Environmental Nexus Approach

Environmental Nexus Approach: Management of Water, Waste, and Soil establishes linkages between environmental resources, such as water, waste, and soil, in order to facilitate sustainable management of these resources. It shows the nexus approach as a policy-relevant means of environmental management by focusing on integrated management of water, waste, and soil resources. It synthesizes interdisciplinary theory, concepts, definitions, models, and findings involved in complex global sustainability problem-solving, making it an essential guide and reference. It includes real-world examples and applications making the book accessible to a broader interdisciplinary readership.

Features:

1. Explores cutting-edge developments in the environmental nexus approach of water, waste, and soil.
2. Introduces the key mechanisms regarding antibiotic resistance genes, microplastics, and other emerging contaminants in the water, waste, and soil nexus.
3. Investigates the fate and behavior of heavy metals, polyaromatic hydrocarbons, plastics, and pesticides in soil systems and their risk assessment.
4. Provides insights into the latest developments, current research perspectives, technology development, critical thinking, and societal requirements of the environmental nexus between water, waste, and soil.

This book is aimed at graduate students and researchers in environmental science and engineering, environmental engineering, and waste management.

AF386382

Environmental Nexus in Waste Management

Series Editors: Vineet Kumar, *Department of Microbiology, School of Life Sciences at the Central University of Rajasthan, Rajasthan, India* and **Sunil Kumar**, *researcher in Environmental Engineering and Science*

The book series on "Environmental Nexus for Waste Management" addresses novel approaches and techniques linked to sustainable development paths with an interdisciplinary focus on waste resources monitoring, assessment, management, reclamation, and recycling, as well as discussing solutions for more sustainable uses of natural resources. It is a broad compendium of research and development related to waste management and discussions of the most up-to-date strategies for tackling waste issues. The series of books constitutes an information source and facilitator for the transfer of knowledge, focusing on practical solutions, and better understanding towards achieving sustainable development.

Solid Waste Treatment Technologies
Challenges and Perspectives
Edited by Pratibha Gautam, Vineet Kumar and Sunil Kumar

Microbial Nexus for Sustainable Wastewater Treatment
Resources, Efficiency, and Reuse
Edited by Vineet Kumar, Sunil Kumar, Pradeep Verma and Sartaj Ahmad Bhat

Environmental Nexus for Resource Management
Edited by Hanuman Singh Jatav, Tatiana Minkina, Satish Kumar Singh, Bijay Singh and Vishnu D. Rajput

Environmental Nexus Approach: Management of Water, Waste, and Soil
Edited by Sartaj Ahmad Bhat, Vineet Kumar, Fusheng Li, Fuad Ameen, and Sunil Kumar

For more information about this series, please visit: www.routledge.com/Environmental-Nexus-in-Waste-Management/book-series/CRCENWM

Environmental Nexus Approach

Management of Water, Waste, and Soil

Edited by
Sartaj Ahmad Bhat, Vineet Kumar, Fusheng Li,
Fuad Ameen, and Sunil Kumar

CRC Press
Taylor & Francis Group
Boca Raton London New York

CRC Press is an imprint of the
Taylor & Francis Group, an **informa** business

Designed cover image: © Shutterstock Images

First edition published 2025
by CRC Press
2385 NW Executive Center Drive, Suite 320, Boca Raton FL 33431

and by CRC Press
4 Park Square, Milton Park, Abingdon, Oxon, OX14 4RN

CRC Press is an imprint of Taylor & Francis Group, LLC

ISBN: 9781032450292 (hbk)
ISBN: 9781032527741 (pbk)
ISBN: 9781003408352 (ebk)

DOI: 10.1201/9781003408352

Typeset in Times
by Newgen Publishing UK

Contents

SECTION II Environmental Nexus in Waste Management 131

SECTION III Environmental Nexus in Contaminated Soil Management 239

Preface

Environmental Nexus Approach: Management of Water, Waste, and Soil addresses the linkages between environmental resources, such as water, waste, and soil, to deal with sustainable management of resources. This book presents an up-to-date and comprehensive collection of chapters contributed by prominent experts, in relevant fields of water, waste, and soil, working in the top institutions around the world. This book promotes the nexus approach as a policy-relevant means of environmental management by focusing on integrated management of water, waste, and soil resources. This book is divided into three sections, namely Section I: Environmental Nexus in Water/Wastewater Management, Section II: Environmental Nexus in Waste Management, and Section III: Environmental Nexus in Contaminated Soil Management. Section I consists of eight chapters covering research on water and wastewater treatment and distribution, wastewater collection/treatment/disposal, and microbial diversity analysis in wastewater treatment systems. This section also discusses advancements in conventional and modern waste treatment systems. As we know, water is essential for all known forms of life and is required for survival on Earth. As a result, an appropriate treatment train is required. However, while conventional domestic waste and wastewater treatment methods remove contaminants of concern up to a certain extent, they may cause environmental problems due to their poor efficacies. Such waste and wastewater must be treated suitably using emerging, low-cost, feasible, and environmentally friendly treatment processes or combinations of processes; these are fully addressed in this section.

Section II consists of five chapters covering research on the behavior and fate of microplastics from municipal solid waste, on the toxicity assessment and bioremediation strategies of antibiotics and antibiotic resistance genes in organic wastes, and on the microbe-assisted enzymatic degradation of emerging contaminants like microplastics. Soil contamination has also generated great interest on a worldwide scale, particularly in waste disposal locations, industrial wasteland, and agricultural farmland. Section III consists of seven chapters covering the areas of emerging contaminants, like microplastics and polyaromatic hydrocarbons, and different technologies for contaminated soil remediation.

In order to provide the theoretical basis for developing remediation technologies that can be a tool to establish a good environmental status, this book defines the current knowledge on water, waste, and soil contamination in relation to emerging contaminants, remediation strategies, improving approaches for remediation efficiency, and the prospects of this remediation technology. The book uses the selected studies to explore the use of the nexus approach in the integrated management of water, waste, and soil systems. Therefore, this book will serve as an invaluable source of basic knowledge on the environmental nexus between environmental resources such as water, waste, and soil to deal with sustainable management.

Contributors

Amal Abdelhaleem
Environmental Engineering Department, Egypt-Japan University of Science and Technology, New Borg El-Arab City, Alexandria, Egypt

Mohamed Gar Alalm
Public Works Engineering Department, Faculty of engineering, Mansoura University, Mansoura, Egypt

Analía Alvarez
Planta Piloto de Procesos Industriales Microbiológicos (PROIMI-CONICET), Avenida Belgrano y Pasaje Caseros, San Miguel de Tucumán, Tucumán, Argentina

Deachen Angmo
Department of Botanical and Environmental Sciences, Guru Nanak Dev University, Amritsar, Punjab, India

Juan Daniel Aparicio
Planta Piloto de Procesos Industriales Microbiológicos (PROIMI-CONICET), Avenida Belgrano y Pasaje Caseros, San Miguel de Tucumán, Tucumán, Argentina

O.O. Ayeleru
Centre for Nanoengineering and Tribocorrosion University of Johannesburg, Johannesburg, South Africa

Faissal Aziz
Laboratory of Water, Biodiversity, and Climate Change, Faculty of Sciences Semlalia, Cadi Ayyad University, Marrakech, Morocco and National Center for Research and Studies on Water and Energy, Cadi Ayyad University, Marrakech, Morocco

Claudia Susana Benimeli
Planta Piloto de Procesos Industriales Microbiológicos (PROIMI-CONICET), Avenida Belgrano y Pasaje Caseros, San Miguel de Tucumán, Tucumán, Argentina

Tanvi Bhatia
ICAR-Central Institute for Research on Buffaloes, Hisar, Haryana

Suman Chaudhary
ICAR-Central Institute for Research on Buffaloes, Hisar, Haryana

Fajun Chen
Department of Entomology, College of Plant Protection, Nanjing Agricultural University, Nanjing, China

Stefanie Bernardette Costa-Gutierrez
Planta Piloto de Procesos Industriales Microbiológicos (PROIMI-CONICET), Avenida Belgrano y Pasaje Caseros, San Miguel de Tucumán, Tucumán, Argentina

Hans-Uwe Dahms
Department of Biomedical Science and Environmental Biology, Kaohsiung Medical University, Kaohsiung City, Taiwan

Saprativ P. Das
Department of Chemical Engineering, Indian Institute of Technology Bombay, Mumbai, Maharashtra, India

Ana Laura De la Colina Martínez
Doctora en Ciencia de Materiales de la Facultad de Química, Universidad Autónoma del Estado de México, Paseo Colón Esquina Paseo Tollocan S/N, Toluca Estado de México, México

David Joaquín Delgado-Hernández
Facultad de Ingeniería, Universidad Autónoma del Estado de México, Avenida Universidad S/N, Cerro de Coatepec, Ciudad Universitaria, Toluca, México

Parvesh Devi
Department of Chemistry, CCS Haryana Agricultural University, Hisar, Haryana, India

Ritu Devi
Department of Chemistry, CCS Haryana Agricultural University, Hisar, Haryana, India

Farzaneh Dianatdar
Department of Cell and Molecular Biology & Microbiology, Faculty of Biological Science and Technology, University of Isfahan, Isfahan, Iran

Dushyant R. Dudhagara
Department of Life Sciences, Bhakta Kavi Narsinh Mehta University, Khadiy, Junagadh, India

Zahra Etemadifar
Department of Cell and Molecular Biology & Microbiology, Faculty of Biological Science and Technology, University of Isfahan, Isfahan, Iran

M. Fathima Sana
Department of Biotechnology, School of Bio Sciences and Technology, Vellore Institute of Technology, Vellore, Tamil Nadu, India

Sougata Ghosh
Department of Physics, Faculty of Science, Kasetsart University, Bangkok, Thailand

Sharareh Harirchi
Swedish Centre for Resource Recovery, University of Borås, Borås, Sweden

M. Hemalatha
Department of Biotechnology, School of Bio Sciences and Technology, Vellore Institute of Technology, Vellore, Tamil Nadu, India

Jiang-Shiou Hwang
Institute of Marine Biology, National Taiwan Ocean University, Keelung, Taiwan

Bhumi M. Javia
Department of Life Sciences, Bhakta Kavi Narsinh Mehta University, Khadiya, Junagadh, India

Sonia Kapoor
University Institute of Engineering & Technology, M. D. University, Rohtak, India

Prerna Kashyap
Department of Biotechnology, School of Bio Sciences and Technology, Vellore Institute of Technology, Vellore, Tamil Nadu, India

Pardeep Kaur
Department of Agriculture, Khalsa College Amritsar, Punjab, India

Mukesh Kumar
Department of Microbiology, CCS Haryana Agricultural University Hisar, Haryana, India

Rakesh Kumar
Department of Agriculture, Khalsa College Amritsar, Punjab, India

Anita Kumari
Department of Botany and Plant Physiology, CCS Haryana Agricultural University Hisar, Haryana, India

Renu Lamba
Department of Botany and Plant Physiology, CCS Haryana Agricultural University Hisar, Haryana, India

Alaa El Din Mahmoud
Environmental Sciences Department, Faculty of Science, Alexandria University, Alexandria, Egypt

H.U. Modekwe
Renewable Energy and Biomass Research, Department of Chemical Engineering, University of Johannesburg, Doornfontein Campus, Johannesburg, South Africa

V. Mohanasrinivasan
Department of Biotechnology, School of Bio Sciences and Technology, Vellore Institute of Technology, Vellore, Tamil Nadu, India.

Monika Moond
Department of Chemistry, CCS Haryana Agricultural University, Hisar, Haryana, India

Reihaneh Moridshahi
Department of Cell and Molecular Biology & Microbiology, Faculty of Biological Science and Technology, University of Isfahan, Isfahan, Iran

Debarshi I. Mukherjee
Department of Biotechnology, Indian Institute of Technology, Kharagpur, Kharagpur, West Bengal, India

Kadarkarai Murugan
Department of Zoology, School of Life Sciences, Bharathiar University, Coimbatore, India

Anuradha S. Nerurkar
Department of Microbiology and Biotechnology Centre, Faculty of Science, The Maharaja Sayajirao University of Baroda, Vadodara, Gujarat, India

B.O. Oboirien
Department of Chemical Engineering, University of Johannesburg, Doornfontein Campus, Johannesburg, South Africa

P.A. Olubambi
Centre for Nanoengineering and Tribocorrosion University of Johannesburg, Johannesburg, South Africa

M.A. Onu
Centre for Nanoengineering and Tribocorrosion University of Johannesburg, Johannesburg, South Africa

Chellasamy Panneerselvam
Faculty of Science, Department of Biology, University of Tabuk, Saudi Arabia

Roshni J. Patel
GSFC University, Vigyan Bhavan, P.O. Fertilizernagar, Vadodara, Gujarat, India

Marta Alejandra Polti
Planta Piloto de Procesos Industriales Microbiológicos (PROIMI-CONICET), Avenida Belgrano y Pasaje Caseros, San Miguel de Tucumán, Tucumán, Argentina

Upasana Priyadarshini
Indian Institute of Technology Bhubaneswar, Argul, Odisha, India

Shokufeh Rafieyan
Department of Cell and Molecular Biology & Microbiology, Faculty of Biological Science and Technology, University of Isfahan, Isfahan, Iran

Enzo Emanuel Raimondo
Planta Piloto de Procesos Industriales Microbiológicos (PROIMI-CONICET), Avenida Belgrano y Pasaje Caseros, San Miguel de Tucumán, Tucumán, Argentina

Rajapandian Rajaganesh
Department of Zoology, School of Life Sciences, Bharathiar University, Coimbatore, India

Jyoti Rani
Department of Chemistry, CCS Haryana Agricultural University, Hisar, Haryana, India

Madhu Rani
University Institute of Engineering & Technology, M. D. University, Rohtak, India

Srinithya Ravinuthala
Biochemical Engineering Lab, Department of Chemical Engineering, National Institute of Technology Tiruchirappalli, Tiruchirappalli, Tamil Nadu, India

Neelancherry Remya
Indian Institute of Technology Bhubaneswar, Argul, Odisha, India

Hazarika Risha
School of Agro and Rural Technology, Indian Institute of Technology Guwahati, North Guwahati, Assam, India

Saravanan S.
Biochemical Engineering Lab, Department of Chemical Engineering, National Institute of Technology Tiruchirappalli, Tiruchirappalli, Tamil Nadu, India

Juliana María Saez
Planta Piloto de Procesos Industriales Microbiológicos (PROIMI-CONICET), Avenida Belgrano y Pasaje Caseros, San Miguel de Tucumán, Tucumán, Argentina

Seema Sangwan
Department of Microbiology, CCS Haryana Agricultural University, Hisar, Haryana, India

Patra Sanjukta
Department of Biosciences and Bioengineering, Indian Institute of Technology Guwahati, North Guwahati, Assam, India

Bishwarup Sarkar
College of Science, Northeastern University, Boston, MA, USA

Reshmi Sasi
School of Biotechnology, National Institute of Technology Calicut, India

Natchiappan Senthilkumar
Division of Chemistry and Bio Prospecting, Institute of Forest Genetics and Tree Breeding, Coimbatore, Tamil Nadu, India

Douglas J.H. Shyu
Department of Biological Science and Technology, National Pingtung University of Science and Technology, Pingtung, Taiwan

Satyavir Singh Sindhu
Department of Microbiology, CCS Haryana Agricultural University, Hisar, Haryana, India

Jaswinder Singh
Department of Zoology, Khalsa College Amritsar, Punjab, India

Joginder Singh
School of Bioengineering and Biosciences, Lovely Professional University Phagwara, Punjab, India

Shubhra Singh
Department of Tropical Agriculture and International Cooperation, National Pingtung University of
  Science and Technology, Pingtung, Taiwan

Sushila Singh
Department of Chemistry, CCS Haryana Agricultural University, Hisar, Haryana, India

Rejeti Venkata Srinadh
Indian Institute of Technology Bhubaneswar, Argul, Odisha, India

C. Subathra Devi
Department of Biotechnology, School of Bio Sciences and Technology, Vellore Institute of
  Technology, Vellore, Tamil Nadu, India

T.V. Suchithra
School of Biotechnology, National Institute of Technology Calicut, India

Gangar Tarun
Department of Biosciences and Bioengineering, Indian Institute of Technology Guwahati, North
  Guwahati, Assam, India

Babita Thakur
School of Bioengineering and Biosciences, Lovely Professional University Phagwara, Punjab, India

Sirikanjana Thongmee
Department of Physics, Faculty of Science, Kasetsart University, Bangkok, Thailand

Anjana K. Vala
Department of Life Sciences, Maharaja Krishnakumarsinhji Bhavnagar University, Bhavnagar,
  Bhavnagar, India

Murugan Vasanthakumaran
Department of Zoology, School of Life Sciences, Bharathiar University, Coimbatore, India

Adarsh Pal Vig
Department of Botanical and Environmental Sciences, Guru Nanak Dev University, Amritsar,
  Punjab, India

Lan Wang
School of Life Science, Shanxi University, Taiyuan, Shanxi Province, China

Leela Wati
Department of Microbiology, CCS Haryana Agricultural University Hisar, Haryana, India

Rui-De Xue
Anastasia Mosquito Control District, St. Augustine, Florida, USA

Mohamed Néjib Daly Yahia
Environmental Sciences Program, Department of Biological and Environmental Sciences, College
  of Arts and Sciences, Qatar University, Doha, Qatar

About the Editors

Sartaj Ahmad Bhat is working as JSPS Postdoctoral Researcher at the River Basin Research Center, Gifu University, Japan. He received his Ph.D. in Environmental Sciences from Guru Nanak Dev University, Amritsar, India in 2017. His primary research focuses on the development and evaluation of treatment technologies for organic waste and wastewater from domestic and industrial outlets as well as organic waste recycling, with a focus on biological and sustainable treatment by earthworms. He has published more than 70 papers in peer-reviewed journals and edited over 15 books published by Elsevier Science, CRC Press (Taylor & Francis Group), International Water Association, Royal Society of Chemistry, and Springer Nature. Dr. Bhat is serving as an Associate/Academic Editor and Editorial Board Member/ Advisory Board Member of more than 20 journals. Dr. Bhat is a recipient of several prestigious awards, such as the JSPS Postdoctoral Fellowship to pursue research at River Basin Research Center, Gifu University, Japan, the Basic Scientific Research Fellowship (BSR JRF, SRF) by the University Grants Commission (UGC) India, the DST-SERB National Postdoctoral Fellowship at CSIR-NEERI, Nagpur, India and Swachhta Saarthi Fellowship by the Govt. of India. He has also received the 2020 Outstanding Reviewer Award by the International Journal of Environmental Research and Public Health, MDPI, and Top Peer Reviewer 2019 award in Environment and Ecology by Web of Science and has more than 800 Verified Reviews and 80 Editor Records to his credit.

Vineet Kumar is presently working as a National Postdoctoral Fellow in the Department of Microbiology, School of Life Sciences at Central University of Rajasthan, Rajasthan, India. He received his M.Sc. and M.Phil. in Microbiology from Ch. Charan Singh University, Meerut, India. Subsequently, he earned his Ph.D. (2018) in Environmental Microbiology from Babasaheb Bhimrao Ambedkar (A Central) University, Lucknow, India. Dr. Kumar's research work mainly focuses on the development of integrated and sustainable treatment techniques that can help in minimizing or eliminating hazardous waste in the environment. He has published more than 40 articles in reputed international peer-reviewed journals, with more than 1600 citations and an h-index of 24. In addition, he is the author/co-author of four proceeding articles, 50 book chapters, and four scientific magazine articles. Moreover, he has published two authored and 20 edited books on different aspects of phytoremediation, bioremediation, wastewater treatment, waste management, omics, genomics, and metagenomics. Dr. Kumar has been serving as a guest editor and reviewer for many prestigious international journals and has served on the editorial board of various reputed journals. He has presented several papers relevant to his research areas at national and international conferences. He is an active member of numerous scientific societies, including the Microbiology Society (UK), the Indian Science Congress Association (India), and the Association of Microbiologists of India (India). He is the founder of the Society for Green Environment, India (website: www.sgeindia.org).

Fusheng Li is Professor in the Division of Water System Safety and Security Studies at the Graduate School of Engineering at Gifu University, Japan. He received his BS degree for environmental engineering from Lanzhou Jiaotong University of China in 1986, MS degree from Kitami Institute of Technology of Japan in 1994, and PhD degree from the Gifu University of Japan in 1998. Dr. Li is directing the Division of Water Quality Studies that covers the fields from water quality to water and wastewater treatment, and recently to resource and energy recovery from organic waste. The ongoing research projects in his lab include adsorption; membrane filtration, enhanced coagulation, disinfection; biological water and wastewater treatment; vermicomposting treatment of vegetable waste and activated sludge; microbial fuel cell; physicochemical water quality assessment; and biological water quality assessment. He has over 350 scholarly publications, including more than 160 in peer reviewed journal papers. As a principal supervisor, he has guided so far 50 masters and 21 doctorate graduate students to the completion of their degrees. Dr. Li is the recipient of awards from several academic societies and associations for his research work on water treatment and water quality dynamics studies.

Fuad Ameen is an associate professor and researcher at the Department of Botany and Microbiology, College of Science, King Saud University, Riyadh, Saudi Arabia. He graduated his Ph.D. in biodegradation of urban waste by mangrove fungi from King Saud University, Saudi Arabia. In addition, he had served as a researcher in different places in the field of biotechnology and applied microbiology. After that, he has been involved in many kinds of projects, many of them dealing with new solutions to treat the polluted sites. He has collected and studied microbial strains from arid and marine ecosystems, indicating its ability to biodegrade the most common pollutants, and his program of research on these organisms has taken him to many places and to examples of every major type of terrestrial and marine ecosystem. He is the author or coauthor of 140 papers and book chapters in peer-reviewed journals.

Sunil Kumar is a well-rounded researcher with more than 22 years of experience in leading, supervising, and undertaking research in the broader field of Environmental Engineering and Science with a focus on Solid and Hazardous Waste Management. Dr. Kumar is a graduate in Environmental Engineering and Management from the Indian Institute of Technology, Kharagpur, India. He completed his Ph.D. in Environmental Engineering from Jadavpur University, Kolkata, India. His primary area of expertise is solid waste management (municipal solid waste, electronic waste etc.) over a wide range of environmental topics including contaminated sites, EIA, and wastewater treatment. His contributions in these fields led to a citation of approx. 10000, h-index of 46, and i10-index of 187 (Google scholar). His contributions since inception at CSIR-National Environmental Engineering Research Institute (NEERI), India in 2000 include 300 refereed publications, five books and 40 book chapters, 10 edited volumes, and numerous project reports to various governmental bodies and private, local, and international academic/research bodies. He is the Associate Editor of peer-reviewed journals of international repute i.e., Environmental Chemistry Letter, International Journal of Environmental Science & Technology, ASCE Journal of Hazardous, Toxic and Radioactive Waste. He also serves on the Editorial Board of Bioresource Technology, Elsevier. He has completed more than 22 research projects as PI with 15 (seven awarded) Ph.D. and 20 M.Phil/M.Tech thesis/dissertations. Dr. Kumar was awarded the most prestigious award Alexander von Humboldt-Stiftung Jean-Paul-Str.12 D-53173 Bonn, Germany as a senior researcher for developing a Global Network and Excellence for more advanced research and technology innovation.

Acknowledgements

The editors would like to express their sincere thanks to the contributors for submitting their work in a timely manner. The editors are also thankful to many anonymous reviewers who took the time to critically review individual chapters of this book; their useful comments were gratefully received and have enhanced the quality of the book.

Dr. Bhat acknowledges the Japan Society for the Promotion of Science (JSPS) for the Postdoctoral Fellowship. Dr. Vineet Kumar gratefully acknowledges the Science and Engineering Research Board, Government of India for providing a Postdoctoral Fellowship (F.No: PDF/2022/000038).

The editors would also like to thank CRC Press (Taylor and Francis Group) for providing the opportunity to accomplish the project and share the knowledge with the scientific and academic fraternity. Particular thanks go to Dr. Gagandeep Singh, Senior Publisher (Engineering) at CRC Press (Taylor & Francis Group India Pvt. Ltd.), for the execution of the publishing agreement, encouragement, support, valuable suggestions, and unconditional support till the submission of manuscripts for production. The editors are also thankful to Anitha AL, project manager at Newgen Knowledge Works P Ltd., Chennai, India, for the skillful organization and management of the entire book project.

The editors also acknowledge the support received from CRC Taylor and Francis Group, for their guidance in finalizing the book.

Sartaj Ahmad Bhat
Gifu, Japan
Vineet Kumar
Rajasthan, India
Fusheng Li
Gifu, Japan
Fuad Ameen
Riyadh, Saudi Arabia
Sunil Kumar
Maharashtra, India

Section I

Environmental Nexus in Water/ Wastewater Management

1 Role of Nanoparticles for Removal of Heavy Metals and Dyes from Wastewater

Monika Moond, Sushila Singh, Seema Sangwan, Jyoti Rani, Ritu Devi, and Parvesh Devi

1.1 INTRODUCTION

One of the biggest global concerns of the twenty-first century is ensuring reliable access to clean water as the multifaceted growth of society depends on access to clean and safe water. Rapid population growth, growing industrialisation, urbanisation, and widespread agricultural activities have led to the production of wastewater that has made the water not only dirty or polluted but also contaminated or lethal (Keerti et al. 2021). Since they represent major dangers to the environment and human health, extreme quantities of dyes and heavy metals in water supplies have long been a source of worry. Dye is a natural and xenobiotic compound that provides colour to substances. Additionally, the dyes are toxic, mutagenic, and teratogenic to a wide variety of microbiological species. The presence of dyes reduces light penetration which might have a negative impact on marine life's ability to photosynthesise. Additionally, they can damage the liver and cause major side effects in individuals, like renal failure. More than 8,000 dyes, both soluble and insoluble, have been produced and employed in a variety of sectors, including the paper, textile, paint, and tannery industries. These dyes may stay stable in the environment for a very long time if they are not handled properly, which could have very bad consequences for human health. Therefore, it is vital to clean and reuse both municipal and industrial waste water. On the other hand, growing industrialisation results in an excessive amount of heavy metals being released into the environment, which raises concerns around the world due to their chronic toxicity. Heavy metals from mining operations, battery manufacturing, metal plating, petroleum refining, pesticides, tanneries, and smelting operations, among others, are frequently found in industrial wastewater. These metals include mercury, chromium, arsenic, manganese, chromium, copper, iron, cobalt, and lead (Deepika et al. 2022).

Due to their significant potential for toxicity, heavy metals are typically seen as a threat to both humans and ecosystems. In extremely small amounts, the human body needs heavy metals including copper, chromium, zinc, manganese, and iron to function. However, if there is a certain level of metals present, they can poison people and have serious effects on their health. Heavy metals are non-biodegradable materials, just like plastic. The presence of heavy metals, even at trace levels, is thought to pose a risk to the environment and human health. The heavy metals should be eliminated since they typically exist in concentrations that are higher than the safe permitted limits. Unlike biological waste, which degrades, heavy metals can accumulate in living organisms and cause a variety of illnesses and disorders (Wan and Hanafiah 2008). For the eradication of all varieties of organic pollutants, numerous technologies including adsorption, biological oxidation, and chemical oxidation have been employed. These methods have a number of drawbacks, such as a high energy need, insufficient pollution removal, and the production of hazardous waste. The employment of more efficient and environmentally friendly procedures have become necessary due to the shortcomings of existing techniques and materials for removing undesirable and hazardous substances from water. A cutting-edge strategy known as nanotechnology has emerged as one of the most flexible and cost-effective ways to remove dyes and heavy metals. It is the molecular or atomic level manipulation of matter to create new structures, devices, and systems with superior electronic, optical, magnetic, conductive, electrochemical, and mechanical properties. Nanotechnology is emerging as a promising technology and has achieved remarkable successes in a number of fields, including wastewater treatment. Because of their high surface-to-volume ratio and tunable pore size, high sensitivity, reactivity, adsorption capacity, catalytic activity, antimicrobial activity, and ease of functionalisation, nanoparticles are well suited for use in wastewater treatment. The size effect of nanoparticles is the main factor responsible for excellent efficiency. Their exceptionally large surface area and surface reactivity are closely related to their size. Due to their intriguing properties that meet the required criteria, nanoparticles are increasingly being used to address environmental challenges. The harmful heavy metals, biological contaminants, organic pollutants, and dyes are easily removed from contaminated water using nanoparticles. There are numerous ways to create nanoparticles, including physical, chemical, and biological pathways. Nanoparticles have received a lot of interest in the removal of heavy metals and dyes from wastewater due to their large active groups, minimal flocculant production, and large surface area Adsorption using nanoparticles is one of the best available technologies since it is simple to use, inexpensive, and very effective. The adsorption process is generally greatly influenced by a number of variables, including pH, adsorbent dose, agitation rate, contact time, temperature, and the physical characteristics of the used sorbent material, such as sorbent surface area and particle size (Azmier et al. 2014). Different adsorption kinetics and isotherms have been proposed so far for establishing the adsorption equilibrium to better understand the adsorption mechanism of nanoparticles with heavy metals and dyes. The usage of nanoparticles for water treatment and purification will increase significantly in the future (Ahmad et al. 2021).

1.2 NANOPARTICLES FOR REMOVAL OF HEAVY METALS

1.2.1 REMOVAL OF CHROMIUM

A very toxic, liquid, and mobile metalloid is called chromium (Cr). Chronic Cr(VI) exposure results in diarrhoea, liver damage, and tumours. DNA damage, itchy skin, and immune system decline are some of its harmful effects. Cr(VI) exposure through inhalation results in acute toxicity and damage to the respiratory tract. Hexavalent chromium (Cr(VI)) and trivalent chromium (Cr(III)) are the two stable forms of the metal chromium. Trivalent chromium is non-toxic and is primarily found in the environment in the forms of $Cr(OH)^{2+}$ and $Cr(OH)_3$. The mobility of Cr(III) in the environment is weak because positively charged $Cr(OH)^{2+}$ can be absorbed onto the colloid and other media with negatively charged surface by electrostatic action, whereas $Cr(OH)_3$ itself exists as precipitation shown in Figure 1.1 (Richmond 1991).

Zero-valent iron nanoparticles (ZVI NPs) can reduce heavy metal Cr(VI) to Cr(III) since they have a reducing potential of primarily −0.44 V as shown in Reaction 1. Adsorption of Cr(VI) onto the surface of the nanoparticles, reduction of Cr(VI) to Cr(III), and coprecipitation of Fe(III)-Cr(III) (oxy) hydroxides are the steps involved in removing Cr(VI).

Reaction 1

$$3Fe^0 + 2CrO_4^{2-} + 16H^+ \longrightarrow 3Fe^{2+} + 2Cr^{3+} + 8H_2O$$

$$3Fe^{2+} + CrO_4^{2-} + 8H^+ \longrightarrow 3Fe^{3+} + Cr^{3+} + 4H_2O$$

$$xCr^{3+} + (1-x)Fe^{3+} + 2H_2O \longrightarrow Cr_xFe_{(1-x)}OOH + 3H^+$$

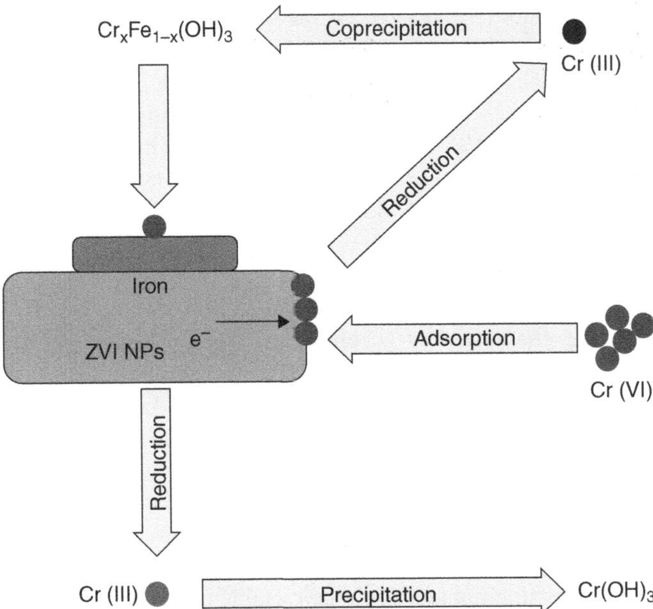

FIGURE 1.1 Mechanism for removal of Cr(VI) using ZVI NPs.

Iron oxide coating, which prohibits further contact with contaminants, is quickly generated on the surface of ZVI NPs due to its strong reactivity, particularly in neutral and alkaline conditions. It is reported that the removal of Cr(VI) is favourable, endothermic, and spontaneous. The elimination of Cr(VI) is mostly accomplished through chemisorption-couple reduction, which is a pseudo-second-order kinetic process. In recent years, it has been discovered that effectively alleviating ZVI NP's passivation requires adding another catalytic metal to its surface. Palladium, nickel, and copper are a few examples of catalytic metals that can increase the number of reactivity sites and promote electron transfer on nanoparticle surfaces. With a copper loading rate of 3%, the ZVI NP's capacity to remove Cr(VI) increases from 29.3% to 84.0%. The initial pH and Cr(VI) concentration of wastewater has an impact on the removal of Cr(VI) by Cu/Fe bimetallic nanoparticles, and removal effectiveness declines as pH and Cr(VI) concentration rise (Jien et al. 2021).

Due to the metallic nanoparticles' improved catalytic efficiency, the conversion of extremely poisonous Cr(VI) to innocuous Cr(III) has attracted growing interest in recent years and is thought to be a successful strategy.

1.2.2 REMOVAL OF MERCURY

Mercury (Hg) in the water environment offers a higher risk to human and animal health than Hg in the air or soil environments because it acts as a longer-term store of the metal. The most prevalent form of inorganic mercury in water is Hg^{2+}, which can undergo biological methylation to form methyl mercury, which is highly poisonous and readily absorbed by animals before moving up the food chain and eventually reaching humans. Therefore, discovering ways to remove Hg^{2+} from water should be a top priority.

ZVI NPs have been successfully used to remove Hg^{2+} from contaminated water. Since the standard redox potential of mercury (E^0(Hg^{2+}/Hg^0) = 0.86 V) is significantly higher than that of iron (E^0(Fe^{2+}/Fe^0) = 0.44 V), reducing Hg^{2+} by ZVI NPs is thermodynamically favourable. ZVI NPs suspension (0.18 g Fe/L) immobilised 98% of Hg^{2+} after 15 min, removal ranging between 94% and 98%. The interaction between Hg^{2+} and ZVI NPs comprises reduction to Hg^0 and adsorption and/or complexation to the nanoparticle shell and is dependent on the dose of ZVI NPs utilised (Gil-Díaz 2021).

Recently, Silver nanoparticles (AgNPs)-based adsorbents have been widely used for the removal of Hg(II) from contaminated water. AgNPs protected by mercaptosuccinic acid (MSA) and supported on activated alumina have a high removal ability of 0.8 g of mercury per gram. There are two proposed mechanisms for removing Hg^{2+} ions, and each depends on the Hg^{2+} concentration. At lower concentrations of Hg (> 25 ppm), the ions interact with the nanoparticle core, leading to reduction. At higher concentrations (<100 ppm), the uptake may be occurring by reduction or adsorption with the carboxylate groups of MSA (Sumesh et al. 2011).

1.2.3 REMOVAL OF LEAD

Lead (Pb) is a very toxic chemical that accumulates over time, mainly in the kidneys, bones, muscles, and brain. Chronic lead poisoning harms the liver, brain, kidneys, nervous system, and reproductive system. The removal mechanism of heavy metals from wastewater using ZVI NPs may differ depending on standard potential (E^0) of heavy metals. In aqueous media, ZVI NPs has a core–shell structure where the core is made up of Fe^0 and the shell is mainly composed of iron oxides and hydroxides that are structurally flawed and result from the oxidation of Fe^0. Due to its low redox standard potential ($E^0 = 0.41$ V), the Fe^0 in the core functions as an electron donor and reductant, while the oxide shell exhibits sorption processes through electrostatic interactions and complexation while also allowing electron transfer from the metallic core. These two sorption and reduction features suggest significant applications for contaminant removal. For instance, the removal mechanism is primarily composed of reduction and sorption for Pb(II), whose E^0 is somewhat more positive than that of Fe(II), and can be outlined in Reaction 2 as follows (Jinyue et al. 2019).

Reaction 2

$$\text{Reduction: } Fe^0 + Pb^{2+} \longrightarrow Fe^{2+} + Pb^0$$

$$\text{Sorption: } FeOOH + Pb^{2+} \longrightarrow FeOOPb^{2+} + H^+$$

Iron nanoparticles (FeNPs) synthesised using tea extract removes Pb(II) pollutants with 97.5% efficiency. The Pb(II) metal ions are first adsorbed on the surface of FeNPs, and then interacts more specifically with the functional groups present there (Figure 1.2). Adsorption of Pb(II) by FeNPs was mostly accomplished through chemisorption and followed pseudo-second-order kinetics.

The Pb(II) removal efficiency declined from 96.4% to 55.2% as pH was changed from 5 to 3. This indicated that at lower pH, H^+ and Pb(II) could compete for the few accessible adsorption sites on FeNPs due to the high concentration of H^+ in the solution (Ze et al. 2020).

1.2.4 REMOVAL OF COPPER

The excessive ingestion of copper (Cu) in humans can cause severe capillary damage, acute mucosal inflammation, hepatic and renal damage, and disturbances to the central nervous system. The ability of metal oxide nanoparticles to produce adsorbents with excellent electronic properties is superior.

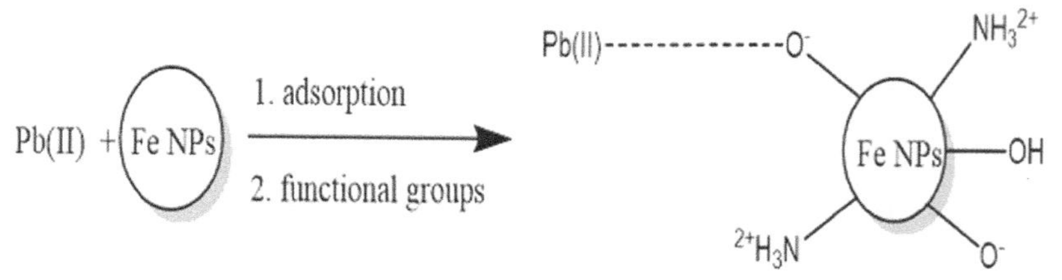

FIGURE 1.2 Mechanism for removal of Pb(II)+using FeNPs.

This is due to the narrow gap between the oxide particles and particle size, resulting in conductivity variation and chemical reactivity. The pH is a common determinant of the adsorption effectiveness of pollutants removal based on metal oxide nanoparticles. For instance, depending on the chemistry of the metal and wastewater or the type of the functional group, the rate of removal of heavy metal ions rises with pH as a result of the formation of metal complexes and electrostatic interactions. Additionally, increasing the negative charge on sorbent materials increases the forces of attraction between positive metal ions and the negative sites. Adsorption took place between the competing metal ions and H^+ ions at low pH values.

Zinc Oxide nanoparticles (ZnO NPs) synthesised using Aloe vera efficiently remove Cu^{2+} metal ions from wastewater. Hydroxyl (OH) groups will form when ZnO NPs are exposed to water; these groups will then operate as adsorptive active sites to remove metal ions by reacting with OH on the ZnO surface (Mahsa et al. 2014). Since the Cu(II) ions interact with the active groups (OH) on the oxide surface in this model of complexing of ion adsorption in hydrated materials shown in Reaction 3, it can be concluded that the mechanism of ion adsorption is as follows:

Reaction 3

$$ZnOH + Cu^{2+} \text{ (aq)} \longrightarrow ZnOCu^+ + H^+\text{(aq)}$$

The amount of active sites on the ZnO surface increases as the pH of the solution rises, making the ZnO surface more conducive to the adsorption of the Cu(II) ion and increasing the removal effectiveness. The precipitation of Cu(II) ions below pH 6 and the formation of complexes that are not adsorbed by ZnO adsorbents can be linked to the decrease in Cu(II) ion removal at pH 6 (Julia de O et al. 2020).

1.2.5 REMOVAL OF CADMIUM

Cadmium (Cd) is hazardous to both humans and animals and can negatively impact health in both the short and long term. It can enter the organism through breathing, eating, or skin absorption. A brief exposure to cadmium can have negative effects on the lungs, circulatory system, liver, and nervous system in addition to causing vomiting, diarrhoea, liver damage, convulsions, shock, and altering kidney function. Cadmium is of toxicological concern because of its bioaccumulation and non-biodegradability property even at trace levels. Because of distinctive characteristics of nanoparticles such as their small size, high specific surface area, and morphological features, these are a promising material in terms of separation, performance, and efficiency for removing heavy metals from water. Tin oxide (SnO_2) is quite intriguing since it is an n-type semiconductor with a direct band gap of 3.6 eV between the bottom tin states in the conduction band and the full oxygen 2p Valence band and offers environmental application for the removal of heavy metal ions. SnO_2 nanoparticles act as effective adsorbents for the removal of Cd(II) with maximum adsorption capacities of 1275.5 mg/g.

A key factor in regulating the adsorption of Cd(II) by SnO_2 nanoparticles is electronegativity. The oxygen atoms on the surfaces of nanoparticles tend to establish stronger covalent connections with the electronegative metal. Metal ion adsorption can occur by a variety of reaction processes, such as surface adsorption, ion exchange, and/or the formation of covalent bonds. The hydroxyl groups that are typically linked to the surfaces of metal oxides in aqueous solution can change form at various pH levels. These groups can generate the necessary conditions for their reaction with both acids and bases because they contain a double pair of electrons and a dissociable hydrogen atom. Because of the dissociation (ionisation) of the surface hydroxyl groups depending on the pH of the solution, the charge on the surface predominates the adsorption or desorption of protons. The starting concentration of metal ions, pH, temperature, and contact time are among the factors that affect the adsorption of Cd(II) ions onto SnO_2 nanoparticles. The adsorption capacity can be successfully promoted by increasing the starting concentration, temperature, and adsorbent dose (Kumar et al. 2016).

Monodentate   **Bidentate complex**

FIGURE 1.3   Complex structures of As(V) on the FeNPs' surface.

1.2.6 REMOVAL OF ARSENIC

One of the most toxic and carcinogenic chemical elements is arsenic (As), which is usually present as either arsenite (As(III)) or arsenate (As(V)). It is one of the most hazardous pollutants present in the environment and therefore its removal, especially from water, becomes essential. Iron nanoparticles (FeNPs) synthesised using eucalyptus leaf extracts are used for the removal of As from contaminated water. The removal mechanism for As(V) is based on the reaction between As(V) and FeNPs that results in the formation of a monodentate chelating ligand first, followed by a bidentate binuclear complex (Figure 1.3). The surface area of FeNPs decreases from 51.14 m^2g^{-1} to 26.03 m^2 g^{-1} after adsorption of As(V) (Zhicheng et al. 2019).

The adsorption of As(V) on to FeNPs followed a pseudo-second-order kinetic model and the Langmuir adsorption model and Liu model, where the maximum single layer As(V) adsorption capacity of FeNPs is 21.59 mg/g.

1.2.7 REMOVAL OF NICKEL

Modern metallurgies use nickel (Ni) in a wide range of metallurgical processes, including alloy creation, electroplating, the manufacture of nickel-cadmium batteries, and usage of the metal as a catalyst in the chemical and food industries. The widespread use of products containing this metal inevitably causes nickel and its byproducts to pollute the water during all phases of production, recycling, and disposal. Undoubtedly, nickel causes cancer in humans. The concentration and duration of exposure affect the potential toxicity of nickel. Contact with nickel can have a number of negative health effects on humans, including allergies, kidney and heart problems, lung fibrosis, and lung and nasal cancer.

Magnetite nanoparticles (Fe_3O_4 NPs) are effective adsorbents for removal of Ni(II) from waste water as they exhibit a high ratio of surface area to volume, mesoporous property, relative huge pore volume, and superparamagnetic characteristics. Additionally, they are non-toxic, recyclable, and easy to synthesise and are highly useful in novel separation processes (Figure 1.4). The removal efficiency of Ni^{2+} from wastewater using Fe_3O_4 NPs is 98.3%. With time, the rate of Ni^{2+} uptake progressively drops and does not seem to get much higher. This is due to the fact that there are more active sites on the Fe_3O_4 NPs adsorbents at the beginning of the adsorption process and, over the course of accumulation time, these sites gradually become saturated. The pH of the medium is still a key regulating factor in the adsorption process because it has an impact on both the active sites of the adsorbent and the chemistry of the metal ion in solution. The pH of the solution is maintained at 5.5 because at pH below 5.5, the medium's proton (H^+) density is higher, and as a result, there may be more competition between H^+ and metal ions for the active adsorption sites on the Fe_3O_4 NPs while at pH greater than 5.5, the removal efficiency of Ni^{2+} decreases with the increasing pH of the solution as Ni^{2+} start precipitating (Congcong et al. 2016). The fractional adsorption becomes independent of the starting metal ion concentration at low metal ion concentrations due to the huge ratio of the number of moles of metal ions to the active sites of the Fe_3O_4 nanoadsorbents. However, as the metal ion concentration rises, there are fewer active sites on the Fe_3O_4 nanoadsorbents available relative to the available moles

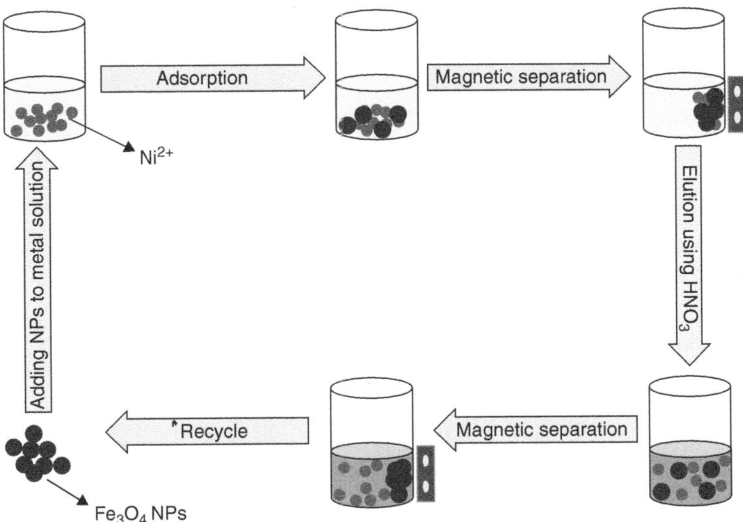

FIGURE 1.4 Removal of Ni^{2+}ions from aqueous medium using Fe$_3$O$_4$ NPs.

of metal, which results in a reduction in the metal ions' ability to adsorb. The adsorption process is chemisorptive, spontaneous, and endothermic. The Fe$_3$O$_4$ nanoadsorbents are shown to be effectively recoverable in dilute nitric acid at pH 1.0 and may preserve almost 85% of their original adsorption capacity after five subsequent adsorption/desorption cycles (Fato et al. 2019).

1.3 NANOPARTICLES FOR REMOVAL OF DYES

Dyes are used extensively in numerous industries, including textile, rubber, paper, plastic, and cosmetics. A significant portion of the world's daily dye production enters the water during the dyeing process. This has major negative impacts on the environment such as an increase in toxicity, chemical oxygen demand (COD), biochemical oxygen demand (BOD), bad smell, and water colour. In recent years, attempts have been made to use nanoparticles, particularly metal and metal oxides (nanoparticles as adsorbent for the treatment of wastewater for dye removal), which have unique properties such as large surface area, high efficiency, chemical stability, and surface functional groups. Coated nanoparticles are better adsorbents for the removal of dye because the coating of nanoparticles causes strong interactions between surface groups of nanoparticles and functional regions of dye. Especially for metal oxides, these materials have a minimal environmental impact, low solubility, and no secondary pollution (Wenqian et al. 2019).

1.3.1 REMOVAL OF METHYLENE BLUE

The blue heterocyclic cationic azo dye known as methylene blue (MB) is widely used in the textile industry and is a recognised water contaminant. Dyes have been removed using a variety of techniques. One effective advanced oxidative process (AOP) for the removal of dyes is photocatalysis. The technique is based on the mineralisation of contaminants in water using reactive oxygen species (ROS). The photocatalytic destruction of dyes has been widely carried out using metal oxide nanoparticles, especially metal tungstates. High photostability and a suitable narrow energy band gap in metal tungstate (MWO$_x$) nanoparticles make them effective photocatalysts for the destruction of dyes. Metal tungstates have the general formula (MWO$_4$), where M are metals such as Ba, Ca, Co, Cd, Sr, Pb, Cu, Zn, and Mn. CoWO$_4$ NPs are popular due to their low cost, narrow energy band gap, and environmental friendliness, making them ideal for photocatalytic research. CoWO$_4$ NPs (cobalt

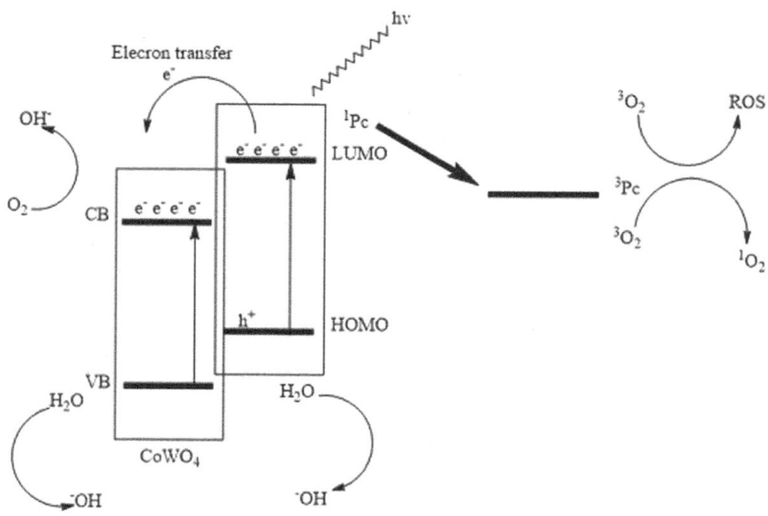

FIGURE 1.5 Mechanism for formation of ROS used for photodegradation of MB.

metal tungstate nanoparticles) are versatile and have been used in a variety of radiations such as solar, visible, and ultra-violet light for the degradation of organic dyes. The zinc(II) phthalocyanine (ZnPc) complexes with $CoWO_4$ NPs nanoparticles are used for photodegradation of MB. The MB's UV-visible spectrum has three distinct bands at 240, 287, and 664 nm and is bright blue in colour. Due to the presence of azo dye, the most significant band is located at 664 nm. The benzene rings are what cause the peaks to appear below 300 nm. In the presence of photocatalysts and visible light, the MB peak at 664 nm is observed to be decreasing. Finally, the vibrant blue MB fades to black. The photodegradation of dyes in the presence of phthalocyanine-semiconductor nanoparticles is caused by photoexcitation of the Pc, which produces electrons in the lowest unoccupied molecular orbital (LUMO) and holes in the highest occupied molecular orbital (HOMO). Conduction band (CB) of $CoWO_4$ receives photo-induced electrons that are transported from the LUMO of the Pc. The vibrational band (VB) of $CoWO_4$ or the HOMO of the Pc complex are where the OH radicals are produced by photo-generated holes. The Pc complex in the excited state may go through intersystem crossing (ISC) to the triplet state, where ROS including singlet oxygen are generated through energy/electron transfer, degrading MB as shown in Figure 1.5 (Sithi and Tebello 2021).

In recent years, adsorption by metal nanoparticles synthesised using green pathway has been the most important approach due to rapid dye removal, no use of toxic chemical compounds, cost effectiveness, and highly flexible simple design and process implementation. The pH of dye solutions, temperature, time of contact of dye, adsorbents, and concentration of adsorbent play an important role in the adsorption process, especially in the adsorption capacity. The pH_{pzc} is a term which expresses ionisation degree of adsorbent surface. When pH is lower than adsorbent pH_{pzc}, adsorbent surface is positively charged and prefers the adsorption of anionic species, whereas when pH is higher than pH_{pzc}, the adsorbent surface is negatively charged and favours the adsorption of cationic species.

The AgNPs synthesised using Tragopogon b. extract is used as an adsorbent for MB and has an adsorption capacity of 45.24887 mg/g based on Langmuir model with $R^2 = 0.9999$ at pH 10 and 65°C. The pH_{pzc} of AgNPs is 6.8, thus adsorption of MB is intensified approximately after pH 7, that is in accordance with the pH_{pzc} and cationic nature of MB. Adsorption of MB on AgNPs increases with increasing pH up to 10 and then remains constant. As the adsorption is an endothermic process, adsorption capacity increases with increases in temperature. The rate of adsorption increases with increases in contact time but the high adsorption rate is initially because of several activated adsorption sites available for dye molecules that increases dye penetration on the adsorption surface. As the adsorbent concentration increases, the adsorption capacity for MB decreases due to the saturation of adsorption sites (Raoof and Nahid 2021). $BiFeO_3$ with a rhombohedrally deformed perovskite structure, is a novel significant visible-light photocatalyst for the degradation of dyes due to its decreasing energy band gap (2.2 eV) and remarkable chemical stability. At neutral pH, the maximum adsorption capacity of $BiFeO_3$ for MB is 166.02 mg/g at 45°C within the contact time of 230 minutes (Tayyebe and Mohammad 2013). Electrostatic interaction, Lewis acid-base interaction, and hydrogen bonding are found to be the main mechanisms for the adsorption of MB depending upon the surface groups of nanoparticles (Wei et al. 2015).

1.3.2 REMOVAL OF RHODAMINE B

Rhodamine B (RhB) dye is an artificial cationic xanthine dye commonly used in textile, paper, and food industries. The exposure of animals and humans to RhB causes malfunctioning of the respiratory system, and skin and eye irritation. Humic acid (HA) modifying Fe_3O_4 nanoparticles (Fe_3O_4/HA NPs) are used as an adsorbent for removal of Rhodamine B from wastewater with maximum adsorption capacities of 161.8 mg/g. The adsorption parameters such as pH, dose of adsorbent, temperature, initial RhB concentration and contact time are need to be optimised for better adsorption of RhB. The Fe_3O_4/HA NPs are able to remove over 98.5% of Rhodamine B in water at optimised pH. The pH_{pzc} of humic acid (HA) modified Fe_3O_4 nanoparticles is 2.3. When the pH is higher than the pH_{pzc} of Fe_3O_4/HA, the surface of adsorbent is negative, which increases the removal efficiency of the RhB cationic form. However, when the pH is higher than 4, the removal efficiency of RhB decreases. Adsorption of the RhB to Fe_3O_4/HA attains equilibrium in less than 15 minutes and follows the Langmuir adsorption model (Liang et al. 2012).

1.3.3 REMOVAL OF CONGO RED

Congo red is a non-biodegradable anionic azo red dye and stable against oxidising agents due to its complex aromatic structure. Thus it can persist in the environment for a longer time. ZnO nanoparticles synthesised using leaf extracts of *Hibiscus rosasinensis* are one of the potential adsorbents which is used for the removal of Congo red dye from wastewater. The adsorption of dye on the surface of adsorbent entirely depends on the surface chemistry of the adsorbent and dye nature. The concept of point of zero charge (pH_{zpc}) provides a clear understanding of this adsorption phenomenon. The adsorbent surface has no charge if the pH_{zpc} is zero. The pH_{pzc} of green synthesised ZnO NPs is 6.71 which means above this particular pH, the surface of nanoparticles is negative and below this pH, the surface is positively charged. Congo red dye is highly sensitive towards pH. When pH is 4 (below pH_{pzc}), there are more electrostatic attractions between the positively charged surface of the ZnO NPs and the negatively charged dye molecules, which significantly improves dye removal; however, when pH is higher than 4, dye removal decreases continuously. It might be caused by an increase in the electrostatic attraction between negatively charged dye molecules and the negatively charged (hydroxyl ions) ZnO nanoparticle surface. Congo red dye removal increases from 53.4% to 78.6% as contact time increases from 15 minutes to 150 minutes. With a temperature increase from 10°C to 70°C, the proportion of dye removal falls from 73% to 50%, demonstrating the exothermic nature of adsorption. It might be because anionic dye molecules become more soluble with increasing temperature, which causes the adsorbent's internal heat energy to force dye molecules out of solution. ZnO NPs have a maximum capacity of 71.43 mg/g for Congo red dye, and their adsorption efficiency constantly improves from 62% to 90% with an increase in adsorbent dose from 0.02 g to 0.10 g (Priyanka and Naba 2020). The possible mechanism of ZnO NPs' interaction with Congo red is shown in Reaction 4. ZnO nanoparticles may interact with amine groups and may also be capable of bonding with the azo groups (-N=N-) of Congo red dye. FTIR spectrum of Congo red dye adsorbed on ZnO NPs is different from FTIR of ZnO NPs. It demonstrates a modest reduction in the broad band around 3300–3600 cm^{-1}, confirming the role of the OH group in dye adsorption. Stretching vibrations at 1053 cm^{-1} caused by the sulfonate group (S=O) are observed following CR dye adsorption, as well as a drop and shift in the peaks at 1600–1700 cm^{-1}, as shown in Reaction 4. (Kataria and Garg 2017).

Reaction 4

1.3.4 REMOVAL OF METHYL ORANGE

An anionic, acidic, and non-biodegradable dye is methyl orange (MO). It produces shin eczema and is toxic. Because of its excellent qualities, such as tiny size, high surface area, large adsorption capacity, a high capacity for regeneration, and availability of a large number of reactive sites, nanoparticles are widely utilised for the removal of MO dye. The removal of MO dye from wastewater is accomplished using nickel oxide nanoparticles (NiO NPs), which have an adsorption capacity of 97.56 mg/g. Due to its effect on the surface characteristics of the adsorbent and on the dissociation or ionisation of adsorbate molecules, pH is a crucial element in the interaction between adsorbate and adsorbent. The pH_{ZPC} of NiO NPs is 6.4. The nanoparticles' surface can be positively/negatively charged depending on acidic/basic conditions or pH_{ZPC} value and can attract/repel anionic/cationic dyes (Shoukat et al. 2017). As the pH rises, the proportion of MO dye removal increases steadily, with the maximum adsorption occurring at pH 4. The increase in negatively charged sites and decrease in positively charged sites on NiO NPs (pH > pH_{ZPC}) cause electrostatic repulsion and decrease MO adsorption, respectively. Additionally, hydroxyl species begin to compete with anionic dye for adsorption sites on NiO NPs, which decreases anionic dye adsorption shown in Reaction 5.

Reaction 5

The reduction in MO removal at extremely acidic pH levels may be caused by the fact that low pH leads to a rise in H^+ ions, and the protons can interact with one of the nitrogens in the $N = N$ bond of MO, making it protonated, which results in electrostatic repulsion between protonated MO and the positively charged NiO NPs (pH< pH_{ZPC}). Because MO is more mobile at higher temperatures, the adsorption of MO onto NiO NPs is endothermic in nature, and removal of MO rises as temperature rises. The adsorption and desorption of MO onto NiO NPs is shown in Figure 1.6. At 45°C and 50°C, more than 94% of MO is removed using NiO NPs. MO uptake is considerably higher at a low adsorbent dose, which reduces linearly with increases in the adsorbent amount. The decline in the adsorption capacity of MO at a high adsorbent dose may be due to low surface area and unavailability of adsorption sites (Qamar et al. 2022).

The Bentonite-supported zero-valent iron nanoparticles (B-ZVI NPs) are used to remove 99.75% MO dye effectively from wastewater. The mechanism for removing MO is explained in Reaction 6. The primary elements of the mechanism are the oxidation of B-ZVI NPs, followed by the adsorption of MO to the surface of B-ZVI NPs, the formation of chelate complexes of Fe(II)-dye, and the breakage of the azo bond. The MO molecule accepts one of the two electrons given by ZVI NPs. The azo link was broken by the reaction between ZVI NPs and H_2O or hydrogen ions, which results in the disappearance of the 464 nm visible absorption peaks. Various parameters such as pH, initial concentration of MO, dosage, and temperature affect the removal of MO. The degradation of MO is an endothermic reaction, since the removal rate increases as temperature increases (Zheng-xian et al. 2011).

Reaction 6

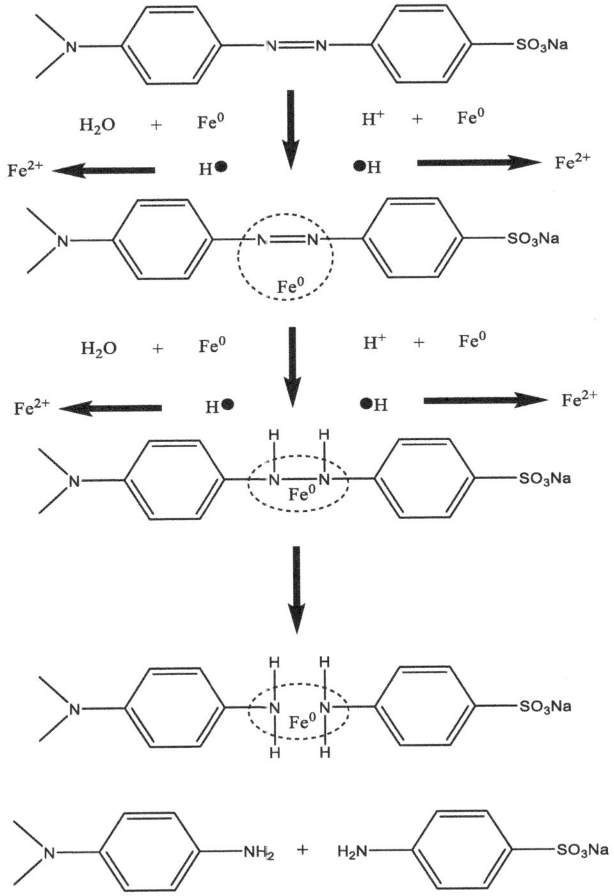

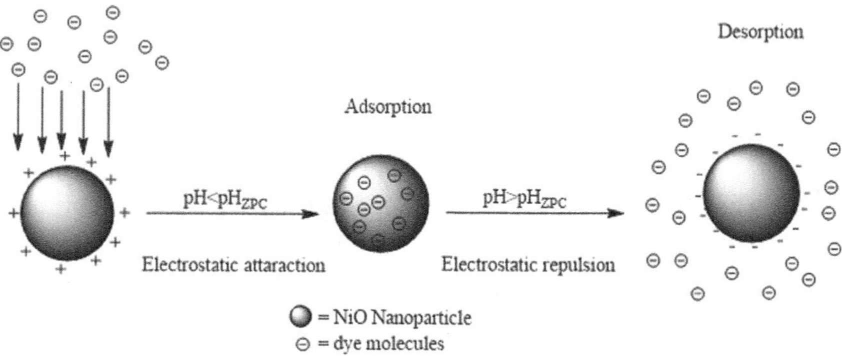

FIGURE 1.6   Adsorption and desorption of MO onto NiO NPs.

1.3.5 REMOVAL OF METHYL RED

Methyl red (MR) is a monoazo dye extensively used in textiles, paper printing, and cosmetic industries. When released into water bodies, MR's carcinogenic and mutagenic properties tend to concentrate in living things, lower the concentration of dissolved oxygen, and negatively impact aquatic life. The dye pollutants are eliminated from the effluent using a number of physical and chemical techniques. However, the dye components are not entirely broken down by these processes, which could result in more pollution. The best approach for removing dye pollutants is photocatalysis utilising nanoparticles since it transforms the dye's components into CO_2, H_2O, and nonharmful byproducts, preventing secondary contamination. As a result of the nanoparticles' high surface-to-volume ratio, there are more available active sites for adsorption, a shorter intraparticle diffusion path, and tunable pore size and surface chemistry. When exposed to sunlight, zinc oxide nanoparticles (ZnO NPs), which are metallic nanoparticles, exhibit photocatalytic activity in the breakdown of MR dye solution. The excited electrons on the photocatalyst's surface travel from the valence band to the conduction band when sunlight strikes a dye solution containing ZnO NPs, leaving a hole in the valence band. While the electrons engage with oxygen to form superoxide anions, the holes interact with water molecules to make reactive hydroxyl radicals. Consequently, the highly reactive hydroxyl radical and superoxide anion are in charge of degrading the methyl red dye shown in Reaction 7 (Anjali et al. 2022).

Reaction 7

$$ZnO\ NPs + hv \longrightarrow h^+\ (valence\ band) + e^-\ (conduction\ band)$$

$$h^+\ (valence\ band) + H_2O \longrightarrow H^+ + OH^-\ (radical)$$

$$e^-\ (conduction\ band) + O_2 \longrightarrow O_2^{\bullet}\ (radical)$$

$$Dye\ molecules + OH^{\bullet}\ (radical) \longrightarrow degraded\ products$$

Magnetic nanoparticles, in particular Fe_3O_4, have received enormous interest in recent years due to their unique magnetic properties, biocompatibility, low toxicity, and simple and inexpensive synthesis procedure. Fe_3O_4-based nanomaterials are also regarded as potential adsorbents for removing contaminants from wastewaters due to the absence of external diffusion resistance, high activity, high surface area, and rapid recovery from liquid phases by an external magnetic field of Fe_3O_4 magnetic nanoparticles. Silica (SiO_2)-coated Fe_3O_4 magnetic nanoparticles remove MR dye molecules from wastewater through adsorption phenomenon by acting as adsorbents. Adsorption is a method for eliminating chemical impurities from wastewater, particularly when typical biological wastewater treatments are ineffective at considerably removing the impurities. This technique is preferred to other ones for wastewater treatment purposes in terms of flexibility and simplicity, cost, ease of use, and sensitivity to toxic compounds. The pH of the solution affects the electrostatic interaction between methyl red and SiO_2-coated Fe_3O_4 magnetic nanoparticles. At low pH levels, partial ionisation of Si-OH could begin, making the surface of the SiO_2-coated Fe_3O_4 magnetic nanoparticles negatively charged. At low pH levels, MR is positively charged which facilitates the electrostatic interaction between dye and SiO_2-coated Fe_3O_4 magnetic nanoparticles. Si-OH ionises on increasing pH which creates more electrostatic attraction sites for dye. The maximum adsorption capacity for the adsorption of MR onto SiO_2-coated Fe_3O_4 magnetic nanoparticles is 49.5mg/g at pH 5 and follows Langmuir adsorption isotherm (Farshid and Soran 2019).

1.3.6 Removal of Brilliant Green

A cationic dye called brilliant green (BG) is frequently released into the water by the textile industry. BG dye causes several health problems such as eye and skin irritation, vomiting, coughing, diarrhoea, and nausea. For the treatment of wastewater, attempts are made to employ metal oxide nanoparticles by utilising techniques like adsorption and photocatalysis. With the use of metal oxide nanoparticles like TiO_2 and ZnO in conjunction with microwave (MW) irradiation, dyes are catalytically degraded. MW technology has attracted interest in recent years due to its various advantages over other advanced oxidation processes (AOPs), including quick and non-selective heating, lower activation energy, improved degradation of organic pollutants, and increased reaction rate. Many applications of MW technology in environmental remediation, such as the treatment of dye or textile effluent, have been proposed. To increase the effectiveness of removing organic contaminants from the wastewater, metal composites are used as MW absorbing materials. Due to its nano size, ZnO can absorb MW energy and can accelerate the breakdown of pollutants. ZnO NPs' ability to absorb MW is enhanced by doping it with composite materials. There have been several attempts to dope non-metallic elements (C, S, and N) into ZnO NPs to enhance their catalytic activity for the breakdown of organic contaminants. Carbon-doped ZnO (C-ZnO) NPs, by absorbing MW energy, may create a hotspot effect leading to OH radical production from water splitting and the degradation of organic dye compounds. OH radical help in intermediate product degradation or the full mineralisation of BG is shown in Reaction 8 (Binay et al. 2020).

Reaction 8

$$MW + C\text{-}ZnO \longrightarrow \text{hotspot effect} + \text{heating effect} \longrightarrow e^- + OH^\cdot \text{ (water spilliting)}$$

$$BG + OH^\cdot \longrightarrow \text{intermediates} \longrightarrow CO_2 + H_2O$$

ZnO nanoparticles alone are also used as adsorbent for removal of BG. The pH has a significant impact on how dye molecules adsorb on the surface of an adsorbent. ZnO nanoparticles have a pH_{pzc} of 6.9. The presence of hydroxyl groups causes the surface of ZnO NPs to be negatively charged above pH 6.9. The cationic BG dye present in solution interacts electrostatically with this negatively charged surface shown in Reaction 9 (Kataria and Garg 2017).

Reaction 9

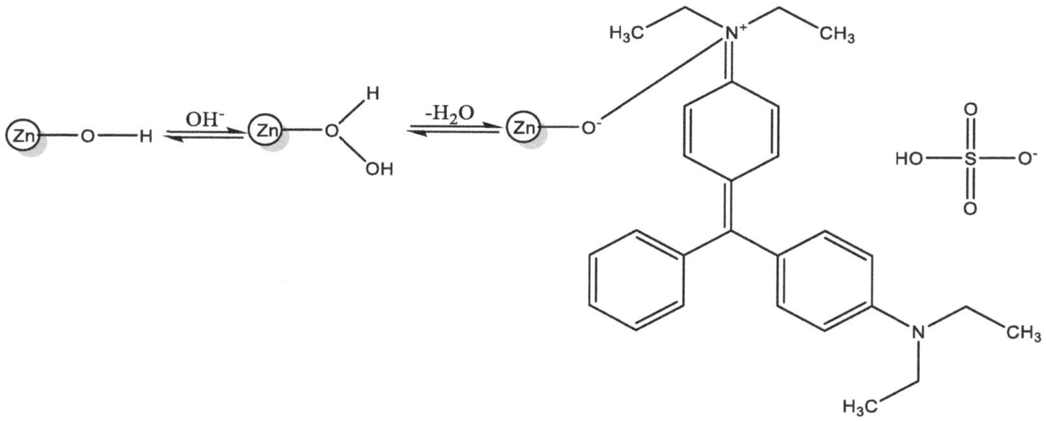

1.3.7 REMOVAL OF INDIGO CARMINE DYE

Indigo carmine dye is an anionic dye, used in the dyeing of clothes. It serves as a microscopic stain in biology and a redox indicator in analytical chemistry. The indigo carmine dye is a member of the highly poisonous indigoid class of dye and its exposure can irritate human skin and eyes. It can also induce permanent injury to conjunctiva and cornea. The photocatalytic activity of nanoparticles in photocatalytic oxidation is influenced by their shape, crystallite size, surface area, band gap, and rate of electron/hole (e^-/h^+) pair recombination. Numerous semiconductor oxides, such as ZnO, TiO_2, and SnO_2 nanoparticles, are utilised as photocatalysts for the breakdown of these toxic dyes. Because of their large band gap, these semiconductor oxides only absorb 5% of sunlight's UV rays. As a result, it is necessary to create a catalyst system capable of enhancing the degradation of water contaminants. Because of its narrow band gap energy and high chemical stability, bismuth ferrite nanoparticles ($BiFeO_3$ NPs) are an active photocatalyst with enhanced degradation activity. Substituting Bismuth ferrite nanoparticles ($BiFeO_3$ NPs) for Gadolinium (Gd^{3+}) reduces the band gap even further, increasing visible-light absorption and water contaminant degradation. When Gd^{3+} substituted $BiFeO_3$ NPs are exposed to visible light, they generate photo-generated holes in the valance band and electrons in the conduction band. The substituent Gadolinium in $BiFeO_3$ NPs acts as an electron trapping agent, and captures the excited electrons. It results in the separation of the e^-/h^+ pair and promotes the transfer of charges from the photocatalyst's bulk to its surface. These electrons, which are transported to the photocatalyst's surface, interact with the adsorbed O_2 to form O_2^-. The water molecules and holes combine to generate the HO radical. The breakdown of dye molecules is facilitated by these e^-, h^+, O_2^-, and OH radical species shown in Reaction 10.

Reaction 10

$$Bi_{1-x}Gd_xFeO_3 + h\nu \longrightarrow Bi_{1-x}Gd_xFeO_3 \ (e^- + h^+)$$

$$e^- + Gd \text{ substituent} \longrightarrow (e^- + Gd \text{ substituent trapped})$$

$$(e^- + Gd \text{ substituent trapped}) + O_2 \longrightarrow O_2^-$$

$$Bi_{1-x}Gd_xFeO_3 \ (h^+) + H_2O \longrightarrow OH^- \text{ (radical)}$$

$$h^+ e^- + OH^- \text{(radical)} + \text{organic dyes} \longrightarrow CO_2 + H_2O + \text{byproducts}$$

An increase in catalyst dosage causes an increase in degradation rate up to a certain point, after which it causes a decrease in photocatalytic rate. This could be due to the presence of nanoparticles, which increase the number of absorbed photons, resulting in the adsorption of more dye molecules. Furthermore, the density and screening properties of nanoparticles act as shields, and the increased catalyst dosage impedes light penetration. Consequently, less surface area is available for light to penetrate, which lowers the rate of degradation (Patil et al. 2018).

A novel adsorbent called cetyltrimethylammonium bromide (CTAB)-modified TiO_2 nanoparticles was used for efficient removal of anionic dyes. Despite having huge surface areas and a hydrophilic character, the metal oxide NPs have a limited ability to adsorb hydrophobic substances. The addition of a surfactant coating to these materials can improve their ability to adsorb organic contaminants. Surfactant forms bilayered micelles on the surface of metal oxide NPs. The surfactant-coated metal oxide surface turns hydrophilic in nature again due to the ionic group of surfactant, which is able to provide both hydrophobic and electrostatic interactions with pollutants. The ability of CTAB (cetyltrimethylammonium bromide)-coated TiO_2 NPs to adsorb Indigo carmine dye is enhanced due to the strong electrostatic interaction between the negatively charged dye and the positively charged surface of the coated nanoparticles (Javad et al. 2014).

1.4 CONCLUSION

The continuous development and advancement of nanotechnology in many forms for wastewater treatment has been driven by the constant demand for fresh water. Although many scientists view nanotechnology as a new era, due to the novelty of the technology, most people are still unaware of many aspects of it. Due to their very profitable properties as adsorbents, nanoparticles have attracted special interest for water treatment over the past decade. Researchers have seen an exponential growth in nanoparticle synthesis due to the exploitation of its high surface area, high rate of adsorption, and simple, affordable synthesis for the purpose of wastewater purification through the removal of heavy metals and dyes. For the purpose of eliminating heavy metals from drinking water, nanoparticles are efficient and much less dangerous tools. Adsorption is a useful method for removing dye pollutants and heavy metals from water because of its high efficacy, low cost, and ease of use. It will be necessary to reduce risks to public and environmental health by reducing potential exposure to nanoparticles and promoting their safer design in order to achieve widespread adoption of nanotechnology in water treatment. This will require overcoming the relatively high costs of nanomaterials through their ability to be reused. To reduce unexpected repercussions and support sustainable water management, the advancement of nanotechnology must coexist alongside research into environmental health and safety. As a result, there is still much to learn and understand about the best ways to combine various treatment technologies in order to develop the finest treatment methods that will allow people to preserve Earth's water supply for future generations.

REFERENCES

Ahmad K., Badawi, M. Abd Elkodous, and Gomaa AM Ali. "Recent advances in dye and metal ion removal using efficient adsorbents and novel nano-based materials: An overview." *RSC Advances* 11, no. 58 (2021): 36528–36553.

Ahmad, Mohd Azmier, Norhidayah Ahmad, and Olugbenga Solomon Bello. "Adsorptive removal of malachite green dye using durian seed-based activated carbon." *Water, Air, & Soil Pollution* 225, no. 8 (2014): 1–18.

Anjali, K. P., R. Raghunathan, Geetha Devi, and Susmita Dutta. "Photocatalytic degradation of methyl red using seaweed mediated zinc oxide nanoparticles." *Biocatalysis and Agricultural Biotechnology* 43 (2022): 102384.

Binay Kumar, Tripathy, Sumit Kumar, Mathava Kumar, and Animesh Debnath. "Microwave induced catalytic treatment of brilliant green dye with carbon doped zinc oxide nanoparticles: Central composite design, toxicity assessment and cost analysis." *Environmental Nanotechnology, Monitoring & Management* 14 (2020): 100361.

Congcong, Ding, Wencai Cheng, Xiangxue Wang, Zhen-Yu Wu, Yubing Sun, Changlun Chen, Xiangke Wang, and Shu-Hong Yu. "Competitive sorption of Pb (II), Cu (II) and Ni (II) on carbonaceous nanofibers: A spectroscopic and modeling approach." *Journal of Hazardous Materials* 313 (2016): 253–261.

Deepika, Thilakan, Jaie Patankar, Srushti Khadtare, Nilesh S. Wagh, Jaya Lakkakula, Khalid Mohamed El-Hady, Saiful Islam et al. "Plant-Derived Iron Nanoparticles for Removal of Heavy Metals." *International Journal of Chemical Engineering* 2022 (2022).

Farshid, Ghorbani, and Soran Kamari. "Core–shell magnetic nanocomposite of Fe3O4@ SiO2@ NH2 as an efficient and highly recyclable adsorbent of methyl red dye from aqueous environments." *Environmental Technology & Innovation* 14 (2019): 100333.

Fato, Fato Patrice, Da-Wei Li, Li-Jun Zhao, Kaipei Qiu, and Yi-Tao Long. "Simultaneous removal of multiple heavy metal ions from river water using ultrafine mesoporous magnetite nanoparticles." *ACS Omega* 4, no. 4 (2019): 7543–7549.

Gil-Díaz, M., J. Rodríguez-Alonso, C. A. Maffiotte, D. Baragano, R. Millan, and M. C. Lobo. "Iron nanoparticles are efficient at removing mercury from polluted waters." *Journal of Cleaner Production* 315 (2021): 128272.

Javad, Zolgharnein, Maryam Bagtash, and Neda Asanjarani. "Hybrid central composite design approach for simultaneous optimization of removal of alizarin red S and indigo carmine dyes using cetyltrimethylammonium bromide-modified TiO$_2$ nanoparticles." *Journal of Environmental Chemical Engineering* 2, no. 2 (2014): 988–1000.

Jien, Ye, Yi Wang, Qiao Xu, Hanxin Wu, Jianhao Tong, and Jiyan Shi. "Removal of hexavalent chromium from wastewater by Cu/Fe bimetallic nanoparticles." *Scientific reports* 11, no. 1 (2021): 1–11.

Jinyue, Yang, Baohong Hou, Jingkang Wang, Beiqian Tian, Jingtao Bi, Na Wang, Xin Li, and Xin Huang. "Nanomaterials for the removal of heavy metals from wastewater." *Nanomaterials* 9, no. 3 (2019): 424.

Julia de O., Primo, Carla Bittencourt, Selene Acosta, Ayrton Sierra-Castillo, Jean-François Colomer, Silvia Jaerger, Verônica C. Teixeira, and Fauze J. Anaissi. "Synthesis of zinc oxide nanoparticles by ecofriendly routes: Adsorbent for copper removal from wastewater." *Frontiers in Chemistry* 8 (2020): 571790.

Kataria, Navish, and V. K. Garg. "Removal of Congo red and Brilliant green dyes from aqueous solution using flower shaped ZnO nanoparticles." *Journal of Environmental Chemical Engineering* 5, no. 6 (2017): 5420–5428.

Keerti, Jain, Anand S. Patel, Vishwas P. Pardhi, and Swaran Jeet Singh Flora. "Nanotechnology in wastewater management: A new paradigm towards wastewater treatment." *Molecules* 26, no. 6 (2021): 1797.

Kumar, K. Yogesh, TN Vinuth Raj, S. Archana, SB Benaka Prasad, Sharon Olivera, and H. B. Muralidhara. "SnO2 nanoparticles as effective adsorbents for the removal of cadmium and lead from aqueous solution: Adsorption mechanism and kinetic studies." *Journal of Water Process Engineering* 13 (2016): 44–52.

Liang, Peng, Pufeng Qin, Ming Lei, Qingru Zeng, Huijuan Song, Jiao Yang, Jihai Shao, Bohan Liao, and Jidong Gu. "Modifying Fe3O4 nanoparticles with humic acid for removal of Rhodamine B in water." *Journal of hazardous materials* 209 (2012): 193–198.

Mahsa, Bagheri Saeid Azizian, Babak Jaleh, and Abdolkarim Chehregani. "Adsorption of Cu (II) from aqueous solution by micro-structured ZnO thin films." *Journal of Industrial and Engineering Chemistry* 20, no. 4 (2014): 2439–2446.

Patil S., Basavarajappa, Bhojya Naik H. Seethya, Nagaraju Ganganagappa, Kumaraswamy B. Eshwaraswamy, and Raghava Reddy Kakarla. "Enhanced photocatalytic activity and biosensing of gadolinium substituted BiFeO3 nanoparticles." *Chemistry Select* 3, no. 31 (2018): 9025–9033.

Priyanka, Debnath, and Naba Kumar Mondal. "Effective removal of congo red dye from aqueous solution using biosynthesized zinc oxide nanoparticles."

Qamar, Riaz, Madiha Ahmed, Muhammad Nadeem Zafar, Muhammad Zubair, Muhammad Faizan Nazar, Sajjad Hussain Sumrra, Iqbal Ahmad, and Ahmad Hosseini-Bandegharaei. "NiO nanoparticles for enhanced removal of methyl orange: Equilibrium, kinetics, thermodynamic and desorption studies." *International Journal of Environmental Analytical Chemistry* 102, no. 1 (2022): 84–103.

Raoof, Jabbari and Nahid Ghasemi. "Investigating methylene blue dye adsorption isotherms using silver nano particles provided by aqueous extract of tragopogon buphthalmoides." *Chemical Methodologies* 5, no. 1 (2021): 21–29.

Richmond J., Bartlett, "Chromium cycling in soils and water: Links, gaps, and methods." *Environmental Health Perspectives* 92 (1991): 17–24.

Shoukat, Sidra, Haq Nawaz Bhatti, Munawar Iqbal, and Saima Noreen. "Mango stone biocomposite preparation and application for crystal violet adsorption: A mechanistic study." *Microporous and Mesoporous Materials* 239 (2017): 180–189.

Sithi, Mgidlana, and Tebello Nyokong. "Asymmetrical zinc (II) phthalocyanines cobalt tungstate nanomaterial conjugates for photodegradation of methylene blue." *Journal of Photochemistry and Photobiology A: Chemistry* 418 (2021): 113421.

Sumesh, E., M. S. Bootharaju, and T. Pradeep. "A practical silver nanoparticle-based adsorbent for the removal of Hg^{2+} from water." *Journal of Hazardous Materials* 189, no. 1–2 (2011): 450–457.

Tayyebe, Soltani, and Mohammad H. Entezari. "Photolysis and photocatalysis of methylene blue by ferrite bismuth nanoparticles under sunlight irradiation." *Journal of Molecular Catalysis A: Chemical* 377 (2013): 197–203.

Wan W.S., Ngah, and MAK Megat Hanafiah. "Removal of heavy metal ions from wastewater by chemically modified plant wastes as adsorbents: A review." *Bioresource Technology* 99, no. 10 (2008): 3935–3948.

Wei, W., L. Yang, W. H. Zhong, S. Y. Li, J. Cui, and Z. G. Wei. "Fast removal of methylene blue from aqueous solution by adsorption onto poorly crystalline hydroxyapatite nanoparticles." *Dig. J. Nanomater. Biostruct* 19 (2015): 1343–1363.

Wenqian, Ruan, Jiwei Hu, Jimei Qi, Yu Hou, Chao Zhou, and Xionghui Wei. "Removal of dyes from wastewater by nanomaterials: A review." *Advanced Materials Letters* 10, no. 1 (2019): 9–20.

Ze, Lin, Xiulan Weng, Gary Owens, and Zuliang Chen. "Simultaneous removal of Pb (II) and rifampicin from wastewater by iron nanoparticles synthesized by a tea extract." *Journal of Cleaner Production* 242 (2020): 118476.

Zheng-xian, Chen, Xiao-ying Jin, Zuliang Chen, Mallavarapu Megharaj, and Ravendra Naidu. "Removal of methyl orange from aqueous solution using bentonite-supported nanoscale zero-valent iron." *Journal of Colloid and Interface Science* 363, no. 2 (2011): 601–607.

Zhicheng, Wu, Xiaobao Su, Zhang Lin, Gary Owens, and Zuliang Chen. "Mechanism of As (V) removal by green synthesized iron nanoparticles." *Journal of hazardous materials* 379 (2019): 120811.

2 Hybrid Treatment Technologies for the Removal of Pharmaceuticals in Wastewater

Amal Abdelhaleem, Faissal Aziz, Mohamed Gar Alalm, and Alaa El Din Mahmoud

2.1 INTRODUCTION

Worldwide water use is expected to rise by 55%, and water stress is already present in about 25% of major cities (Salehi 2022). More recently, the world's finite freshwater resources have been further strained by factors like global warming, water scarcity, demand growth, and inadequate management, leaving over 4 billion people with acute water shortages for at least one month each year (van Vliet, Jones et al. 2021, Dotaniya, Meena et al. 2022).

The majority of the research on water scarcity concentrates on the amount of water. However, if safety procedures are not in place, the water quality given to consumers during a water shortage could be in danger (Mahmoud, Umachandran et al. 2021). In order to preserve reliable access to sufficient clean water as droughts brought on by global warming become more frequent, efficient water management and distribution planning are crucial. Due to the increasing consumption of water sources and insufficient infrastructure many water treatment plants have failed to fulfill typical water demand (Salehi 2022).

Due to improvements in research and development, the expanding world population, and improved access to healthcare and medications, consumption of pharmaceuticals and personal care products (PPCPs) increased dramatically over the past decade (Adeleye, Xue et al. 2022). These compounds are considered as emerging contaminants and have gained attention after the traditional priority pollutants (such as polychlorinated biphenyls and polycyclic aromatic hydrocarbons) (Xiang, Wu et al. 2021). Pharmaceuticals in drinking water have become a significant issue worldwide, particularly in developed countries, according to the World Health Organization (WHO) (Lu and Astruc 2020). Furthermore, the emergence of COVID-19 led to higher levels of PPCPs in the aquatic environment and the usage of therapeutic drugs (Gwenzi, Selvasembian et al. 2022).

Pharmaceuticals can enter the environment in various ways, such as wastewater from pharmaceutical industries and solid waste. Drugs' non-metabolized active pharmaceutical ingredients (APIs) are expelled in the urine and feces, and topical medications wash off the skin when you take a bath. These nominal leftovers greatly impact the general status of environmental contamination. The body alters 90% of medications taken orally that aren't for local use; as a result, 7.3 units of the active ingredient end up in the wastewater system (Vambol, Vambol et al. 2021). Accordingly, municipal wastewater treatment facilities produce solid sludge containing many pharmaceutically active chemicals. Table 2.1 illustrates the usual constituents of pharmaceutical wastewater effluents.

DOI: 10.1201/9781003408352-3

TABLE 2.1
Usual Constituents of Pharmaceutical Wastewater Effluents

Parameters	Effluent values (mg L^{-1}, except pH)
Total dissolved solids	1350–7250
Total suspended solids	500–2000
pH	1.5–7
Chemical oxygen demand	2000–6000
Biological oxygen demand	400–900

The ecosystem, human health, and the accessibility of drinking water in aquatic areas are all adversely affected by pharmaceuticals (Dhangar and Kumar 2020). One of the primary problems with pharmaceuticals is antibiotic residues, which are regularly found in surface waters and may lead to the formation of antibiotic-resistant microorganisms. The hormone imbalance due to its bioaccumulation in the aquatic environment has a number of impacts, including reduced fertility, reproductive problems, and a higher risk of hormone-related cancer (Dhangar et al. 2020). A channel and a sink for pharmaceutical release into the environment are possible with wastewater treatment plants (WWTPs). However, pharmaceuticals are not currently targeted or removed by WWTPs, which are made to remove nutrients, pathogens, and particulates from industrial and municipal wastewater. Therefore, it is crucial and helpful for future attempts to increase their removal efficiency by WWTPs to have a thorough understanding of the degradation mechanisms and contribution of each treatment step (Adeleye, Xue et al. 2022).

Advanced oxidation processes (AOPs) are efficient techniques for treating wastewater that contains biorecalcitrant pharmaceuticals. The most promising AOP is heterogeneous photocatalysis. Although numerous AOPs, such as photocatalysis and electrochemical oxidation, have been used for water purification, most of these methods are expensive, time-consuming, inadequate, and demand significant energy inputs. Hence, hybridizing AOPs with other treatment technologies could be the best option to fill these gaps.

With an emphasis on innovative AOPs, the main goal of this chapter is to explore hybrid wastewater treatment technologies for eliminating pharmaceutical pollutants from wastewater effluents.

2.2 PHARMACEUTICAL COMPOUNDS IN WASTEWATER

Pharmaceutical compounds are often produced using static methods, which results in the diversity of PPCPs in the wastewater produced by various procedures. Huge amounts of liquid cleaners are needed for solid washing cake, extraction, and equipment cleanup. Pharmaceutical industry manufacturing processes and the widespread consumption of PPCPs are the two main contributors to the presence of pharmaceutical compounds in drinking water.

The effluents created by various pharmaceutical and medication production procedures contain various chemicals. On the one hand, pharmaceutical commerce is overwhelmed by high-value, low-volume multiproduct factories that are predominantly static processes in which liquid waste is processed. On the other hand, factories are set aside for large-scale, continuous, and semi-batch pharmaceutical production. These facilities employ a variety of reaction mixes, (homogeneous) catalysts, solvent-based solids, and water, all under the control of specialist equipment. Instead of purity, the type of impurity affects the price of the medication in these units the most.

As a result, separation procedures are crucial to the operation of that industry. The environmental quotient or E-factor of PPCPs manufacturing ranges from 50 to 100 kg/t due to the multistep procedures (5 to 30 steps). In the medicine industry, ultrapure water is also used as an extractant or solvent, or to repeatedly wash the solid cake. This water is not reused due to the stringent limitations outlined in the government's drug master file (DMF) protocols. Pharmaceutical residues in aquatic ecosystems provide difficulties in terms of their presence, impacts, and hazardous effects. Therefore, as a treatment alternative for this problem, the recovery of highly valuable APIs and pharmaceutical medicines from diluted streams should be investigated. It has a high concentration of non-biodegradable organic substrate, such as spent solvents, beta-lactamides, reproductive hormones, animal and plant steroids, over-the-counter drugs, anti-inflammatories, antibiotics, anti-depressants, lipid regulators, detergent metabolites, personal care items, byproducts of oil use and combustion, flame retardants, and other commonly used substances (Vuppala, Motappa et al. 2012, Ramola and Singh 2013). The complete dissemination and consequences of such particles' presence in the environment are frequently unidentified and nonspecific. Trace amounts of these compounds, typically a few parts per trillion, have been found in the aquatic environment, including lakes, rivers, and groundwater. Therefore, it is commonly believed that drugs will not hurt the environment. It is challenging to conduct a thorough risk assessment due to a lack of approved analysis techniques, relevant monitoring data, and supporting knowledge regarding the toxicity, the destination, and/or the metabolites of pharmaceutical substances in aquatic ecosystems. Due to the large range of wastewater features, the predicted effects of wastewater on fauna and flora are highly variable. Pharmaceutical chemicals provide an increased risk to both human and livestock health due to their genotoxicity and mutagenic properties, in addition to their acute toxicity. Because these contaminants build up in the environment through the food chain, when they are dumped on land or into bodies of water, they have an impact on the health of people and other living organisms.

2.3 TREATMENT OF PHARMACEUTICALS IN WASTEWATER

In the production of pharmaceuticals and chemicals, water is a crucial feedstock. Additionally, other processes like production, cooling, and material processing depend on stable and high-quality water sources. In addition to being a prerequisite for sterilizing reactors or medical equipment in various medical applications, such as injection water, process water quality control is of utmost significance in pharmaceutical manufacture (Gadipelly, Pérez-González et al. 2014). Process wastewaters are the wastewaters produced by internal processes in any industry. As a result, water that interacts with various unit operations or processes, including raw materials, finished goods, intermediates, or waste products during production or processing, is referred to as a process wastewater. The effluent from PPCPs reactors are different in composition and characteristics, but it is rarely given particular treatment because the volumes are always so limited and various products are produced using the same processors and filters. Reusing water reduces waste, discarding expenses and feedwater needs, offsetting the operational costs of waste reuse (Gadipelly, Pérez-González et al. 2014).

The pharmaceutical industry makes use of numerous different wastewater treatment and disposal methods. Based on the raw ingredients and manufacturing techniques used to create various drugs, the substance and quantity of these companies' wastewaters varies by location, season, and even time period. The value of readily available water is also influenced by a plant's location. As a result, it is quite challenging to pinpoint a single treatment plan for a sector of the economy as diverse as pharmaceuticals. Numerous treatment options are available to handle the enormous variety of wastes produced by this company, but they are industry- and waste-specific.

2.3.1 RECOVERY PROCESSES

An integrated waste management strategy for pharmaceutical plants entails the pre-treatment and recovery of numerous valuable byproducts, including solvents, acids, heavy metals, and various key APIs that find their way into industrial effluents. The fermentation broth of crops used for fermentation is a rich source of mycelia and solvent. Some of the solvents are not biodegradable and have a high biochemical oxygen demand (BOD) concentration. Recovering the pharmaceutical chemicals can immediately balance operational costs for waste treatment and improve the processes' economics by lowering or reducing the cost of waste disposal for the main process system and the need for raw water for the intermediate unit process. The process can recycle the waste stream that can be recovered, and the water can be utilized as a boiler supply or as a cooling tower intake. In reality, following processing, hot waste streams have the potential to feed additional heat exchangers (heat pinching) or boilers, decreasing water and energy expenditures. In general, if the product is in the stream alone and has a molecular weight greater than 250 Da, it can be retrieved utilizing efficient membrane techniques. Using reverse/forward osmosis and nano/ultra filtration can provide significant economic benefits. The filtrate can then go through additional processing steps.

The most recent pressure-driven membrane separation technique, known as nanofiltration (NF), has seen its applications quickly grow over the past ten years (Mei, Shixi et al. 2000, Zhang, He et al. 2003). NF can recover over 80% of a highly contaminated water at a higher grade than feedwater for high operating efficiency and security of products. Ultrafiltration has also been utilized effectively to extract organic molecules from various synthetic media from the fermentation process effluent. A significant component of fermentation is alkaline protease, which accounts for around 60% of all enzyme sales. Using an ultrafiltration process with an optimum transmembrane pressure of 90 kPa and a feed flow of 714 L/(h/m^2), 83% of the alkaline protease activity was recovered.

2.3.2 Biological Treatment

Since they efficiently convert the majority of waste into innocuous gases and get rid of sludge, biological treatment technologies are typically utilized to treat the light streams from industrial units. The treatments are offered for the anaerobic hybrid reactor, carbon-attached activated sludge process, trickling filtering, and activated sludge method. Advanced oxidation systems and membrane approaches exist in addition to the aforementioned standard treatment methods (Deegan, Shaik et al. 2011). Traditionally, biological treatment technologies (aerobic and anaerobic) have remediated pharmaceutical compounds.

2.3.2.1 Aerobic Treatment

Some of the most popular aerobic treatment techniques include membrane bioreactors, extended aeration activated sludge (EAS), activated sludge (AS), and activated sludge with granular activated carbon (AS-GAC) (Peng, Li et al. 2004, Helmig, Fettig et al. 2005). Activated sludge (AS) is the most prevalent aerobic treatment that has been demonstrated to be effective for a variety of pharmaceutical wastewater types (El-Gohary, Abou-Elela et al. 1995). The two key elements that affect how efficiently and affordably the traditional activated sludge treatment operates are temperature and hydraulic retention time. The effectiveness of the AS approach is also impacted by the chemical oxygen demand (COD), suspended solids (SS), BOD, pH, and the presence of biorecalcitrant products. In a previous study (Tekin, Bilkay et al. 2006), the authors used an aerobic sequential batch reactor to remove 98% of COD from industrial wastewater that had undergone Fenton pre-treatment. In other studies, ibuprofen, naproxen, bezafibrate, ethinylestradiol, and several other estrogens were efficiently degraded by aerobic treatment, whereas sulfa medications including sulfamethoxazole, carbamazepine, and diclofenac exhibited lower removal efficiencies (Clara, Strenn et al. 2005, Deegan, Shaik et al. 2011).

Due to the fact that membrane bioreactors (MBRs) are a technically feasible alternative for wastewater treatment, particularly due to the high sludge retention period, they have attracted a lot of attention in the last ten years. The MBR may have a microorganism concentration of up to 20 mg/L suggesting faster degradation of organic pharmaceuticals (Radjenovic, Petrovic et al. 2007). This high biomass content enhances the degradability of bigger organic compounds. Separating suspended particles by membranes, which is not restricted by the settling properties of the sludge, is another advantage of membrane treatment (Urase, Kagawa et al. 2005). An MBR coupled with a conventional activated sludge reactor achieved removal efficiencies of 98.7% in a study with wastewaters containing antibiotics (ofloxacin, sulfisoxazole, and erythromycin), analgesics, and anti-inflammatory drugs (ibuprofen, naproxen, indomethacin, and acetaminophen). It is uncommon for MBR or a single procedure to eliminate all medications. In a different study, MBR effectively reduced the levels of more than 10 estrogens, including 17-estradiol (E2), 17-estradiol, 17-dihydroequilin, medrogestone, estriol (E3), norgestrel, estradiol valerate, and trimegestone, to levels that were close to or below analytical detection. Nevertheless, a specific serotonin reuptake inhibitor could not completely eliminate various pharmaceutical effluents from the system. It is therefore necessary to apply a mix of several pre-and post-treatment procedures (Urase, Kagawa et al. 2005).

The molecularly imprinting technique (MIP) has various benefits over standard immunosorbent (IS), including excellent affinity and selectivity, ease of preparation, and high stability. Compared to biological receptors, the MIPs may be employed repeatedly without losing their function, have high mechanical strength, and resist severe chemical media, heat, and pressure. MIP targeting tetracycline (TC) and oxytetracycline (OTC) to remove the antibiotics selectively was developed, along with various tetracycline derivatives from pig kidney tissue. Suedee et al. described the use of molecularly imprinted polymers from a combination of tetracycline and its breakdown products to generate affinity membranes for removing tetracycline from water (Suedee, Srichana et al. 2004). The pharmaceutical industry will utilize membrane processes to implement a global approach based on the contaminants' sizes, starting with microbial particles (microfiltration), moving on to biomolecules and pathogens (ultrafiltration), divalent ions (NF), and finally reverse osmosis for monovalent ions.

2.3.2.2 Anaerobic Treatment

Fluidized bed reactors, up-flow anaerobic digester reactors, continuously stirred tank reactors (anaerobic digestion), etc. have all been used for treating wastewater (Beun, Hendriks et al. 1999, Enright, McHugh et al. 2005, Oktem, Ince et al. 2008, Tang, Zheng et al. 2011). Anaerobic composite systems, a combination of suspended culture and connected culture reactors, have drawn a lot of attention lately. The merits of anaerobic treatment versus aerobic systems cover the capacity to handle considerable quantities of wastewater with reduced energy requirements, a low biomass release, better pricing, and the economic expansion of biogas as a valued source of energy as a reaction product (Deegan, Shaik et al. 2011). Large volumes of medicines can be removed from wastewater with great efficiency using the up-flow anaerobic batch reactor (USAR). Even at a very high organic material concentration of 9 kg COD/(m3/day), a USAR operating at higher temperatures of roughly 55°C demonstrated a high removal of COD (65–75%) and BOD (80.94%) (Sreekanth, Sivaramakrishna et al. 2009). Antibiotic-effluent wastewater can be shown to be capable of removing 75% COD and more than 95% Tylosin, making USAR a suitable use for such wastewater. High COD vitamin process effluent was treated using a combination of anaerobic biological oxidation and catalytic moist air oxidation units (Kang, Zhan et al. 2011). With this combination, greater than 94.66% COD elimination was achieved, and the organic content was completely biodegradable. For chemical synthesis effluent with a COD between 400 and 6000 mg/L, better than 60.65% removal was achieved by a composite vertical anaerobic sludge blanket system. The methanogenic investigation did not reveal any inhibitory effect, and the generated biosolids was extremely cost-effective (Oktem, Ince et al. 2008). Recently, Sponza and Celebi introduced an oxytetracycline-laced antibiotic effluent to an anaerobic multi-chamber bed reactor (AMCBR) and continuous stirred tank reactor (CSTR) (Sponza and Çelebi 2012). The combination of an anaerobic AMCBR and aerobic CSTR treatment system shown effectiveness in removing OTC from synthetic wastewater with high yields (>95%) at OTC loadings (177.78 g of OTC/(m3/day)). Colistin sulfate and kitamycin could not be eliminated by Tang et al. from wastewater containing antibiotics (Tang, Zheng et al. 2011).

2.3.3 ADVANCED TREATMENT PROCESSES

To improve the effectiveness of the secondary treatment, advanced treatment of pharmaceuticals in wastewater may be regarded as the primary treatment or pre-treatment phase. Standard advanced treatment technologies include membrane technology, membrane distillation, activated carbon adsorption, and advanced oxidation techniques.

2.3.3.1 Membrane Technology

Recently, the use of membranes in water filtration has steadily increased. It is reported that low-pressure membranes may eliminate microbiological components without raising disinfection byproducts, allowing compliance with the requirements enacted in response to the Surface Water Treatment Rule Amendments of 1986 (Snyder, Adham et al. 2007). As reverse osmosis (RO) pre-treatment procedures, low-pressure membrane systems play a significant role in desalination and water reuse. In one study, 95% diclofenac rejection was achieved with RO membranes. Nanofiltration/ultrafiltration (NF/UF) methods are particularly recommended in the case of small spaces and/or unpredictable feedwater quality (Adham, Chiu et al. 2005). NF and UF are highly effective for the removal of micropollutants and natural organic materials (NOMs) from reclaimed wastewater and drinking water. The NF membrane held more personal care products (PCPs), especially endocrine-disrupting chemicals (EDCs), than the UF membrane, indicating the high influence of the pore size. Additionally, the chemistry of the source water showed an impact on the retention of EDC/PCPs (Yoon, Westerhoff et al. 2007). As a result, it can be said that both RO and NF are more effective at getting rid of particular organic drugs, but the disposal problem with retentate and concentrates persists.

2.3.3.2 Activated Carbon

Due to its substantial surface area (approximately 1000 m^2/g), well-developed surface morphology, and physical–chemical properties, activated carbon (AC) is good for removing diclofenac and nimesulide from aqueous solutions (dos Reis, Bin Mahbub et al. 2016). Ibuprofen elimination using activated carbon generated from trash was recently proven (Mestre, Pires et al. 2007). Thus, the AC method offers the benefit of a simple raw material input for carbon synthesis. The AC method employs powdered activated carbon (PAC) or granulated activated carbon (GAC). PAC offers an advantage over GAC since it is often new, whereas GAC is typically recycled in fixed bed columns (Westerhoff, Yoon et al. 2005, Stoquart, Servais et al. 2012). PAC is more efficient than GAC, but it is more expensive, and it can be difficult to regenerate or dispose of saturated GAC columns (Snyder, Adham et al. 2007). Depending on the physicochemical characteristics of each molecule, PAC could remove target compounds to variable degrees, according to research on the adsorption of EDCs and PCPs by PAC in various water sources (Cyr, Suri et al. 2002). The most difficult part of using PAC is separating the treated water from the adsorbent; as a result, it must be used in conjunction with a filtration device. Activated carbon could be utilized as a pre-treatment in several recent pieces of research on employing AC in conjunction with other treatment methods (Westerhoff, Yoon et al. 2005).

2.3.3.3 Membrane Distillation

Membrane distillation is a vital purification method with intriguing merits. First, membrane distillation can be employed for the production of mineralized water (Hausmann, Sanciolo et al. 2011, Gryta 2012). Second, the membrane distillation method uses very little heat and operates at atmospheric pressure (Hausmann, Sanciolo et al. 2011). The method has successfully recovered process waters utilizing the heat generated by industrial operations, which makes its future applications very promising (Song, Li et al. 2007). Thirdly, membrane distillation produces extremely pure water, although membrane fouling is a significant drawback. Finally, membrane distillation has successfully recovered acid from fermentation broths (Gryta, Markowska-Szczupak et al. 2013).

2.3.3.4 Advanced Oxidation Processes (AOPs)

Because many pharmaceuticals have a limited capacity for biodegradation, current treatment methods fall short of completely eliminating these species, and discharging treated effluents into receiving waters runs the risk of contaminating the water with these contaminants. Additionally, these products may be discharged to the surface waters at amounts that are harmful to the surrounding species. AOPs are thus a general term for aqueous-phase redox processes that rely on the involvement of strong oxidizing species, predominantly but not solely hydroxyl radicals, in the procedures that result in the degradation of the predicted contaminant. Based on the type of pharmaceutical wastewater and the intended target of removal or transformation, these procedures may be used independently or in combination with other treatments.

2.4 HYBRID SYSTEMS FOR THE REMOVAL OF PHARMACEUTICALS

The types and ranges of pollutants that physical and chemical water and wastewater treatment processes can remove from water are still subject to limitations (Mahmoud, Fawzy et al. 2022). Nowadays, combining NF membranes with membrane technology is practical since it is a useful tool for removing pharmaceuticals. NF commercial membrane removed caffeine, ibuprofen, dipyrone, diclofenac, and acetaminophen (NF90). NF90 successfully eliminated the five pharmaceutical medicines with a more than 88% drug rejection rate. At a pressure of 20 bar and a pH of 5, rejection rates for ibuprofen and diclofenac were greater than 90% (Licona, Geaquinto et al. 2018).

For the attenuation of pharmaceuticals from wastewater, there are many created wetlands designs. The total discharge of pharmaceuticals can be significantly decreased with the combination of various constructed wetlands methods. For instance, pharmaceutical reductions of over 70% were achieved using three different real-scale hybrid treatment combinations (Escolà Casas and Matamoros 2021). The filler material selection mostly influences the adsorption ability of the constructed wetlands for eliminating pharmaceuticals. A possible method to increase the removal of pharmaceuticals appears to be using biochar as filler material in constructed wetlands. Filters based on biochar caused biodegradation and adsorption of pharmaceuticals such as carbamazepine (73–99%) and caffeine (97–99%) (Dalahmeh, Ahrens et al. 2018). For sulfamethazine, removal was 52% (Chen, Wei et al. 2016). Recently, Chand et al. showed that vertical flow constructed wetland could remove various pharmaceuticals with an inlet concentration of 1000 µg/L with removal efficacy of 87.53% (caffeine), 79.93% (ibuprofen), and 75.50% (amoxicillin) in the wetland setup with plant stand and biochar amended substrate (Chand, Suthar et al. 2022). Adding biochar to the support matrix of artificial wetlands may be a sustainable strategy to expand their capacity since it often increases carbon delivery but also enhances the mass transfer through adsorbing and enteric bacteria degradations (Wu, Xu et al. 2019).

Delgado et al. developed a configuration of a column packed with powder-activated carbon and sand ($\geq$150 µm) with a bed height of 20 cm and flow rate of 0.4 cm^3/(cm^2.min) (Delgado, Marino et al. 2022). It was found that this design efficiently removed carbamazepine and sildenafil from aqueous solutions during three months of operation, with removal efficiencies $\geq$90%. Combining adsorption columns with PAC may lessen the degree of the detrimental effects that medications can have on water resources, aquatic life, and human health.

Significant levels of organic contaminants, ranging from reagents to intermediates to finished products, are frequently included in the wastewater from chemical synthesis processes. To treat such a matrix, many researchers have explored technology and biological therapy methods. A continuously mixed tank system and a vertical anaerobic sludge blanket clarifier, which serve as the acidogenic and methanogenic phases, respectively, make up Chen et al.'s two-phase anaerobic digestion (TPAD) system (Chen, Ren et al. 2008). The combined pilot plant removed 99% of the COD, while the MBR balanced the pH. Therefore, TPAD-MBR may be effectively used for wastewater resulting from chemical production. In order to achieve removal efficiencies of 86% COD and 90% turbidity, Boroski et al. used electrocoagulation (EC) followed by heterogeneous photocatalysis (TiO_2), where the removal rose from 70% (EC) to 76% with UV/H_2O (Boroski, Rodrigues et al. 2009). The combination is most effective for wastewater with significant concentrations of refractory/non-biodegradable compounds. The following parameters were evaluated for a hybrid up-flow anaerobic sludge blanket reactor for wastewater: TDS, 8500–9000 mg/L; TSS, 2800–3000 mg L^{-1}; COD, 13000–15000 mg L^{-1}; BOD, 7000–7500 mg L^{-1}; BOD: COD ratio, 0.45–0.60 (Sreekanth, Sivaramakrishna et al. 2009). The biological oxidation of this effluent is extremely prone to it. The removal efficiencies are: BOD: 80–90%; COD: 65–75%. The method provides a high biomass production rate, making it economically viable.

2.5 ADVANCED OXIDATION PROCESSES IN HYBRID SYSTEMS

2.5.1 FUNDAMENTAL OF ADVANCED OXIDATION PROCESSES

Advanced oxidation processes (AOPs) are highly efficient for degrading persistent organic micropollutants by generating reactive radical species (e.g. OH$^\bullet$). The main advantage of AOPs is this technology's capability to complete the mineralization of the micropollutants and the generated byproducts. OH$^\bullet$ is a strong oxidant with high efficiency due to its non-selectivity, high standard oxidation potential, and high reaction rate constants. The reactivity of OH$^\bullet$ is 10^6 to 10^{12} times higher than other oxidizing agents such as ozone (O_3). In addition, it possesses a high redox potential ($E°(OH^\bullet/H_2O) = 2.8$ V vs NHE) that can non-selectively react with most organic contaminants until their total mineralization. The OH$^\bullet$ attack on the organic molecules can occur through various processes, including electron-transfer reactions, hydrogen atom abstraction, and OH$^\bullet$ addition to double bonds in alkenes aromatic compounds (Eq. 2.3) as demonstrated in the reactions (Eqs. 2.1–2.3).

$$R - H + OH^\bullet \rightarrow R^\bullet + H_2O \qquad (2.1)$$

$$CH_3COOH + OH^\bullet \rightarrow CH_2^\bullet COOH + H_2O \qquad (2.2)$$

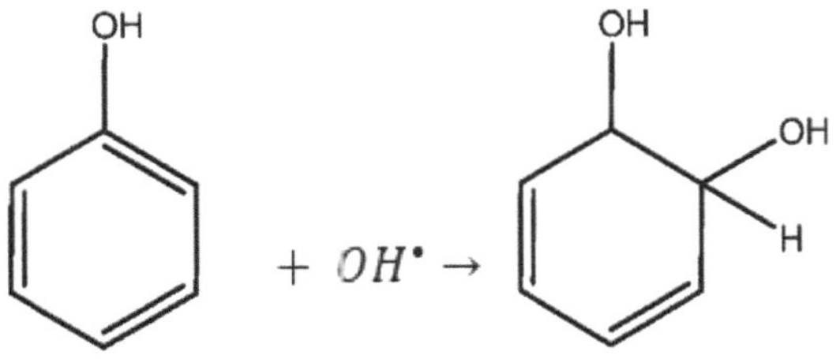

$$\text{(2.3)}$$

AOPs involve photolysis, ozonation, heterogenous photocatalysis, Fenton and electrochemical oxidation, and sonolysis. The following sections describe the fundamentals and mechanism of various approaches to AOPs.

2.5.2 AOPs Technologies

2.5.2.1 Photolysis

Photolysis occurs when an organic pollutant is exposed to UV light irradiation or natural sunlight leading to the degradation of the pollutant via photochemical reactions. UV photolysis, however, can only destroy organic contaminants; it is not capable of achieving the complete mineralization of the pollutant. One of the most effective AOPs for degrading organic contaminants is Vacuum UV (VUV) photolysis, which could achieve higher mineralization degrees than UV photolysis. While UV light irradiation is between 200 and 400 nm, VUV light irradiation utilizes wavelengths less than 190 nm. Water undergoes hemolysis and photochemical reactions upon VUV exposure, yielding reactive reducing and oxidizing species for degrading organic pollutants (Eqs. 2.4–2.5).

$$H_2O + h\upsilon\left(< 190\,nm\right) \rightarrow H^{\cdot} + OH^{\cdot} \tag{2.4}$$

$$H_2O + h\upsilon\left(< 190\,nm\right) \rightarrow H^{+} + e^{-} + OH^{\cdot} \tag{2.5}$$

The coupling of UV photolysis with additional oxidants such as H_2O_2 has also been adopted in wastewater treatment to guarantee the effective oxidation of organic contaminants. Upon UV light irradiation, H_2O_2 absorbs UV light leading to the breakdown of H_2O_2 into $OH^{\cdot}$ (Eq. 2.6).

$$H_2O_2 + h\upsilon \rightarrow 2OH^{\cdot} \tag{2.6}$$

Although UV/H_2O_2 can be used to completely mineralize a number of organic pollutants, this method has some drawbacks, including the need for a special reactor design, the limited ability to use solar radiation, and poor UV absorption.

2.5.2.2 Ozonation

Ozone is a unique oxidizing product that dissociates in water to produce OH•, that is an even more potent oxidizer than ozone itself. This causes certain functional groups of organic compounds to be attacked via an electrophilic process (Dantas, Contreras et al. 2008). Ozonation is a promising method for removing biodegradation-resistant organic pollutants from pharmaceutical effluents, especially for antibiotic elimination (Balcıoğlu and Ötker 2003, Dantas, Contreras et al. 2008). The elimination of excessive quantities of penicillin and the enhancement of the fermentation effluent biodegradability were investigated (Arslan-Alaton and Dogruel 2004). Various processes can be used for ozonation to accelerate the oxidation process, such as O_3/H_2O_2, $O_3/H_2O_2/UV$, and O_3/UV. The generation of $OH^{\cdot}$ can be induced through ozone breakdown in an aqueous medium in the presence of sole H_2O_2 (Eq. 2.7), sole UV (Eq. 2.8), UV/H_2O_2 (Eq. 2.9), and OH^{-} in basic medium (Eqs. 2.10–2.11) (Rosman, Salleh et al. 2018). Nevertheless, $O_3/H_2O_2/UV$ is the most effective process among various processes.

$$2O_3 + H_2O_2 \rightarrow 2OH^{\cdot} + 3O_2 \tag{2.7}$$

$$O_3 + H_2O + h\upsilon \rightarrow 2OH^{\cdot} + O_2 \tag{2.8}$$

$$H_2O_2 + O_3 + H_2O + h\mho \rightarrow 4OH^{\bullet} + O_2 \qquad (2.9)$$

$$O_3 + OH^- \rightarrow O_2 + HO_2^- \qquad (2.10)$$

$$O_3 + HO_2^- \rightarrow OH^{\bullet} + O_2^{\bullet -} + O_2 \qquad (2.11)$$

The main drawbacks of applying the ozonation process in wastewater treatment include the high cost required for ozone production and its poor solubility in aqueous solutions.

2.5.2.3 Heterogeneous Photocatalysis

In this process, heterogeneous photocatalysts such as TiO_2, WO_3, Fe_2O_3, and ZnO absorb the required photon energy depending on their bandgap energy for the excitation of the electron from the valence band (VB) to the conduction band (CB). Generally, solar light irradiation can be an economical and available light source for treating pharmaceuticals using photocatalysis. The excitation of electrons results in the generation of holes at the valence band. The generated electrons (e^-) and holes (h^+) initiate redox reactions on the active sites of the photocatalyst leading to the destruction of organic molecules. As shown in Figure 2.1, the generated h^+ reacts with H_2O in the aqueous medium yielding $OH^{\bullet}$, whereas e^- reacts with dissolved O_2 to produce superoxide radicals $OH^{\bullet}$ and $O_2^{\bullet -}$. TiO_2 is the most commonly used photocatalyst in most pharmaceutical photocatalytic research. Photocatalysis is the most convenient technology for effluents with a high COD and for entirely converting extremely refractory organic pollutants into levels compatible with biological treatment.

2.5.2.4 Fenton and Photo-Fenton Oxidation

Fenton and photo-Fenton oxidation processes are common AOPs for degrading organic pollutants. Fenton-based AOPs mainly depend on the reaction of ferrous salts with H_2O_2 in acidic conditions leading to H_2O_2 decomposition into $OH^{\bullet}$ (Eqs. 2.12–2.13).

$$Fe^{+2} + H_2O_2 \rightarrow Fe^{+3} + OH^{\bullet} + OH^- \qquad (2.12)$$

$$Fe^{+3} + H_2O_2 \rightarrow Fe^{+2} + HO_2^{\bullet} + H^+ \qquad (2.13)$$

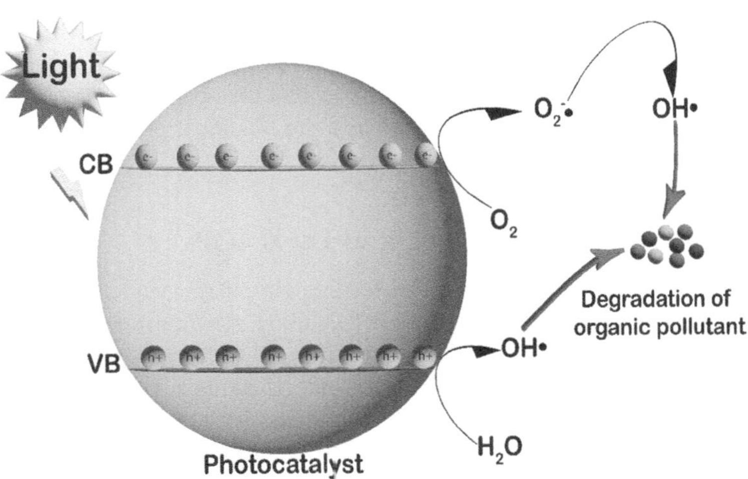

FIGURE 2.1 The degradation of organic pollutants by the heterogenous photocatalysis.

To speed up the breakdown and mineralization of organic pollutants, photo-Fenton-based AOPs under UV irradiation have been adopted further. In general, Fenton-based AOPs have some problematic issues regarding managing produced sludge. Therefore, the Fenton reaction may be employed as a pre-treatment step for converting non-biodegradable compounds into biodegradable compounds, making biological treatment of the effluent more efficient. A wastewater effluent containing chloramphenicol and paracetamol COD at concentrations that exceed 12000 mg/L was treated using Fenton-biological treatment processes, and thus, COD removal rates of greater than 95% could be achieved (Badawy, Wahaab et al. 2009).

2.5.2.5 Sonolysis

Sonolysis is a chemical-free AOP at which the destruction of organic contaminants relies mainly on the production of $OH^{\bullet}$ through acoustic cavitation when a liquid is exposed to ultrasound irradiation with high intensity (Eq. 2.14) (Kıdak and Doğan 2018). Acoustic cavitation is mainly caused by the tension resulting from the change in pressure. The main flaws of sonolysis are the high energy requirements and limited mineralization. Recently, biological treatment and hydrodynamic cavitation have been utilized to remove pharmaceutical substances from wastewater. Integrating the attached-growth culture, hydrodynamic cavitation/H_2O_2, and UV treatment technologies could achieve about >90% degradation for clofibric acid and >98% for carbamazepine and diclofenac (Zupanc, Kosjek et al. 2013).

$$H_2O + ultrasound\ irradiation \rightarrow OH^{\bullet} + H^{\bullet} \tag{2.14}$$

2.5.2.6 Electrochemical Oxidation/Degradation

An electrochemical reaction is based on the degradation of organic pollutants by $OH^{\bullet}$ produced at the surface of an anode electrode (i.e., boron-doped diamond (BDD)). Simulated wastewater containing diclofenac, carbamazepine, propranolol, ibuprofen, and ethinylestradiol showed complete degradation when treated using the electrochemical method. The electrooxidation performance can be improved by involving dissolved iron, i.e., the electro-Fenton process, which facilitates the decomposition of H_2O_2 to $OH^{\bullet}$. Additionally, the electro-Fenton reaction using a doped BDD electrode could minimize the toxicity of the generated byproducts (Sirés, Arias et al. 2007).

2.5.2.7 Wet Air Oxidation

Wet air oxidation is a thermochemical reaction that involves the formation of $OH^{\bullet}$ and other reactive oxygen species at extreme temperatures (200–320°C) and pressures (220 MPa) (Debellefontaine and Foussard 2000). A recent study has demonstrated that this technique can effectively eradicate COD. Therefore, this method may be employed as a pre-treatment process of wastewater prior to activated sludge systems.

2.5.3 REMOVAL EFFICIENCY OF PHARMACEUTICALS USING AOPS

Various treatment processes have been applied for treating pharmaceuticals, including adsorption and electrochemical technologies. However, these technologies suffer from inherent drawbacks that limit their application to remove pharmaceuticals efficiently. The adsorption technique can only reduce the concentrations of pharmaceuticals without achieving complete mineralization (Choi, Kim et al. 2008). On the other hand, electrochemical processes have major limitations regarding high energy requirements and high cost of reactors (He, Dong et al. 2019). AOPs are one of the most promising technologies that could address the aforementioned gaps since they can utilize solar light, minimize energy consumption, and efficiently degrade pharmaceuticals in a short time. Table 2.2

TABLE 2.2
The Removal Efficiency of Common Pharmaceuticals Using Various Photocatalysts

Pharmaceutical	System	Maximum removal efficiency/time	Reference
Tetracycline	Fe-MOF/TiO$_2$/solar light	92.76% in 10 min	(He, Dong et al. 2019)
	Fe$_3$O$_4$/PS/visible light	78.50% in 90 min	(Zhang, Mei et al. 2022)
	CuFe$_2$O$_4$@g-C$_3$N$_4$/H$_2$O$_2$/visible light	94% in 100 min	(Sun, Huang et al. 2022)
Sulfamethazine	W-TiO$_2$/visible light	100% in 30 min	(Fouad, Alalm et al. 2020)
	Co-TiO$_2$/PMS	95.8% in min	(Zhang, He et al. 2022)
	AgCl/MIL-100(Fe)/visible light	99.9% in 120 min	(Ning, Pang et al. 2022)
Sulfamethoxazole	TiO$_2$@rGO/UV	92% in 3 h	(Lin, Wang et al. 2017)
	Ag$_3$PO$_4$@g-C$_3$N$_4$/visible light	99% in 1 h	(Zhou, Zhang et al. 2017)
	Cu$_2$O@SCG/visible light	80% in 120 min	(Zheng, Yang et al. 2022)
Acetaminophen	CuO@C/H$_2$O$_2$/visible LED	95% in 60 min	(Abdelhaleem, Abdelhamid et al. 2022)
	TiO$_2$/UV	95% in 80 min	(Yang, Liya et al. 2008)
Naproxen	g-C$_3$N$_4$/visible light	92.9% in 210 min	(Mafa, Malefane et al. 2022)
	MOF-derived BiFeO$_3$/PMS/ visible light	95.5% in 40 min	(Yao, Chen et al. 2022)
Diphenhydramine	UVC/H$_2$O$_2$	Nearly 100% in 60 min	(López, Plaza et al. 2017)
	UV254/sulfite	Nearly 100% in 6 min	(So, Wang et al. 2022)

shows the removal efficiency of some common pharmaceuticals using AOPs based on previous investigations.

2.5.3.1 Tetracycline

Tetracycline (TC) is a low-cost broad-spectrum antibiotic frequently used to treat human and animal diseases (Zhou, Liu et al. 2017). The wide application of TC caused its ubiquity in many rivers. For instance, the surface water's average TC levels reached 0.357 µg L−1 (Chen, Jing et al. 2018). Hence, the TC treatment is significant for abating its spread in the waterways and minimizing its associated risks to humans and the environment. TC could be photodegraded effectively using Fe-based metal-organic framework/TiO$_2$ (Fe-MOF/TiO$_2$) magnetic composite. This composite is an eco-friendly photocatalyst that could degrade about 92.76% of TC in 10 mins under solar light (He, Dong et al. 2019). A photocatalyst of Fe$_3$O$_4$ coated with carbon (Fe$_3$O$_4$@C) was successfully used to degrade 78.50% of TC within 90 min under visible light in the presence of potassium persulfate (PS) at both strong acid and strong alkali pH conditions (Zhang, Mei et al. 2022). A photo-Fenton AOP was utilized in order to efficiently degrade TC under visible light using a magnetic CuFe$_2$O$_4$/ g-C$_3$N$_4$ photocatalyst in the presence of interfering components such as NO$_3^-$, HCO$_3^-$, HPO$_4^{2-}$, SO$_4^{2-}$, humic acid (HA), and high salt levels (Sun, Huang et al. 2022). The heterojunction design of CuFe$_2$O$_4$/g-C$_3$N$_4$ photocatalyst can reduce the recombination of photogenerated electron–hole pairs and thereby increase the degradation efficiency. The magnetic photocatalyst in the photo-Fenton process exhibited 94% decay of TC within 100 min (Sun, Huang et al. 2022). Interestingly, the toxicity evaluation of the generated byproducts at the end of the photo-Fenton reaction confirmed the safety and reliability of the proposed technology for TC removal.

2.5.3.2 Sulfamethazine

Sulfamethazine (SMZ) is a common antimicrobial drug broadly used for aquaculture and livestock (Lin and Wu 2018). The livestock's organs do not entirely absorb antibiotics, and some are consequently released into the environment (Yu, Kiwi et al. 2019). As a result, SMZ has been detected in marine environments at concentrations between ng^{-1} and mg L^{-1} (Fouad, Alalm et al. 2020). Although SMZ exists in surface and groundwater at significant levels, its removal is challenging using conventional biological treatment technologies due to its resistance to biodegradation (Tang and Wang 2019). Numerous gram-negative and gram-positive bacteria can have their growth hampered by the sulfonamide group found in the SMZ structure (Zhang, He et al. 2022). SMZ was completely degraded using W-TiO$_2$ photocatalyst under visible irradiation, where nearly 100% of SMZ was degraded in 30 mins (Fouad, Alalm et al. 2020). AOPs based on peroxymonosulfate have shown to be highly effective at removing organic pollutants that are resistant to biological treatment. An amorphous Co-TiO$_2$ heterojunction photocatalyst was fabricated for SMZ degradation through peroxymonosulfate (PMS) activation. The Co-TiO$_2$/PMS system could achieve 95.8% of SMZ within 30 mins (Zhang, He et al. 2022). Additionally, the system revealed an outstanding activity for SMZ decay over a broad pH range, even in the presence of coexisting organic and inorganic chemicals. A step scheme (S-scheme) AgCl/MIL-100(Fe) heterojunction nanocomposite was fabricated to accelerate the transfer of electrons and holes and showed 99.9% degradation of SMZ in 120 mins under visible light irradiation (Ning, Pang et al. 2022). The findings of the toxicity assessment revealed that the degradation byproducts are less harmful than SMZ.

2.5.3.3 Sulfamethoxazole

As one of the most widely used antibiotics in aquatic environments, sulfamethoxazole (SMX) has the potential to lead to new threats to aquatic ecosystems. SMX can be released into wastewater treatment plants through the release of urine and feces and ultimately end up in surface water. According to reports, the SMX level in surface water could reach 15 µg L^{-1} (Tran, Hoang et al. 2019). Due to the sulfonamide group's presence in the SMX structure, it possesses antimicrobial characteristics that may boost its resistance to the environment (Ioannidou, Frontistis et al. 2017). Hence, developing effective treatment technologies for its removal is crucial to abate its accumulation in the environment. The TiO$_2$@rGO/UV system was able to eliminate about 92% of SMX in 3 h (Lin, Wang et al. 2017). It was reported that the Ag$_3$PO$_4$@g-C$_3$N$_4$ heterojunction catalyst could degrade 99% of SMX in 1 h under the presence of visible light (Zhou, Zhang et al. 2017). On the other hand, Cu$_2$O nanoparticles were incorporated with biochar produced from coffee grounds (Cu$_2$O@SCG) and employed for SMX removal under visible light irradiation. The Cu$_2$O@SCG reduced the charge carrier recombination, accelerated interfacial electron transfer kinetics, and improved visible light absorption. The photocatalytic degradation of SMX using the Cu$_2$O@SCG/visible light system was higher than 80% in 120 min (Zheng, Yang et al. 2022).

2.5.3.4 Acetaminophen (Paracetamol)

It has long been common practice to use paracetamol (PCM) as an analgesic to treat body aches, fevers, and headaches. It was reported that the usage of PCM has grown significantly during the COVID-19 pandemic by 198%, indicating its potential ubiquity in wastewater treatment plants (Galani, Alygizakis et al. 2021). PCM has been found in natural streams in the USA at concentrations of 10 µg/L and 6 µ g/L in European sewage treatment plant effluents (Skoumal, Cabot et al. 2006). PCM is resistant to degradation by conventional treatment methods in wastewater treatment plants (Ahmed, Zhou et al. 2017). A photo-Fenton-like AOP was employed for PCM degradation using a metal-Organic Framework-derived CuO@C photocatalyst under energy-efficient visible light emitting diode (LED) lamps as a light source. The proposed system could degrade 95% of PCM in 60 mins with a mineralization degree of 68% (Abdelhaleem, Abdelhamid et al. 2022). A considerable decrease in PCM concentration was achieved using TiO$_2$ under UV irradiation, where 95% was decayed in 80 mins with a mineralization degree up to 85% (Yang, Liya et al. 2008).

2.5.3.5 Naproxen

Naproxen (NPX) is a non-steroidal anti-inflammatory (NSAID) medicine that has been found in numerous waterways at levels ranging from the ng/L to the µg/L and applied for treating pain and inflammation caused by menstrual cramps and arthritis (Arany, Szabó et al. 2013, Regmi, Kshetri et al. 2018, Mlunguza, Ncube et al. 2019). The US Environmental Protection Agency categorized NPX as a priority drug due to its massive global consumption, toxicity, stability, and persistence (De Voogt, Janex-Habibi et al. 2009). About 92.9% of NPX was eliminated in 210 mins using a highly stable multi-elemental doped polymeric graphitic carbon nitride (g-C_3N_4) photocatalysts under visible light irradiation (Mafa, Malefane et al. 2022). The NPX degradation performance reached 95.5% within 40 mins when NPX was degraded in a MOF-derived $BiFeO_3$/PMS/visible light system (Yao, Chen et al. 2022). Additionally, the combined system UV_{254}/Vacuum UV was proven efficient for the photodegradation and mineralization of NPX (Arany, Szabó et al. 2013).

2.5.3.6 Diphenhydramine

Diphenhydramine (DPH) was extensively used as a drug for relieving symptoms of allergy and the common cold (Reznikov, Norris et al. 2021). Additionally, it showed antiviral properties against SARS-CoV-2 (Reznikov, Norris et al. 2021). Based on several investigations, DPH was found in soil at levels of 20 to 50 µg/kg and in main streams at levels of 0.01 to 0.10 µg/L (Ferrer, Heine et al. 2004). The UVC/H_2O_2 (López, Plaza et al. 2017), UV_{254}/sulfite (So, Wang et al. 2022), and Fe^{+2}/H_2O_2 (López, Plaza et al. 2017) homogenous systems were reported in previous studies for the DPH removal. Additionally, DPH was removed by the TiO_2/UVA heterogeneous system (López, Plaza et al. 2017).

2.5.3.7 Ciprofloxacin

According to reports, the antibiotic ciprofloxacin had the highest concentration of more than 200 different pharmaceuticals found in river waters around the world, at 6.5 mg/L. Acute toxicity tests on a single substance revealed that the median adequate levels for many organic micropollutants were less than 1 mg/L.s (Pathak, Tran et al. 2020). The photocatalytic destruction of this biotoxic molecule was observed with S-doped g-C_3N_4/ZnO. The S-doped g-C_3N_4/ZnO material could achieve about 98.8% and 75.8% decay under UV and visible light, respectively (Gupta, Gupta et al. 2021). Nano-zero valent iron nanoparticles combined with UV, and H_2O_2 were studied and the results revealed 100% Ciprofloxacin removal occurred in 30–40 min (Mondal, Saha et al. 2018).

2.5.3.8 Carbamazepine

Integrated MBR and the TiO_2 photocatalysis method was adopted to remove non-biodegradable pharmaceuticals like carbamazepine from pharmaceutical effluents. According to the obtained results, carbamazepine has a 95% removal as well as COD being reduced (Pal 2018).

2.5.4 Hybrid Systems

AOPs have been integrated with other technologies in a hybrid system to accelerate the degradation and mineralization rate, scale up the commercialization in large-scale plants, and reduce energy usage.

2.5.4.1 AOP/membrane Hybrid Systems

Membrane technologies have been broadly used for water remediation due to their great efficiency, compact design, and durability. The hybridization of membranes with AOP as a pre-treatment process plays a positive role in reducing membrane fouling. Hybrid ozone/membrane systems are one of the most common technologies for removing organic pollutants, including pharmaceuticals. The levels of dicloxacillin and ceftazidime antibiotics in surface water samples decreased to below the detection limits using a hybrid ozone/membrane system compared to membrane filtration only (Alpatova, Davies et al. 2013). A combined system using electro-Fenton-based AOP (EF) and membrane distillation (MD) technology was also used to simultaneously decompose organic contaminants and separate the purified water from the mixture (Omi, Rastgar et al. 2022). The model pollutant ibuprofen (IBF) was chosen to assess the effectiveness of the EF/MD hybrid system. The effectiveness of the IBF degradation was investigated after adjusting the temperature of the feed and permeate sides of the MD unit at 60°C and 20°C to ensure sufficient dewatering while permitting small concentrations of volatile organic molecules to cross the membrane and reach the permeate side. In comparison to sole EF technology, the EF/MD hybrid system could reduce energy consumption and enhance IBF degradation by nearly 19.3% within 3 hours, suggesting the merit of hybrid processes.

2.5.4.2 AOP/biological Hybrid Systems

A viable strategy for lowering the operating costs required to cure pharmaceuticals could be linking a biological process with an improved oxidation process in a system. The anaerobic/aerobic/oxic system was combined with ozonation to remove PCM from wastewater (Rosal, Rodríguez et al. 2010). Additionally, integrating a membrane bioreactor (MBR) technology with TiO2 photocatalysis was found to be effective for removing carbamazepine (Laera, Chong et al. 2011). The integrated MBR-TiO$_2$ photocatalysis system could remove 95% of Carbamazepine. In another study, an integrated MBR/Ozone system showed higher cost-effectiveness over sole MBR in terms of the degradation of pharmaceuticals in hospital wastewater when 156 mg/l of ozone was used for 20 min (Nielsen, Hastrup et al. 2013). Moreover, the application of ozonation as a post-treatment technology, integrated with a moving bed biofilm reactor (MBBR), was proven to be efficient for removing pharmaceuticals in hospital wastewater (Hansen, Spiliotopoulou et al. 2016). The Fenton-based AOP could improve the degradation performance of amoxicillin to 80% after the hybridization with the activated sludge process in a combined system (Guo, Xie et al. 2015). The use of a sono-photo-Fenton-based AOP in combination with a traditional biological treatment led to a significant degradation (removal: up to 91.13%) of 15 pharmaceuticals in hospital wastewater, including loratadine, ciprofloxacin, norfloxacin, acetaminophen, diclofenac, valsartan, irbesartan, carbamazepine, and venlafaxine (Serna-Galvis, Silva-Agredo et al. 2019).

Fermentation broth, mycelia, and nutrients used for cell culture make up the majority of the fermentation process' waste water. In addition, several organic solvents are added to the extraction process for the active pharmaceutical ingredient. Helmig et al. handled estrogen-containing API formulation waste with a hybrid approach, including pre-treatment ozonation and aerobic treatment, i.e. membrane bioreactor technology (Helmig, Fettig et al. 2005). As a result, removing more than 90% of COD and TSS was possible. In addition, the MBR resulted in total wastewater treatment. Cokgor et al. investigated wash water waste from penicillin manufacture (Cokgor, Alaton et al. 2004). They used biological activated sludge treatment in conjunction with ozonation (pre-treatment) and synthetic biomass containing 30% COD. Ozone ozonation successfully removed 34% COD and 24% total organic carbon, and activated sludge thereafter revealed excellent COD removal and improved biodegradability. In certain instances, penicillin formulation effluents contain contaminants, such as tylosin, that have a refractory effect on biological treatment systems; thus, implementing a hybrid technique results in their total elimination. Tylosin and avilamycin-containing wastewater was treated using a hybrid vertical anaerobic stage reactor (Chelliapan and Sallis 2011). For avilamycin macrolide and tylosin antibiotic waste streams, UASR may be employed as a pre-treatment alternative with a COD removal of 70–75%; thus, tylosin can be decomposed effectively under anaerobic conditions. Tekin et al. examined the production process and wash fluids comprising traces of organic compounds, iodine, and metal salts with concentrations ranging from 900 to 6800 mg/L for COD and 85 to 3600 mg/L for BOD (Tekin, Bilkay et al. 2006). They used an AOP with biological treatment to deal with this effluent. In a sequencing batch reactor, the Fenton oxidation (pre-treatment) coagulation step, followed by aerobic biological degradation, removed 45 to 50% of the COD, with the biological treatment reducing the COD to 98%.

2.5.4.3 AOP/AOP Hybrid Systems

The mineralization of recalcitrant compounds can be improved by applying simultaneous AOP hybrid systems. For instance, integrating sonolysis with other AOPs in a hybrid system is desirable to enhance the mineralization degree and minimize the operation costs. A combined AOP process was used for the degradation of IBF, including photo-Fenton and sonolysis under solar light irradiation in a hybrid system (Mendez-Arriaga, Torres-Palma et al. 2009). The sono-photo-Fenton hybrid system might, according to the results, boost IBF mineralization and degradation by as much as 60% and 95%, respectively.

2.5.4.4 AOP/adsorption Hybrid Systems

Adsorption was used in conjunction with AOPs to reduce the organic load further. A hybrid strategy of ozone-based AOP and adsorption by activated char was applied to remove pharmaceuticals. Chemical oxygen demand (COD) was removed with an efficiency of 75–88.5% using ozonation in the integrated ozonation/activated char process. The COD content was further reduced to 85.4–92.7% using subsequent activated char adsorption (Patel, Mondal et al. 2020).

2.6 CONCLUSION

Wastewater is a pathway for drugs to enter our rivers, as current treatments are inappropriate for these compounds. Hospitals are a source to be considered, given their wastewater is loaded with pharmaceutical compounds. Additionally, the majority of how medications end up in the environment comes through poor manufacturing facility disposal. Chemical synthesis and fermentation wastewaters make up the majority of the pharmaceuticals that end up in the environment from different production sites. Numerous academics are interested in the unsettling presence of these compounds in our aquatic habitats.

In recent decades, a number of papers on the management of pharmacological drugs and endocrine disruptors have been released. Effluent from chemical and biological processes is what most wastewater treatment systems deal with. The pre-treatment stage, which frequently uses advanced oxidation procedures to get rid of recalcitrant/refractory compounds that are frequently non-biodegradable, is the most frequent treatment method used to both wastewater streams. The trash with improved biodegradability can then be effectively treated using biological treatment techniques. Among the advanced biological treatment technologies, the membrane biological reactor (MBR) is a proven feasible choice.

Furthermore, anaerobic reactors are widely utilized as a byproduct, i.e., biogas from the process may be economically utilized by the agriculture industry alongside the treated sludge. However, current research aims to limit the contamination of the environment by these micropollutants by proposing hybrid advanced oxidation technologies. Furthermore, hybrid methods have been used to treat specific substances not removed by single-stage therapy. Therefore, the hybrid advanced oxidation process is highly potential for the degradation of pharmaceutical micropollutants; it has shown tremendous effectiveness in the removal of these compounds.

REFERENCES

Abdelhaleem, A., et al. (2022). "Photocatalytic degradation of paracetamol using photo-Fenton-like metal-organic framework-derived CuO@ C under visible LED." *Journal of Cleaner Production*: 134571.

Adeleye, A. S., et al. (2022). "Abundance, fate, and effects of pharmaceuticals and personal care products in aquatic environments." *Journal of Hazardous Materials* 424: 127284.

Adham, S., et al. (2005). *"Development of a microfiltration and ultrafiltration knowledge base."* Awwa Research Foundation, US Department of the Interior, Bureau of Reclamation, Pasadena.

Ahmed, M. B., et al. (2017). "Progress in the biological and chemical treatment technologies for emerging contaminant removal from wastewater: a critical review." *Journal of Hazardous Materials* 323: 274–298.

Alpatova, A. L., et al. (2013). "Hybrid ozonation-ceramic membrane filtration of surface waters: the effect of water characteristics on permeate flux and the removal of DBP precursors, dicloxacillin and ceftazidime." *Separation and Purification Technology* 107: 179–186.

Arany, E., et al. (2013). "Degradation of naproxen by UV, VUV photolysis and their combination." *Journal of Hazardous Materials* 262: 151–157.

Arslan-Alaton, I. and S. Dogruel (2004). "Pre-treatment of penicillin formulation effluent by advanced oxidation processes." *Journal of Hazardous Materials* 112(1–2): 105–113.

Badawy, M. I., et al. (2009). "Fenton-biological treatment processes for the removal of some pharmaceuticals from industrial wastewater." *Journal of Hazardous Materials* 167(1–3): 567–574.

Balcıoğlu, I. A. and M. Ötker (2003). "Treatment of pharmaceutical wastewater containing antibiotics by O_3 and O_3/H2O2 processes." *Chemosphere* 50(1): 85–95.

Beun, J., et al. (1999). "Aerobic granulation in a sequencing batch reactor." *Water Research* 33(10): 2283–2290.

Boroski, M., et al. (2009). "Combined electrocoagulation and TiO2 photoassisted treatment applied to wastewater effluents from pharmaceutical and cosmetic industries." *Journal of Hazardous Materials* 162(1): 448–454.

Chand, N., et al. (2022). "Removal of pharmaceuticals by vertical flow constructed wetland with different configurations: Effect of inlet load and biochar addition in the substrate." *Chemosphere* 307: 135975.

Chelliapan, S. and P. J. Sallis (2011). "Application of anaerobic biotechnology for pharmaceutical wastewater treatment." *IIOAB J* 2(1): 13–21.

Chen, H., et al. (2018). "Characterization of antibiotics in a large-scale river system of China: occurrence pattern, spatiotemporal distribution and environmental risks." *Science of the Total Environment* 618: 409–418.

Chen, J., et al. (2016). "Removal of antibiotics and antibiotic resistance genes from domestic sewage by constructed wetlands: Optimization of wetland substrates and hydraulic loading." *Science of the Total Environment* 565: 240–248.

Chen, Z., et al. (2008). "A novel application of TPAD–MBR system to the pilot treatment of chemical synthesis-based pharmaceutical wastewater." *Water Research* 42(13): 3385–3392.

Choi, K.-J., et al. (2008). "Removal of antibiotics by coagulation and granular activated carbon filtration." *Journal of Hazardous Materials* 151(1): 38–43.

Clara, M., et al. (2005). "Removal of selected pharmaceuticals, fragrances and endocrine disrupting compounds in a membrane bioreactor and conventional wastewater treatment plants." *Water Research* 39(19): 4797–4807.

Cokgor, E. U., et al. (2004). "Biological treatability of raw and ozonated penicillin formulation effluent." *Journal of Hazardous Materials* 116(1–2): 159–166.

Cyr, P. J., et al. (2002). "A pilot scale evaluation of removal of mercury from pharmaceutical wastewater using granular activated carbon." *Water Research* 36(19): 4725–4734.

Dalahmeh, S., et al. (2018). "Potential of biochar filters for onsite sewage treatment: Adsorption and biological degradation of pharmaceuticals in laboratory filters with active, inactive and no biofilm." *Science of the Total Environment* 612: 192–201.

Dantas, R. F., et al. (2008). "Sulfamethoxazole abatement by means of ozonation." *Journal of Hazardous Materials* 150(3): 790–794.

De Voogt, P., et al. (2009). "Development of a common priority list of pharmaceuticals relevant for the water cycle." *Water Science and Technology* 59(1): 39–46.

Debellefontaine, H. and J. N. Foussard (2000). "Wet air oxidation for the treatment of industrial wastes. Chemical aspects, reactor design and industrial applications in Europe." *Waste Management* 20(1): 15–25.

Deegan, A., et al. (2011). "Treatment options for wastewater effluents from pharmaceutical companies." *International Journal of Environmental Science & Technology* 8(3): 649–666.

Delgado, N., et al. (2022). "Pharmaceutical compound removal using down-flow fixed bed filters with powder activated carbon: A novel configuration." *Journal of Environmental Chemical Engineering* 10(3): 107706.

Dhangar, K. and M. Kumar (2020). "Tricks and tracks in removal of emerging contaminants from the wastewater through hybrid treatment systems: A review." *Science of the Total Environment* 738: 140320.

dos Reis, G. S., et al. (2016). "Activated carbon from sewage sludge for removal of sodium diclofenac and nimesulide from aqueous solutions." *Korean Journal of Chemical Engineering* 33(11): 3149–3161.

Dotaniya, M. L., et al. (2022). "Reuse of poor-quality water for sustainable crop production in the changing scenario of climate." *Environment, Development and Sustainability*.

El-Gohary, F., et al. (1995). "Evaluation of biological technologies for wastewater treatment in the pharmaceutical industry." *Water Science and Technology* 32(11): 13–20.

Enright, A.-M., et al. (2005). "Low-temperature anaerobic biological treatment of solvent-containing pharmaceutical wastewater." *Water Research* 39(19): 4587–4596.

Escolà Casas, M. and V. Matamoros (2021). Novel Constructed Wetland Configurations for the Removal of Pharmaceuticals in Wastewater. *Removal and Degradation of Pharmaceutically Active Compounds in Wastewater Treatment*. S. Rodriguez-Mozaz, P. Blánquez Cano and M. Sarrà Adroguer. Cham, Springer International Publishing: 163–190.

Ferrer, I., et al. (2004). "Combination of LC/TOF-MS and LC/ion trap MS/MS for the identification of diphenhydramine in sediment samples." *Analytical Chemistry* 76(5): 1437–1444.

Fouad, K., et al. (2020). "A novel photocatalytic reactor for the extended reuse of W–TiO2 in the degradation of sulfamethazine." *Chemosphere* 257: 127270.

Gadipelly, C., et al. (2014). "Pharmaceutical industry wastewater: review of the technologies for water treatment and reuse." *Industrial & Engineering Chemistry Research* 53(29): 11571–11592.

Galani, A., et al. (2021). "Patterns of pharmaceuticals use during the first wave of COVID-19 pandemic in Athens, Greece as revealed by wastewater-based epidemiology." *Science of the Total Environment* 798: 149014.

Gryta, M. (2012). "Effectiveness of water desalination by membrane distillation process." *Membranes* 2(3): 415–429.

Gryta, M., et al. (2013). "The study of membrane distillation used for separation of fermenting glycerol solutions." *Journal of Membrane Science* 431: 1–8.

Guo, R., et al. (2015). "The degradation of antibiotic amoxicillin in the Fenton-activated sludge combined system." *Environmental Technology* 36(7): 844–851.

Gupta, B., et al. (2021). "A multivariate modeling and experimental realization of photocatalytic system of engineered S–C3N4/ZnO hybrid for ciprofloxacin removal: Influencing factors and degradation pathways." *Environmental Research* 196: 110390.

Gwenzi, W., et al. (2022). "COVID-19 drugs in aquatic systems: a review." *Environmental Chemistry Letters* 20(2): 1275–1294.

Hansen, K. M., et al. (2016). "Ozonation for source treatment of pharmaceuticals in hospital wastewater–ozone lifetime and required ozone dose." *Chemical Engineering Journal* 290: 507–514.

Hausmann, A., et al. (2011). "Direct contact membrane distillation of dairy process streams." *Membranes* 1(1): 48–58.

He, L., et al. (2019). "A novel magnetic MIL-101 (Fe)/TiO2 composite for photo degradation of tetracycline under solar light." *Journal of Hazardous Materials* 361: 85–94.

Helmig, E. G., et al. (2005). *API removal from pharmaceutical manufacturing wastewater–results of process development, pilot-testing, and scale-up*. WEFTEC 2005, Water Environment Federation.

Ioannidou, E., et al. (2017). "Solar photocatalytic degradation of sulfamethoxazole over tungsten–Modified TiO2." *Chemical Engineering Journal* 318: 143–152.

Kang, J., et al. (2011). "Integrated catalytic wet air oxidation and biological treatment of wastewater from Vitamin B6 production." *Physics and Chemistry of the Earth, Parts A/B/C* 36(9–11): 455–458.

Kıdak, R. and Ş. Doğan (2018). "Medium-high frequency ultrasound and ozone based advanced oxidation for amoxicillin removal in water." *Ultrasonics Sonochemistry* 40: 131–139.

Laera, G., et al. (2011). "An integrated MBR–TiO2 photocatalysis process for the removal of Carbamazepine from simulated pharmaceutical industrial effluent." *Bioresource Technology* 102(13): 7012–7015.

Licona, K. P. M., et al. (2018). "Assessing potential of nanofiltration and reverse osmosis for removal of toxic pharmaceuticals from water." *Journal of Water Process Engineering* 25: 195–204.

Lin, C.-C. and M.-S. Wu (2018). "Feasibility of using UV/H2O2 process to degrade sulfamethazine in aqueous solutions in a large photoreactor." *Journal of Photochemistry and Photobiology A: Chemistry* 367: 446–451.

Lin, L., et al. (2017). "Immobilized TiO2-reduced graphene oxide nanocomposites on optical fibers as high performance photocatalysts for degradation of pharmaceuticals." *Chemical Engineering Journal* 310: 389–398.

López, N., et al. (2017). "Treatment of Diphenhydramine with different AOPs including photo-Fenton at circumneutral pH." *Chemical Engineering Journal* 318: 112–120.

Lu, F. and D. Astruc (2020). "Nanocatalysts and other nanomaterials for water remediation from organic pollutants." *Coordination Chemistry Reviews* 408: 213180.

Mafa, P. J., et al. (2022). "Multi-elemental doped g-C3N4 with enhanced visible light photocatalytic activity: insight into naproxen degradation, kinetics, effect of electrolytes, and mechanism." *Separation and Purification Technology* 282: 120089.

Mahmoud, A. E. D., et al. (2021). 26 – Water resources security and management for sustainable communities. *Phytochemistry, the Military and Health*. A. G. Mtewa and C. Egbuna, Elsevier: 509–522.

Mahmoud, A. E. D., et al. (2022). Technical Aspects of Nanofiltration for Dyes Wastewater Treatment. *Membrane Based Methods for Dye Containing Wastewater: Recent Advances*. S. S. Muthu and A. Khadir. Singapore, Springer Singapore: 23–35.

Mei, S., et al. (2000). "Application of nanofiltration membrane in the purification process of tylosin." *Chinese Journal of Antibiotics* 25(3): 172–174.

Mendez-Arriaga, F., et al. (2009). "Mineralization enhancement of a recalcitrant pharmaceutical pollutant in water by advanced oxidation hybrid processes." *Water Research* 43(16): 3984–3991.

Mestre, A., et al. (2007). "Activated carbons for the adsorption of ibuprofen." *Carbon* 45(10): 1979–1988.

Mlunguza, N. Y., et al. (2019). "Adsorbents and removal strategies of non-steroidal anti-inflammatory drugs from contaminated water bodies." *Journal of Environmental Chemical Engineering* 7(3): 103142.

Mondal, S. K., et al. (2018). "Removal of ciprofloxacin using modified advanced oxidation processes: Kinetics, pathways and process optimization." *Journal of Cleaner Production* 171: 1203–1214.

Nielsen, U., et al. (2013). "Removal of APIs and bacteria from hospital wastewater by MBR plus O3, O3+H2O2, PAC or ClO2." *Water Science and Technology* 67(4): 854–862.

Ning, R., et al. (2022). "An innovative S-scheme AgCl/MIL-100 (Fe) heterojunction for visible-light-driven degradation of sulfamethazine and mechanism insight." *Journal of Hazardous Materials* 435: 129061.

Oktem, Y. A., et al. (2008). "Anaerobic treatment of a chemical synthesis-based pharmaceutical wastewater in a hybrid upflow anaerobic sludge blanket reactor." *Bioresource Technology* 99(5): 1089–1096.

Omi, F. R., et al. (2022). "Synergistic effect of thermal dehydrating on the emerging contaminants removal via Electro-Fenton." *Journal of Cleaner Production* 356: 131880.

Pal, P. (2018). "Treatment and Disposal of Pharmaceutical Wastewater: Toward the Sustainable Strategy." *Separation & Purification Reviews* 47(3): 179–198.

Patel, S., et al. (2020). "Treatment of a pharmaceutical industrial effluent by a hybrid process of advanced oxidation and adsorption." *ACS Omega* 5(50): 32305–32317.

Pathak, N., et al. (2020). "Removal of Organic Micro-Pollutants by Conventional Membrane Bioreactors and High-Retention Membrane Bioreactors." 10(8): 2969.

Peng, Y., et al. (2004). "Nitrogen removal from pharmaceutical manufacturing wastewater with high concentration of ammonia and free ammonia via partial nitrification and denitrification." *Water Science and Technology* 50(6): 31–36.

Radjenovic, J., et al. (2007). "Analysis of pharmaceuticals in wastewater and removal using a membrane bioreactor." *Analytical and Bioanalytical Chemistry* 387(4): 1365–1377.

Ramola, B. and A. Singh (2013). "Heavy metal concentrations in pharmaceutical effluents of Industrial Area of Dehradun (Uttarakhand), India." *Journal of Environmental and Analytical Toxicology* 3(173): 2161–0525.1000173.

Regmi, C., et al. (2018). "Visible-light-driven S and W co-doped dendritic BiVO4 for efficient photocatalytic degradation of naproxen and its mechanistic analysis." *Molecular Catalysis* 453: 149–160.

Reznikov, L. R., et al. (2021). "Identification of antiviral antihistamines for COVID-19 repurposing." *Biochemical and Biophysical Research Communications* 538: 173–179.

Rosal, R., et al. (2010). "Occurrence of emerging pollutants in urban wastewater and their removal through biological treatment followed by ozonation." *Water Research* 44(2): 578–588.

Rosman, N., et al. (2018). "Hybrid membrane filtration-advanced oxidation processes for removal of pharmaceutical residue." *Journal of Colloid and Interface Science* 532: 236–260.

Salehi, M. (2022). "Global water shortage and potable water safety; Today's concern and tomorrow's crisis." *Environment International* 158: 106936.

Serna-Galvis, E. A., et al. (2019). "Effective elimination of fifteen relevant pharmaceuticals in hospital wastewater from Colombia by combination of a biological system with a sonochemical process." *Science of the Total Environment* 670: 623–632.

Sirés, I., et al. (2007). "Degradation of clofibric acid in acidic aqueous medium by electro-Fenton and photoelectro-Fenton." *Chemosphere* 66(9): 1660–1669.

Skoumal, M., et al. (2006). "Mineralization of paracetamol by ozonation catalyzed with Fe^{2+}, Cu^{2+} and UVA light." *Applied Catalysis B: Environmental* 66(3–4): 228–240.

Snyder, S. A., et al. (2007). "Role of membranes and activated carbon in the removal of endocrine disruptors and pharmaceuticals." *Desalination* 202(1–3): 156–181.

So, H. L., et al. (2022). "Insights into the degradation of diphenhydramine–an emerging SARS-CoV-2 medicine by UV/Sulfite." *Separation and Purification Technology*: 122193.

Song, L., et al. (2007). "Direct contact membrane distillation-based desalination: novel membranes, devices, larger-scale studies, and a model." *Industrial & Engineering Chemistry Research* 46(8): 2307–2323.

Sponza, D. T. and H. Çelebi (2012). "Removal of oxytetracycline (OTC) in a synthetic pharmaceutical wastewater by sequential anaerobic multichamber bed reactor (AMCBR)/completely stirred tank reactor (CSTR) system: biodegradation and inhibition kinetics." *Journal of Chemical Technology & Biotechnology* 87(7): 961–975.

Sreekanth, D., et al. (2009). "Thermophilic treatment of bulk drug pharmaceutical industrial wastewaters by using hybrid up flow anaerobic sludge blanket reactor." *Bioresource Technology* 100(9): 2534–2539.

Stoquart, C., et al. (2012). "Hybrid Membrane Processes using activated carbon treatment for drinking water: A review." *Journal of Membrane Science* 411: 1–12.

Suedee, R., et al. (2004). "Use of molecularly imprinted polymers from a mixture of tetracycline and its degradation products to produce affinity membranes for the removal of tetracycline from water." *Journal of Chromatography B* 811(2): 191–200.

Sun, X., et al. (2022). "Efficient degradation of tetracycline under the conditions of high-salt and coexisting substances by magnetic CuFe2O4/g-C3N4 photo-Fenton process." *Chemosphere* 308: 136204.

Tang, C.-J., et al. (2011). "Enhanced nitrogen removal from pharmaceutical wastewater using SBA-ANAMMOX process." *Water Research* 45(1): 201–210.

Tang, J. and J. Wang (2019). "MOF-derived three-dimensional flower-like FeCu@ C composite as an efficient Fenton-like catalyst for sulfamethazine degradation." *Chemical Engineering Journal* 375: 122007.

Tekin, H., et al. (2006). "Use of Fenton oxidation to improve the biodegradability of a pharmaceutical wastewater." *Journal of Hazardous Materials* 136(2): 258–265.

Tran, N. H., et al. (2019). "Occurrence and risk assessment of multiple classes of antibiotics in urban canals and lakes in Hanoi, Vietnam." *Science of the Total Environment* 692: 157–174.

Urase, T., et al. (2005). "Factors affecting removal of pharmaceutical substances and estrogens in membrane separation bioreactors." *Desalination* 178(1–3): 107–113.

Vambol, S., et al. (2021). Comprehensive insights into sources of pharmaceutical wastewater in the biotic systems. *Pharmaceutical Wastewater Treatment Technologies: Concepts and Implementation Strategies.* N. A. Khan, S. Ahmed, V. Vambol and S. Vambol, IWA Publishing: 0.

van Vliet, M. T. H., et al. (2021). "Global water scarcity including surface water quality and expansions of clean water technologies." *Environmental Research Letters* 16(2): 024020.

Vuppala, V., et al. (2012). "Photocatalytic degradation of methylene blue using a zinc oxide-cerium oxide catalyst." *European Journal of Chemistry* 3(2): 191–195.

Westerhoff, P., et al. (2005). "Fate of endocrine-disruptor, pharmaceutical, and personal care product chemicals during simulated drinking water treatment processes." *Environmental Science & Technology* 39(17): 6649–6663.

Wu, Z., et al. (2019). "Highly efficient nitrate removal in a heterotrophic denitrification system amended with redox-active biochar: A molecular and electrochemical mechanism." *Bioresource Technology* 275: 297–306.

Xiang, Y., et al. (2021). "A review of distribution and risk of pharmaceuticals and personal care products in the aquatic environment in China." *Ecotoxicology and Environmental Safety* 213: 112044.

Yang, L., et al. (2008). "Degradation of paracetamol in aqueous solutions by TiO2 photocatalysis." *Water Research* 42(13): 3480–3488.

Yao, J., et al. (2022). "New insight into the regulation mechanism of visible light in naproxen degradation via activation of peroxymonosulfate by MOF derived BiFeO3." *Journal of Hazardous Materials* 431: 128513.

Yoon, Y., et al. (2007). "Removal of endocrine disrupting compounds and pharmaceuticals by nanofiltration and ultrafiltration membranes." *Desalination* 202(1–3): 16–23.

Yu, J., et al. (2019). "Evidence for a dual mechanism in the TiO2/CuxO photocatalyst during the degradation of sulfamethazine under solar or visible light: critical issues." *Journal of Photochemistry and Photobiology A: Chemistry* 375: 270–279.

Zhang, H., et al. (2022). "Efficient activation of persulfate by C@ Fe3O4 in visible-light for tetracycline degradation." *Chemosphere* 306: 135635.

Zhang, W., et al. (2003). "Development and characterization of composite nanofiltration membranes and their application in concentration of antibiotics." *Separation and Purification Technology* 30(1): 27–35.

Zhang, Y., et al. (2022). "Amorphous Co@ TiO2 heterojunctions: A high-performance and stable catalyst for the efficient degradation of sulfamethazine via peroxymonosulfate activation." *Chemosphere* 307: 135681.

Zheng, M.-W., et al. (2022). "Mechanisms of biochar enhanced Cu2O photocatalysts in the visible-light photodegradation of sulfamethoxazole." *Chemosphere* 307: 135984.

Zhou, L., et al. (2017). "Z-scheme mechanism of photogenerated carriers for hybrid photocatalyst Ag3PO4/ g-C3N4 in degradation of sulfamethoxazole." *Journal of Colloid and Interface Science* 487: 410–417.

Zhou, Y., et al. (2017). "Modification of biochar derived from sawdust and its application in removal of tetracycline and copper from aqueous solution: adsorption mechanism and modelling." *Bioresource Technology* 245: 266–273.

Zupanc, M., et al. (2013). "Removal of pharmaceuticals from wastewater by biological processes, hydrodynamic cavitation and UV treatment." *Ultrasonics Sonochemistry* 20(4): 1104–1112.

3 Bioelectrochemical Systems for Remediation and Treatment of Wastewater

Srinithya Ravinuthala, Saravanan S., and Saprativ P. Das

3.1 INTRODUCTION

At present, major global concerns that require immediate attention include the exponentially rising global population, in turn increasing water and energy demands, environmental and climate change, depleting fossil fuel resources, and degradation of the environment due to pollution. All life forms on Earth depend on water as a natural resource, and the quantity and quality of that water available determines the many factors of their lifestyle. For the establishment and sustenance of various human activities, such as housing, agriculture, and industry, clean water must be available. Due to the growth in global population and industrial activity, freshwater is becoming one of the fastest dwindling resources on Earth.

Various industries have grown dramatically over the past few centuries as technology advanced, especially after the industrial revolution in the 19th century. The rise of the industrial sector has led to an increase of harmful industrial effluent release into bodies of water, as well as environmental pollution from hazardous and toxic substances polluting soil, sediments, and the air. Almost 80% of wastewater is dumped directly into bodies of water without being treated. Ground water contamination is becoming more of an issue as the use of pesticides and fertilizers in agricultural operations is increasing uncontrollably. Wastewater not only permanently damages the delicate ecological balance, but it also contributes to the decline of freshwater supplies, harming future generations. As a result of industrialization, the use of water in production has been severely strained. Increased requirements of output in terms of products have led to an upsurge in the generation of large volumes of industrial wastewater. As a result, wastewater treatment is essential for use and restoration, since its accessibility is closely related to the availability of clean and sufficient water supplies.

Wastewater contains a highly complex mixture of organic, as well as inorganic pollutants, depending on its origin. According to the source of the water, water contamination can be in the forms of chemical, microbiological, and/or nutrient contamination, suspended matter, oxygen depletion, and surface water and groundwater contamination. Wastewater effluents from various sources must be treated before being recycled or reused. Wastewater is high in nutrients, and releasing it into the environment may promote eutrophication and endanger flora and animals. Rising amounts of micropollutants such as pharmaceuticals, organic polymers, and suspended particles are contaminating water reservoirs. Wastewater may also be the source of harmful pathogens and toxic chemicals that could cause digestive or skin issues when ingested or come into direct contact with the skin.

Water is contaminated by toxic heavy metals (such as mercury (Hg), lead (Hb), chromium (Cr), zinc (Zn)), hazardous pollutants, and waste due to a variety of processes. Traditional methods are being utilized to partially treat this water before it is dumped into bodies of water. Presently,

DOI: 10.1201/9781003408352-4

industries are employing chemical methods such as electro-chemical, oxidation processes, and electro-flocculation, as well as other physical methods such as grit and flotation in order to treat wastewater effluents. This reduces the toxicity of the effluents, therefore protecting the downstream users of these effluents from health risks. However, mostly the technologies being used are expensive and result in sludge formation and subsequent water contamination. Especially low-cost effectiveness is proving problematic for industrial sectors, hence offering researchers a cause and space to develop better wastewater clean-up methods and materials (Verma et al. 2021).

Despite efforts being made at local, national, and international levels either to prevent industrial wastewater discharge by treating it prior to release into water bodies or to remediate already polluted waters, water bodies are becoming more and more contaminated. This has negative effects on both the environment and humans. As per the United Nations Water Integrated Monitoring Initiative on SDG 6 (IMI-SDG6) Progress Report, Water Target 6.3 is to "By 2030, improve water quality by reducing pollution, eliminating dumping and minimizing release of hazardous chemicals and materials, halving the proportion of untreated wastewater and substantially increasing recycling and safe reuse globally."

Treating wastewater using biological methods is a low-energy, economical technology that has gained popularity in recent years. When compared to traditional approaches, using microorganisms for wastewater treatment has demonstrated to be more successful in degrading a variety of contaminants, including persistent and emergent pollutants. The microbes employed for treatment may be in free state, i.e. suspended in the solution, or in immobilized/encapsulated form. As compared to using immobilized cells, the use of free cells for wastewater treatment has some drawbacks. This includes the possibility of variations in the biomass concentrations and loss of the microbes during washing, as well as more susceptibility to external factors. On the other hand, immobilized/encapsulated form not only have the advantage of overcoming these drawbacks, but also are known to be more stable and do not lose their activity as quickly (Mehrotra et al. 2021).

Recently, there has been a rise in focus towards deploying microalgae for wastewater treatment. Microalgae are organisms that have photosynthetic capabilities of utilizing the organic matter in effluents and converting it into biomass. They can also be used to reduce the chemical oxygen demand (COD) and biological oxygen demand (BOD), as well as for the removal of nutrients such as nitrogen, phosphorus, and heavy metals. The biomass produced can be used for extraction of biofuels (Bhat et al. 2021).

However, there are several limitations to biological wastewater treatment. The performance efficiency of any biological-based system is largely reliant on pH and temperature. On a more practical scale, if the developers opt for aerobic digestion, which therefore necessitates aeration, the developers will end up using more electricity. It also produces biological sludge, which, if not adequately managed, causes eutrophication of aquatic bodies. Anaerobic digestion, on the other hand, targets fewer contaminants than aerobic digestion and takes longer to operate as efficiently. Efforts are being undertaken to address these constraints so that microbial wastewater treatment can become more widely used.

According to an estimate, up to ten times the energy necessary for municipal wastewater treatment may be recoverable. Exoelectrogen microorganisms (or exoelectrogens), which are extracellular electron transferors (EETs) capable of transferring (donating) or uptaking (accepting) electrons, are one of the important sustainable technologies beneficial for recovering energy from waste. This method yields two benefits: wastewater treatment and energy generation from waste. EETs, like organisms involved in biological waste remediation, digest soluble organic or inorganic substances from wastewater effluents to create metabolic products like hydrogen, bioethanol, or methane. The generation of metabolites from biowaste is extremely beneficial to the environment and also energy saving. As a result of the relevance of these electron transfer events, electrochemists have used several natural bacterial mechanisms for energy generation and conservation at the anode and/or cathode (Yan et al. 2020). Technologies that employ such microbes with EET capabilities are known as bioelectrochemical systems (BES). These are again broadly divided into two major types – microbial fuel cells (MFCs) and microbial electrolysis cells (MECs). Roughly differentiating these two, MFCs produce electricity while degrading organic compounds, while MECs produce biohydrogen when an external voltage is supplied. This chapter discusses in detail the various types and working of these BES, and their application in treating wastewater effluents.

3.2 DESIGN AND WORKING MECHANISM OF VARIOUS TYPES OF BIOELECTROCHEMICAL SYSTEMS (BES)

MFCs and MECs have more or less the same major components – two electrodes (anode and cathode), and a membrane separating them. Oxidative reactions take place at the anode, and reductive reactions take place at the cathode. The catalyst used at the anode is generally a biocatalyst such as singular or multicellular organisms (bacteria, algae, fungi, etc.), enzymes, or cell organelles. To speed up the reduction reactions at the cathode, a catalyst, which may be inorganic or organic (bio-based), is required to be applied on the cathode. The working of each of these BES, however, tends to differ – while MFCs degrade organic matter to produce bioelectricity, an external voltage is needed to be applied to MECs with biohydrogen being the end product. The following sub-sections not only elaborate about the exact working mechanisms of both the BESs, but also explain a few of the different types of designs used.

3.2.1 MICROBIAL FUEL CELLS (MFC)

In 1911, Potter constructed an MFC, in which *Escherichia coli* and *Saccharomyces cerevisiae* were employed as biocatalysts, platinum as the electrode material, and sugar solution as the substrate. This is the first reported publication of an MFC for bioelectricity production. During his studies, he discovered that the maximum voltage production was independent of the capacity of the container as well as the thickness of the electrode (Potter 1911). In 1931, Cohen Barnett arranged a series of bacterial cultures to create an MFC with a potential of 35 V but only 2 mA of current. He also proposed using potassium ferricyanide or benzoquinone to boost the performance of the cells. Such chemicals, which shuttle electrons between the microbe and the electrode surface, are now known as mediators (Cohen 1931).

Following these discoveries, NASA launched a space mission in 1963 to convert human waste to power during space travels, creating hype around work in MFCs. The next notable contribution in the field of MFCs is the work by Bennetto and his team. They researched extensively on several classes of mediators for improving BFC performance in the 1980s, taking the works of Cohen (1931) into account (Delaney et al. 1984). In 1991, Habermann and Pommer exhibited an MFC that could operate for five years without any need of maintenance or without facing any malfunctions using sulphate reducing bacteria. Their work pioneered the use of MFCs for concomitant wastewater treatment and bioelectrogenesis (Habermann and Pommer 1991). A major breakthrough in the working of MFCs was discovered by Kim and his co-researchers when they showed that *Shewanella putrefaciens* can accomplish electron transfer without the use of external mediators (Kim et al. 1999). According to a bibliometric analysis of MFC publications by Naseer et al. (2021) between 1970 and 2020, there has been an exponential growth in the number of publications on MFCs from fewer than 50 publications in 2000 to more than a thousand in 2020.

From the electro-chemical perspective, MFCs have a theoretical maximum potential that can be reached in a cell of any scale. Considering a cell with glucose as the substrate, and oxygen as the terminal electron acceptor at the cathode, the reactions occurring at the anode and cathode are equations (3.1) and (3.2) respectively.

$$C_6H_{12}O_6 + H_2O \rightarrow 6CO_2 + 24\,e^- + 24\,H^+ \tag{3.1}$$

$$O_2 + 4e^- + 4H^+ \rightarrow 2H_2O \tag{3.2}$$

TABLE 3.1

Standard Potentials of Some Reactions in Anode and Cathode in MFC

Half-Cell Reaction on Anode		Half-Cell Reaction on Cathode	
Substrate & Reaction	Standard Potential E_{anode} (Volt)	Terminal Electron Acceptor & Reaction	Standard Potential $E_{cathode}$ (Volt)
Glucose $C_6H_{12}O_6 + H_2O \rightarrow 6CO_2 + 24\,e^- + 24\,H^+$	-0.428	Oxygen $O_2 + 4e^- + 4H^+ \rightarrow 2H_2O$ $2O_2 + 4H^+ + 4e^- \rightarrow 2H_2O_2$	0.816 0.295
Acetate $C_2H_3O_2^- + 4H_2O \rightarrow 2HCO^{3-} + 9H^+ + 8e^-$	-0.296	Ferricyanide $Fe(CN)_6^{3-} + e^- \rightarrow Fe(CN)_6^{4-}$	0.361
Butyrate $C_4H_8O_2 + 2H_2O \rightarrow 2C_2H_4O_2 + 4H^+ + 4e^-$	-0.28	Permanganate $MnO^{4-} + 4H^+ + 3e^- \rightarrow MnO_2 + 2H_2O$	1.68
Formate $HCOO + 2H_2O \rightarrow HCO_3^- + 3H^+ + 2e^-$	-0.43	Persulfate $S_2O_8^{2-} + 2e^- \rightarrow 2SO_4^{2-}$	2.01

As per Table 3.1, standard potential of anodic and cathodic reactions are –0.428 V and 0.816 V respectively. Applying the Nernst equation to find the overall cell potential (Equation 3.3):

$$Ecell = Ecathode - Eanode = 0.816\ V - (-0.428\ V) = 1.254\ V \qquad (3.3)$$

Hence from this, it can be implied that the theoretical maximum potential that could be generated by complete oxidation of glucose is 1.254 V. However, in practical applications, the output value is much lower, mostly ranging around 0.4 V to 0.7 V due to internal resistance losses, the microbes used, etc.

As shown in Table 3.1, oxidation reactions take place at the anode, and reduction reactions take place at the cathode. When a single substrate is used at the anode, the theoretical maximum voltage potential of the half-cell reaction can be found. On the cathodic side, the terminal oxygen acceptor determines the half-cell potential. The values given for the half-cell reactions in the table are standardized values.

The sub-sections below will describe three basic designs and working of MFCs developed over the years.

3.2.1.1 Double-chamber MFC

Double-chamber or dual-chamber MFCs are the simplest type of MFC of all. In general, two chambers are connected, separated by a membrane, known as the proton exchange membrane (PEM), which selectively allows the movement of protons across the chambers. One chamber acts as the anode chamber while the other as the cathode chamber. The anodic chamber is in anaerobic conditions, while the cathode, in general, is in aerobic conditions (if the terminal electron acceptor is oxygen, else it may be in anaerobic conditions as well in some cases). The substrate (organic matter, which may be wastewater, organic wastes, etc.) is usually placed in the anode. When the organic matter is degraded by the microbes, protons and electrons are released. The protons diffuse through the PEM to the cathode chamber. Electrons flow through the external circuit, hence creating a voltage potential and current.

The main advantage of double-chamber MFCs is the possibility of eliminating oxygen diffusion from the cathode chamber to the anode chamber. When oxygen diffuses to the anode chamber, it not only competes with the anode but also decreases the overall working efficiency of the MFC.

However, the main drawback of double-chamber MFCs is that they cannot be effectively scaled up due to design and cost restraints.

3.2.1.2 Single-chambered MFC

MFCs consisting of only one chamber (normally the anaerobic anode chamber) are known as single-chamber MFC (Figure 3.1) and include both the anode and cathode. Park and Zeikus (2003) reported the first usage of single-chambered MFC. As compared to double-chambered MFCs, these MFCs are simpler in design, and generate higher power densities. Internal ohmic resistance can be reduced by shortening the distance between electrodes, and hence the higher power densities. However, in single-chamber MFCs, microbial contamination and reverse-flow of oxygen are some problems that arise. But these MFCs also exhibit low internal resistance, and do not require membranes, therefore being more cost-effective as compared to their double-chamber counterparts.

3.2.1.3 Sediment MFC

An anode buried in anoxic sediment and a cathode submerged in aerobic water make up a sediment MFC (S-MFC). The oxidizable organic compounds used by the electro-microorganisms in the sediment transmit electrons to the anode. Electrons passing through the external circuit interact with protons that diffuse from sediment and dissolved oxygen in water to create water and produce electricity. The redox potential difference between sediment containing the reductant and water comprising the oxidant is the primary driver of electricity production. The activity of the sediment microbes is related to the redox potential difference and determines the output.

Microbial population changes as sediment depth increases, moving from aerobic to anoxic and even anaerobic species. The redox potential difference between the anode and cathode is the consequence of microorganisms that are present in the bottom layer of sediment using the organic matter and producing the reductant simultaneously. Due to the redox potential difference, electrons move from low potential to high potential, or in other words, S-MFCs produce current.

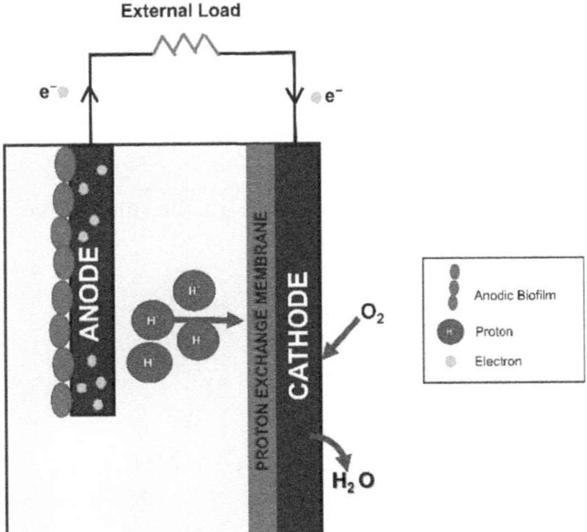

FIGURE 3.1 Schematic representation of a single-chambered MFC.

Note: A single-chambered MFC contains only the anodic chamber with the anode and anolyte contained within it. Generally, the cathode is an air cathode, with oxygen being the terminal electron acceptor. Although, in this figure, a proton exchange membrane is depicted separating the anode chamber and cathode, membrane-less MFCs are also being used.

S-MFCs have the ability to generate a sizeable amount of electrical energy that may be utilized as the power source for low-power consumption equipment. On the other hand, S-MFC is an alternative to in situ bioremediation technique since organic contaminants are degraded during the power generation. Presently, S-MFC technology is being focused for bioremediation with simultaneous powering of low-power consuming devices (Zhao et al. 2021).

3.2.2 Microbial Electrolysis Cells (MEC)

Previously, H_2 was produced utilizing a different electrolytic, photolytic, and/or thermochemical processes. In thermochemical operations, pressure along with heat are utilized to break down molecular bonds. Electrolysis is the use of electricity to separate water into its constituent elements. Photolytic processes are used to extract H_2 with the help of bacteria. However, thermochemical processes need the use of fossil fuels as raw materials, whereas electrolytic and photolytic processes require a great deal of energy and are thus highly expensive. Fossil fuel combustion produces greenhouse gases (CO_2, SO_2, and NOx) as well as hazardous pollutants such as polycyclic hydrocarbons, mercury, and volatile compounds, which contribute to global warming and have a detrimental impact on human health.

Biohydrogen generation is necessary to address thermodynamic and environmental concerns by employing wastewater for hydrogen synthesis while also treating the effluent. When using biomass as a raw material, the organic molecules dissolved in wastewater have a higher energy state as compared to ones that are not, which makes mechanical combustion difficult. The raw materials used in biological hydrogen generation, including wastewater effluents, organic compounds in the form of biodegradable wastes, and lignocellulosic biomass, are easily accessible, are cost-effective, and use waste from other industries. Although several types of water electrolysis technologies have been developed, more research is required before they can be integrated into large-scale, cost-effective energy networks. A recent techno-economic research study, for example, found that water electrolysis using solar energy is still not economically viable when compared to hydrogen creation from fossil sources. Within this paradigm, microbial electrolysis is seen as a beneficial and new method. MECs were first proposed in 2005. In these conditions, microbial electrolysis cell technology delivers a combined benefit of energy production in the form of biohydrogen as well as organic waste treatment. MECs are one of the most alluring green biohydrogen production technologies since they not only yield more hydrogen than other biological technologies, but also need a lot less external energy input (theoretically >0.14 V), especially when compared to water electrolysis (>1.2 V).

For example, considering acetate as the substrate in the anode, the following overall reaction takes place in the MEC (Equation 3.4):

$$C_6H_{12}O_6 + 12H_2O \rightarrow 6CO_2 + 12H_2 \tag{3.4}$$

The reaction at the anode is as follows (Equation 3.5):

$$C_6H_{12}O_6 + 12H_2O \rightarrow 6CO_2 + 24H^+ + 24e^- \tag{3.5}$$

The reaction at the cathode is as follows (Equation 3.6):

$$8H^+ + 8e^- \rightarrow 4H_2 \tag{3.6}$$

In order to achieve hydrogen evolution reaction (HER) in MECs, overcoming the endothermic barrier is necessary. The electric potentials of anodic oxidation and cathodic reduction that are determined using the Nernst equation may be used to calculate HER. The reaction at the anode has

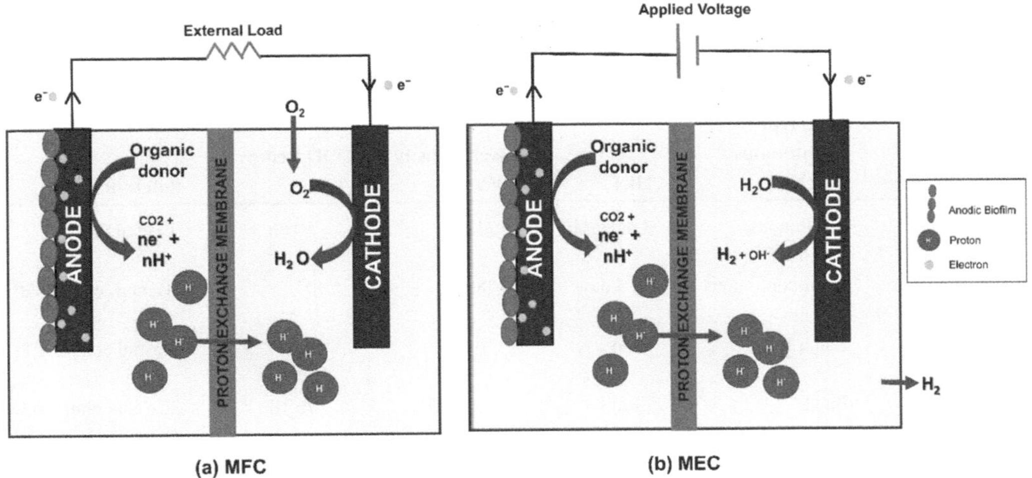

FIGURE 3.2 Working of MFC vs. MEC.

Note: While the design and anodic reactions do not vary much between MFCs and MECs, the major difference is the use of an applied voltage in the case of an MEC. This use of an applied voltage further influences the reactions that take place at the cathode. As seen in the figure, in an MFC, oxygen becomes the terminal electron acceptor. In an MEC, due to the applied voltage, hydrogen gas is generated.

a redox potential value of −0.3 V, while the cathodic reaction redox potential is −0.414 V. Using the Nernst equation, and substituting these values (Equation 3.7):

$$\text{Ecell} = \text{Ecathode} - \text{Eanode} = -0.414\ \text{V} - (-0.3\ \text{V}) = -0.114\ \text{V} \tag{3.7}$$

The negative value indicates that the reaction cannot take place spontaneously, but requires an external voltage of 0.114 V or more to be applied. This applied voltage is required to overcome losses such as ohmic and activation as well as in the limitation of mass transport.

The hydrogen produced by a MEC employing exoelectrogens generally requires an applied potential in the range of 0.6 to 0.9 V, as per literature. The large range is due to several factors – microbial activity, the configuration of the cell, properties of the electrode employed, the substrate type and concentration of organic substrate, and selectivity of the PEM used (Yang et al. 2021).

Configurations of MFCs and MECs are found to be similar – there are many studies using the typical double- and single-chambered designs. The materials used do not vary much either, although suitable arrangements should be made for the collection of hydrogen gas generated, and an external voltage source. Figure 3.2 compares the differences in the working of MFCs and MECs.

3.3 BES FOR WASTEWATER TREATMENT

The most cited use of BES is the simultaneous generation of power along with wastewater treatment. Some systems have also been designed to recover metals and nutrients as well. In this section, BES applications that have focused on treating wastewater will be discussed.

3.3.1 WASTEWATER TREATMENT WITH SIMULTANEOUS BIOELECTRICITY PRODUCTION

MFCs are known for their ability to treat wastewater and also produce electricity, giving these technologies an extra edge over typical or even biological treatment. In MFCs inoculated with anaerobic sludge or other effluent sources, a consortium of exoelectrogens, pollutant-degrading bacteria, and

TABLE 3.2

MFCs Reported for the Application in Treatment of Various Wastewaters

Wastewater Type	MFC Type (Continuous / Batch)	HRT	Power Density (mW/m³)	COD Removal (%)	Reference
Animal carcass wastewater	Continuous	3 days	2190	50.66	(Li et al. 2013)
Dairy wastewater	Semi-continuous flow	10.5 days	4800	93	(Marassi et al. 2020)
Municipal solid wastewater	Batch	4 days	2010	80.5	(Rashid et al. 2021)
Rice mill wastewater	Batch	7 days	656.10	76.18	(Raychaudhuri and Behera 2020)
Pulp/paper wastewater	Batch	70 h	9450	65.6	(Chen et al. 2020)
Sanitary wastewater	Batch	36 h	61*	87.29	(Das et al. 2020)

Note: *power generated expressed in terms of mW.

several other non-exoelectrogenic microorganisms perform in a complicated interspecific coordination. While pollutant-degrading bacteria first break down the pollutants, exoelectrogens and other minor bacteria may be engaged in bioprocesses such as electron transfer, nutrient absorption, nitrification, denitrification, biomass synthesis, and degradation. It has also been discovered that non-exoelectrogens retain an anaerobic state, allowing for optimal conditions towards increased power production. This is the reason why mixed microbial consortiums and communities outperform single pure cultures (Guo et al. 2020; Ilamathi and Jayapriya 2018). MFCs have been used to treat a variety of different wastewater, with a versatile use of materials. Table 3.2 discusses some of the reported MFCs that have been used for wastewater treatment.

Table 3.2 shows that due to the ability of microbes to use a wide range of substrates, MFCs can be used to treat effluents from versatile sources. The power output is influenced by many factors, including the design of the MFC, material used, and effluent used.

3.3.2 Wastewater Treatment with Simultaneous Valuable Product Recovery

BES also have the ability to recover valuable products along with their power generation capacities. As compared to the conventional recovery technologies being used, despite being primarily still in the bench-scale study phase, BES have a great deal of potential to become energy-efficient techniques for recovery of valuable products like metals and nutrients.

3.3.2.1 Nutrient Recovery

Two elements that are necessary for life to exist on Earth are nitrogen and phosphorus. For nitrogen-containing biomolecules, such as amino acids that may be utilized to create various types of proteins, nitrogen is one of the most crucial components. Phosphorus, which is mostly found in the form of phosphate in bones, cells, and proteins, participates in a number of vital metabolic processes in a variety of living species. The nitrogen cycle and the phosphorus cycle, respectively, have been responsible for maintaining the balance of nitrogen and phosphorus in nature for thousands of years. The requirement for chemical fertilizers in modern agricultural practices to restore soil fertility and preserve soil nutrient reserves has lately had an impact on the balances.

According to reports, only 42–47% of the nitrogen supplied to croplands worldwide is used to produce crops, and the remainder adds to the eutrophication of water bodies, which deteriorates water quality and causes the atmosphere to become more acidic, both of which are dangerous for humans. Such circumstances are made worse by the effluent discharge from wastewater treatment facilities with poor ammonium nitrogen removal. As a result, the importance of nitrogen removal from wastewater before disposal has become a matter of concern.

3.3.2.1.1 NITROGEN REMOVAL

BES has lately grown in popularity with nitrogen removal and recovery due to its many benefits over conventional methods. First off, there are no extra chemicals required to raise the catholyte pH. The catholyte pH can be raised by both oxygen reduction and hydrogen evolution, which leads to the conversion of ammonium to ammonia. Second, without the use of alkali or buffers, recycling ammonia to the anode from the cathode can prevent the anolyte's pH from becoming acidic and maintaining current (for example, pH 5.5 significantly hinders current production). Thirdly, the BES offers energy-efficient ammonia recovery along with power generation and wastewater treatment, in contrast to conventional energy-intensive ammonia recovery methods (such as stripping or electro-dialysis). In addition, the hydrogen produced in the cathode chamber of MEC can be utilized to strip ammonia out that then can be absorbed in acid bottles (Cheng, Kaksonen, and Cord-Ruwisch 2013).

There are several ammonia removal mechanisms found in BES. Ammonium could be transported through the membrane of the BES through diffusion (passively) or migration (actively). Because of the cathode's higher pH than the surrounding environment, ammonia can escape the system by evaporating and turning into a gas. Independent of the electrode processes, it can be (biologically) oxidized by oxygen at the cathode chamber and denitrified by the microorganisms upon the cathode or in the solution. Although it has been hypothesized that microorganisms near the anode can directly nitrify or denitrify ammonia to nitrogen gas, there is still no conclusive evidence to support this. Finally, either the anode or the cathode chamber can integrate ammonium into biomass for growth (Rodríguez Arredondo et al. 2015).

3.3.2.1.2 PHOSPHORUS REMOVAL

Phosphorus is a non-renewable material found in nature in the form of phosphate rocks, and it attracts public attention as a necessary nutrient for living creatures as well as significant aquatic pollution. The most practical method for obtaining phosphorus for fertilizer manufacture is mining from phosphate reserves. If there is no appropriate management, it is estimated that commercial phosphate rock deposits would be exhausted in another 70–140 years. On the other side, phosphorus release causes eutrophication, which damages freshwater ecosystems and makes it more difficult to reclaim water from lakes, reservoirs, estuaries, and seas. A large portion of the phosphorus that is mined is released into the environment from point sources (such as ineffective phosphorus removal in wastewater treatment facilities) and nonpoint sources (such as soil erosion and agricultural runoff with leftover chemical fertilizers). As a result, it is vital to think of phosphorus recovered from wastewater streams as a possible valuable phosphorus resource for business and agriculture.

In contrast to nitrogen removal in BES, phosphorus recovery has not received as much attention. Phosphorus, however, has sparked considerable attention in its removal as well as recovery in BES since it is both a valuable resource and a waste. The most frequent kind of phosphorus recovered in BES is struvite mineral, which forms when magnesium, ammonium nitrogen, and phosphorus are present in an alkaline environment (pH >9.2). Because of the decreased oxygen in the cathode compartment, which would naturally raise the catholyte pH, it is thought that BES can promote the development of struvite.

3.3.2.2 Metal Recovery

Microorganisms are known to remove metals from their environments through processes such as bioaccumulation, biosorption, biomineralization, and bioreduction. BES have recently been utilized to recover metals from effluents. In MFCs, the substrate that is oxidized at the anode releases

TABLE 3.3
Metal Recovery in BES

Reactor Type	Metal Removed & Reaction	Power Density (mW/m²) / Voltage Applied (V)	Removal Efficiency (%)	Reference
MFC	Copper(II) $Cu^{2+} + 2e^- = Cu$	5500 mW/m²	90% in 24 h	(Motos et al. 2015)
MFC	Selenium(IV) $Se^{4+} + 4e^- = Se$	2900 mW/m²	99% in 48 h	(Catal, Bermek, and Liu 2009)
MFC	Vanadium(V) $VO_2^+(aq) + 2H^+ + e^- = VO_2^+(aq) + H_2O$	529 mW/m²	~100% in 168 h	(Qiu et al. 2017)
MEC	Lead(II) $Pb^{2+} + 2e^- = Pb$	0.34 V	47.5%	(Modin et al. 2012)
MEC	Nickel(II) $Ni^{2+} + 2e^- = Ni$	1.1 V	67 ± 5.3%	(Qin et al. 2012)
MEC	Zinc(II) $Zn^{2+} + 2e^- = Zn$	-1.0 V	60%–99% in 23 hours	(Modin, Fuad, and Rauch 2017)

electrons that go through the circuit and reach the cathode. The metal or metalloid ions serve as the terminal electron acceptors at the cathode, where they are reduced by the electrons and accumulate as deposits or precipitates upon the cathode, while concurrently generating bioelectricity. For the precipitation of these metals or metalloids to take place, the redox potential of the cathodic half-cell must be higher than the anode potential of the cell. Examples of metals that have positive redox potentials and have been precipitated in MFCs include Ag(I), Cu(II), Au(II), Hg(II), Se(IV), and V(V) (Nancharaiah, Venkata Mohan, and Lens 2015).

For metals and metalloids having low or closer redox potentials to the potential of the anode, such as Zn(II), Pb(II), Cd(II), Ni(II), and Co(II), an additional potential should be supplied. This way, it ensures the electron flow is directed towards the cathode and not the reverse direction. MECs are used in this case since these BES require an external power source for their functioning (Wang and Ren 2014). Table 3.3 shows the reactions involved in the removal of metals, as well as the power densities generated (MFCs) or the applied external voltage (MEC). Depending upon the type of metal that is to be removed from the effluent, there may be or may not be a requirement for an external voltage to be applied. In general, to remove metals that have higher redox potentials than the anodic potential, MFCs are used. For metals that have a lower redox potential, MECs are used to recover the same.

3.3.3 WASTEWATER TREATMENT COUPLED WITH HYDROGEN GAS PRODUCTION

The majority of laboratory research on MECs started when it was discovered that hydrogen could be created by biocatalysed electrolysis. The studies concentrated on creating novel electrodes and new cell designs, examining the performance limitations of MECs, or evaluating the impact of various biological variables. Although the researchers in these investigations were able to alter the substrate composition by using synthetic media, it has long been understood that laboratory testing with actual wastewater is necessary to determine the technology's viability in the real world. The difficulties of employing actual wastewater as a substrate for hydrogen synthesis in MECs are covered in this section.

In 2007, the first MEC with the ability to treat domestic wastewater was reported. It was a double-chamber MEC with a 300 ml anodic chamber (Ditzig, Liu, and Logan 2007). The cell exhibited good BOD removal (BOD 5 ≤ 7 mg l⁻¹) and virtually a complete elimination of COD (87–100%) while operating at applied voltages of 0.2–0.6 V. The cell was batch-fed and had retention durations ranging from 30 to 108 hours. The low hydrogen output (less than 10% of the theoretical maximum) contrasted with the comparatively high treatment efficiency and was explained by the low substrate conversion into energy as well as hydrogen leakage through the tubing.

Analysis of literature can help in concluding several factors that should be taken into account when employing an MEC, including the voltage to be applied, hydraulic retention time (HRT), and COD of the effluent. According to studies, a minimum applied voltage of 0.4–0.6 V is necessary to create a considerable quantity of hydrogen, and increasing applied voltage beyond 0.8 V simply lowers the efficiency of energy recovery without producing any appreciable improvements in current density.

High HRTs (>30 h, batch) are needed to produce high-quality effluent, although HRTs of 3–10 h (continuous) can be utilized to reduce energy use and maximize hydrogen synthesis while maintaining a respectable COD removal rate. The organic loading rate (OLR), as well as the COD content of the wastewater, plays crucial roles in the reactor's overall performance and the optimal applied voltage. Low current densities and coulombic efficiencies, insignificant hydrogen generation, and relatively high energy consumption are typical side effects of low strength wastewater (below 250 mg COD l⁻¹), but COD concentrations over 360–400 mg COD l⁻¹ markedly improved overall performance.

In a study by Tenca et al. 2013, the effect of various wastewater types and cathodic catalyst used affecting MEC performance has been described. They discovered that, contrary to predictions, systems that used industrial wastewater produced a better quality of gas effluent along with a greater quantity of biogas than as compared to systems that used food processing wastewater. Nonetheless, they concluded that a considerable portion of the organic matter was decomposed by non-anode-dependent pathways.

Due to their flexibility and possible capacity to overcome thermodynamic obstacles, MECs have been proposed as an acceptable second stage method in the treatment of dark fermentation effluents. Sewage sludge, a by-product of wastewater treatment plants (WWTPs), may also be utilized as a substrate for MECs that generate hydrogen. Table 3.4 summarizes some studies which have used MECs for treating various types of wastewaters.

TABLE 3.4
MECs Reported for the Application in Treatment of Various Wastewaters

Wastewater Type	MEC Type (Continuous / Batch)	HRT	Applied Voltage (V)	Hydrogen Production (L H₂ L⁻¹ day⁻¹)	COD Removal (%)	Columbic Efficiency (%)	Reference
Distillery wastewater	Batch	70 h	1.0	0.035	77.5	17.25	(Samsudeen et al. 2020)
Domestic wastewater	Continuous	85 days	1.1	0.015	34%	55%	(Heidrich et al. 2013)
Food processing wastewater	Batch	150 h	0.7	0.35	67	35	(Tenca et al. 2013)
Potato wastewater	Batch	48 h	0.9	0.74	79	80	(Kiely et al. 2011)
Swine wastewater	Batch	184 h	0.5	1.0	75	70	(Wagner et al. 2009)
Winery wastewater	Batch	72 h	0.9	0.17	47	50	(Cusick, Kiely, and Logan 2010)

MECs have the ability to use substrates such as wastewater at the anode chamber. At the cathode chamber, depending on the applied voltage and degradation of the substrate at the anode, hydrogen gas is produced.

3.4 CHALLENGES AND LIMITATIONS OF BES IN WASTEWATER TREATMENT

While both MFCs and MECs present great potential as sustainable technologies for wastewater treatment along with green energy production, they have their own boundaries and bottlenecks to overcome for becoming reliable systems at a large commercial scale. Majorly, limitations of BES stem from the microbial design and/or economical restraints. This section addresses the challenges that must be considered and overcome in order to optimize BES towards commercialization.

3.4.1 MICROBIAL ACTIVITY

In terms of microbial activity, methanogenesis – an undesirable process which happens often in MFCs as well as MECs – poses a barrier to MECs for hydrogen generation. Methanogens and exoelectrogens thrive in comparable settings and compete for limited substrates. Methanogens actively develop in BES by taking up the substrate, resulting in methane generation instead of bioelectricity and hydrogen production, therefore lowering overall BES performance. In terms of hydrogen generation, one mole of methane represents 8 moles of electron loss, which is equal to 4 moles of hydrogen gas.

Another issue is the fact that anodic exoelectrogens exhibit limited biodegradability toward complex carbon sources (wastewater, cellulosic compounds, organic wastes, etc.); the range of substrates that may be degraded in BES is also an issue. MECs produce a lot of hydrogen when given simple carbon sources like acetate and glucose, but less hydrogen when given complex carbon sources. Similarly, MFCs give a lower power output.

3.4.2 MATERIAL AND COSTS

Reduction reactions at the cathode (in the case of MFCs) and HER (in MECs) is extremely slow on ordinary cathode materials, necessitating a large overpotential to induce hydrogen generation. A catalyst is utilized on the cathode surface to increase hydrogen production and minimize cathode overpotential. Because of its strong catalytic activity, platinum is the best-studied cathode catalyst to date. However, due to its high cost, rapid cation poisoning, and shorter durability, its practical applicability is limited. As a result, for the actual implementation of MEC, a highly active catalyst which is cost-effective as well as scalable is required. Because of their widespread availability, low cost, and minimal toxicity to microorganisms, transition metals are gaining popularity as an efficient alternative catalyst for cathodic HER.

The most common base materials used in electrodes nowadays are carbonaceous in nature, with good conductivity, a large surface area, and great biocompatibility. Granular graphite, carbon fabric, carbon felt, and carbon paper are among the materials used. Rozendal et al. (2008) have calculated that catalyst-loaded cathodes account for more than 47% of the entire cost of MECs, and the low-cost effectiveness and poor wettability of these materials are problems. To overcome this issue, biocathodes are being used as an alternative and cheap electrode material; microorganisms use the cathode surface as an electron acceptor to catalyse the reduction reactions of HER. The results depicted a large reduction in overpotential, as well as a dramatic increase in cell performance in terms of start-up time and, either bioelectricity or hydrogen output. Despite this material's potential, the large-scale effectiveness of biocathodes as an efficient and low-cost replacement electrode material remains yet to be comprehensively evaluated.

3.4.3 CELL DESIGN AND ARCHITECTURE

Additionally, BES lacks an appropriate standardized cell architecture and scale-up measurements. Proton exchange membranes (PEMs) are used in two-chambered configurations to reduce hydrogen diffusion from the cathode to the anode and prevent the movement of wastewater solution from the anode to the cathode chambers. This increases the purity of the hydrogen generated at the cathode. Even though the PEM is essential, research towards membrane-less systems has been instigated keeping in mind PEM's high internal resistance as well as economic feasibility. A single-chamber BES is strongly preferable from an economic standpoint since it avoids the need for a costly PEM that accounts for more than 25% of MEC capital costs. Nevertheless, there is still no obvious solution for the best cell design towards real-world applications that fully accounts for efficient performance, unwanted substrate exchanges as well as gas diffusion, and the initial investment required.

3.4.4 EXTERNAL VOLTAGE REQUIREMENT

In the case of MECs specifically, these BES sustaining technology is hampered by the requirement for an external power source due to the thermodynamically unpredictable nature of hydrogen evolution and numerous types of potential loss. The theoretically necessary voltage to generate hydrogen gas at the cathode chamber is 0.114 V (in the case of acetate), although prior research has clearly shown that 0.25 V is the absolute minimum needed to produce actual current densities and hydrogen production rates of MECs (Ivanov et al. 2017). According to an estimate, when a MEC operated with an applied voltage of 0.8 V, it was predicted that this external power input might occupy more than 94% of the entire energy need, followed by around 5% of the pumping system (Zou and He 2018).

3.5 FUTURE PROSPECTIVE OF BES

Despite limitations in optimizing BES, over the past decade, there are instances where these technologies are being employed for wastewater treatment at a large scale.

Recently, a municipal wastewater treatment facility in Chateauneuf, Switzerland has constructed the largest MFC system, with a 1000 L capacity. A 1000 L system was created by joining together 64 MFC units that were separated into four sub-stacks. Reticulated vitreous carbon was used as the electrode material. With a continuous wastewater flow rate of 20 L/h and a catholyte flow rate of 10 L/h, acclimation was performed by adding a load of 1 kohm of external resistance. There were voltage reversals found in the channels occasionally, plus variations in the rainfall frequency and seasonal temperature, causing some issues in the performance. However, despite these difficulties, the 1000 L system was a successful arrangement for producing power, treating wastewater, and allowing CO_2 from exhaust gas to be recycled (Blatter et al. 2021).

The Ecovolt reactor from Cambrian Innovation has accomplished the feat of becoming the first commercial application of MEC for high-strength wastewater treatment. Their unique internal architecture claims to produce high-quality biogas (>70% methane) while performing high-rate, bulk BOD removal (80% elimination). An installation's typical net power output through combined heat and power cogeneration is estimated to be between 30 and 200 kW.

A start-up company called "JSP Enviro" in Chennai, India uses MFC technology to treat wastewater from the country's dyeing, dairy, printing, and leather sectors while also producing electricity. This technology outperforms the existing traditional ones on the market in terms of cost and maintenance requirements, making it affordable for small businesses as well.

3.6 CONCLUSION

While BES do provide a promising solution for sustainable renewable technologies, as well as pollution elimination, at a theoretical level, it is to be remembered that at a practical level, there are many performance- as well as economic-based challenges still to be overcome. There is a further need of developing materials that serve the purpose of improving performance of BES, but also are scalable as well. Research is now focused on using composite materials and integrating nanotechnology, especially for cathode catalysts, membranes, and electrode efficiency. There is also active awareness and focus upon the life cycle and techno-economic analyses of these technologies, as compared to the existing ones. However, more emphasis should be laid on the kind of materials used, such that they are environmentally friendly as well as economical. Also, since BES are bio-based systems, there are possibilities of performance ambiguity each time. Detailed study of microbial niches, and the dynamics of the microbial population, needs to be observed. Only then can the long-term working of BES be assessed and understood more clearly. Nevertheless, in spite of several research gaps still needing to be addressed, these technologies are evolving continuously. Start-ups are now offering services and products based on these BES, and moving towards mass commercialization as future sustainable technologies.

REFERENCES

Bhat, Sartaj Ahmad, Guangyu Cui, Wenjiao Li, Yongfen Wei, Fusheng Li, Sunil Kumar, and Fuad Ameen. 2021. "Challenges and Opportunities Associated with Wastewater Treatment Systems." *Current Developments in Biotechnology and Bioengineering: Strategic Perspectives in Solid Waste and Wastewater Management*, 259–83. https://doi.org/10.1016/B978-0-12-821009-3.00008-7.

Blatter, Maxime, Louis Delabays, Clément Furrer, Gérald Huguenin, Christian Pierre Cachelin, and Fabian Fischer. 2021. "Stretched 1000-L Microbial Fuel Cell." *Journal of Power Sources* 483: 229130. https://doi.org/10.1016/j.jpowsour.2020.229130.

Catal, Tunc, Hakan Bermek, and Hong Liu. 2009. "Removal of Selenite from Wastewater Using Microbial Fuel Cells." *Biotechnology Letters* 31(8): 1211–6. https://doi.org/10.1007/s10529-009-9990-8.

Chen, Fu, Siyan Zeng, Zhanbin Luo, Jing Ma, Qianlin Zhu, and Shaoliang Zhang. 2020. "A Novel MBBR–MFC Integrated System for High-Strength Pulp/Paper Wastewater Treatment and Bioelectricity Generation." *Separation Science and Technology (Philadelphia)* 55 (14): 2490–99. https://doi.org/10.1080/01496 395.2019.1641519.

Cheng, Ka Yu, Anna H. Kaksonen, and Ralf Cord-Ruwisch. 2013. "Ammonia Recycling Enables Sustainable Operation of Bioelectrochemical Systems." *Bioresource Technology* 143: 25–31. https://doi.org/10.1016/ j.biortech.2013.05.108.

Cohen, Barnett. 1931. "The Bacterial Culture as an Electrical Half-Cell." *Journal of Bacteriology* 21: 18–19.

Cusick, Roland D., Patrick D. Kiely, and Bruce E. Logan. 2010. "A Monetary Comparison of Energy Recovered from Microbial Fuel Cells and Microbial Electrolysis Cells Fed Winery or Domestic Wastewaters." *International Journal of Hydrogen Energy* 35 (17): 8855–61. https://doi.org/10.1016/ j.ijhydene.2010.06.077.

Das, Indrasis, M. M. Ghangrekar, Rajiv Satyakam, Piyush Srivastava, Swarup Khan, and H. N. Pandey. 2020. "On-Site Sanitary Wastewater Treatment System Using 720-L Stacked Microbial Fuel Cell: Case Study." *Journal of Hazardous, Toxic, and Radioactive Waste* 24 (3): 1–7. https://doi.org/10.1061/(asce)hz.2153-5515.0000518.

Delaney, Gerard M., H. Peter Bennetto, Jeremy R. Mason, Sibel D. Roller, John L. Stirling, and Christopher F. Thurston. 1984. "Electron-Transfer Coupling in Microbial Fuel Cells. 2. Performance of Fuel Cells Containing Selected Microorganism-Mediator-Substrate Combinations." *Journal of Chemical Technology and Biotechnology. Biotechnology* 34 (1): 13–27. https://doi.org/10.1002/jctb.280340104.

Ditzig, Jenna, Hong Liu, and Bruce E. Logan. 2007. "Production of Hydrogen from Domestic Wastewater Using a Bioelectrochemically Assisted Microbial Reactor (BEAMR)." *International Journal of Hydrogen Energy* 32 (13): 2296–2304. https://doi.org/10.1016/j.ijhydene.2007.02.035.

Guo, Yajing, Jiao Wang, Shrameeta Shinde, Xin Wang, Yang Li, Yexin Dai, Jun Ren, Pingping Zhang, and Xianhua Liu. 2020. "Simultaneous Wastewater Treatment and Energy Harvesting in Microbial Fuel Cells: An Update on the Biocatalysts." *RSC Advances* 10 (43): 25874–87. https://doi.org/10.1039/d0r a05234e.

Habermann, W., and E. H. Pommer. 1991. "Biological Fuel Cells with Sulphide Storage Capacity." *Applied Microbiology and Biotechnology* 35: 128–133. https://doi.org/10.1007/BF00180650.

Heidrich, E. S., J. Dolfing, K. Scott, S. R. Edwards, C. Jones, and T. P. Curtis. 2013. "Production of Hydrogen from Domestic Wastewater in a Pilot-Scale Microbial Electrolysis Cell." *Applied Microbiology and Biotechnology* 97 (15): 6979–89. https://doi.org/10.1007/s00253-012-4456-7.

Ilamathi, R., and J. Jayapriya. 2018. "Microbial Fuel Cells for Dye Decolorization." *Environmental Chemistry Letters* 16 (1): 239–50. https://doi.org/10.1007/s10311-017-0669-4.

Ivanov, Ivan, Yong Tae Ahn, Thibault Poirson, Michael A. Hickner, and Bruce E. Logan. 2017. "Comparison of Cathode Catalyst Binders for the Hydrogen Evolution Reaction in Microbial Electrolysis Cells." *International Journal of Hydrogen Energy* 42 (24): 15739–44. https://doi.org/10.1016/j.ijhyd ene.2017.05.089.

Kiely, Patrick D., Roland Cusick, Douglas F. Call, Priscilla A. Selembo, John M. Regan, and Bruce E. Logan. 2011. "Anode Microbial Communities Produced by Changing from Microbial Fuel Cell to Microbial Electrolysis Cell Operation Using Two Different Wastewaters." *Bioresource Technology* 102 (1): 388–94. https://doi.org/10.1016/j.biortech.2010.05.019.

Kim, Byung-Hong, Hyung-Joo Kim, Moon-Sik Hyun, and Doo-Hyun Park. 1999. "Direct Electrode Reaction of Fe(III)-Reducing Bacterium, Shewanella Putrefaciens." *Journal of Microbiology and Biotechnology* 9 (2): 127–131.

Li, Xiaohu, Nengwu Zhu, Yun Wang, Ping Li, Pingxiao Wu, and Jinhua Wu. 2013. "Animal Carcass Wastewater Treatment and Bioelectricity Generation in Up-Flow Tubular Microbial Fuel Cells: Effects of HRT and Non-Precious Metallic Catalyst." *Bioresource Technology* 128: 454–60. https://doi.org/10.1016/j.biort ech.2012.10.053.

Marassi, Rodrigo J., Lucas G. Queiroz, Daniel C.V.R. Silva, Fabiana S. dos Santos, Gilmar C. Silva, and Teresa C.B. de Paiva. 2020. "Long-Term Performance and Acute Toxicity Assessment of Scaled-up Air–Cathode Microbial Fuel Cell Fed by Dairy Wastewater." *Bioprocess and Biosystems Engineering* 43 (9): 1561–71. https://doi.org/10.1007/s00449-020-02348-y.

Mehrotra, Tithi, Subhabrata Dev, Aditi Banerjee, Abhijit Chatterjee, Rachana Singh, and Srijan Aggarwal. 2021. "Use of Immobilized Bacteria for Environmental Bioremediation: A Review." *Journal of Environmental Chemical Engineering* 9 (5): 105920. https://doi.org/10.1016/j.jece.2021.105920.

Modin, Oskar, Nafis Fuad, and Sebastien Rauch. 2017. "Microbial Electrochemical Recovery of Zinc." *Electrochimica Acta* 248: 58–63. https://doi.org/10.1016/j.electacta.2017.07.120.

Modin, Oskar, Xiaofei Wang, Xue Wu, Sebastien Rauch, and Karin Karlfeldt Fedje. 2012. "Bioelectrochemical Recovery of Cu, Pb, Cd, and Zn from Dilute Solutions." *Journal of Hazardous Materials* 235–236: 291–97. https://doi.org/10.1016/j.jhazmat.2012.07.058.

Motos, Pau Rodenas, Annemiek ter Heijne, Renata -van der Weijden, Michel Saakes, Cees J.N. Buisman, and Tom H.J.A. Sleutels. 2015. "High Rate Copper and Energy Recovery in Microbial Fuel Cells." *Frontiers in Microbiology* 6: 527. https://doi.org/10.3389/fmicb.2015.00527.

Nancharaiah, Y. V., S. Venkata Mohan, and P. N.L. Lens. 2015. "Metals Removal and Recovery in Bioelectrochemical Systems: A Review." *Bioresource Technology* 195: 102–14. https://doi.org/10.1016/ j.biortech.2015.06.058.

Naseer, Muhammad Nihal, Asad A. Zaidi, Hamdullah Khan, Sagar Kumar, Muhammad Taha bin Owais, Juhana Jaafar, Nuor Sariyan Suhaimin, Yasmin Abdul Wahab, Kingshuk Dutta, Muhammad Asif, S.F. Wan Muhamad Hatta, and Muhammad Uzair. 2021. "Mapping the Field of Microbial Fuel Cell: A Quantitative Literature Review (1970–2020)." *Energy Reports* 7: 4126–38. https://doi.org/10.1016/j.egyr.2021.06.082.

Park, Doo Hyun, and J. Gregory Zeikus. 2003. "Improved Fuel Cell and Electrode Designs for Producing Electricity from Microbial Degradation." *Biotechnology and Bioengineering* 81 (3): 348–55. https://doi. org/10.1002/bit.10501.

Potter, MC. 1911. "Electrical Effects Accompanying the Decomposition of Organic Compounds." In *Proceedings of the Royal Society of London. Series B, Containing Papers of a Biological Character*, 84:260–76. https://doi.org/10.1098/rspb.1911.0073.

Qin, Bangyu, Haiping Luo, Guangli Liu, Renduo Zhang, Shanshan Chen, Yanping Hou, and Yong Luo. 2012. "Nickel Ion Removal from Wastewater Using the Microbial Electrolysis Cell." *Bioresource Technology* 121: 458–61. https://doi.org/10.1016/j.biortech.2012.06.068.

Qiu, Rui, Baogang Zhang, Jiaxin Li, Qing Lv, Song Wang, and Qian Gu. 2017. "Enhanced Vanadium (V) Reduction and Bioelectricity Generation in Microbial Fuel Cells with Biocathode." *Journal of Power Sources* 359: 379–83. https://doi.org/10.1016/j.jpowsour.2017.05.099.

Rashid, Tazien, Farooq Sher, Abu Hazafa, Rizwan Qureshi Hashmi, Ayesha Zafar, Tahir Rasheed, and Sadiq Hussain. 2021. "Design and Feasibility Study of Novel Paraboloid Graphite Based Microbial Fuel Cell for Bioelectrogenesis and Pharmaceutical Wastewater Treatment." *Journal of Environmental Chemical Engineering* 9 (1): 104502. https://doi.org/10.1016/j.jece.2020.104502.

Raychaudhuri, Aryama, and Manaswini Behera. 2020. "Comparative Evaluation of Methanogenesis Suppression Methods in Microbial Fuel Cell during Rice Mill Wastewater Treatment." *Environmental Technology and Innovation* 17: 100509. https://doi.org/10.1016/j.eti.2019.100509.

Rodríguez Arredondo, M., P. Kuntke, A. W. Jeremiasse, T. H.J.A. Sleutels, C. J.N. Buisman, and A. Ter Heijne. 2015. "Bioelectrochemical Systems for Nitrogen Removal and Recovery from Wastewater." *Environmental Science: Water Research and Technology* 1 (1): 22–33. https://doi.org/10.1039/c4ew00066h.

Rozendal, René A., Hubertus V.M. Hamelers, Korneel Rabaey, Jurg Keller, and Cees J.N. Buisman. 2008. "Towards Practical Implementation of Bioelectrochemical Wastewater Treatment." *Trends in Biotechnology* 26 (8): 450–59. https://doi.org/10.1016/j.tibtech.2008.04.008.

Samsudeen, N, Joshua Spurgeon, Manickam Matheswaran, and Jagannadh Satyavolu. 2020. "Simultaneous Biohydrogen Production with Distillery Wastewater Treatment Using Modified Microbial Electrolysis Cell." *International Journal of Hydrogen Energy* 45 (36): 18266–74. https://doi.org/10.1016/j.ijhydene.2019.06.134.

Tenca, Alberto, Roland D. Cusick, Andrea Schievano, Roberto Oberti, and Bruce E. Logan. 2013. "Evaluation of Low Cost Cathode Materials for Treatment of Industrial and Food Processing Wastewater Using Microbial Electrolysis Cells." *International Journal of Hydrogen Energy* 38 (4): 1859–65. https://doi.org/10.1016/j.ijhydene.2012.11.103.

Verma, Priyanka, Achlesh Daverey, Ashok Kumar, and Kusum Arunachalam. 2021. "Microbial Fuel Cell – A Sustainable Approach for Simultaneous Wastewater Treatment and Energy Recovery." *Journal of Water Process Engineering* 40: 101768. https://doi.org/10.1016/j.jwpe.2020.101768.

Wagner, Rachel C., John M. Regan, Sang Eun Oh, Yi Zuo, and Bruce E. Logan. 2009. "Hydrogen and Methane Production from Swine Wastewater Using Microbial Electrolysis Cells." *Water Research* 43 (5): 1480–88. https://doi.org/10.1016/j.watres.2008.12.037.

Wang, Heming, and Zhiyong Jason Ren. 2014. "Bioelectrochemical Metal Recovery from Wastewater: A Review." *Water Research* 66: 219–32. https://doi.org/10.1016/j.watres.2014.08.013.

Yan, Xuejun, Hyung-Sool Lee, Nan Li, and Xin Wang. 2020. "The Micro-Niche of Exoelectrogens Influences Bioelectricity Generation in Bioelectrochemical Systems." *Renewable and Sustainable Energy Reviews* 134: 110184. https://doi.org/10.1016/j.rser.2020.110184.

Yang, Euntae, Hend Omar Mohamed, Sung Gwan Park, M. Obaid, Siham Y. Al-Qaradawi, Pedro Castaño, Kangmin Chon, and Kyu Jung Chae. 2021. "A Review on Self-Sustainable Microbial Electrolysis Cells for Electro-Biohydrogen Production via Coupling with Carbon-Neutral Renewable Energy Technologies." *Bioresource Technology* 320: 124363. https://doi.org/10.1016/j.biortech.2020.124363.

Zhao, Qing, Min Ji, Hongmei Cao, and Yanli Li. 2021. "Recent Advances in Sediment Microbial Fuel Cells." *IOP Conference Series: Earth and Environmental Science* 621: 012010. https://doi.org/10.1088/1755-1315/621/1/012010.

Zou, Shiqiang, and Zhen He. 2018. "Efficiently 'Pumping out' Value-Added Resources from Wastewater by Bioelectrochemical Systems: A Review from Energy Perspectives." *Water Research* 131: 62–73. https://doi.org/10.1016/j.watres.2017.12.026.

4 Advancements in Conventional and Modern Waste Treatment Approaches

Rejeti Venkata Srinadh, Upasana Priyadarshini, and Neelancherry Remya

4.1 INTRODUCTION

As the global population expands, solid waste generation is anticipated to hit 3.4 billion MT annually by the year 2050. Asia now generates one-third of the world's solid waste, with India having a share of 0.5–0.9 kg/capita/d and China with 0.44–4.3 kg/capita/d. Population growth is crucial to the production of a substantial amount of solid waste, which poses a serious threat to sustainable development. The escalating quantity and heterogeneity of industrial and urban garbage have been identified as one of the potential threats to humanity. Rapid growth in industrialization along with the population is leading to excessive dependency on fossil fuels for energy requirements. These non-renewable fuels are on the verge of depletion due to the excessive burden. Hence, solar energy, wind energy, geothermal energy, tidal energy, biomass as biofuels, and many more renewable energy technologies are all at various stages of development. These alternative energy applications have different boons and drawbacks, which in turn affect their commercial usage. One of the recently growing yet promising alternative energy sources is biomass, which is renewable and can quench the demand for carbon-based fuels and chemicals in all three phases viz. solid (biochar), liquid (bio-oil), and gas (syngas).

A sudden rise in the quantity of waste generated during the outbreak of COVID-19 was observed across the globe, which shattered the current waste management operations of many developing and under-developed countries. A notable increase was observed in emerging contaminants (ECs), which are unregulated, new compounds that could have far-reaching effects on ecosystems and human health. Apart from ECs, other bio-wastes, like pharmaceutical industry sludge, antibiotic-resistant genes, and micro-pollutants, have also been accumulating at a higher pace in wastewater. Modern treatment practices like vermiremediation (IndraKumar Singh et al., 2022), phytoremediation (Kumar et al., 2022), and microbial fuel cells are being relied upon for the effective removal of the above-mentioned pollutants. These advanced technologies are proven to be both economical and environmentally sound for wastewater treatment (Bhat et al., 2021). On the other hand, a huge increase in plastic waste (PPE kits, single-use plastics, facemasks) and hazardous waste (hospital waste, used syringes and medicine covers, bottles, glassware, etc.) was observed (Ali & Parvin, 2022). The global waste composition as of the year 2021 is represented in Figure 4.1.

The majority of solid waste constitutes biomass waste originating from various sources, like agricultural waste, municipal solid waste (MSW), and other industrial organic byproducts. These biomass wastes predominantly consist of lignocellulosic components embedded within a variety of minerals, proteins, and sugars. Hence, these biomass wastes carry a huge potential to act as an efficient energy source (an efficient alternative to fossil fuels), subjected to proper treatment. Different treatment strategies can be adopted to transform these waste sources based on their characteristics.

DOI: 10.1201/9781003408352-5

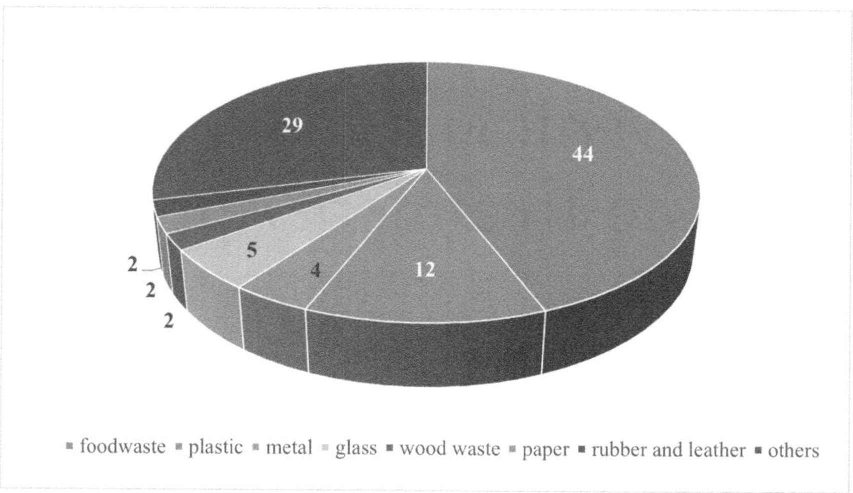

FIGURE 4.1 Global waste composition as per 2021 statistics.

Understanding the characteristics and corresponding constituents of solid waste are necessary for careful management and the application of appropriate technology.

4.2 SOLID WASTE CHARACTERISTICS

4.2.1 PHYSICAL CHARACTERISTICS

The physical characteristics of solid waste include moisture content, particle size and density, calorific value, and field capacity, which have a significant impact on determining the fate of the waste. The extent of compostability, combustibility, or recyclability of the solid waste is determined using these characteristics. The particle size and density of biomass fuels are crucial because they affect the combustion characteristics, specifically the drying and heating rate throughout the combustion process. The low bulk density of biomass is primarily responsible for its poor volume-based heating value. In general, biomass is densified into condensed feedstock to enhance its energy density, hence decreasing the cost of bioenergy conversion. On the other hand, field capacity and moisture content are significant in determining the ease of biological conversion processes. The higher the calorific value of the biomass, the better the combustibility.

4.2.2 CHEMICAL CHARACTERISTICS

Chemical characteristics determine the presence of different chemical components in the waste through two major analyses, namely proximate analysis and ultimate analysis. The former helps in the determination of moisture, volatile matter, ash content, and fixed carbon content, and ultimate analysis determines the availability of different elemental components like carbon (C), oxygen (O), hydrogen (H), sulphur (S), and nitrogen (N). Table 4.1 depicts the physical and chemical characteristics of some of the biomass waste used in energy conversion.

Biomass waste also consists of lignocellulosic materials like hemicellulose, cellulose, and lignin along with sugars, starch, lipids, proteins, and carbohydrates. Lignocellulosic biomass is comparably less effective than others in biochemical conversion due to the presence of lignin that requires a lot of time and energy to break down . Hence, lignocellulosic biomass is preferably used in thermo-chemical conversion technologies.

TABLE 4.1
Physical and Chemical Characteristics of Biomass

Feedstock	Ultimate Analysis (%)				Proximate Analysis (%)				Density (Kg/m³)	HHV (MJ/Kg)
	C	H	N	O	Volatile Matter	Fixed Carbon	Ash Content	Moisture Content		
Sewage Sludge	23.10–42.30	3.10–6.00	3.20–8.30	17.70–34.40	38.30–73.70	0.30–9.41	24.08–52.00	37.32–35.16	370–560	13.05
Rice Husk	39.37–41.12	5.13–5.28	0.32–0.39	53.18–53.58	61.20–70.60	12.87–16.47	16.53–20.91	0.84–0.94	90–150	12.2–16.58
Sugarcane Bagasse	44.70–44.90	5.91–5.97	0.08–0.12	48.79–48.99	76.85–77.65	13.25–13.37	9.59–9.61	0.86–0.94	80–120	15.7–19.5
Switch Grass	42.33–48.80	5.30–6.99	0.03–1.16	37.58–43.68	81.20	13.81	2.54–5.51	2.65–12.00	115–182	17.90–19.06
Corn Stover	43.65	5.56	0.61	43.31	75.17	19.25	5.58	6.2–12.00	131–158	16.18–20.40
Douglas Fir	47.90	6.55	0.08	45.57	76.08	18.89	0.21	4.82	530	19.5–21.5

Source: Arpia et al., 2021; Barskov et al., 2019.

Notes:
C: carbon; H: hydrogen; N: nitrogen; O: oxygen; HHV: higher heating value
Physical characteristics include HHV and density.
Chemical characteristics are determined by proximate analysis and ultimate analysis

4.3   NEED FOR SOLID WASTE MANAGEMENT

The major objective of solid waste management is a reduction of the quantum of waste to be disposed of in a landfill. There are various important steps in order to achieve this objective involving source reduction, proper collection and segregation, and waste processing for recycling or reuse before dumping at the landfill sites. The waste processing facility involves two major functions: waste-volume reduction, and material and energy recovery from waste.

4.3.1   WASTE-VOLUME REDUCTION

Waste-volume reduction aims at minimizing the quantity of waste (reducing the size occupied by waste) to be transported to landfill. It involves a range of processes, from size reduction, compaction, segregation, and recycling to open burning and incineration/combusting. The source reduction of waste plays a significant role in the waste-volume reduction. This can be achieved through general sensitive practices (limiting the usage of packaging material, usage of recycled products, reusing resources in a greater efficient manner during their full lifespan, etc.). Once the waste is collected, the volume reduction of waste occurs using any of the three technologies, namely chemical waste-volume reduction, mechanical waste-volume reduction, and thermal waste-volume reduction. Chemical waste-volume reduction uses processes like hydrolysis or chemical conversions, which induce certain chemical changes within the volume, thereby decreasing the quantum of waste to be dumped. Certain mechanical machinery like shredders and compactors are used in the size reduction and densification of solid waste. Solid waste collection vehicles are generally equipped with compactor machines to effectively collect larger volumes of waste per trip. It is to be noted that the presence of biodegradable waste reduces the recyclable performance of wastes like paper and cardboard after compaction. Size reduction helps in balling the wastes like agricultural residues, which increases the density of the waste. Thermal conversion processes like incineration are generally practised in urban areas and industries to convert the byproducts into ash. This procedure reduces the volume of waste produced by almost 90% (Nanda & Berruti, 2021). The byproduct of incineration is then transferred to the dumping sites or secured landfills. This process of volume reduction also increases the lifespan of the landfill.

4.3.2   ENERGY AND RESOURCE RECOVERY

Certain wastes contain valuable components that are to be recovered for reuse or recycling into other materials. The primary objective of energy and resource recovery is to recover recyclable materials and generate alternative energy sources from the trash. In this way, the recovered material may be converted into new goods, reducing the need for fresh raw materials. Considerably less power is needed to make the same thing out of recycled resources than it would out of raw ones. Moreover, it reduces the amount of energy used in the various phases of the product life cycle, which includes extraction, production, utilization, and disposal. Energy recovery shall be done from homogeneous waste sources that have a higher heating value (Prajapati et al., 2021). The energy recovery from waste is an environmentally sustainable and economically feasible waste management technique, which results in the production of newer energy sources like char, syngas, bio-oil, and methane. These technologies are briefly discussed in the upcoming section of this chapter.

4.4   WASTE TO ENERGY (WTE) CONVERSION TECHNOLOGIES

4.4.1   THERMOCHEMICAL CONVERSION

Thermochemical conversion involves the usage of heat energy to initiate chemical reactions within biomass waste in order to create value-added products like syngas, biochar, and liquid fuel like bio-oil

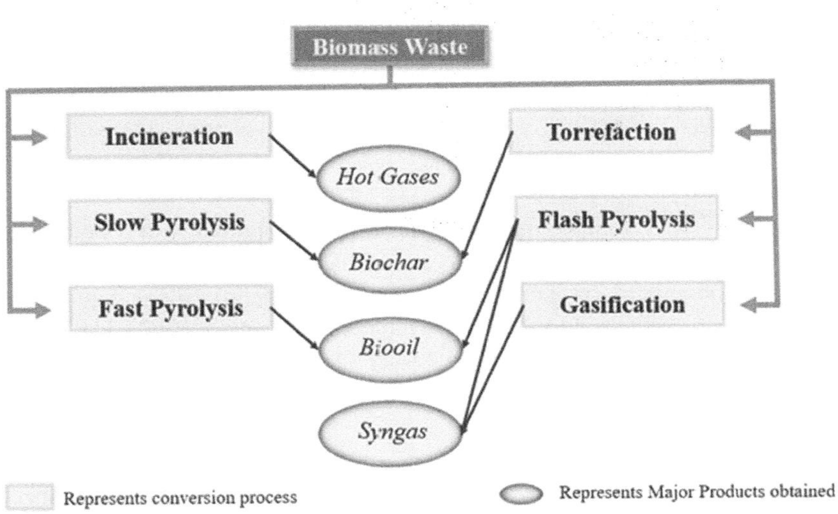

FIGURE 4.2   Different thermochemical waste to energy conversion technologies.

(Nanda & Berruti, 2021). There are four major thermochemical technologies involved in conventional WtE conversion, namely (1) combustion; (2) pyrolysis; (3) gasification; and (4) torrefaction (Figure 4.2).

4.4.1.1   Combustion / Incineration

Incineration is a familiar and established WtE method that retrieves high-temperature energy due to the burning of solid waste for thermal and power facilities. In addition to generating energy, incineration may also be utilized for electricity generation. Incineration of MSW has been favoured over landfills, which are odorous and space-intensive, and which interfere with domestic or agricultural regions. Incineration often decreases the weight and volume of MSW by 80–85% and 95–96%, respectively. The waste properties that affect the efficiency of the incineration process are calorific value, moisture content, particle size of feedstock, elemental composition, ash fusion temperature, and bulk density. Moisture, CO, CO_2, CH_4, H_2, light hydrocarbons, volatile organic carbons (VOCs), dioxins, furans, tars, and polycyclic aromatic hydrocarbons (PAH) are the majority constituents among the volatile matter emissions from incineration. Temperature, mineral matter, heating rate, and inorganic materials of municipal solid waste govern the production of gases, slag, ashes, and heat during incineration (Nanda & Berruti, 2021). Co-combustion of MSW and coal overcomes a number of underlying difficulties, such as improved boiler operations and decreased ash volume, smoke, CO, and particle emissions. In addition, the coal boilers might be converted to utilize MSW as a co-fuel, thereby improving the economics of the process. Co-combustion of MSW and coal could be an effective way to generate environmentally friendly electricity and reduce waste.

4.4.1.2   Pyrolysis

Pyrolysis is considered a viable replacement for incineration since it enables a high rate of energy as well as resource recovery. Recently, pyrolysis has attracted substantial interest due to its greater energy generation capability and capacity to solve incineration's current disadvantages, like transportation constraints and the difficulty in locating new sites for landfilling. It is fundamentally a thermochemical process undertaken at 400–600 °C, devoid of oxygen. The production of gases, bio-oil, and char from pyrolysis is dependent on the process's pressure, temperature, and heating rate.

Major issues associated with bio-oil produced in this process are high water content, high acid content, and greater instability both thermally and oxidatively. Separate bio-oil upgradation processes, like deoxygenation, emulsification, and thermal cracking, are done using different catalysts. Recently, the pyrolysis process has evolved with different reactor types (fixed bed reactor, thermal tubular reactor, fluidized bed reactor, etc.), which tends to increase product yield and cost efficiency.

4.4.1.3　Gasification

There are several ways and technologies for gasification. The optimal gasification system is chosen based on the nature and availability of biomass as well as the desired output. The high volatile component of biomass (about 80% for a dry and ash-free (DAF) fuel) results in a quick transformation into a gaseous product. Gasifiers generally run at temperatures between 800 and 950 °C for biomass. In terms of environmental effects and energy recovery opportunities, gasification technology is preferably superior to waste incineration technology (Ongen et al., 2016).

4.4.1.4　Torrefaction

Typically, torrefaction occurs at comparably modest pyrolysis temperatures in the range of 200–300 °C in an oxygen-free environment. The feedstock is heated at a heating rate below 50 °C/min, over a duration of time ranging from hours to days, thus the volatiles are released and carbon's structure is maintained. In the first phase, water is lost, which is a component that might reduce the calorific value of a fuel. This is followed by the loss of small amounts of CO_2, CO, H_2, and CH_4. This results in the retention of around 70% of the bulk and 90% of the total energy. The solid substance has a low affinity for water (hydrophobic) and may be preserved for a prolonged amount of time (Thengane et al., 2022). During this process, biomass tends to lose a higher amount of oxygen (O_2) and hydrogen (H) over carbon, enhancing the calorific value of the char product. The elimination of O_2 from biomass as a mixture of gaseous compounds (CO_2, H_2O, and CO) and condensable volatile derivatives (like furans, acids, and phenols) is primarily responsible for the enhanced qualities of torrefied biomass. This process is also applicable as a pre-treatment method for increasing the productivity of thus generated biofuels (Arpia et al., 2021).

4.4.2　BIOCHEMICAL CONVERSION

The waste raw material for biological conversion shall have components like dry solids, moisture, volatile solids, organic content, carbon/nitrogen (C/N) ratio, micronutrients and macronutrients for microbial growth, and the absence of pollutants. The relevant biochemical conversion processes, such as anaerobic digestion, vermicomposting, and composting, are selected based on the metabolic activities of the involved microbes.

4.4.2.1　Anaerobic Digestion

Anaerobic digestion is a method that tries to digest organic waste, converting it to a gaseous product and a stabilized carbon-rich effluent, through the growth and digestion activities of a different microorganisms (facultative and stringent anaerobes) under non-toxic circumstances. A collection of microorganisms undertake the biochemical process, which is often separated into four stages: hydrolysis, acidogenesis, acetogenesis, and methanogenesis. The initial phase, hydrolysis, is characterized by the breakdown of complex components like lipids, carbohydrates, and proteins into soluble compounds like short-chain sugars, amino acids, and fatty acids. This is the limiting step of the procedure, as the absorption of these molecules by the microorganisms is dependent on the pace of substrate breakdown by the enzymes. In the next step, these obtained substrates achieved from the hydrolysis step are absorbed by microbes and utilized in the energy metabolism of these microorganisms and volatile organic acids (VOCs) like lactic, propionic, acetic, and formic acids, certain alcohols, and gaseous substances like CO_2 and H_2. During acetogenesis, organic acids produced during acidogenesis are converted to acetic acid. A specific set of homoacetogenic bacteria accomplish this activity. In the last phase, methanogenesis, the acetotrophic bacteria utilize acetic acid to make methane, whereas the hydrogenotrophic bacteria use carbon dioxide and hydrogen (Cremonez et al., 2021).

4.4.2.2 Composting

Organic material is prone to microbial degradation under regulated, damp, hot, oxidative, and non-oxidative settings throughout the composting process. Composting is the most prevalent and economical method for treating the organic part of municipal solid waste. It is the finest technique for the degradation of organic content present in municipal solid waste, including dairy, vegetable, culinary, and slaughterhouse waste (Prajapati et al., 2021). There are two primary composting processes: aerobic composting and anaerobic composting. Microbes decompose organic waste in the oxygen-rich atmosphere during aerobic composting, but oxygen is non-existent during anaerobic composting. The compost obtained is usually free of pathogens and may be used as an effective fertilizer on farms. Some 50–85% of garbage may be reduced with composting (Pujara et al., 2019). Composting's capacity to operate as a good soil conditioner, improve solids quality, and serve as an organic resource to farmland is a key advantage. It also reduces the burden on landfills. Co-composting is a technology that combines multiple source components to enhance the former process. In this technique, the biomass is combined with a specialized thickening substance that provides proper particle stability and density to enhance the growth of the aerobic microorganisms, hence accelerating the breakdown rate. In addition, the co-composting method may be used to reclaim depleted soils from indoor crops, as well as polluted soils. A correct mixture of soil and organic matter can restore the structure of the soil, increase the productivity of nutrient-dependent fields, and treat polluted grounds through the breakdown of different heavy metals (Hu et al., 2020).

4.4.2.3 Vermicomposting

During the vermicomposting process, microbes, red worms, and earthworms alter the biochemical complexity of the organic portion of waste to generate fertilizer (Bhat et al., 2017; 2018). This process consists of three distinct phases: mixing, digesting, and maturation. In the stage of mixing, the waste component is typically digested by microbes and worms. During this phase, a temperature of around 15 °C and at least 2–5 days are necessary. However, throughout this process, the temperature rises by around 50 °C due to the energy produced. During the digestion phase, fungi, actinomycetes, and earthworms are used to biodegrade semi-complex chemicals into a substrate. This stage demands a temperature of around 60 °C and a least of 10–30 days. During the last and final step, maturation, complex chemicals are biodegraded into a substrate over a period of 10–20 days. It is a highly helpful technique since the right management of biomass waste results in the generation of a higher-quality bio-fertilizer (Prajapati et al., 2021).

4.4.3 Challenges of WtE

The major challenges associated with traditional WtE conversion technologies are high input power requirements, greater retention time, low-quality biofuel yield, the requirement for post-treatment technologies, and waste generation. Biological conversion technologies have certain drawbacks that are discussed in Table 4.2, which depicts the challenges faced by conventional WtE conversion technologies. Similarly, thermochemical conversion technologies like combustion / incineration produce greater quantities of bottom ash, which require further waste management techniques for safe disposal or reuse. This process involves a lot of heat energy and the establishment of air pollution control devices to remove pollutants from the gases evolved during the process, which constitutes a huge economic burden in the setting up and maintenance of the technology.

TABLE 4.2
Challenges for Waste to Energy (WtE) Conversion Technologies

Sl No	Conversion Process	Challenges Involved
1.	Incineration	Ash generation and release of toxic gases like dioxins and furans
		Requirement of air pollution control devices (APCD)
		Expensive operation and maintenance of APCDs
		Require further waste management techniques for ash disposal
2.	Pyrolysis	Product yield reduces due to the presence of metals in solid waste
		Feedstock pre-treatment is required
		High power requirement
		Time-consuming process
		Feedstock properties influence product yield
3.	Gasification	Plugging of the reactor due to tar formation
		Waste heat recovery issues
		Presence of salts and metals in feedstock potentially corrodes the reactor
		Requires high-temperature conditions and prolonged retention time
4.	Hydrothermal Liquefaction	Presence of metals may corrode the reactor
		High-pressure conditions must be maintained
		Require catalysts for effective yields
5.	Torrefaction	Consumes ample amount of time
		Requires higher temperature ranges compared to MW-assisted process for similar yield
		Comparatively lower HHV of the yielded biochar
6.	Anaerobic Digestion	Time-taking process due to slower microbial metabolism
		Anaerobic conditions must be maintained throughout
		Chance of leakage creates a foul odour
		Single-stage AD is less efficient
		Little control over product yield
7.	Composting	Anaerobic conditions might develop if not properly mixed
		Proper C/N ratio must be maintained throughout
		Little control over temperature
		Slow microbial metabolism leading to more time requirement

Source: Khan et al., 2022; Narayan et al., 2021.

Note: MW: microwave; AD: anaerobic digestion; C/N: carbon to nitrogen ratio

Apart from incineration, Pyrolysis and gasification processes require larger residence time due to a lesser heating rate involving surface heating mechanisms. These processes yield better biofuel output compared to other thermal processes, yet high energy consumption and greater residence time are the major setbacks. The addition of catalysts and feedstock pre-treatment sums up to the additional cost required for improving the biofuel yield. To overcome these challenges, a novel thermochemical process emerged that uses microwaves for converting waste to energy, which is extensively discussed in further sections.

4.5 ADVANCEMENT IN SWM

4.5.1 Microwave Application in SWM

One of the major lacunas of conventional thermal bioenergy conversion technologies is process efficiency owing to the properties of heat transfer. The thermal conversion of biomass using microwave-assisted processes has garnered a great deal of interest in recent years because of its ability to enhance conversion efficiency and product yield and quality. The location of microwaves in the electromagnetic scale is denoted by a range of 300 MHz–300 GHz of frequencies with wavelength ranging between 1 mm and 1 m. There are three possible interactions between microwaves and matter: reflection, absorption, and transmission (in terms of heat generated due to the electrical aspect of the microwave system) or any mixture of the three processes thus described. Consequently, the following categories may be used to classify materials:

(i)   Absorbers, in which the microwaves are absorbed by the substance (e.g., water), resulting in an accelerated process of heating the material.

(ii) Microwave-transparent material, in which microwaves flow through or propagate with little or no loss (such as Teflon, quartz, and ceramics).

(iii) Microwave-reflective material, in which the microwave propagation is prohibited and rebounded (such as metals), having negligible energy coupling to the process.

Dielectric heating is reliant on the electromagnetic field properties and feedstock characteristics. The primary molecular concepts behind the heating mechanism generated by microwave (MW) radiation are interfacial polarization, dipole reorientation, and conduction.

Microwave radiation has been used in a wide range of industrial and commercial applications over several years (Asomaning et al., 2018). Transformation of electromagnetic energy to thermal energy occurs in MW heating due to ionic conduction and dipole rotation (Simonetti, Martín, and Dionisi, 2022). The polar particles present in the feedstock get targeted by MW. Production of heat, cell lyses, and breakage of the polymeric network occurs due to the wavering of polar molecules (Uthirakrishnan et al., 2022). Numerous benefits of MW heating make it a desirable option, including instant initiation and ceasing and also the creation of immediate heating of the material. Compared to traditional approaches, this quick heating speeds up a lot of organic reactions that are assisted by microwaves. Greater yields and selectivity of the substance can be attained in less time throughout the reaction. Furthermore, because microwaves pass through the vessel wall, no immediate interaction exists between the energy source and the reactants. For direct interaction with the components of the reaction mixture, the container wall is generally transparent to the microwaves (Ethaib et al., 2020). So, MW heating is a very efficient process to extract valuable products from waste.

4.5.2 Microwave in Pre-treatment

Pre-treatments are effective in enhancing the decomposability of organic materials and growth in the production of methane. Pre-treatment can be classified into chemical, biological, and physical, each of which can be used alone or as a mix. Microwave pre-treatment (MWP) falls under physical pre-treatment. As compared to thermal pre-treatments, MW heating is fast and has a feature of selective heating because of its different heating mechanism. Typically, water, organic solvents, alkalis, and dilute-acid based solutions have been used to perform MW pre-treatment for bioethanol synthesis. Pre-treatment can facilitate the hydrolysis of polymers into smaller molecules during the anaerobic digestion (AD) process. As a pre-treatment for the AD process, microwave-assisted heating is combined with H_2O_2, alkalis, and acids (Arpia et al. 2021). The application of MW in various pre-treatment processes is depicted in Figure 4.3.

An experiment with varying temperature (50–96 °C) shows that the biogas production is increased by 14±4% at the temperature of 96 °C due to MWP. Hence, at higher temperature better performance occurs (Uthirakrishnan et al., 2022). Table 4.3 summarizes different MW pre-treatment process undertaken in recent years. Studies on MWP of food waste shows that in comparison to residence time, temperature has more effect. Lignocellulosic biomass solubilization increases up to 40% due to the MWP. Instead of heat, electrical energy is utilized in MW, hence renewable energy like wind

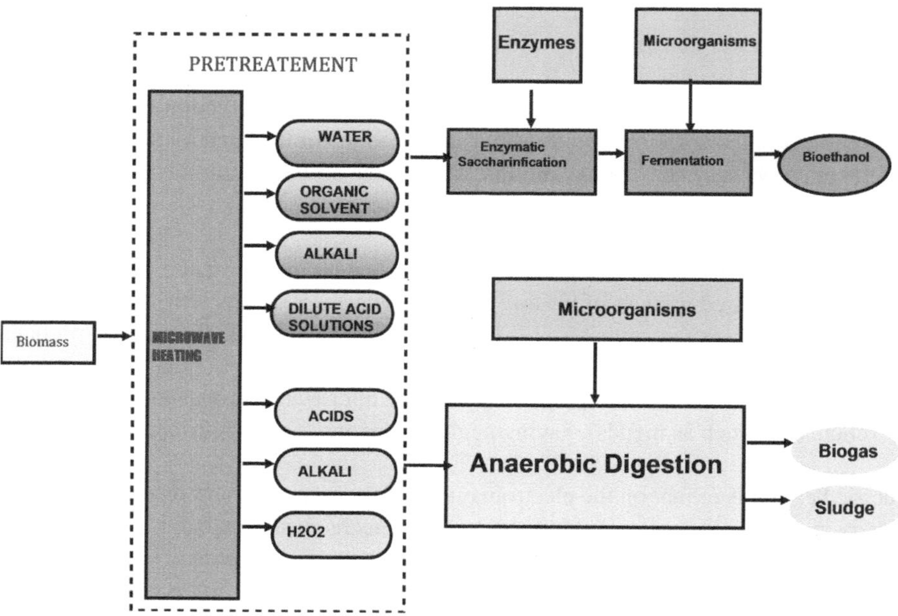

FIGURE 4.3 Microwave application in biomass pre-treatment processes.

TABLE 4.3
MW Pre-treatment in Bioenergy Production

Feedstock	Operating Conditions	Treatment	Yield	References
Cassava Pulp	90 °C, 30min	MW-alkali pre-treatment	Bioethanol	(Hoang et al., 2021)
Rice Straw	680 W, 24 min	MW-pre-treatment	Bioethanol	(Hoang et al., 2021)
Rape Straw	900 W, 1 min	MW-pre-treatment	Bioethanol	(Hoang et al., 2021)
Sewage Sludge	1000 W, 10 min	MW-NaOH pre-treatment	Compared to raw sludge, the solubilization degree is 18 times higher	(Arpia et al., 2021)
Sewage Sludge	600 W, 2 min	MW-alkali pre-treatment	COD solubilization is around 46%	(Asomaning et al., 2018)

Note: MW: microwave.

or solar energy can be used as a source of power. This makes MWP a complete sustainable process (Simonetti et al., 2022). For enhanced solid biofuel, MWP is directly related to the nature of feedstock. The power requirement varies from 200 to 2200 W, temperature variation from 110 °C to 200 °C, and retention time ranges from 2 to 5 minutes (Angulo-Mosquera et al., 2021).

4.5.3 Microwave-assisted Pyrolysis (MAP)

Microwave pyrolysis is the technique of directly heating and pyrolysing materials using microwaves as an indirect source of heat and carbon compounds as the microwave receptor. MAP is a revolutionary approach for effective in-situ treatment of biomass. This makes the microwave-assisted process an approachable method for recovering energy from biomass, thereby converting those wastes into valuable goods. Figure 4.4 represents the application of microwave in different bioenergy production processes. Compared to conventional pyrolysis, this method yields oils that include less harmful byproducts and produces a variety of chemicals with industrial relevance. It is equally desirable and helpful that the gas portion generates greater quantities of syngas. In contrast, MAP enables a huge opportunity to shift the garbage from traditional, environmentally harmful disposal techniques like landfilling and incineration. It also offers a practical way to recover commercially useful goods from waste (Ethaib et al., 2020)

In general, different MAP conditions will result in variable product yields. The operational conditions of this process ultimately determine the effectiveness of microwave pyrolysis. The rate of temperature change, addition of microwave absorbers (like silicon carbide, carbon-based materials, and metal oxides), moisture content, and initial inert gas flow rate and residence time are the variables that have major effects on the production efficiency of MAP. Understanding how these variables affect and interact with one another throughout the microwave-assisted pyrolysis process is crucial. Biomass often absorbs microwaves at a low level. However, adding the microwave-absorbing elements improves the treated materials' ability to absorb microwaves. The presence of inorganic materials and relatively high moisture levels may affect the final product. It has been noted that the initial moisture content of the biomass has a significant influence on the yields of MAP products (bio-oil, syngas, and biochar). At a higher temperature, the drier feedstock promoted the yield of high viscosity oil (Ethaib et al., 2020).

4.5.4 Microwave-assisted Torrefaction (MAT)

Torrefaction is a thermal pre-treatment for biomass waste residues at mild temperature that takes place in an oxygen-free environment. The main yield of torrefaction is biochar, which can be comparable to coal in terms of combustible properties. During this process, the calorific value and energy density

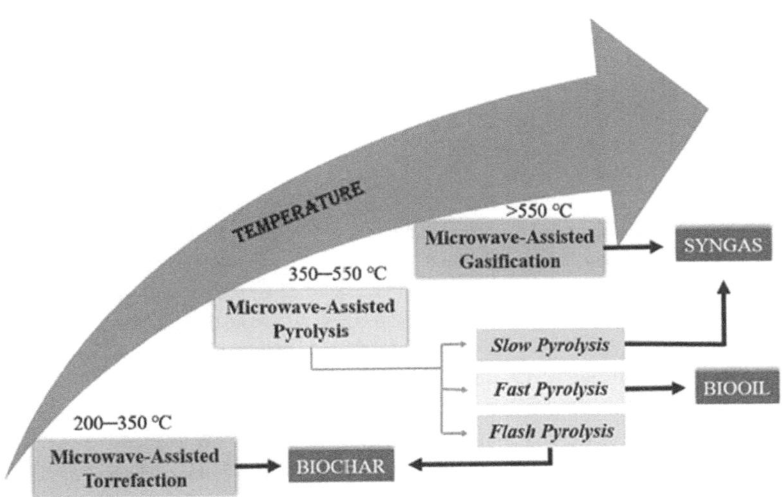

FIGURE 4.4	Different microwave-assisted bioenergy production processes.

of feedstock are significantly ameliorated due to the thermal breakdown of the hemicelluloses and the partial degradation of cellulose and lignin. A high degree of degradation of hemicellulose and cellulose occurs in microwave-assisted torrefaction (Jiao et al., 2022). Microwave-assisted heating is preferable to conventional heating due to its better heat transfer efficiency inside the sample bulk with shorter thermal latency, volumetric heating, cheaper energy requirement, and enhanced command over the process. Here, conduction accompanies radiation inside the sample, in contrast to the convection and conduction modes of heat transmission in traditional heating. Despite the high conversion loss in microwave torrefaction, net energy balance and processing efficiency are better than in conventional torrefaction, in addition to the microwave-assisted process's shorter time span (Prashanth et al., 2022).

4.5.5 Microwave-assisted Gasification (MAG)

Gasification is the conversion of carbon-rich compounds, such as biomass, into syngas, which is a flammable gas. Apart from synthesis gas, it also produces a little quantity of oil and tar (liquid fraction) and biochar (solid fraction). In microwave-assisted gasification, maximum biochar to syngas conversion occurs. In addition, the gasification efficiency is enhanced due to the microwave application. In MAG, at shorter durations of reaction better fuel properties are achieved. As compared to conventional gasification, MAG has an excellent biochar conversion efficiency of 49% (Arpia et al., 2021). Different MW-assisted treatment techniques for biofuel production are presented as Table 4.4.

TABLE 4.4
Different MW-assisted Treatment Techniques for Biofuel Production

Feedstock	Process Involved	Operating Conditions	Output	Remarks	References
Corn Stover	MAP	800 W 400–500 °C 10 min	Bio-oil: 38.1 wt% Biochar: 34.4 wt% Biogas: 27.5 wt%	$MgCl_2$ used as catalyst	(Arpia et al., 2021)
Soapstock	MAP	1000 W 500–600 °C 10 min	Bio-oil: 46% (with bentonite)	HZSM-5 zeolite, Bentonite used as catalyst	(Asomaning et al., 2018)
Sewage Sludge and Leucaena Wood	MAT	250 W	Biochar with higher carbon content and better absorption of CO_2	Leucaena optimization increases adsorptive capabilities of biochar	(Huang et al., 2019)
Waste Oil and Oil Palm Waste	MAT	250–300 °C, 5–8 min	Energy and mass yield	At temperature 300 °C, 100% energy yield and 85% mass yield were reported	(Arpia et al., 2021)
Oil Palm Shell (OPS)	MAG	920 °C, 5 min, 800 W	Biogas yield: 1.98 m³/kg Gasification efficiency: 69.09%	Activated carbon is utilized as MW susceptor. HHV of syngas produced by MAG is 7.32 MJ/kg	(N. Ismail, G. S. Ho, N. A. S. Amin, 2015)
Rice Straw Biochar	MAG	550 °C, 60 min	Syngas yield: 1.32 m³/kg	$Ca(OH)_2$ was recognized as a superior catalyst for both CO_2 absorption and MW-assisted gasification, which decreased the reaction time from 90 min to 60 min	(Xiao et al., 2015)

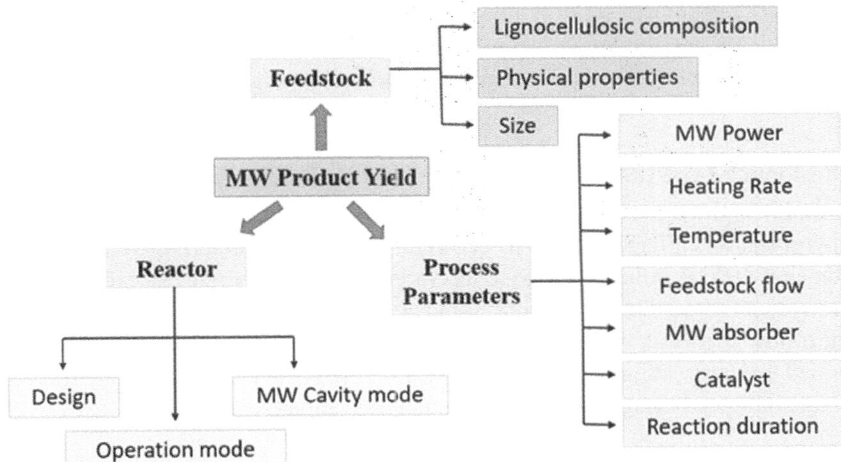

FIGURE 4.5 Factors affecting product yield in microwave bioprocessing.

4.5.6 Factors Affecting MW-assisted Processes

Some critical parameters need to be addressed in microwave-assisted thermal conversion of biomass because they significantly affect the final product quality and distribution. There are various factors involved in product yield through MW-assisted bioenergy conversion processes (Figure 4.5). The major process/operating parameters that affect the microwave-assisted thermal conversion are listed below.

4.5.6.1 Composition and Size of Incoming Biomass

The heating performance of microwave-assisted thermal processes can be controlled by the content of the biomass feedstock. Due to the combustible property of input material the temperature and the heating rate increases. The material gets heated up from inside out on exposure to microwave radiations, hence particle size is a crucial phenomenon. Compared to conventional pyrolysis, microwave-assisted pyrolysis significantly reduces the need for particle size reduction for the purpose of quick heating. This is because microwave radiation has a fast, volumetric heating characteristic, which gradually reduces the energy required for feedstock size reduction. As a result, the cost of pre-treatment, like the pulverization of wood, may be decreased. The appropriate kind and size of the feedstock is required to reduce microwave pyrolysis energy and time, further adding to the commercial viability of the technology (Ge et al., 2021).

4.5.6.2 Moisture Content of Incoming Biomass

The heating efficiency of the feedstock during microwave heating is determined by its water and moisture content. Microwave energy utilization efficiency is improved due to a specific level of moisture in the feedstock, since this moisture can elevate the rate of heating during MAP. Sometimes water is the only component of the feedstock that helps in sufficient absorption of microwaves during microwave heating (Tripathi et al., 2016).

4.5.6.3 Temperature of the Reaction

The reaction temperature affects a material's dielectric characteristics, therefore during solvent heating, the dielectric constant typically falls with the rise in temperature. The reaction temperature has a big impact on how pyrolysis products are distributed and how they behave. Greater liquid yields are produced in the intermediate temperature range whereas char yields are typically reduced and gas yields are increased on increasing the pyrolysis temperature (Xiao et al., 2015).

4.5.6.4 Residence Time (Reaction Time) and Heating Rate

Depending on low or high heating rates and short or long reaction times, the liquid to solid fraction ratio after pyrolysis gets affected. The type of biomass pyrolysis is determined by heating rates. The bio-oil or biochar yield is influenced due to the variation in heating rates and reaction temperature. Fast pyrolysis, usually at a shorter residence time and a high heating rate, has been used to produce bio-oil, while slow pyrolysis with a long residence time at a comparatively moderate temperature shall be applied for the creation of biochar. Fast heating rates provide more gas and less char because they encourage the speedy fragmentation of the biomass (Ethaib et al., 2020).

4.5.6.5 Output Power of Microwave

The heating rate and final pyrolysis temperature are influenced by the microwave output power; low power lengthens the process duration at a slow heating rate. On the other hand, a stronger microwave can offer a faster heating rate and shorter processing time. Therefore, in order to deliver the appropriate heating rate and process temperature, optimization of the system design and process for microwave power regulation is essential (Ge et al., 2021).

4.5.7 Scope and Opportunities

Microwave pyrolysis can convert a wide range of biomass into renewable biofuels (biochar, bio-oil, and syngas), making it a potentially game-changing method for producing clean energy. There is a lot of research focused on the scaling up of the MW technology to commercial-scale bioenergy production processes. Certain factors that influence the scalability of MW processes are: (1) the need for high power density, (2) MW penetration depth, (3) the continuous feed process (to improve energy efficiency and biofuel yield), (4) the selection of a MW absorber/catalyst, and (5) the cost of catalyst synthesis and regeneration (Ge et al., 2021). Power density refers to the quantity of power that is absorbed by the unit volume of feedstock. The power density of biomass that has been subjected to the effect of an electric field in MW will be determined by the dielectric characteristics of the biomass. In most cases, the dielectric constant (that denotes the potential of a substance to accumulate electrical energy) and dielectric loss factor (that reflects the capacity of a substance to uptake the electric energy) are utilized in order to ascertain whether or not a material is capable of being heated by microwaves. It has been proposed that the pyrolysis process might have been triggered by the superheating of bound water inside the feedstock, resulting in a huge pressure increase due to the greater power density resulting in better biofuel yield (Beneroso et al., 2017). Understanding the fluctuation of dielectric characteristics with frequency and temperature is critical for designing an effective MW treatment system. MW heating is strongly reliant on the dielectric characteristics of the material. Despite the fact that the dielectric characteristics of the raw biomass materials are almost identical, the dielectric properties change throughout MWP. When biomass starts to turn into char with the increase in temperature, a significant rise in loss tangent levels could be noticed. At this stage, the carbon content in the biomass increases, and as a result, the material's MW absorption qualities alter dramatically. Thus, biochar has different dielectric characteristics than the related biomass (Arpia et al., 2021). Therefore, biomass processing in an up-scaled MW system will be difficult, as MW power delivery may need to be regulated for improved efficiency of the process. The penetration depth is largely reliant on the biomass loading rate and MW frequency; typically, a lower frequency yields a deeper penetration. This parameter directly affects the efficiency and uniformity of biomass heating. If the feedstock dosage is too large, then it is difficult for microwaves to permeate the feedstock in the reactor. Therefore, depth of penetration plays a significant part in the commercialization of MW biomass bioconversion processes. The majority of commercial MW processing systems are based on a conveyor belt, which is capable of providing effective MW penetration depth (Siddique et al., 2022).

With the help of catalytic loading for MAP, the final product of higher quality can be achieved. Figure 4.6 represents the improvement in product yield using catalysts in MAP of different feedstock. Using a carbon-based catalyst, one may achieve rapid heating. Only by employing zeolite was the liquid fraction able to produce a high-quality product, albeit in a smaller amount. Alternatively, the outcomes of using metal oxides were different (Vignesh et al., 2022). Certain catalysts lower the working temperature, resulting in energy savings. This is a benefit for the MW procedure, but it is still required to overcome certain disadvantages. The extraction and regeneration of the catalyst from the char constitute additional operational expenses for the process, and the catalyst must be stable over time, mechanically robust, and simple to recover. Therefore, the quantity and quality of the products formed are influenced by microwave power, temperature, the chemical properties and physical properties of the feedstock, MW catalyst utilized, and chemical characteristics of the catalyst, all of which are required to be taken into account while feeding the catalyst during MAP. Hence, this parameter must be thoroughly investigated before contemplating commercialization.

Batch-scale microwave processes are extensively studied on a lab scale compared to the continuous mode of operation. Continuous MW processing offers various advantages, like optimum residence time of feedstock in the reactor, proper penetration of the microwaves, maximized product yield, and better MW power and temperature control, thereby reducing the fouling tendency of the reactor. Hence, the continuous mode of operation tends to be more efficient in commercial or pilot-scale MW bioprocessing units.

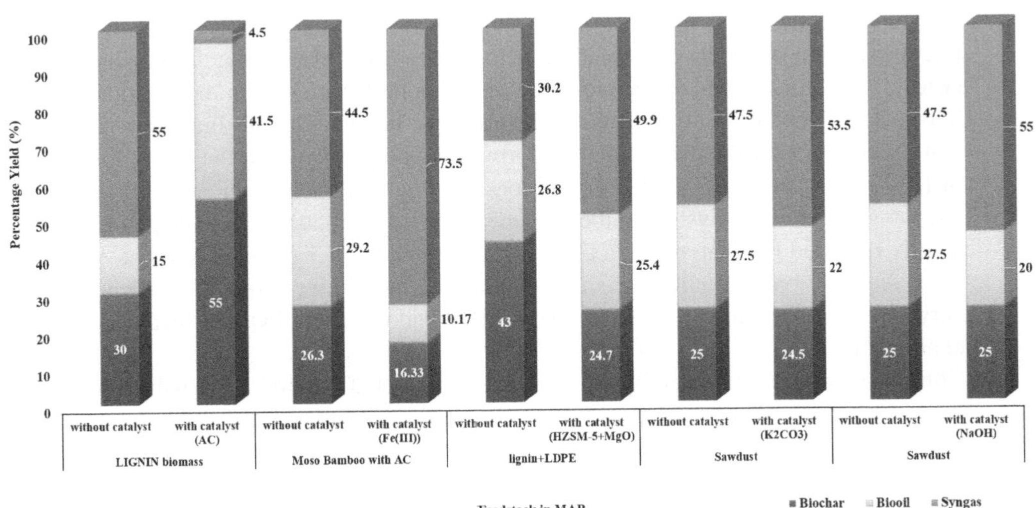

FIGURE 4.6 Variation in biofuel yield with respect to catalyst addition in microwave-assisted pyrolysis.

Source: Bu et al., 2014; Dong et al., 2018; Fan et al., 2017; Shang et al., 2015.

4.5.7.1 Scalable Concepts for MW Bioenergy Production

Apart from the above-discussed factors, one of the initial steps involved in the creation of a scalable MW processing system should be the determination of the most suitable system for handling materials and their electromagnetic compatibility. The major challenges involved in up-scaling MW processes are the dielectric properties of feedstock, hotspot formation, energy conversion efficiency of the MW system, and the reactor design. The problems associated with energy conversion and dielectric properties can be addressed by the addition of suitable MW susceptors and/or catalysts. In order to overcome the challenges, Beneroso et al. (2017) reviewed five probable chemical industry processes that may be utilized in the MAP scale-up procedure. Some of these procedures were tested in relation to the equipment's biomass flow rate.

(a) Concept of rotary kiln: Typically, the feed is supplied from one side of the kiln, and the spinning action transports the treated products to the exit along the span of the kiln. The rotational action topples the feedstock without compacting it, which improves the release of volatile compounds from biomass. A waveguide supplies microwave energy to one of the static portions. The revolving cavity consists of a central body, a SiC layer that acts as a microwave absorber, and an insulating layer formed of non-microwave-absorbing materials amidst the main body and the layer that is absorbing. A key concern noted is the requirement for attentive construction of the exit port to permit evacuation of processed substance and prevent microwave leaks. Because the electric field strength is greatest in the middle of the kiln and gradually decreases outward, the biomass (which accumulates on the outside edges due to the rotational motion) becomes inaccessible to strong electric field intensity. The rotational motion of the kiln permits the feedstock to pass from one location of higher power density at least, where two hot spots are noticeable in the feed inlet region (Buttress et al., 2016).

 Integrating the microwave input system and the system to distinguish biochar and volatile compounds during continuous operation is a particularly difficult aspect of the rotary kiln because it is essential to prevent microwave radiation losses to the environment.

(b) Concept of conveyor transport through tunnel applicator: This design proposed transferring biomass feedstock via a tunnel on a microwave-transparent conveyor belt where the microwaves are transmitted perpendicular to the feedstock movement. The dielectric properties of the biomass and the conveyor belt play an important role in the positioning of the hotspots with high power density along the conveyor belt length. This limited power dissemination provides for greater thermal uniformity in comparison with the rotary kiln concept of MW processing. The residence durations may be extremely brief, which is crucial for encouraging rapid pyrolysis, and the produced volatile compounds from the pyrolysis can be readily removed by a fan through a number of specifically engineered holes at the cavity's apex. An additional benefit of this microwave processing idea is that the thermal inertia is very low; therefore, the process may be initiated and terminated within seconds (Robinson et al., 2010).

(c) Concept of rotating ceramic-based disc: It is a patented method that enables biomass to be processed within a channel made by stationary metal walls connected to a circular disc that may be isolated and decontaminated beyond the processing region, hence eliminating contamination difficulties. The biomass can be isolated from the circular disc by a spinning window constructed of microwave-transparent material, such as alumina, which acts as a protection against magnetron contamination. Microwave feeding can be undertaken beneath the microwave-transparent spinning glass to avoid waveguide contamination due to the volatile compounds. In addition, a waveguide mounted on the top of the reactor would prevent the volatile compounds from being extracted from the region where the majority of them are produced. To minimize arcing effects, a thorough electromagnetic model must be implemented.

(d) Concept of MW fluidized bed reactor: This unique gasifier design, integrating microwave heating and a dual fluidized bed gasifier, divides the fluidized bed into two zones: a gasification zone and a heating zone. A bed material (MW absorber) circulating loop is established between these two phases. The bed material along with the catalyst is circulated to the heating zone, where the MW absorbers absorb the microwaves and attain the required temperature. The hot bed material supplies the energy for the endothermic gasification of steam. The bed material is carried back to the gasification zone after gas–solid separation. Furthermore, limestone can be added to the reactor and combined with the bed material for CO_2 absorption on-site. This approach enables the production of an H_2-rich gas with minimal tar concentration that is suitable for cogeneration plants and biofuel synthesis (Xie et al., 2014).

(e) Concept of auger reactor: Extrusion-based systems, like the auger reactor, have received attention for pyrolysis applications due to their ability to run progressively with essentially minimal carrier gas. This movement improves the heat exchange between solids, liquids, and gases, as well as the movement of pyrolysed products towards the reactor exit. As char particles exit the reactor, the additional biomass feedstock can be supplied and pyrolysed, allowing for continuous operation. It has already been demonstrated that extrusion-based methods are compatible with microwave processing. Albeit the speed of rotation of screws may be adjusted to match the solids retention time criteria for microwave pyrolysis, this approach requires ceramic materials for the walls that are resistant to contamination from the generated bio-oils and can endure temperatures above 500 °C. Utilizing liquid additives to biomass may assist in the scalability of the extrusion idea for microwave pyrolysis (Álvarez-Chávez et al., 2021). The University of Minnesota's Center for Biorefining developed a 4.5 kW microwave pyrolysis reactor capable of processing 10 kg/h of biomass using an auger transport mechanism (Beneroso et al., 2017).

Luo et al., 2021 examined the MAP of biomass at 250–300 °C on a bench scale and compared it to conventional pyrolysis. The fundamental factors and connections related to the dielectric characteristic of biomass, rapid hotspots, and a diminishing energy barrier were uncovered by analysing these novel results. To prove the viability of scaling up, a continuous 80 kg/h microwave-assisted auger reactor at pilot size has also been developed (Luo et al., 2021).

These aforementioned concepts can be further explored for up-scaling the application of microwave technology for commercial bioenergy production.

4.6 CONCLUSION

Recovery of valuable products and energy from waste is necessary due to the increased rate of waste generation, thereby reducing the waste volume, curbing unwanted pollution, and ultimately achieving "Zero Waste" in a landfill. The scientific advancement led to a paradigm shift from open burning, composting, and dumping in landfill to incineration, pyrolysis, anaerobic composting, etc. The thermochemical conversion of waste is a widely studied technology platform to convert waste to solid, liquid, and gas replacements for fossil-derived counterparts. Over the past few years, a dramatic increase has been observed in the microwave-assisted thermal conversion of wastes to fuels and chemicals due to numerous advantages that it could potentially provide over conventional processes. Microwave pyrolysis can convert a wide range of biomass into renewable biofuels (biochar, bio-oil, and syngas), making it a potentially game-changing method for producing clean energy. The several factors that affect the yield in microwave bioprocessing were briefly discussed. Yet, there is a lot of research underway on the scaling up of the MW technology to commercial-scale bioenergy production processes. The major challenges involved in up-scaling of MW processes are the dielectric properties of feedstock, hotspot formation, the energy conversion efficiency of the MW system, and the reactor design. Therefore, the scalable techniques can be investigated further in order to scale up the implementation of microwave technology for commercial bioenergy production.

REFERENCES

Ali, S. A., & Parvin, F. (2022). Examining challenges and multi-strategic approaches in waste management during the COVID-19 pandemic: A systematic review. In *Waste Management and Research* (Vol. 40, Issue 9, pp. 1356–1380). https://doi.org/10.1177/0734242X221079303

Álvarez-Chávez, B. J., Godbout, S., & Raghavan, V. (2021). Optimization of microwave-assisted hydrothermal pretreatment and its effect on pyrolytic oil quality obtained by an auger reactor. *Biofuel Research Journal*, 8(1), 1316–1329. https://doi.org/10.18331/BRJ2021.8.1.3

Angulo-Mosquera, L. S., Alvarado-Alvarado, A. A., Rivas-Arrieta, M. J., Cattaneo, C. R., Rene, E. R., & García-Depraect, O. (2021). Production of solid biofuels from organic waste in developing countries: A review from sustainability and economic feasibility perspectives. In *Science of the Total Environment* (Vol. 795). https://doi.org/10.1016/j.scitotenv.2021.148816

Arpia, A. A., Chen, W. H., Lam, S. S., Rousset, P., & de Luna, M. D. G. (2021). Sustainable biofuel and bioenergy production from biomass waste residues using microwave-assisted heating: A comprehensive review. *Chemical Engineering Journal*, 403(July 2020), 126233. https://doi.org/10.1016/j.cej.2020.126233

Asomaning, J., Haupt, S., Chae, M., & Bressler, D. C. (2018). Recent developments in microwave-assisted thermal conversion of biomass for fuels and chemicals. *Renewable and Sustainable Energy Reviews*, 92(April 2017), 642–657. https://doi.org/10.1016/j.rser.2018.04.084

Barskov, S., Zappi, M., Buchireddy, P., Dufreche, S., Guillory, J., Gang, D., Hernandez, R., Bajpai, R., Baudier, J., Cooper, R., & Sharp, R. (2019). Torrefaction of biomass: A review of production methods for biocoal from cultured and waste lignocellulosic feedstocks. *Renewable Energy*, 142, 624–642. https://doi.org/10.1016/J.RENENE.2019.04.068

Beneroso, D., Monti, T., Kostas, E. T., & Robinson, J. (2017). Microwave pyrolysis of biomass for bio-oil production: Scalable processing concepts. *Chemical Engineering Journal*, 316, 481–498. https://doi.org/10.1016/J.CEJ.2017.01.130

Bhat, S. A., Cui, G., Li, W., Wei, Y., Li, F., Kumar, S., & Ameen, F. (2021). Challenges and opportunities associated with wastewater treatment systems. *Current Developments in Biotechnology and Bioengineering: Strategic Perspectives in Solid Waste and Wastewater Management*, 259–283. https://doi.org/10.1016/B978-0-12-821009-3.00008-7

Bhat, S.A., Singh, J. & Vig, A.P. (2017). Amelioration and degradation of pressmud and bagasse wastes using vermitechnology. *Bioresource technology*, 243, 1097–1104.

Bu, Q., Lei, H., Wang, L., Wei, Y., Zhu, L., Zhang, X., Liu, Y., Yadavalli, G., & Tang, J. (2014). Bio-based phenols and fuel production from catalytic microwave pyrolysis of lignin by activated carbons. *Bioresource Technology*, 162, 142–147. https://doi.org/10.1016/J.BIORTECH.2014.03.103

Buttress, A. J., Binner, E., Yi, C., Palade, P., Robinson, J. P., & Kingman, S. W. (2016). Development and evaluation of a continuous microwave processing system for hydrocarbon removal from solids. *Chemical Engineering Journal*, 283, 215–222. https://doi.org/10.1016/j.cej.2015.07.030

Cremonez, P. A., Teleken, J. G., Weiser Meier, T. R., & Alves, H. J. (2021). Two-Stage anaerobic digestion in agroindustrial waste treatment: A review. *Journal of Environmental Management*, 281(December 2020). https://doi.org/10.1016/j.jenvman.2020.111854

Dong, Q., Niu, M., Bi, D., Liu, W., Gu, X., & Lu, C. (2018). Microwave-assisted catalytic pyrolysis of moso bamboo for high syngas production. *Bioresource Technology*, 256, 145–151. https://doi.org/10.1016/J.BIORTECH.2018.02.018

Ethaib, S., Omar, R., Kamal, S. M. M., Radiah, D., Biak, A., & Zubaidi, S. L. (2020). Microwave-Assisted Pyrolysis of Biomass Waste: A Mini Review. *Processes*, 8, 1–17. https://doi.org/10.3390/pr8091190

Fan, L., Chen, P., Zhang, Y., Liu, S., Liu, Y., Wang, Y., Dai, L., & Ruan, R. (2017). Fast microwave-assisted catalytic co-pyrolysis of lignin and low-density polyethylene with HZSM-5 and MgO for improved bio-oil yield and quality. *Bioresource Technology*, 225, 199–205. https://doi.org/10.1016/J.BIORTECH.2016.11.072

Ge, S., Yek, P. N. Y., Cheng, Y. W., Xia, C., Wan Mahari, W. A., Liew, R. K., Peng, W., Yuan, T. Q., Tabatabaei, M., Aghbashlo, M., Sonne, C., & Lam, S. S. (2021). Progress in microwave pyrolysis conversion of agricultural waste to value-added biofuels: A batch to continuous approach. *Renewable and Sustainable Energy Reviews, 135*(January 2020), 110148. https://doi.org/10.1016/j.rser.2020.110148

Hoang, A. T., Nižetić, S., Ong, H. C., Mofijur, M., Ahmed, S. F., Ashok, B., Bui, V. T. V., & Chau, M. Q. (2021). Insight into the recent advances of microwave pretreatment technologies for the conversion of lignocellulosic biomass into sustainable biofuel. *Chemosphere, 281*(March). https://doi.org/10.1016/j.chemosphere.2021.130878

Hu, J., Yang, Z., Huang, Z., Li, H., Wu, Z., Zhang, X., Qin, X., Li, C., Ruan, M., Zhou, K., Wu, X., Zhang, Y., Xiang, Y., & Huang, J. (2020). Co-composting of sewage sludge and Phragmites australis using different insulating strategies. *Waste Management, 108*, 1–12. https://doi.org/10.1016/J.WASMAN.2020.04.012

Huang, Y. F., Chiueh, P. Te, Lo, S. L., Sun, L., Qiu, C., & Wang, D. (2019). Torrefaction of sewage sludge by using microwave heating. *Energy Procedia, 158*, 67–72. https://doi.org/10.1016/J.EGYPRO.2019.01.047

IndraKumar Singh, S., Singh, W. R., Bhat, S. A., Sohal, B., Khanna, N., Vig, A. P., Ameen, F., & Jones, S. (2022). Vermiremediation of allopathic pharmaceutical industry sludge amended with cattle dung employing Eisenia fetida. *Environmental Research, 214*, 113766. https://doi.org/10.1016/J.ENVRES.2022.113766

Ismail, N. G. S. Ho, N. A. S. Amin, F. N. A. (2015). Microwave Plasma Gasification of Oil Palm Biochar. *Jurnal Teknologi (Science & Engineering)*, 7–13. https://doi.org/10.11113/jt.v74.4827

Jiao, L., Li, J., Yan, B., Chen, G., & Ahmed, S. (2022). Microwave torrefaction integrated with gasification: Energy and exergy analyses based on Aspen Plus modeling. *Applied Energy, 319*. https://doi.org/10.1016/J.APENERGY.2022.119255

Khan, A. H., López-Maldonado, E. A., Khan, N. A., Villarreal-Gómez, L. J., Munshi, F. M., Alsabhan, A. H., & Perveen, K. (2022). Current solid waste management strategies and energy recovery in developing countries – State of art review. *Chemosphere, 291*(November 2021). https://doi.org/10.1016/j.chemosph ere.2021.133088

Kumar, V., Agrawal, S., Bhat, S. A., Américo-Pinheiro, J. H. P., Shahi, S. K., & Kumar, S. (2022). Environmental impact, health hazards, and plant-microbes synergism in remediation of emerging contaminants. *Cleaner Chemical Engineering, 2*, 100030. https://doi.org/10.1016/J.CLCE.2022.100030

Luo, H., Zhang, Y., Zhu, H., Zhao, X., Zhu, L., Liu, W., Sun, M., Miao, G., Li, S., & Kong, L. (2021). Microwave-assisted low-temperature biomass pyrolysis: From mechanistic insights to pilot scale. *Green Chemistry, 23*(2), 821–827. https://doi.org/10.1039/d0gc03348k

Nanda, S., & Berruti, F. (2021). A technical review of bioenergy and resource recovery from municipal solid waste. *Journal of Hazardous Materials, 403*, 123970. https://doi.org/10.1016/J.JHAZMAT.2020.123970

Narayan, A. S., Marks, S. J., Meierhofer, R., Strande, L., Tilley, E., Zurbrügg, C., & Lüthi, C. (2021). Advancements in and Integration of Water, Sanitation, and Solid Waste for Low- And Middle-Income Countries. In *Annual Review of Environment and Resources* (Vol. 46, pp. 193–219). https://doi.org/10.1146/annurev-environ-030620-042304

Ongen, A., Ozcan, H. K., & Ozbas, E. E. (2016). Gasification of biomass and treatment sludge in a fixed bed gasifier. *International Journal of Hydrogen Energy, 41*(19), 8146–8153. https://doi.org/10.1016/j.ijhyd ene.2015.11.159

Prajapati, P., Varjani, S., Singhania, R. R., Patel, A. K., Awasthi, M. K., Sindhu, R., Zhang, Z., Binod, P., Awasthi, S. K., & Chaturvedi, P. (2021). Critical review on technological advancements for effective waste management of municipal solid waste – Updates and way forward: Advancements in Municipal Solid Waste Management. *Environmental Technology and Innovation, 23*, 101749. https://doi.org/10.1016/j.eti.2021.101749

Prashanth, P. F., Gurrala, L., Mohan, R. V., Sarvanakumar, K., & Vinu, R. (2022). Microwave-assisted torrefaction and pyrolysis of rice straw pellets for bioenergy. In *IET Renewable Power Generation*. https://doi.org/10.1049/rpg2.12445

Pujara, Y., Pathak, P., Sharma, A., & Govani, J. (2019). Review on Indian Municipal Solid Waste Management practices for reduction of environmental impacts to achieve sustainable development goals. *Journal of Environmental Management, 248*, 109238. https://doi.org/10.1016/J.JENVMAN.2019.07.009

Robinson, J. P., Kingman, S. W., Snape, C. E., Bradshaw, S. M., Bradley, M. S. A., Shang, H., & Barranco, R. (2010). Scale-up and design of a continuous microwave treatment system for the processing of oil-contaminated drill cuttings. *Chemical Engineering Research and Design*, *88*(2), 146–154. https://doi.org/10.1016/j.cherd.2009.07.011

Shang, H., Lu, R. R., Shang, L., & Zhang, W. H. (2015). Effect of additives on the microwave-assisted pyrolysis of sawdust. *Fuel Processing Technology*, *131*, 167–174. https://doi.org/10.1016/J.FUPROC.2014.11.025

Siddique, I. J., Salema, A. A., Antunes, E., & Vinu, R. (2022). Technical challenges in scaling up the microwave technology for biomass processing. *Renewable and Sustainable Energy Reviews*, *153*(October 2021), 111767. https://doi.org/10.1016/j.rser.2021.111767

Simonetti, S., Martín, C. F., & Dionisi, D. (2022). Microwave Pre-Treatment of Model Food Waste to Produce Short Chain Organic Acids and Ethanol via Anaerobic Fermentation. *Processes*, *10*(6). https://doi.org/10.3390/pr10061176

Thengane, S. K., Kung, K. S., Gomez-Barea, A., & Ghoniem, A. F. (2022). Advances in biomass torrefaction: Parameters, models, reactors, applications, deployment, and market. *Progress in Energy and Combustion Science*, *93*(November 2021). https://doi.org/10.1016/j.pecs.2022.101040

Tripathi, M., Sahu, J. N., & Ganesan, P. (2016). Effect of process parameters on production of biochar from biomass waste through pyrolysis: A review. *Renewable and Sustainable Energy Reviews*, *55*, 467–481. https://doi.org/10.1016/j.rser.2015.10.122

Uthirakrishnan, U., Godvin Sharmila, V., Merrylin, J., Adish Kumar, S., Dharmadhas, J. S., Varjani, S., & Rajesh Banu, J. (2022). Current advances and future outlook on pretreatment techniques to enhance biosolids disintegration and anaerobic digestion: A critical review. *Chemosphere*, *288*. https://doi.org/10.1016/J.CHEMOSPHERE.2021.132553

Vignesh, N. S., Soosai, M. R., Chia, W. Y., Wahid, S. N., Varalakshmi, P., Moorthy, I. M. G., Ashokkumar, B., Arumugasamy, S. K., Selvarajoo, A., & Chew, K. W. (2022). Microwave-assisted pyrolysis for carbon catalyst, nanomaterials and biofuel production. *Fuel*, *313*(September 2021), 123023. https://doi.org/10.1016/j.fuel.2021.123023

Xiao, N., Luo, H., Wei, W., Tang, Z., Hu, B., Kong, L., & Sun, Y. (2015). Microwave-assisted gasification of rice straw pyrolytic biochar promoted by alkali and alkaline earth metals. *Journal of Analytical and Applied Pyrolysis*, *112*, 173–179. https://doi.org/10.1016/J.JAAP.2015.02.001

Xie, Q., Borges, F. C., Cheng, Y., Wan, Y., Li, Y., Lin, X., Liu, Y., Hussain, F., Chen, P., & Ruan, R. (2014). Fast microwave-assisted catalytic gasification of biomass for syngas production and tar removal. *Bioresource Technology*, *156*, 291–296. https://doi.org/10.1016/j.biortech.2014.01.057

5 From Linear to Circular Bioeconomy

A Paradigm Shift in Wastewater Management

M.A. Onu, O.O. Ayeleru, H.U. Modekwe, B.O. Oboirien, and P.A. Olubambi

5.1 INTRODUCTION

Water, a key production component, requires sustainable management of freshwater resources and wastewater generated from usage. The global standards for wastewater treatment have evolved; while protecting aquatic ecosystems and human health remain important priorities, there is a growing desire to increase access to clean water and energy by producing valuable goods from wastewater (Bhat et al., 2021). The perception of wastewater as a "nuisance" is progressively giving way to that of a "resource" (WWAP (United Nations World Water Assessment Programme), 2017). Reducing the amount of polluted water and its treatment for reuse and safe discharge back into the environment is necessary for sustainable development and environmental conservation. Wastewater is a by-product of home, industrial, commercial, or agricultural operations. Depending on the source, it may contain various components, including organics, inorganics, nutrients, heavy metals, toxins, and pathogens (Gupta et al., 2021). To ensure that wastewater's intrinsic potential is realized, the World Bank collaborates with partners worldwide for effective wastewater management (World Bank, 2019). A common method of disposing of wastewater from many sources over the years has been to release it into neighboring surface water bodies without any treatment. These undesirable substances have the potential to alter the ecosystem of any body of water and negatively impact aquatic life. For instance, effluent with high levels of organics and nutrients tends to lower the water's dissolved oxygen (DO) content. The organics and other oxygen-consuming materials act as food supplies for aquatic bacteria, who use the oxygen in the water for their metabolism and so reduce the amount of DO available to higher life forms (Gupta et al., 2021). The sustainable development goals (SDGs) can be achieved by using wastewater to produce energy, fresh water, fertilizers, and nutrients (World Bank, 2019). As a result, reuse and resource recovery have significantly replaced end-of-pipe treatment internationally (Zvimba et al., 2021). The "5R" (Reduce, Reuse, Recycle, Recover, and Restore) principle serves as additional support for the requirement to shift the perception of wastewater from one of a "nuisance" to one of a "resource" (Zvimba et al., 2021). In a linear economy, energy, chemicals, and other important goods and services are produced using finite natural resources. As a result, a significant amount of waste may be produced in the industrial and household sectors, which would then be dumped in landfills or wastewater, increasing pollution (Ye et al., 2022). The circular economy has drawn a lot of attention recently as a way to meet the current production and consumption demands. The goal of the circular economy is to increase resource efficiency, resource recovery, and recycling in order to reduce the exploitation of virgin resources (Ubando et al., 2020). The term "bioeconomy," which is used to describe economic activities resulting from advancements in biosciences and a rise in scientific knowledge, is relatively new. Georgescu-Roegen coined the term "bioeconomics" to describe the biological basis of the economic process and highlight the issue of mankind's existence with a finite supply of resources that are asymmetrically distributed and inequitably utilized (Vargas-Hernández, 2020). The European Commission describes the bioeconomy as the economy that employs renewable biological resources from land and sea (such as animals, crops, fish, forests, and microbes) to produce energy, food, and materials (D'Adamo et al., 2021). A "circular economy" system seeks to limit waste production while preserving the economic worth of materials, goods, and resources for as long as it is feasible (Smol et al., 2020). According to Feleke et al. (2021), the circular economy aims to "keep the added value in products for as long as it is practicable and to reduce waste disposal" (Feleke et al., 2021). The definitions provided emphasize the importance of waste management and more sensible use of resources. It should be recognized that all resource and waste groups across all industrial sectors should be the focus of improvement initiatives. More judicious use of water resources (which are limited) and more sustainable wastewater procedures are envisaged as a means of attaining a circular economy in the water and wastewater sector (Smol et al., 2020). Materials in bioeconomy are, in some ways, circular by nature. The usage of biomaterials can also be relatively linear, though. Recently, efforts to move the world toward a circular economy have been emphasized (Achillas & Bochtis, 2020). The circular bioeconomy's economic model takes into account not just the traditional bioeconomy's sectors but also, in specific terms, those of bio-based chemicals and fuels as well as waste management (Kircher, 2022). The bioeconomy integrates the relationships between biological and economic systems from a novel viewpoint that marks the paradigm change (Vargas-Hernández, 2020). The goal of this chapter is to give an overview of global wastewater resources management, the circular economy in the water sector, bioeconomy, and sustainable development; the economic and environmental benefits of effective wastewater management, challenges and limitations to water reuse, resource recovery, and effective wastewater treatment technologies, and organizational and societal attitudinal changes toward wastewater resources and awareness creation are all highlighted.

5.2   OVERVIEW OF GLOBAL WASTEWATER RESOURCES MANAGEMENT

Water is an essential component for life on Earth and is a crucial part of the global political and economic evolutions; the growth of urban population and an expansion in water use have concurrently led to the massive production of greywater and wastewater (Kundu et al., 2022). Wastewater sources are industrial activities, sewage (domestic) systems, and agricultural runoff. About 380 billion cubic meters of wastewater are produced each year globally, and by the end of 2030 and 2050, those numbers are expected to rise by ~24% and ~52%, respectively, over the current levels (Qadir et al., 2020). North America is estimated to generate the most wastewater per capita annually, at 232 m^3, followed by Europe (124 m^3), the Middle East and North Africa (114 m^3), Oceania (88 m^3), Asia (82 m^3), Latin America and the Caribbean (65 m^3), and Sub-Saharan Africa (46 m^3) (Qadir et al., 2020). Conventional wastewater treatment methods are expensive, energy-intensive, and maintenance-intensive. These water treatment amenities have a number of drawbacks, including poor facility design, electrical problems, low facility maintenance, and a lack of skilled and experienced staff, which causes the treatment facilities to remain non-operational (Kundu et al., 2022). Conventional wastewater treatment uses physical, chemical, and biological techniques to remove sediments, organic materials, and nutrients. The treated wastewater is typically discharged into neighboring lakes and rivers, which has a detrimental effect on the aesthetic value and quality of aquatic life. A good relationship between natural resources, sustainable resource usage that complies with laws and regulations, market opportunities, consumer outreach, and product disposal plan are all aspects of the circular economy that attempt to solve every issue. Furthermore, the circular economy can more effectively meet societal demands to protect the environment (Kundu et al., 2022). According

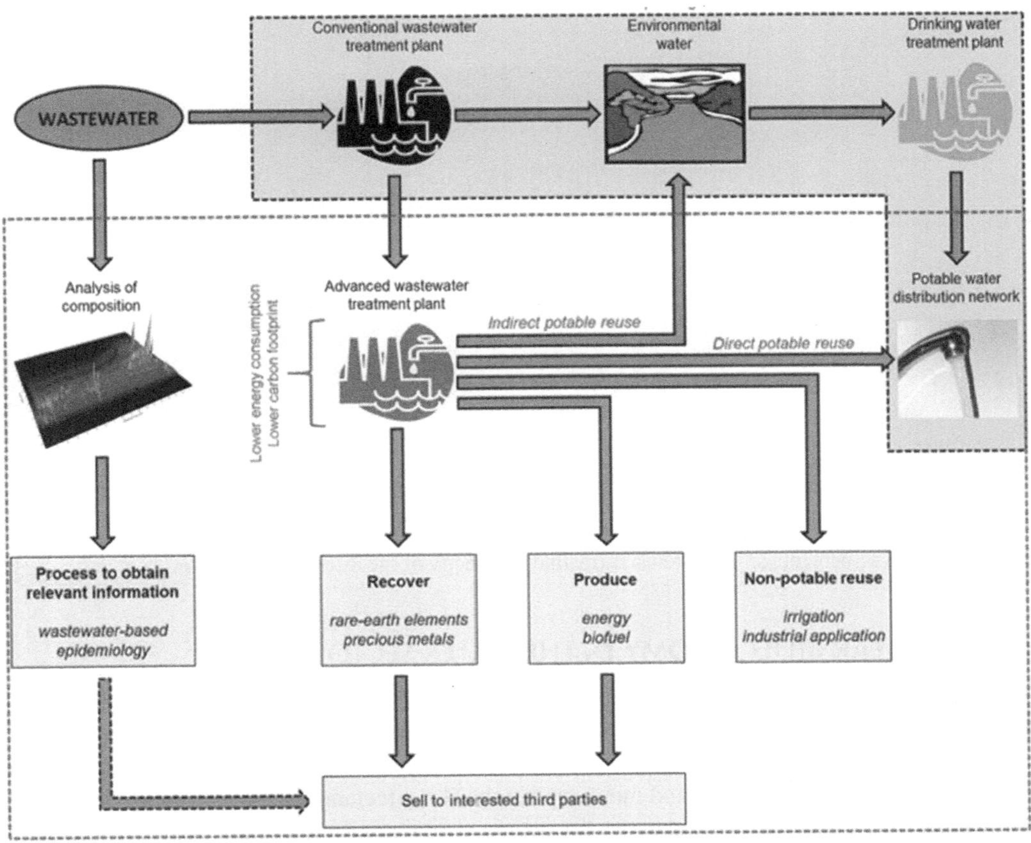

FIGURE 5.1 Overview of the leading wastewater management concepts.

Source: (Villarín & Merel, 2020), accessed 22 October 2022. This work is shared under a Creative Commons Attribution-Non-Commercial-No Derivative Works License and copyright has been granted by the authors.

to Villarín and Merel (2020), reclaimed water is frequently used for agricultural irrigation worldwide; such water reuse calls for more specialized treatment to lessen wastewater pollutants and reduce population exposure (Villarín & Merel, 2020). Figure 5.1 shows an overview of the leading wastewater management concepts.

Wastewater study is influenced by conceptual modifications brought about by the necessities of contemporary society, which also alter how wastewater is seen. For example, as urban agglomerations' population grows, the need for water increases, worsening the water trajectory of big cities. This approach promotes the concept shift, transforming wastewater from an undesirable material to a helpful resource (Villarín & Merel, 2020). Consequently, wastewater management has undergone various fundamental developments due to the circular bioeconomy concept, societal enlightenment, and environmental considerations. Figure 5.2 shows the wastewater treatment capacity of various income levels.

Wastewater can be an inexpensive and viable source of energy, nutrients, and other valuable by-products like organic and organic-mineral fertilizer. The advantages of obtaining such elements from wastewater go beyond those of environmental and human health. They contribute to reducing climate change and ensuring food and energy security (WWAP (United Nations World Water Assessment Programme), 2017).

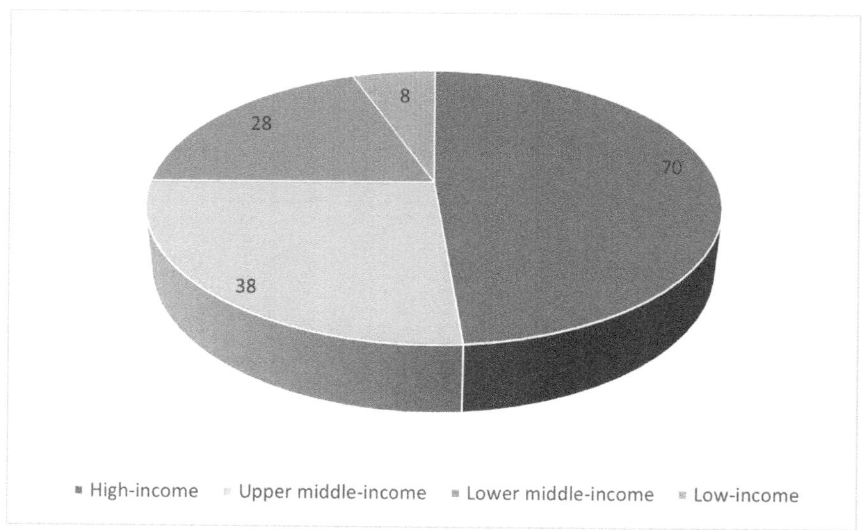

FIGURE 5.2 The percentage of wastewater treatment capacity of the different income levels.

5.3 CIRCULAR BIOECONOMY IN THE WATER SECTOR

Wastewater can be used as a resource since it is a rich source of valuable materials, such as organic matter, nutrients, metals, and energy, due to the depletion of natural resources. By illustrating the idea of resource, re-make, and re-think to create new value-added commodities from trash, resource reuse is becoming more and more appealing. These techniques represent the current paradigm change in wastewater management for resource mining (Kundu et al., 2022). The creation of food, energy, and water are all interdependent. Food production uses one-fourth of the world's energy and is the world's largest consumer of freshwater resources (Mbavarira & Grimm, 2021). Therefore, it is crucial to develop strategies to reduce the quantity of water and energy used in food production. One way to address the global water scarcity issue is through wastewater treatment for reuse, which allows limited freshwater supplies for other purposes or preservation (World Bank, 2019). Figure 5.3 shows resources that are recoverable from wastewater. Water quality improvement through removing elements like pathogens or micro-pollutants that pose a risk to the environment and public health has traditionally been the primary goal of wastewater treatment. However, the circular economy idea represents a significant paradigm change by transforming wastewater constituents like nutrients, previously considered contaminants, into valuable materials to recapture (Villarín & Merel, 2020). Additionally, wastewater treatment plants can become more environmentally and financially viable by using their by-products for energy production and agriculture (World Bank, 2019).

5.3.1 Nutrient Recovery

Large amounts of nutrient-rich wastewater are produced from various sources, including urban sewerage, industrial effluent, and agricultural activities. The release of nutrient-rich wastewater has boosted algal bloom and phytoplankton growth, decreased dissolved oxygen levels, and decreased water transparency, all of which increased the likelihood that water bodies would become overly eutrophic and endanger aquatic life (Kundu et al., 2022). A significant problem for the circular economy of the future is recovering nutrients from wastewater and recycling nutrients as soil fertilizers (Saliu & Oladoja, 2021). Uncontrolled wastewater discharge, limited

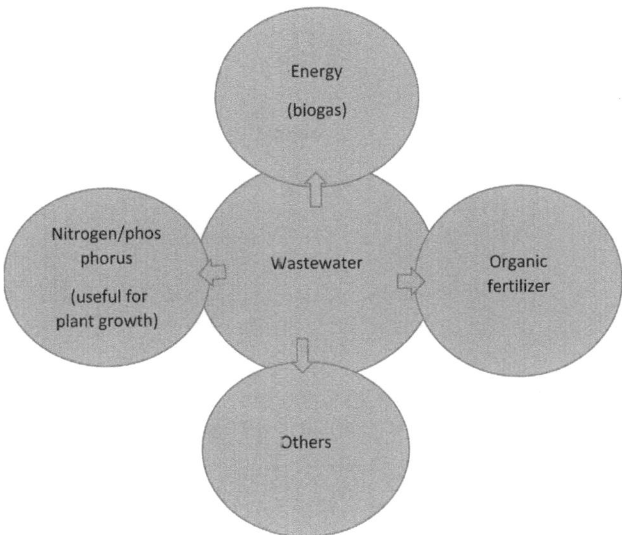

FIGURE 5.3 The recoverable resources present in wastewater.

access to fertilizers in underdeveloped nations, and excessive fertilizer prices are the driving forces behind this. Some studies have presented the techniques for recovering nutrients from primary wastewater sources, including home, industrial, and agricultural wastewater, and the transformation of recovered nutrients into agricultural fertilizers (Saliu & Oladoja, 2021; Van der Hoek et al., 2016). Various methods of nutrient recovery from wastewater have been successfully developed. Postulations are frequently used to determine the recovered nutrient's potential for reuse as fertilizer. In a few experimental circumstances, plant growth and yield potentials with recovered nutrients are comparable to or superior to standard fertilizer (Saliu & Oladoja, 2021; Van der Hoek et al., 2016). This paradigm shift creates a new source of nutrients while preventing the discharge of chemicals that cause the eutrophication of natural waters, energy use, and the emission of greenhouse gases that are often connected with biological nutrient removal (Villarín & Merel, 2020).

5.3.2 ENERGY PRODUCTION

Energy recovery at wastewater treatment facilities is a vital sustainability policy tool. It can be accomplished by creating biogas, using heat pumps in treatment plant effluents and using heat exchangers to recover energy from various high-temperature streams (Neczaj & Grosser, 2018). Wastewater is a source of chemical energy that can be transformed into multiple energy products due to its abundance of organic materials. As a result, researchers are developing cutting-edge technology to produce energy from wastewater treatment plants (Kundu et al., 2022). Wastewater treatment produces biogas, a renewable energy source with several benefits. For example, it is dispatchable and, once upgraded, can be stored and delivered using the existing gas infrastructure network; it is a clean energy source that doesn't rely on essential raw materials; it doesn't disturb animals; and it doesn't compete with agriculture or put a strain on water bodies. Anaerobic digestion is one method that has been used for a long time to convert wastewater streams that are high in organic matter into bioenergy. Several technologies, including biohydrogen, biodiesel, bioethanol, and microbial cell fuels, are currently being developed to turn organic matter into bioenergy (Puyol et al., 2017).

TABLE 5.1
Potential Areas for Wastewater Reuse

Agriculture	Recreational	Industry	Environmental
Edible crops	Landscaping features	Cleaning and washing (beverage, food)	Groundwater rejuvenation
Edible crops utilizing trickle irrigation	Boating lakes	Cooling (power production, paper, and textile)	Flow enhancement
Nonedible crops	Swimming lakes	Process water, boiler feed water (all industries)	Drought amelioration
Farm animals drinking water	Snowmaking		

Credit: (Rodriguez et al., 2020), accessed 22 October 2022.] [CC BY 3.0 IGO public license: https://creativecommons.org/licenses/by/3.0/igo/ (with copyright permission from the World Bank Group).

5.3.3 Water Reuse

Water stress has been brought on by the disparity between water supply and demand in various parts of the world, mainly where water is rare. Because of this, it is crucial to recover water after using it (water reclamation) (Duque et al., 2021). Using treated wastewater for specific purposes can significantly lower potable water usage because it is one of the most readily accessible water resources produced continuously (Angelakis & Snyder, 2015). After applying the appropriate treatment chain (tertiary treatments, advanced oxidation process, ultrafiltration, reverse osmosis, etc.) to remove pathogens, reclaimed water can be used for a variety of operations, including industrial processes, agriculture (irrigation), and even drinking water (Alkhudhiri et al., 2019; Duque et al., 2021). It has long been known that using treated wastewater for irrigation in agriculture can reduce local water stress and enhance agricultural production (Neczaj & Grosser, 2018). The safe reuse of water resources, which entails using them more than once, represents a substantial change from the prior water resource management model, which virtually never took the value of reclaimed wastewater and its reuse applications into account (Tortajada, 2020). Wastewater can be purified to various specifications to meet industry and agriculture demands, among other areas. It can also be recycled as drinking water to keep the ecosystem flowing. One way to address the global water shortage issue is through wastewater treatment for recycling, which saves limited freshwater supplies for other purposes or conservation (Rodriguez et al., 2020). Table 5.1 shows potential areas for wastewater reuse. Reclaimed water can be effectively used in agriculture, recreation, industry, and the environment.

5.4 TECHNOLOGICAL INNOVATION IN WASTEWATER TREATMENT

The modern global water sector is facing significant challenges due to the deterioration in water quality and the stringent wastewater discharge rules (Edokpayi et al., 2017). As a result, wastewater treatment and water urgently need novel technology to ensure economic and environmental sustainability (Kumari, 2019). Despite the widespread reporting of technological advancements in wastewater treatment, these technologies are nevertheless linked with actual prices, both for the assets themselves and for running and maintaining the facility where they are used (Morris et al., 2021). Wastewater treatment makes it possible to dispose of industrial and human effluents without endangering human health or causing unacceptable environmental harm. Water from wastewater is used for irrigation, an efficient disposal method (Kumari, 2019). The common approaches used in wastewater treatment include physical, biological, chemical, and sludge treatment methods (Dutta et al., 2021). In order to attain the treatment goal, physical and chemical treatment methods are characterized as treatment technologies that do not use biomass in the process. Wastewater solids are removed when it passes through filters or screens, or they might be eliminated through air flotation or gravity settling. It is possible to remove air-entrapped particles since they float to the top. In order to make contaminants into a form that makes it easier to remove them, chemicals are utilized in wastewater treatment. To improve removal by physical processes, adjustments can include creating floc or a heavier particle mass. Therefore, to treat wastewater, chemical addition and physical processes are typically used jointly (U.S. EPA, 2013).

Furthermore, biological treatment utilizes microorganisms to remove organic pollutants from wastewater. Removing organic matter and nutrients in wastewater treatment has resulted in the containment and acceleration of natural biodegradation processes. The microbes produce treated wastewater by metabolizing nutrients, colloids, and dissolved organic materials. The treated wastewater undergoes physical methods that eliminate excess microbiological development (U.S. EPA, 2013). Figure 5.4 shows a system flow diagram of the wastewater treatment process. The two most popular techniques for treating wastewater are coagulation and flocculation. Chemical and electrocoagulation are the two different kinds of coagulation techniques that can be applied. Both techniques can remove pollutants from wastewater but work in distinct ways. By separating liquid from solid particles, coagulation and flocculation are extremely important in the treatment of wastewater. Following coagulation, floc will increase throughout the floating process, allowing for sedimentation filtering to take place (Mehta et al., 2021).

Other wastewater treatment processes include membrane filtration processes, which are one of the most feasible and effective methods of wastewater management due to several advantages, such as relatively low energy consumption and compactness. Membrane filtration techniques generally include microfiltration, ultrafiltration (UF), nanofiltration (NF), and reverse osmosis (RO) (Mehta et al., 2021). Adsorption is one of the most effective ways to remove organic pollutants from wastewater. Adsorption is more dependable than electrochemical or biological techniques because it is easier to use, costs less to operate, requires less maintenance, and requires less monitoring (Mehta et al., 2021).

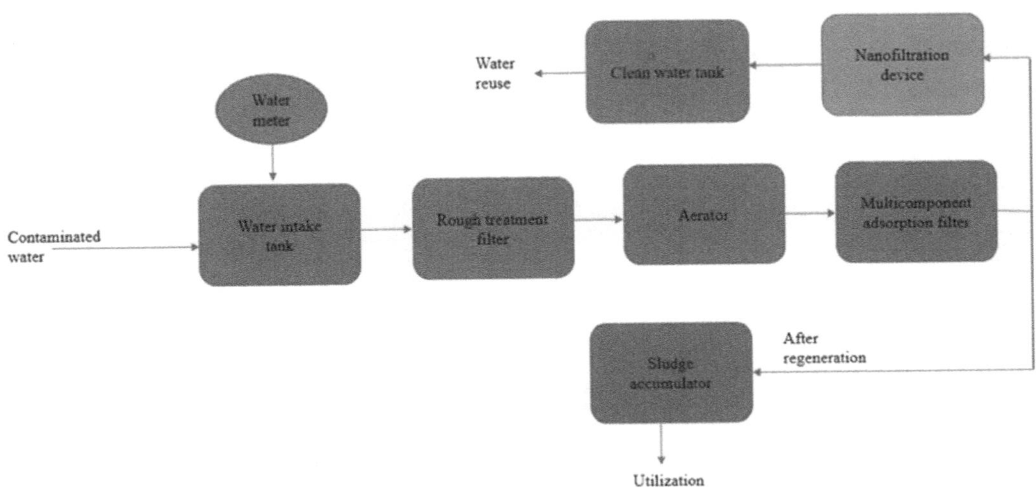

FIGURE 5.4 The designing of a technological, industrial wastewater treatment facility.

5.5 BIOECONOMY AND SUSTAINABLE DEVELOPMENT

The term "circular bioeconomy" refers to blending the bioeconomy and circular economy objectives with different degrees of emphasis on biotechnology (Kershaw et al., 2021). Bioeconomy can be used to achieve SDGs (Anand, 2016); the circular bioeconomy must encompass various information, views, multidimensional issue diagnoses, solution ideas, and noneconomic values to follow a sustainable trajectory (Kershaw et al., 2021). Water technologies that help regeneration can be brought through a circular economy concept. As one of the most crucial resources for manufacturing, water has attracted the focus of circular economy discourse. Almost all industries, including aquaculture and agriculture, rely on it. Because of this, several assessments about the function of water in a circular economy have been made (Flores et al., 2018). According to Delgado Martin et al. (2021), access to clean water and sanitary facilities is essential for society to function well and be healthy. Ecosystem health and biodiversity are supported by water. Slower economic growth results from a lack of water availability because it is also essential for producing food, energy, and most industrial operations (Delgado Martin et al., 2021). The circular economy demonstrates how to deal with issues like undervaluing water, financial and operational inefficiencies, pollution and deteriorated ecosystems, equity, sustainable urban water supply, sanitation services, and growing urban demand that is becoming more and more complicated. Individuals and organizations must abide by the rules and operate in the public interest to improve the water quality and protect water resources (WWAP (United Nations World Water Assessment Programme), 2017). Through the development of fresh business strategies, new funding sources, and assistance in filling the current funding deficit, circular economy projects can also aid in luring the private sector (Delgado Martin et al., 2021). Figure 5.5 shows how the bioeconomy is connected to the SDGs.

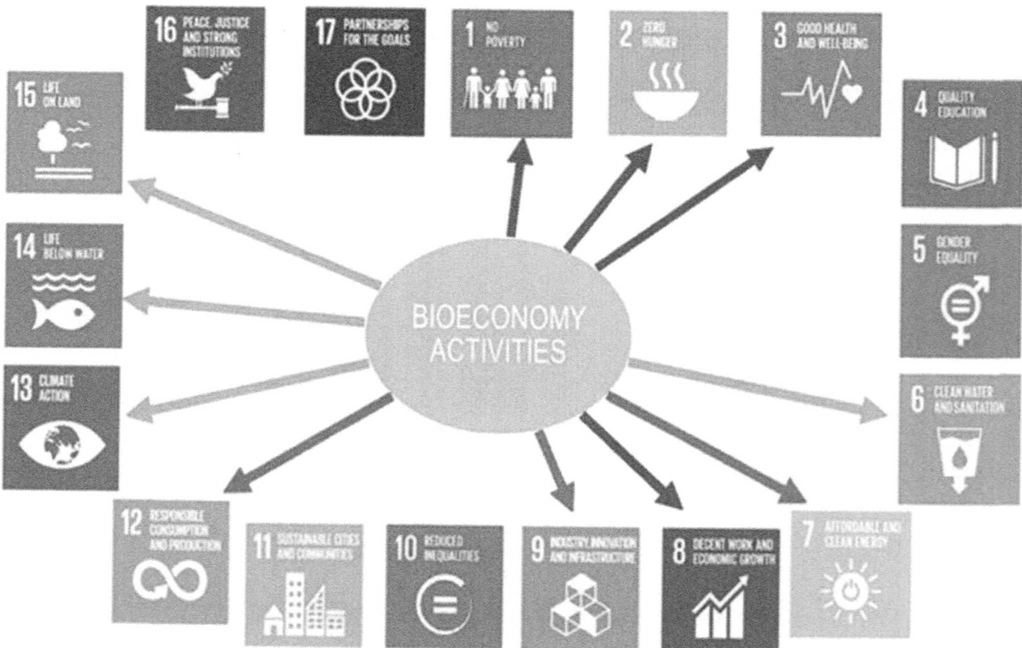

FIGURE 5.5   The relationship between bioeconomy and the sustainable development goals.

Source: (Heimann, 2019), accessed 22 October 2022. This work is shared under a Creative Commons Attribution-Non-Commercial-No Derivative works License and the authors have granted copyright.

Achieving the SDGs is linked with a sustainable bioeconomy; approximately half of the 17 SDGs directly connect to the bioeconomy. Some similarities exist between bioeconomy concepts and the SDGs, and some of the goals of bioeconomy activities are complimentary to or even the same as the SDG targets. A variety of opportunities are presented by the developments in new technologies, such as biotechnology, to assist in addressing some of the development concerns, such as access to clean drinking water, energy provision, efficient health care, and food security (Anand, 2016; Heimann, 2019). The bioeconomy will use technology to boost resource circularity and raise production sustainability. This type of economy aims to accomplish the SDGs established in the UN 2030 Agenda for Sustainable Development, including SDG6, by fostering sustainable economic growth that focuses on enhancing social equity and human well-being while conserving resources and regenerating ecosystems (Thame, 2022).

5.6 DRIVERS OF CIRCULAR BIOECONOMY

Many organizations and stakeholders are engaged in creating and executing a circular bioeconomy that is affected by these drivers in terms of their roles and course of action. Concerns about climate change, diminished biodiversity, resource depletion, food, clean water security, and energy supply are the key motivators behind the idea (Marvik & Philp, 2020). It is crucial to have an ecosystemic approach to the circular bioeconomy, which focuses on different stages in the value chains of the goods connected and controlled by connected stakeholders to create and carry out an effective plan. Wastewater treatment is an essential component of the infrastructure, but in some circumstances, particularly in developing areas, communities are forced to find economical and ecologically friendly solutions to these problems. The progress of wastewater infrastructure may be hampered by a lack of knowledge and technical information and a government concentration on more advanced areas (Poškus et al., 2021).

5.6.1 POLICIES

Legislation may be a powerful tool for directing the bioeconomy (Kardung et al., 2021). The overarching goal of policies to encourage waste stream conversion is to promote a cascading or continuous use of resources (Marvik & Philp, 2020). The objectives of wastewater management, or policy intents, are translated into laws and regulations, with roles given to various actors (WWAP (United Nations World Water Assessment Programme), 2017). By fostering synergies between several policy domains, there is a strong chance to advance a circular bioeconomy (circular economy, climate, agriculture, energy, and industrial policy) (Muscat et al., 2021). Transition management, a term referring to the administration of societally motivated transitions, can be used to study policies to promote a bioeconomy (Marvik & Philp, 2020). Several academic contributions have investigated policy programs toward changes in socio-technical procedures, including energy, mobility, food, and water, by drawing on a variety of disciplines, including evolutionary economics, institutional theory, and invention studies (Davies & Evans, 2019; Diercks et al., 2019; Rogge et al., 2017; Sovacool & Axsen, 2018). Single policy instruments are unlikely to support such changes due to the complexity of sustainability transition processes. As a result, innovation studies emphasize the significance of focusing on policy mixtures, or the blend of policy strategies, policy instruments, and the methods through which such systems and tools evolve (Scordato et al., 2022). The required policy, institutional, and regulatory (PIR) frameworks must be in place to support the paradigm change. PIR motivations are necessary to promote sustainable wastewater investments that consider resource recovery and circular economy standards (World Bank, 2019). According to Fund et al. (2015), 45 nations in all have created national policy plans that have had a substantial impact on the growth of the bioeconomy, eight of which may be categorized as focused or comprehensive bioeconomy plans (the European Union, Finland, Germany, Japan, Malaysia, South Africa, the

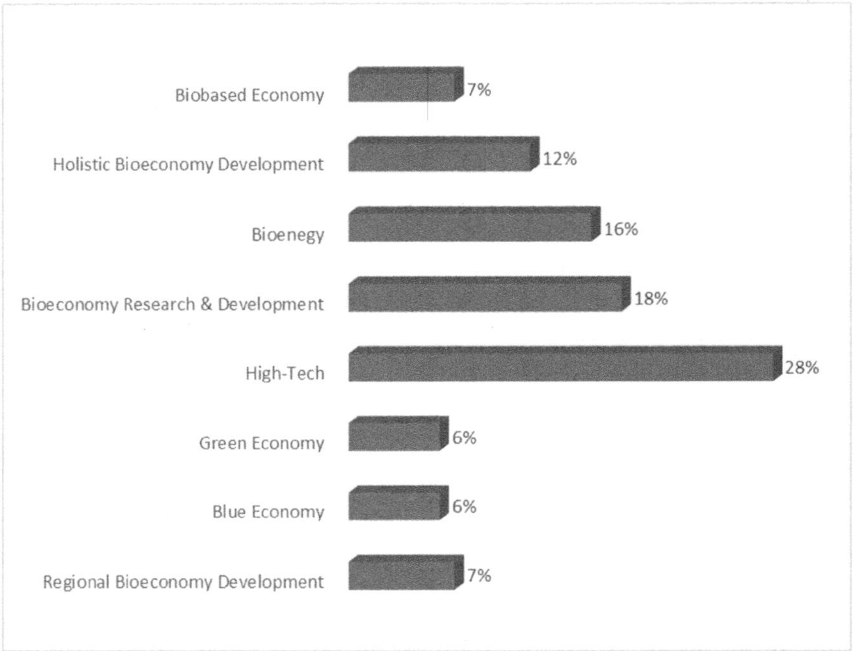

FIGURE 5.6　Policy/strategy of global bioeconomy-related initiatives "modified".

Source: (Fund et al., 2015), accessed 22 October 2022. Permission for use has been granted by the authors.

USA, and the West Nordic Countries). The analysis found other policies for the other nations that expressly mention the growth of the bioeconomy. However, these policies don't take a comprehensive approach; instead, they see the development of the bioeconomy through the prism of particular policy sectors. The research identifies 11 bioenergy strategies and 19 high-tech (biotechnology and convergent technologies) strategies related to the bioeconomy. The authors also list 12 initiatives for research and innovation, with five strategies focused on the bio-based economy. Four green and four blue economy strategies with bioeconomy-related priorities are also identified in their study. Lastly, five regional bioeconomy strategies have been identified (Fund et al., 2015). Figure 5.6 gives an overview of global bioeconomy-related policy/strategy initiatives. Current knowledge and experience of the bioeconomy must be shared and widely disseminated in order to build a stronger consensus on efficient investments and regulations. This will make it easier for it to be developed both nationally and internationally (Chavarria et al., 2020).

5.6.2　Technologies

New technologies and the redesign of old ones are significant forces behind the shift to a circular bioeconomy (Muscat et al., 2021). The innovation approach strongly emphasizes the commercialization of scientific findings, especially by large private enterprises, and is fundamentally driven by technology and science. This strategy involves distributing labor among public actors, private companies, and scientists. Each is accountable for the following: advancing scientific discovery, turning discoveries into innovations, and controlling potential externalities (Scordato et al., 2022). The circular technologies should be practical from an economic, environmental, and engineering standpoint (Mehta et al., 2021). Technology is essential in shifting to a circular bioeconomy (Salvador et al., 2022). Technologies that enable maximizing resource value at the lowest possible cost are preferred.

Additionally, systems prioritizing material recovery over disposal should be emphasized (Salvador et al., 2022). For instance, anaerobic digestion (AD) is a commercial technology that transforms organic wastewater streams from municipalities and industries into renewable energy from methane-rich biogas. Despite energy recovery, AD also offers other significant benefits, such as high organic matter removal efficiency, low excess sludge formation, and minimal space requirements (Puyol et al., 2017).

5.6.3 Organizations

Systems at the organizational level must strike a balance between maintaining profitability while adhering to the law's standards for their goods and services and staying competitive by attracting customers. Nevertheless, these systems can sway their audience by launching new services and goods and attractively branding them. Organizational structures can also influence their political superstructure (Poškus et al., 2021). The advantages of group efforts, enterprises, and corporate innovation are required for various specialties to build cooperative organizations and networks, consortia, and varied types of expertise (Salvador et al., 2022). Government subsidies (financial incentives and tax exemptions) have been cited as a way to "promote more circular activities," though not necessarily under this label (i.e., without necessarily using the terms "circular economy" or "bioeconomy"). Regarding the government's advantages, regulation—particularly the "regulation of items with more ecologically responsible practices"—has been identified as a critical factor in circular bioeconomy activities (Salvador et al., 2022).

5.6.4 Social Behavior

Beyond having access to information and understanding sustainable practices, research increasingly emphasizes the significance that cultural and behavioral factors have in motivating people to behave more sustainably (Muscat et al., 2021). Principles and cultural and behavioral barriers will need to be addressed to change consumption patterns and maintain them within the bounds of the planet to comply with the necessary safeguard (Muscat et al., 2021). One strategy to investigate social implications and reflect on their acceptability while innovations are still being developed is to involve stakeholders. This enables innovators to consider the values and norms at play in stakeholders' assessments of the societal elements of their (intended) product, which is predicted to result in better, more palatable innovations and facilitate the eventual acceptance of these innovations (Hoes et al., 2021). The lack of widespread social acceptance of the usage of items made from recycled human waste is a significant barrier to the market's development for resource recovery. Additionally, many farmers who presently utilize untreated wastewater are opposed to treating it because they think doing so will reduce their agricultural production by removing effluent nutrients (World Bank, 2019).

5.6.5 Markets

Resource optimization makes it possible to extract and utilize resources more effectively, extending their life cycles and, as a result, slowing the rate at which they escape the system. Additionally, increased resource efficiency may result in businesses spending less or making more money (Salvador et al., 2022). Market incentives may hasten the transition from non-circular and fossil-based goods to those made from biological materials. The economic viability of circular products or business models is currently hindered or reduced by market barriers, such as market access restrictions and anti-competitive subsidies that hurt the environment (Muscat et al., 2021). Promoting resource recovery programs will be challenging without an accurate assessment of water resources.

Inadequate water valuation also results in incorrect pricing for water resources and services, discouraging resource recovery initiatives (World Bank, 2019).

5.6.6 AWARENESS CREATION

There is still more to be done to improve the knowledge base and understanding of the circular bioeconomy in wastewater management. Water scarcities and extremes (such as floods and droughts) lead to moments of increased awareness, which expands the window of opportunity for considering circular bioeconomy management alternatives. Wastewater reuse frequently encounters solid public opposition because of an overall lack of knowledge and assurance about the risks to human health. Education and awareness-raising are the best ways to eliminate societal, cultural, and consumer barriers (Cisneros, 2014; Mu'azu et al., 2020). Civil society can be better educated, which is essential in shaping investment and policy environments. Small and medium-sized businesses must become more knowledgeable and engaged because of their significant cumulative influence (WWAP (United Nations World Water Assessment Programme)/UN-Water, 2018).

5.7 ECONOMIC AND ENVIRONMENTAL BENEFITS OF ADEQUATE WASTEWATER TREATMENT

There are significant environmental, health, and financial advantages to investing in wastewater treatment technology that increases resource recovery and water recyclability. In reality, the objective is to seek value from wastewater to go beyond reducing pollution and, if nothing else, as a further means of funding wastewater management and strengthening the system's financial viability. Reducing water withdrawals, resource loss in production systems, and economic values are all factors that assist the shift to a circular economy. These factors include improved wastewater treatment, increased water reuse, and the recovery of valuable by-products (WWAP (United Nations World Water Assessment Programme), 2017). Health and quicker access to clean water are among the advantages of adequate wastewater management. The global benefit–cost ratio of return on investment in dollars for every 1 US dollar invested for sanitation, water, sanitation, and water combined was estimated to be $5.5, $2.0, and $4.3, respectively (Hutton, 2013). In the arid region and the Middle East, a wise approach to wastewater utilization has helped turn wastewater from a risk to the environment and human health into a resource that is both profitable and environmentally responsible (Alawode et al., 2019). The provision of clean, high-quality drinking water, water conservation, and waste reduction would result from effective wastewater management. Sludge can be used to generate renewable energy; treated wastewater could also be used to recharge water bodies (like aquifers, lakes and rivers), which can be used as tourism or recreation destinations, which are further financial advantages of proper wastewater treatment (Levický & Fiľa, 2020).

5.8 LIMITATIONS AND CHALLENGES OF CIRCULAR BIOECONOMY

The bioeconomy's rapid global spread has been one of its most remarkable characteristics. However, because of how quickly technology has developed, its expectations and problems have frequently surpassed those that initially existed. This fact is advantageous. The bioeconomy, however, could become disconnected from the initial goals if those expectations are not met, and the subsequent loss of trust poses a significant risk to the bioeconomy's future (Patermann & Aguilar, 2021). Various possibilities are equally provided by the adoption of bioeconomy initiatives in Europe and elsewhere. Still, there are also difficulties, trade-offs, and possible problems associated with the sustainable production, use, and processing of biomass and bioresources, as well as with ecological responsibility, environmental management, and inclusive development (Hoff et al., 2018). Internal alignment across business functions is one of the main challenges facing implementing circularity programs (Esposito et al., 2018). Executives acknowledge the strategic importance of circularity but differ on how to approach it concerning other projects (Esposito et al., 2018). If the bioeconomy is to acquire more traction, comprehensive technology may still be in its infancy. There is still more work to be done now and in the long run. There is a need for appropriate solutions, which must focus on enabling recovery (rather than "separate for disposal") and be technically and economically practical (Salvador et al., 2022). The limitations and challenges of bioeconomy differ for different countries. For instance, in Africa, according to studies by Salvador et al. (2022), the challenge of finding economic technology, the unpreparedness of local technology, and the lack of a clear and compelling political framework to create a more circular economy is also related to the slowness of administrative procedures for the implementation of circular bioeconomy practices as well as the lack of government support or incentives through policies. Local firms frequently emphasize high investment costs, whether for initial investments or for scaling up, which is commonly followed by discussions that mention "lack of sufficient finances for the realization of the initiative" and "expensive investment expenses" (Salvador et al., 2022). The idea of recovering water resources and how to put it into effect are not generally well understood. Instead of being viewed as a resource, wastewater is still seen as a burden or something that needs to be handled and disposed of. As a result, there is a lack of political will to create laws and policies that encourage and promote resource recovery and wastewater reuse. Besides, regulations are frequently created without taking into account the costs associated with their implementation (World Bank, 2019).

5.9 MITIGATIVE MEASURES

Innovative finance strategies are required to stimulate the creation and investment in wastewater systems, which will help ensure operations' viability and the well-being of regional ecosystems. Also, there is a necessity for new regulatory frameworks that mainly motivate all parties to view wastewater practices as resource recovery operations (World Bank, 2019). Building trust and overcoming unfavorable opinions require public awareness and education programs, for example on the evaluation of synergies and trade-offs between ecosystem services resulting from different bioeconomy strategies, or, to put it another way, the effects of the bioeconomy on various ecosystem services. It could help address challenges crucial to sustainable science and practice, such as legitimacy and acceptability (D'Adamo et al., 2021). The SDGs were put forth by the United Nations in 2015 to increase social equality and human well-being while enhancing environmental quality. The SDGs that highlight the significance of resource efficiency and waste management for a circular economy are SDGs 6 and 12: clean water and sanitation, SDG 8: decent work and economic growth, SDG 11: sustainable cities and communities, and SDG 12: responsible consumption and production. Therefore, cutting-edge research and development are needed for circular technology. For the integration process in the future, an environmental impact evaluation and economic analysis should be undertaken using a variety of approaches. Establishing a new business model to implement the bioeconomy concept globally is the key to its effective adoption (Mehta et al., 2021).

Furthermore, having a defined national policy objective is one of the essential aspects that can promote the growth of wastewater reuse and resource recovery initiatives. Resource recovery efforts have to overcome internal obstacles in addition to external ones. Organizational behavior must be changed for resource recovery projects to succeed, and a committed leadership and team must be built to support and promote resource recovery at all organizational levels (Rodriguez et al., 2020).

5.10 CONCLUSION

The bioeconomy should be seen as a component of extensive economic system changes involving the social, technological, and economic transformation. The heart of these transformation techniques involves institutional innovations for enabling environments and long-term incentives at the level of both businesses and global policy, behavior modification (modified consumption), and institutional innovations for long-term incentives. Additionally, both individual bioeconomy operations and bioeconomy concepts can and should use the SDGs as a suitable sustainability baseline. This analysis shows that the development of the bioeconomy can have significant implications, which are admirably represented by the broad scope of the SDGs. Furthermore, in the transition to bioeconomy, various incremental and radical improvements must be created to accomplish the paradigm shift.

REFERENCES

Achillas, C., & Bochtis, D. (2020). Toward a Green, Closed-Loop, Circular Bioeconomy: Boosting the Performance Efficiency of Circular Business Models. *Sustainability*, *12*(23), 10142.

Alawode, O. I., Ogedengbe, K., Afolayan, A. F., & Adewuyi, O. B. (2019). Prospects of wastewater reclamation and reuse for water scarcity mitigation and environmental pollution control in sub-saharan Africa. *Environmental Management and Sustainable Development*, *8*(4), 1–20.

Alkhudhiri, A., Darwish, N. B., & Hilal, N. (2019). Analytical and forecasting study for wastewater treatment and water resources in Saudi Arabia. *Journal of Water Process Engineering*, *32*, 100915.

Anand, M. (2016). *Innovation and sustainable development: A bioeconomic perspective. Brief for global sustainable development report, GSDR. 2016. In.* New Delhi: The Energy and Resources Institute.

Angelakis, A. N., & Snyder, S. A. (2015). Wastewater treatment and reuse: Past, present, and future. *Water*, *7*(9), 4887–4895.

Atamanova, O. V., Tichomirova, E. I., Politayeva, N. A., Podolsky, A. L., & Istrashkina, M. V. (2019). Innovative technologies for industrial wastewater treatment. *IOP*.

Bhat, S. A., Cui, G., Li, W., Wei, Y., Li, F., Kumar, S., & Ameen, F. (2021). Challenges and opportunities associated with wastewater treatment systems. *Current Developments in Biotechnology and Bioengineering*, 259–283.

Chavarria, H., Trigo, E., Pablo, F. V., & Piñeiro, V. (2020). Bioeconomy: A sustainable development strategy. In *G20-Insights. Policy Area: Sustainable Energy-Water-Food-Systems.* Available www.semantic scholar.org/paper/Bioeconomy%3A-A-Sustainable-Development-Strategy-G20-Chavarr%C3%ADa/ f9285699cc0426e3e9838b4d4b345ec909c33fa3 . Accessed 30 August 2022.

Cisneros, B. J. (2014). *Water recycling and reuse: An overview. Water Reclamation and Sustainability*, 431–454.

D'Adamo, I., Gastaldi, M., Morone, P., Rosa, P., Sassanelli, C., Settembre-Blundo, D., & Shen, Y. (2021). Bioeconomy of sustainability: Drivers, opportunities and policy implications. *Sustainability*, *14*(1), 200.

Davies, A., & Evans, D. (2019). Urban food sharing: Emerging geographies of production, consumption and exchange. *Geoforum*, *99*, 154–159.

Delgado Martin, A., Rodriguez, D. J., Amadei, C. A., & Makino, M. (2021). *Water in Circular Economy and Resilience (WICER).* https://openknowledge.worldbank.org/handle/10986/36254

Diercks, G., Larsen, H., & Steward, F. (2019). Transformative innovation policy: Addressing variety in an emerging policy paradigm. *Research Policy*, *48*(4), 880–894.

Duque, A. F., Campo, R., Val del Rio, A., & Amorim, C. L. (2021). Wastewater valorization: practice around the world at pilot-and full-scale. *International Journal of Environmental Research and Public Health*, *18*(18), 9466.

Dutta, D., Arya, S., & Kumar, S. (2021). Industrial wastewater treatment: Current trends, bottlenecks, and best practices. *Chemosphere*, *285*, 131245.

Edokpayi, J. N., Odiyo, J. O., & Durowoju, O. S. (2017). Impact of wastewater on surface water quality in developing countries: a case study of South Africa. *Water Quality*, *10*(66561), 10.5772.

Esposito, M., Tse, T., & Soufani, K. (2018). Introducing a circular economy: New thinking with new managerial and policy implications. *California Management Review*, *60*(3), 5–19.

Feleke, S., Cole, S. M., Sekabira, H., Djouaka, R., & Manyong, V. (2021). Circular bioeconomy research for development in sub-Saharan Africa: innovations, gaps, and actions. *Sustainability*, *13*(4), 1926.

Flores, C. C., Bressers, H., Gutierrez, C., & de Boer, C. (2018). Towards circular economy–a wastewater treatment perspective, the Presa Guadalupe case. *Management Research Review*, *41*(5), 554–571 https://doi.org/10.1108/MRR-02-2018-0056

Fund, C., El-Chichakli, B., Patermann, C., & Dieckhoff, P. (2015). *Bioeconomy policy (part II): Synopsis of national strategies around the world* (German Bioeconomy Council: Berlin, Germany, Issue. www.surea qua.no/Sureaqua/library/Bioeconomy-Policy_Part-II%20-%20Synopsis%20of%20National%20Strateg ies%20around%20the%20World,%202015.pdf

Gupta, S., Mittal, Y., Panja, R., Prajapati, K. B., & Yadav, A. K. (2021). Conventional wastewater treatment technologies. In A. Pandey) (Ed.), *Current Developments in Biotechnology and Bioengineering* (pp. 47–75). Elsevier. https://doi.org/10.1016/B978-0-12-821009-3.00008-7

Heimann, T. (2019). Bioeconomy and SDGs: Does the bioeconomy support the achievement of the SDGs? *Earth's Future*, *7*(1), 43–57.

Hoes, A., Van Der Burg, S., & Overbeek, G. (2021). Transitioning responsibly toward a circular bioeconomy: Using stakeholder workshops to reveal market dependencies. *Journal of Agricultural and Environmental Ethics*, *34*(4), 1–21.

Hoff, H., Johnson, F. X., Allen, B., Biber-Freudenberger, L., & Förster, J. J. (2018). *Sustainable bio-resource pathways towards a fossil-free world: The European bioeconomy in a global development context. Policy paper produced for the IEEP Think2030 conference.* Institute for European Environmental Policy (IEEP), Brussels.

Hutton, G. (2013). Global costs and benefits of reaching universal coverage of sanitation and drinking-water supply. *Journal of Water and Health*, *11*(1), 1–12. https://doi.org/10.2166/wh.2012.105

Kardung, M., Cingiz, K., Costenoble, O., Delahaye, R., Heijman, W., Lovrić, M., van Leeuwen, M., M'barek, R., van Meijl, H., & Piotrowski, S. (2021). Development of the circular bioeconomy: Drivers and indicators. *Sustainability*, *13*(1), 413.

Kershaw, E. H., Hartley, S., McLeod, C., & Polson, P. (2021). The sustainable path to a circular bioeconomy. *Trends in Biotechnology*, *39*(6), 542–545.

Kircher, M. (2022). The bioeconomy needs economic, ecological and social sustainability. *AIMS Environmental Science*, *9*(1), 33–50.

Kumari, S. (2019). Innovative Technologies for Waste Water Treatment. *IJFMR-International Journal for Multidisciplinary Research*, *1*(3), 636–642.

Kundu, D., Dutta, D., Samanta, P., Dey, S., Sherpa, K. C., Kumar, S., & Dubey, B. K. (2022). Valorization of wastewater: A paradigm shift towards circular bioeconomy and sustainability. *Science of the Total Environment*, 157709.

Levický, M., & Fiľa, M. (2020). *The Level of Circular Economy in Slovakia on the Basis of Selected Indicators*. Virtual Conference.

Marvik, O. J., & Philp, J. (2020). The systemic challenge of the bioeconomy: A policy framework for transitioning towards a sustainable carbon cycle economy. *EMBO reports*, *21*(10), e51478. https://doi. org/10.15252/embr.202051478

Mbavarira, T. M., & Grimm, C. (2021). A systemic view on circular economy in the water industry: learnings from a Belgian and Dutch case. *Sustainability*, *13*(6), 3313.

Mehta, N., Shah, K. J., Lin, Y. J., Sun, Y., & Pan, S. Y. (2021). Advances in circular bioeconomy technologies: from agricultural wastewater to value-added resources. *Environments*, *8*(3), 20.

Morris, J. C., Georgiou, I., Guenther, E., & Caucci, S. (2021). Barriers in implementation of wastewater reuse: identifying the way forward in closing the loop. *Circular Economy and Sustainability*, *1*(1), 413–433.

Mu'azu, N. D., Abubakar, I. R., & Blaisi, N. I. (2020). Public acceptability of treated wastewater reuse in Saudi Arabia: Implications for water management policy. *Science of the Total Environment*, *721*, 137659.

Muscat, A., de Olde, E. M., Ripoll-Bosch, R., Van Zanten, H. H. E., Metze, T. A. P., Termeer, C. J. A. M., van Ittersum, M. K., & de Boer, I. J. M. (2021). Principles, drivers and opportunities of a circular bioeconomy. *Nature Food*, *2*(8), 561–566.

Neczaj, E., & Grosser, A. (2018). *Circular economy in wastewater treatment plant—challenges and barriers*. Multidisciplinary digital publishing institute proceedings.

Patermann, C., & Aguilar, A (2021). A bioeconomy for the next decade. *EFB Bioeconomy Journal*, *1*, 100005.

Poškus, M. S., Jovarauskaitė, L., & Balundė, A. (2021). A Systematic Review of Drivers of Sustainable Wastewater Treatment Technology Adoption. *Sustainability*, *13*(15), 8584.

Puyol, D., Batstone, D. J., Hülsen, T., Astals, S., Peces, M., & Krömer, J. O. (2017). Resource recovery from wastewater by biological technologies: opportunities, challenges, and prospects. *Frontiers in Microbiology, 7*, 2106.

Qadir, M., Drechsel, P., Jiménez Cisneros, B., Kim, Y. K., Pramanik, A., Mehta, P., & Olaniyan, O. (2020). *Global and regional potential of wastewater as a water, nutrient and energy source.* Natural resources forum.

Rodriguez, D. J., Serrano, H. A., Delgado, A., Nolasco, D., & Saltiel, G. (2020). *From Waste to Resource: Shifting paradigms for smarter wastewater interventions in Latin America and the Caribbean.* World Bank. https://openknowledge.worldbank.org/handle/10986/33436

Rogge, K. S., Kern, F., & Howlett, M. (2017). Conceptual and empirical advances in analysing policy mixes for energy transitions. *Energy Research & Social Science, 33*, 1–10.

Saliu, T. D., & Oladoja, N. A. (2021). Nutrient recovery from wastewater and reuse in agriculture: A review. *Environmental Chemistry Letters, 19*(3), 2299–2316.

Salvador, R., Barros, M. V., Donner, M., Brito, P., Halog, A., & Antonio, C. (2022). How to advance regional circular bioeconomy systems? Identifying barriers, challenges, drivers, and opportunities. *Sustainable Production and Consumption, 32*, 248–269.

Scordato, L., Bugge, M. M., Hansen, T., Tanner, A., & Wicken, O. (2022). Walking the talk? Innovation policy approaches to unleash the transformative potentials of the Nordic bioeconomy. *Science and Public Policy, 49*(2), 324–346.

Smol, M., Adam, C., & Preisner, M. (2020). Circular economy model framework in the European water and wastewater sector. *Journal of Material Cycles and Waste Management, 22*(3), 682–697. https://doi.org/10.1007/s10163-019-00960-z

Sovacool, B. K., & Axsen, J. (2018). Functional, symbolic and societal frames for automobility: Implications for sustainability transitions. *Transportation Research Part A: Policy and Practice, 118*, 730–746.

Thame, D. (2022). *Water in the context of a sustainable circular bioeconomy1 Water—from oceans to taps, Southampton School of Law.*

Tortajada, C. (2020). Contributions of recycled wastewater to clean water and sanitation Sustainable Development Goals. *Nature Partner Journals Clean Water, 3*(1), 1–6.

U.S. EPA. (2013). *Emerging technologies for wastewater treatment and in-plant wet weather management* (United States Environmental Protection Agency, Issue. www.epa.gov/sites/default/files/2019-02/documents/emerging-tech-wastewater-treatment-management.pdf

Ubando, A. T., Felix, C. B., & Chen, W. (2020). Biorefineries in circular bioeconomy: A comprehensive review. *Bioresource Technology, 299*, 122585.

Van der Hoek, J. P., de Fooij, H., & Struker, A. (2016). Wastewater as a resource: Strategies to recover resources from Amsterdam's wastewater. *Resources, Conservation and Recycling, 113*, 53–64.

Vargas-Hernández, J. G. (2020). Bio-economy at the crossroads of sustainable development. In *Advanced Integrated Approaches to Environmental Economics and Policy: Emerging Research and Opportunities* (Vol. 15, pp. 23–48). IGI Global.

Villarín, M. C., & Merel, S. (2020). Paradigm shifts and current challenges in wastewater management. *Journal of Hazardous Materials, 390*, 122139.

World Bank. (2019). *From Waste to Resource-Shifting Paradigms for Smarter Wastewater Interventions in Latin America and the Caribbean: Background Paper II. Showcasing the River Basin Planning Process through a Concrete Example-The Río Bogotá Cleanup Project.* https://elibrary.worldbank.org/doi/abs/10.1596/33381

WWAP (United Nations World Water Assessment Programme). (2017). *The United Nations World Water Development Report 2017. Wastewater: The Untapped Resource.* . https://wedocs.unep.org/handle/20.500.11822/20448

WWAP (United Nations World Water Assessment Programme)/UN-Water. (2018). *2018 UN World Water Development Report, Nature-based Solutions for Water.* UNESCO. http://repo.floodalliance.net/jspui/handle/44111/2726

Ye, Y., Ngo, H. H., Guo, W. S., Chang, S. W., Nguyen, D. D., Varjani, S., Liu, Q., Bui, X. T., & Hoang, N. B. (2022). Bio-membrane integrated systems for nitrogen recovery from wastewater in circular bioeconomy. *Chemosphere, 289*, 133175.

Zvimba, J. N., Musvoto, E. V., Nhamo, L., Mabhaudhi, T., Nyambiya, I., Chapungu, L., & Sawunyama, L. (2021). Energy pathway for transitioning to a circular economy within wastewater services. *Case Studies in Chemical and Environmental Engineering, 4*, 100144.

6 Biofilm-based Denitrifying Moving Bed Biofilm Reactor (dMBBR) for the Removal of Nitrate from Wastewater

Roshni J. Patel, Debarshi I. Mukherjee, and Anuradha S. Nerurkar

6.1 INTRODUCTION

Nitrate is a common pollutant endangering human health and the environment, and it is still a major issue throughout the world. Nitrogen-based fertilizers, the dairy industry and landfill leachate are the major sources of nitrate contamination in water (Toolabi et al. 2021). High levels of nitrate in drinking water can lead to infant cyanosis syndrome or "blue baby syndrome," methemoglobinemia, cancer, teratogenesis, thyroid disorders, mutagenesis and abortions (Wu et al. 2018). The presence of nitrate in water bodies can also cause eutrophication, fish hypoxia, toxic algal blooms and toxin production (Briki et al. 2017). The most widely used methods for nitrate removal are ion exchange, reverse osmosis, chemical reduction and biological denitrification. However, due to the generation of waste brine, high capital cost and energy cost, the use of physico-chemical methods has been limited. Whereas low cost, high removal efficiency and low environmental footprint make biological heterotrophic denitrification one of the most widely used. It is a process where denitrifiers reduce NO_3^- into harmless N_2. (Equation 1)

$$(NO_3^- \rightarrow NO_2^- \rightarrow NO \rightarrow N_2O \rightarrow N_2) \tag{1}$$

Over the years, a large number of denitrifying bacteria have been identified as having potential for the denitrification of various effluents and as an inoculum for denitrifying MBBR (Hong et al. 2020). Recently, various microorganisms were used in the MBBR for bioaugmentation purposes, such as a special microbial seed of biofilm-forming denitrifying bacteria (which involved a consortium of *Diaphorobacter* sp. R4, *Pannonibacter* sp. V5, *Thauera* sp. V9, *Pseudomonas* sp.V11 and *Thauera* sp.V14) where *Thauera* was the most efficient genus (Patel et al. 2021; Patel and Nerurkar 2024), *Pseudomonas* sp. SZF15 (Su et al. 2019). *Corynebacterium pollutisoli* SPH6 (Liu et al. 2018), *Pseudomonas mendocina* IHB602 (Hong et al. 2020) and *Acinetobacter* sp.CN86 (Su et al. 2019) efficiently remove nitrate from wastewater. The biological denitrification process is of two types: autotrophic denitrification and heterotrophic denitrification. Autotrophic denitrifiers derive their energy from hydrogen, iron or sulfur compounds (Karanasios et al. 2010). Nitrate serves as the terminal electron acceptor in the microbial mediated heterotrophic denitrification process, while various carbon sources such as sucrose, acetate and methanol serve as electron donors (Rajmohan et al. 2018).

DOI: 10.1201/9781003408352-7

6.1.1 Biofilms in a Wastewater Treatment System

The most extensively used treatment system is the biological wastewater treatment system. Microorganisms are commonly found in wastewater treatment systems as microbial biofilms or flocs. Microorganisms in a biofilm form require less space, have a higher concentration of relevant organisms and do not require biomass return (Ødegaard 2006), which are important parameters for the performance of wastewater treatment plants. Extracellular polymeric substances (EPS) retain microbial cells together in biofilms. Biofilms are known as the "city of microbes" where microorganisms are present and stick to each other and the EPS matrix is "house of the biofilm cells" in which microorganisms are embedded (Flemming and Wingender 2010). Microbe-produced EPS accounts for 90% of biofilm content and is composed of extracellular polysaccharides, proteins, lipids and extracellular DNA (eDNA) (Flemming and Wingender 2010). Biofilm provides high microbial resistance against various chemical agents, antimicrobial agents and various environmental stresses such as UV damage, shear stress and high temperature. Among different biofilm reactors, trickling filters are not volume-effective, mechanical failures occur in rotating biological contactors, granular media biofilters require backwashing and fluidized bed reactors show instability. To overcome operational problems faced by biofilm reactors, the Moving Bed Biofilm Reactor (MBBR) was envisioned. In the MBBR, the biofilm forms on the specifically engineered carriers that move freely, making it highly effective in its performance (Rusten et al. 2006). Moreover, existing conventional activated sludge (CAS) processes can be easily upgraded to MBBR.

6.1.2 Moving Bed Biofilm Reactor (MBBR)

MBBR development began in the 1980s in Norway (Ødegaard 2006). It combines the best features of suspended processes and biofilm-based processes. Biofilm forms on the surface of the carriers that are suspended in the reactor and move freely in the MBBR. Figure 6.1 shows a bench-scale schematic diagram of denitrifying MBBR (dMBBR) for the removal of nitrate from wastewater. The carriers in MBBR "carry" the microorganisms in the reactor as the microbes adhere to the surfaces of the carriers in biofilm form (Ødegaard 2006). Thin and evenly distributed biofilm is the salient feature of efficient MBBR (Rusten et al. 2006). Biofilm formed on the exterior and interior surfaces of carriers can remove nitrate from the wastewater. The carriers suspended in the reactor move freely in the reactor due to the aerator (in aerobic reactors) and mixer (in anoxic and anaerobic reactors). MBBR has been widely used for the breakdown of organic matter and nitrogen and applied in municipal and industrial wastewater treatment as well as in drinking water treatment. It has also been widely used for the detoxification of various industrial effluents. In this chapter, we discussed various parameters important for the optimal removal of nitrate in dMBBR.

6.1.3 Operational Parameters Influencing the Performance of Denitrifying MBBR

Previous studies have proven that the optimization operation parameters enhance MBBR performance. Biofilm is the major element in dMBBR. It is affected by various operating conditions such as carrier type, filling ratio of carriers, the physico-chemical surface of carriers, hydraulic retention time (HRT) and hydrodynamics inside dMBBR. Optimization of these parameters increased nitrate removal efficiencies in the reactor by directly affecting microorganisms and the speed of biofilm development inside dMBBR (Chu et al. 2014). Since biological denitrification is used by bacteria to meet energy requirements, its rate increases proportionately with the growth rate of the microorganisms and biofilm formation.

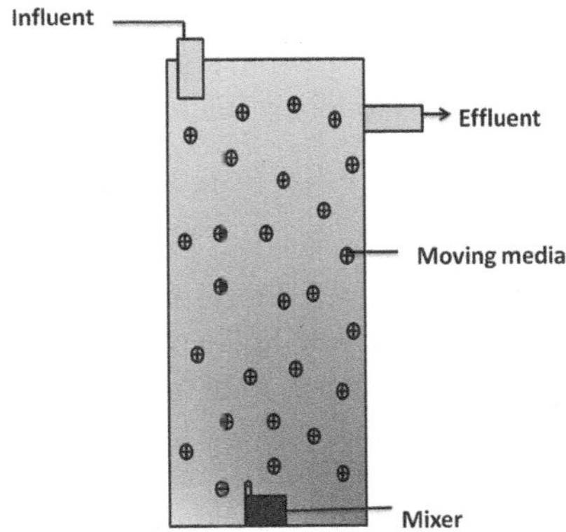

FIGURE 6.1 Schematic diagram of bench-scale dMBBR.

6.1.3.1 Carriers

The carrier media used in dMBBR is a key element that influences its performance. In MBBR, carriers with high suspendability, surface roughness and high mechanical strength showed high performance (Yuan et al. 2015). Carrier shape, structure, porosity and surface properties protect the microbial community developed on the carriers. Biofilm develops more thickly inside the carriers, whereas the abrasion-exposed surface forms thin biofilms (Mahendran et al. 2012; Zhang et al. 2017). The shape of the carrier can also create gradients of varying DO and shear stress that can promote a stable and diverse biofilm of aerobic nitrifiers and anaerobic denitrifiers growing on the same carrier (Wu et al. 2022). Yuan et al. (2015) recommended polyethylene (PE) as the best carrier while demonstrating that polypropylene (PP), polyurethane foam (PUF) and haydite carriers gave relatively lower and less stable denitrification efficiency.

Suspended ceramic products, diatomaceous earth, bioplastic-based carriers, polymeric nano-fibrous material and activated carbon can also be used as carriers in dMBBR (Dong et al. 2011). Other carrier media, such as cellulose, woodchips, plastic media, non-woven carriers and modified carriers, were also employed for MBBR, which increase the biofilm formation on the carriers (Young et al. 2017).

6.1.3.2 Filling Ratio

The filling ratio of carriers is the ratio of the volume of carriers added to the reactor to the total internal volume of the reactor. It should be below 70% to allow the carriers to move freely in suspension. Wang et al. (2005) reported that increasing the filling ratio reduced the biofilm thickness on each carrier, which allows for more mass exchange and biofilm regeneration in the dMBBR, resulting in a higher denitrification rate. They also claim that a high filling ratio causes more carrier collisions, which leads to the selection of bacteria that can grow on the carrier under the selected reactor conditions, making the system more efficient. But too high (>50%) filling ratio reduced the anoxic zone due to the thinner biofilm, which increased aerobic microorganisms, which in turn compete with denitrifying bacteria. These aerobic microorganisms can also consume a significant amount of organic matter present in the water, lowering the denitrification rate of anaerobic bacteria by depleting electron donors. Thus, a balance needs to be struck between the free mobility of carriers, biofilm biomass, mass transfer rate and anoxia to achieve maximal denitrification efficiency. Therefore, it is suggested that the filling ratio should be optimized for proper biofilm development and efficient nitrate removal in dMBBR.

At various filling ratios, biofilm formed on the surface of carriers is composed of EPS, filamentous bacteria, bacilli and cocci that together form biofilm, adhered on the surface of carriers.

Studies have shown that at low filling ratios cocci and bacilli were found to be present on the biofilm of the carriers and formed compact biofilm structures. At increased filling ratios, the filamentous bacteria population was high, which leads to looser microbial biofilm structures (Wang et al. 2005; Yuan et al. 2015).

6.1.3.3 Modification of Carriers

Biofilm development on the surface of the dMBBR carrier and functional microorganisms present on the carriers of dMBBR play an important role in the removal of nitrate. Modification of the carrier surface area and carrier is also an important feature to achieve appropriate biofilm development and initial attachment of the cells to the surface and is a promising option for wastewater treatment. Microbial cells associated with EPS form an initial biofilm on carriers and facilitate the formation of biofilm, which ultimately increases the pollutant removal efficiency due to faster attachment of the bacterial cells on the surface of the carriers (Morgan-Sagastume 2018). Therefore, commonly used plastic carriers are often modified via polymer blending, polymer grafting, wet chemical oxidation, film coating, etc. To increase positive charges and hydrophobicity, Chu et al. (2014) used polymer blending with foam stabilizer, polyether polyol, toluene diisocyanate and other ingredients. These modifications are designed to make the carrier surface more hydrophilic and electrophilic. Hydrophilicity and electrophilicity increase the contact area between the uneven surfaces of the carrier and the influent water. This increased interface allows more bacterial cells to reach the carrier for colonization while also allowing the greater exchange of organic carbon and nitrates for denitrification to proceed.

6.1.3.4 Shear Stress

Shear stress and collisions induce sloughing from carriers and prevent excess biomass build-up. The structure of the carrier and the degree of shear stress determine the size of the hydrodynamic boundary layer in which biofilms form. Due to this hydrodynamic control, carriers in dMBBR do not clog and thin biofilm is maintained, which could easily denitrify water (Dupla et al. 2006). At lower shear forces, thicker but less compact biofilms were formed, and biofilm surface roughness increased, which favored aerobic microorganisms (Wang et al. 2005; Ødegaard 2006) whereas higher shear stress in MBBR supports more stable, thinner, stronger and denser biofilm with increased EPS production (Dezotti et al. 2018). Another mechanism for shear stress-induced compaction of biofilms is the enhanced proton translocation across cell membranes. This increases the hydrophobic nature of the cell membrane and increases cell-to-cell interactions. The resulting denser biofilms are more efficient in denitrification. Therefore, higher hydrodynamic stress should be maintained to develop a stronger biofilm, which ultimately increases the performance of the dMBBR.

6.1.3.5 Hydraulic Retention Time (HRT)

The contact time of the influent wastewater with the microbial biomass inside the reactor, i.e. HRT, is also an important parameter for the denitrification process. An appropriate HRT improves nitrate reduction inside the dMBBR. If the HRT is too long, it may consume high energy and if the HRT is too short it reduces contact time between microbial biomass and the pollutant (Wang et al. 2009). Therefore, optimization of the appropriate HRT is also required for the performance of the dMBBR.

6.1.4 ENVIRONMENTAL AND NUTRITIONAL FACTORS IMPORTANT FOR THE EFFICIENT DENITRIFICATION PROCESS IN dMBBR

Apart from carrier type and filling ratio, environmental and nutritional factors also influence the denitrification efficiency and rates in dMBBR.

6.1.4.1   Carbon Source

In the heterotrophic denitrification process, acetate, methanol, glucose and ethanol have proven more effective carbon sources. These are simple carbon sources with high nitrate reduction efficiency because they are easy to utilize and provide electrons for denitrification process (Mohan et al. 2016). Acetate is more effective in denitrification than other carbon sources such as glucose, ethanol and methanol due to its metabolic properties. It can be utilized by denitrifiers as it forms acetyl-CoA (Onnis-hayden and Gu 2008). Carbon sources are also known to influence the community structure and biofilm structure of bacteria, according to Srinandan et al. (2012).

6.1.4.2   C/N Ratio

The C/N ratio is an important factor of whether sewage is suitable for biological denitrification. Changes in C/N ratios affect heterotrophic denitrifying bacteria. Carbon acts as the electron donor for heterotrophic denitrifying bacteria, while nitrate becomes the terminal electron acceptor. The balance of electron donor and electron acceptor is essential in biological denitrification. Influent with very low carbon content should be amended with appropriate carbon sources because low C/N ratio limits denitrification efficiency by accumulating denitrification intermediates such as NO_2 and N_2O by limiting the electron supply (Yuan et al. 2018). This results in incomplete denitrification and the presence of toxic nitrogen oxides in the water. This toxicity harms the bacterial population and further lowers the denitrification efficiency. Alternatively, high C/N ratio increases the COD of effluents due to the retention of organic matter in the effluent (Patel et al. 2021). This also promotes the growth of undesirable non-denitrifying heterotrophs. Therefore, appropriate C/N ratio should be maintained for the efficient denitrification process.

6.1.4.3   Nitrate Concentration

Nitrate concentration varies in different wastewaters. High nitrate concentration strongly affects bacterial denitrification because the bacterial activity is inhibited by high nitrate concentrations (Dhamole et al. 2007). As a result, bacterial denitrification with nitrate concentrations of more than 100 mM has been studied in only a few investigations (Dhamole et al. 2007).

6.1.4.4   Metal Ions

The presence of metal ions in wastewater induces biofilm formation. Moderate concentrations of Mg^{2+}, Ca^{2+}, Fe^{2+}, Fe^{3+} and Mn^{2+} induce biofilm formation (Srinandan et al. 2010). The addition of metal ions increases the EPS content at moderate metal ion concentrations. Metal ions absorb EPS and become trapped in the EPS matrix network, forming metal-EPS complexes. This makes the biofilm denser and more anoxic and promotes denitrification. Thus, the addition of metal ions can induce biofilm formation inside dMBBR and increase the performance of dMBBR. Different metalloenzymes involved in denitrification can show tolerance or even higher activity in the presence of small amounts of heavy metals. It enhances the activity of nitrite reductase, nitrate reductases and nitric oxide reductase that ultimately increase overall denitrification (Chen et al. 2019).

6.1.4.5   pH, DO and Temperature

pH, DO and temperature all have an impact on denitrifier growth, metabolism, denitrification gene expression and, as a result, denitrification rate.

pH also influences bacterial denitrification ability because it has a significant impact on nitrate removal as well as bacterial growth and metabolism. pH can have an impact on the bacterial surface charge. The ideal pH for denitrification is 7–8, which is also the ideal pH for the majority of environmental denitrifying bacteria (Glass and Silverstein 1998). Without an optimal pH, the activity of reductases and electron transporters decreases, and this leads to lower denitrification efficiency. Further, an acidic pH promotes the formation of toxic nitrous acid from nitrite, which lowers the denitrifying bacterial population. Meanwhile, alkalinity maintains the proton gradient and increases the precipitation of bicarbonate and Ca^{2+} and nutrient availability. Nitrite reduction itself increases pH of the medium; consequently, it gets down regulated in basic media to prevent excess alkalinity. In other words, high pH causes the microbial community to carry out partial denitrification.

DO is also an important parameter in the efficient denitrification of dMBBR. At high DO concentrations the denitrification rate was reduced due to the shortening of the anoxic zone in MBBR (Pochana and Keller 1999). Most denitrifying bacteria are facultative anaerobic and preferentially utilize oxygen as the terminal electron acceptor because it is more energetically favorable than nitrate. The presence of oxygen in high-strength denitrification reactors may induce nitrite build-up, according to Glass and Silverstein (1998). DO levels have a significant impact on the microbial community of biofilm in biological treatment systems (Feng et al. 2012). Temperature has an impact on bacterial survival, growth and the denitrification process. The optimal temperature for denitrification is between 25 and 37 °C, which is also the best temperature for bacterial activity. Low temperatures influence microorganism activity and the composition of the biofilm community (Young et al. 2017). At low temperatures, biofilm formation was higher compared to high temperatures (Morimatsu et al. 2012).

6.1.5 BIOFILM DEVELOPMENT PROCESS

Biofilm formation is a cyclic process. It comprises five distinct stages: (1) initial attachment; (2) irreversible attachment; (3) early development of biofilm; (4) maturation and (5) dispersion. Detached cells return back to planktonic growth, bringing the developmental life cycle of biofilm to a close. Figure 6.2 shows steps involved in biofilm formation.

6.1.5.1 Stage 1: Initial Attachment

The adsorption of molecules such as proteins, lipids, nucleic acids, humic acids and polysaccharides on the surfaces initiates biofilm formation on the carrier material. Absorbed molecules form films, which can have a variety of effects such as changing the surface physico-chemical properties, acting as a nutrient source for microbes, releasing toxic surface metal ions, and detoxifying the bulk (Lewandowski 2011). Bacterial adhesion can occur on the surface via two mechanisms: the first is bacterial adhesion and aggregation to carriers, which is mediated by high-affinity adhesion factors such as polysaccharide, protein and extracellular DNA or accessory structures such as bacterial

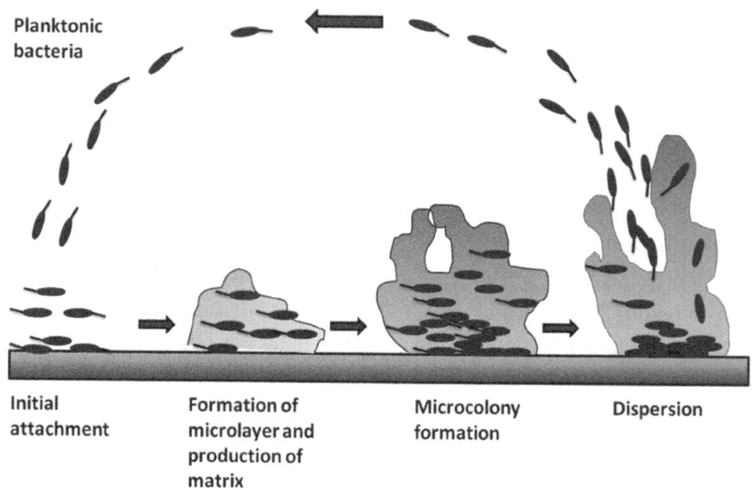

FIGURE 6.2   Steps involved in biofilm formation.

flagella or pili of the bacterial surface; the second is activated by the recognition of specific glyco-protein and adhesion proteins on the surface of bacteria (Flemming et al. 2016).

6.1.5.2 Stage 2: Irreversible Adhesion

During this stage, cells become irreversibly adsorbed. Cells adhere to the substrate and form cell clusters where all cells are in contact. The hydrophobic–hydrophilic characteristics of the contact surface have a significant impact on microbial adhesion to the surface (Liu et al. 2004).

6.1.5.3 Stages 3 & 4: Early Development of Biofilm and Biofilm Maturation

Biomass can persist in mature biofilms for several years (Biswas et al. 2014). The mature biofilm forms a complex architecture, channels and pores inside the biofilm. Zhu et al. (2015) also claim that bacterial adhesion force is related to the amount of EPS on MBBR carriers. Components affecting biofilm characteristics and microbial attachment/detachment interactions become more prominent in mature biofilm EPS.

6.1.5.4 Stage 5: Biofilm Dispersion

When a single cell actively escapes from the biofilm, the sessile matrix-encased biofilm returns to a planktonic mode and this process is known as dispersion. This process is driven by the release of chemical gradient of oxygen, release of waste products and nutrient resources from developed biofilm (Rumbaugh and Sauer 2020). In MBBRs, the collision of carriers expedites the detachment of biomass (Dezotti et al. 2018). Reactor operational parameters such as hydrodynamics, shear stress, HRT and organic loading rates, environmental parameters of the wastewater, microbial physiology and metabolism, microbial interactions and microbial cell properties are all major factors affecting biofilm development in MBBR and they also affect biofilm structure and composition (Flemming et al. 2016).

6.2 CONCLUSIONS AND PROSPECTS

This book chapter discussed various factors affecting dMBBR performance and biofilm formation on carriers for its efficiency. Optimization of various process parameters such as different types of carriers, filling ratio of carriers, surface modifications of carriers, HRT, C/N ratio, carbon sources, DO, temperature and pH can be used as an effective strategy to enhance dMBBR performance for wastewater treatment. More emphasis is needed on the aspect of the microbial component of dMBBR. Further investigations are still needed to characterize the dMBBR biofilm using metaproteomics and metagenomic approaches to analyze the microbial community. Novel dMBBR hybridization strategies are required to generate interest in order to meet the increasing demands for nitrate removal from wastewater with less energy consumption.

REFERENCES

Biswas, K., Taylor, M. W., & Turner, S. J. (2014). Successional development of biofilms in moving bed biofilm reactor (MBBR) systems treating municipal wastewater. *Applied Microbiology and Biotechnology*, 98(3), 1429–1440.

Briki, M., Zhu, Y., Gao, Y., Shao, M., Ding, H., & Ji, H. (2017). Distribution and health risk assessment to heavy metals near smelting and mining areas of Hezhang, China. *Environmental Monitoring and Assessment*, 189(9), 1–19.

Chen, Y., Wang, Q., Zhao, S., Yang, W., Wang, H., & Jia, W. (2019). Removal of nutrients and emission of nitrous oxide during simultaneous nitrification, denitrification and phosphorus removal process with metal ions addition. *International Biodeterioration & Biodegradation*, 142, 143–150.

Chu, L., Wang, J., Quan, F., Xing, X. H., Tang, L., & Zhang, C. (2014). Modification of polyurethane foam carriers and application in a moving bed biofilm reactor. *Process Biochemistry*, 49(11), 1979–1982.

Dezotti, M., Lippel, G., & Bassin, J. P. (2018). Advanced biological processes for wastewater treatment. *Springer, İsviçre. doi*, 10, 978–3.

Dhamole, P. B., Nair, R. R., D'Souza, S. F., & Lele, S. S. (2007). Denitrification of high strength nitrate waste. *Bioresource Technology*, 98(2), 247–252.

Dong, Z., Lu, M., Huang, W., & Xu, X. (2011). Treatment of oilfield wastewater in moving bed biofilm reactors using a novel suspended ceramic biocarrier. *Journal of Hazardous Materials*, 196, 123–130.

Dupla, M., Comeau, Y., Parent, S., Villemur, R., & Jolicoeur, M. (2006). Design optimization of a self-cleaning moving-bed bioreactor for seawater denitrification. *Water Research*, 40(2), 249–258.

Feng, L. J., Xu, J., Xu, X. Y., Zhu, L., Xu, J., Ding, W., & Luan, J. (2012). Enhanced biological nitrogen removal via dissolved oxygen partitioning and step feeding in a simulated river bioreactor for contaminated source water remediation. *International Biodeterioration & Biodegradation*, 71, 72–79.

Flemming, H. C., Wingender, J., Szewzyk, U., Steinberg, P., Rice, S. A., & Kjelleberg, S. (2016). Biofilms: an emergent form of bacterial life. *Nature Reviews Microbiology*, 14(9), 563–575.

Flemming, H., & Wingender, J. (2010). The biofilm matrix. *Nature Publication Group*, 8, 623–633.

Glass, C., & Silverstein, J. (1998). Denitrification kinetics of high nitrate concentration water: pH effect on inhibition and nitrite accumulation. *Water Research*, 32(3), 831–839.

Hong, P., Wu, X., Shu, Y., Wang, C., Tian, C., Wu, H., & Xiao, B. (2020). Bioaugmentation treatment of nitrogen-rich wastewater with a denitrifier with biofilm-formation and nitrogen-removal capacities in a sequencing batch biofilm reactor. *Bioresource Technology*, 303, 122905.

Karanasios, K. A., Vasiliadou, I. A., Pavlou, S., & Vayenas, D. V. (2010). Hydrogenotrophic denitrification of potable water: a review. *Journal of Hazardous Materials*, 180(1–3), 20–37.

Lewandowski, Z., & Boltz, J. P. (2011). Biofilms in water and wastewater treatment. In *Water-Quality Engineering* (pp. 529–570). Elsevier. 10.1016/B978-0-444-53199-5.00095-6

Liu, X., Wang, L., & Pang, L. (2018). Application of a novel strain Corynebacterium pollutisoli SPH6 to improve nitrogen removal in an anaerobic/aerobic-moving bed biofilm reactor (A/O-MBBR). *Bioresource Technology*, 269, 113–120.

Liu, Y., Yang, S. F., Li, Y., Xu, H., Qin, L., & Tay, J. H. (2004). The influence of cell and substratum surface hydrophobicities on microbial attachment. *Journal of Biotechnology*, 110(3), 251–256.

Mahendran, B., Lishman, L., & Liss, S. N. (2012). Structural, physicochemical and microbial properties of flocs and biofilms in integrated fixed-film activated sludge (IFFAS) systems. *Water Research*, 46(16), 5085–5101.

Mohan, T. K., Nancharaiah, Y. V., Venugopalan, V. P., & Sai, P. S. (2016). Effect of C/N ratio on denitrification of high-strength nitrate wastewater in anoxic granular sludge sequencing batch reactors. *Ecological Engineering*, 91, 441–448.

Morgan-Sagastume, F. (2018). Biofilm development, activity and the modification of carrier material surface properties in moving-bed biofilm reactors (MBBRs) for wastewater treatment. *Critical Reviews in Environmental Science and Technology*, 48(5), 439–470.

Morimatsu, K., Eguchi, K., Hamanaka, D., Tanaka, F., & Uchino, T. (2012). Effects of temperature and nutrient conditions on biofilm formation of Pseudomonas putida. *Food Science and Technology Research*, 18(6), 879–883.

Ødegaard, H. (2006). Innovations in wastewater treatment:–the moving bed biofilm process. *Water Science and Technology*, 53(9), 17–33.

Onnis-Hayden, A., & Gu, A. Z. (2008). Comparisons of organic sources for denitrification: biodegradability, denitrification rates, kinetic constants and practical implication for their application in WWTPs. *Proceedings of the Water Environment Federation*, 2008(17), 253–273.

Patel, R. J., & Nerurkar, A. S. (2024). Thauera sp. for efficient nitrate removal in continuous denitrifying moving bed biofilm reactor. *Bioprocess and Biosystems Engineering*, 47(3), 429–442.

Patel, R. J., Patel, U. D., & Nerurkar, A. S. (2021). Moving bed biofilm reactor developed with special microbial seed for denitrification of high nitrate containing wastewater. *World Journal of Microbiology and Biotechnology*, 37(4), 1–13.

Pochana, K., & Keller, J. (1999). Study of factors affecting simultaneous nitrification and denitrification (SND). *Water Science and Technology*, 39(6), 61–68.

Rajmohan, K. S., Gopinath, M., & Chetty, R. (2018). Bioremediation of nitrate-contaminated wastewater and soil. In *Bioremediation: Applications for Environmental Protection and Management* (pp. 387–409). Springer, Singapore.

Rumbaugh, K. P., & Sauer, K. (2020). Biofilm dispersion. *Nature Reviews Microbiology*, 18(10), 571–586.

Rusten, B., Eikebrokk, B., Ulgenes, Y., & Lygren, E. (2006). Design and operations of the Kaldnes moving bed biofilm reactors. *Aquacultural Engineering*, 34(3), 322–331.

Srinandan, C. S., Jadav, V., Cecilia, D., & Nerurkar, A. S. (2010). Nutrients determine the spatial architecture of Paracoccus sp. *biofilm. Biofouling*, 26(4), 449–459.

Srinandan, C. S., D'souza, G., Srivastava, N., Nayak, B. B., & Nerurkar, A. S. (2012). Carbon sources influence the nitrate removal activity, community structure and biofilm architecture. *Bioresource Technology*, 117, 292–299.

Su f, Xue J., lin Huang, T., Wei, L., yu Gao, C., & Wen, Q. (2019). Performance and microbial community of simultaneous removal of NO3--N, Cd2+ and Ca2+ in MBBR. *Journal of Environmental Management*, 250, 109548.

Toolabi, A., Bonyadi, Z., Paydar, M., Najafpoor, A. A., & Ramavandi, B. (2021). Spatial distribution, occurrence, and health risk assessment of nitrate, fluoride, and arsenic in Bam groundwater resource, Iran. *Groundwater for Sustainable Development*, 12, 100543.

Wang, R. C., Wen, X. H., & Qian, Y. (2005). Influence of carrier concentration on the performance and microbial characteristics of a suspended carrier biofilm reactor. *Process Biochemistry*, 40(9), 2992–3001.

Wang, Y., Peng, Y., & Stephenson, T. (2009). Effect of influent nutrient ratios and hydraulic retention time (HRT) on simultaneous phosphorus and nitrogen removal in a two-sludge sequencing batch reactor process. *Bioresource Technology*, 100(14), 3506–3512.

Wu, J., Yin, Y., & Wang, J. (2018). Hydrogen-based membrane biofilm reactors for nitrate removal from water and wastewater. *International Journal of Hydrogen Energy*, 43(1), 1–15.

Wu, Y., Wan, A., Zhao, B., Xue, S., & Xu, A. (2022). Single-stage MBBR using novel carriers to remove nitrogen in rural domestic sewage: The effect of carrier structure on biofilm morphology and SND. *Journal of Environmental Chemical Engineering*, 10(5), 108267.

Young, B., Delatolla, R., Kennedy, K., Laflamme, E., & Stintzi, A. (2017). Low temperature MBBR nitrification: Microbiome analysis. *Water Research*, 111, 224–233.

Yuan, Q., Wang, H., Hang, Q., Deng, Y., Liu, K., Li, C., & Zheng, S. (2015). Comparison of the MBBR denitrification carriers for advanced nitrogen removal of wastewater treatment plant effluent. *Environmental Science and Pollution Research*, 22(18), 13970–13979.

Yuan, Q.; Wang, H.Y.; Chua, Z.-S.; Hang, Q.-Y.; Liu, K.; Li, C.M (2018) Influence of C/N ratio on MBBR denitrification for advanced nitrogen removal of wastewater treatment plant effluent. *Desalin Water Treat* 66, 158–165:. https://doi.org/10.5004/dwt.2017.0263

Zhang, X., Song, Z., Guo, W., Lu, Y., Qi, L., Wen, H., & Ngo, H. H. (2017). Behavior of nitrogen removal in an aerobic sponge based moving bed biofilm reactor. *Bioresource Technology*, 245, 1282–1285.

Zhu, Y., Zhang, Y., Ren, H. qiang, Geng, J. ju, Xu, K., Huang, H., & Ding, L. li. (2015). Physicochemical characteristics and microbial community evolution of biofilms during the start-up period in a moving bed biofilm reactor. *Bioresource Technology*, 180, 345–351.

7 Mosquito Vector Management in Clean, Stagnant, and Sewage Water Ecosystems

Kadarkarai Murugan, Rajapandian Rajaganesh,
Murugan Vasanthakumaran, Douglas J.H. Shyu,
Jiang-Shiou Hwang, Hans-Uwe Dahms, Fajun Chen,
Rui-De Xue, Lan Wang, Natchiappan Senthilkumar,
Mohamed Néjib Daly Yahia, and
Chellasamy Panneerselvam

7.1 INTRODUCTION

For a person to survive, they need water. In fact, it may be argued that water is the source of all life because it makes up 60% of the human body. No aspect of human activity is complete without water. The world is currently arguing whether information flow or energy flow is more crucial. That is a valid query. However, water flow continues to be more significant. It is essential to human equity, the environment, and both the economy and ecology. Global warming and environmental factors mean the issue of water is becoming ever more urgent (Kumar, 2018). Mosquitoes are breeding in different water ecosystems, like clean water, stagnant water, and sewage water.

The main environmental pollutant harming Earth's ecosystem is water contamination, and proper water sanitation is essential to avoid water pollution. Heavy metal contamination in industrial wastewaters from industries including mining, electroplating, and textiles is one of the main offenders (Bolisetty et al., 2019). Higher concentrations of heavy metals from effluents are released into the aquatic ecosystem, which harm not only aquatic life but also people who drink the water, causing heavy metal poisoning, infections, and even death. Various chemical technologies are common ways to remove heavy metals from wastewater (Fu and Wang, 2011; Barakat, 2011). Because of the abundance of major functional groups of various amino acids that have the ability to bind metals, proteins among biological materials provide a viable approach for heavy metal adsorption (Yamashita et al., 1990; Soon et al., 2022). Therefore, in the current study, we try to employ plant extract to remediate contaminated water that threatens ecosystem functioning.

Mosquitoes spread terrible diseases (WHO, 2020). Every year, there are over 500 million cases of illness and 2.7 million fatalities from diseases spread by mosquitoes (Bustamante, 2020). One of the most dangerous illnesses spread by female mosquitoes is malaria, which kills more than one million people annually and is caused by the *Plasmodium* protist. Malaria is one of the diseases that female mosquitoes transfer the most to humans. Additionally, the *Aedes aegypti* mosquito can spread additional serious illnesses (Uraki et al., 2019). The likelihood of outbreaks is rising because of climate change, lack of economy, poor immunizations and the concurrent spread and reemergence of arthropod-borne viral illnesses (Leta et al., 2018; Tjaden et al., 2018; Shearer et al., 2018; Iwamura et al., 2020; Anoopkumar and Aneesh, 2022). Additionally, through *Culex sp.* about

DOI: 10.1201/9781003408352-8

100 million people are thought to be afflicted with lymphatic filariasis (*Wuchereria bancrofti*), of which 43 million have severe cases (Hamama et al., 2022).

Insecticide resistance is a developing issue for managing problematic mosquito populations globally and has the potential to render use of current chemical control methods ineffective (Griffin et al., 2016; Hemingway et al., 2016). The situation outlined above suggests that the use of synthetic compounds containing organophosphorus and pyrethroids as adult mosquitocidal is not having the preferred impact on mosquito vector populations, most likely as a result of the acquired resistance phenomena (Smith et al., 2019; Demok et al., 2019). Insecticide residues end up in ecosystems and human populations due to increased different life stages of mosquito resistance to chemical insecticides, which forces responsible national control bodies to increase fumigation procedure doses and negatively affect humans and other living things (Zhang et al., 2020; Rahman et al., 2020). This also causes the decline of biodiversity, emergence of insect resistance, and pollution of environmental resources and the food chain (Benelli et al., 2017). As a result, scientists have concentrated on cutting-edge, possibly environmentally benign control techniques.

As a result of their insecticidal qualities, biodegradability, and affordability, botanicals and their components have become effective alternatives for the management of disease vectors (Pratheeba et al., 2019). Terpenes and iridoids are mosquitocidal repellant, and oviposition deterrent effects in mosquitoes have been identified as the major components (Castillo et al., 2017; Chellappandian et al., 2018). Because of their characteristic aroma, traditional medicine has utilized *C. cyminum* to treat a number of illnesses (Ramya et al., 2022). Since ancient times, the tiny cumin (*Cuminum cyminum*) seeds have been employed in numerous cultures for both culinary and medicinal uses (Gohari and Saeidnia, 2011). Earlier investigations revealed that the seed extract showed antibacterial and other biological, biochemical, and enzymatic properties (Al-Snafi, 2016), and found that a cluster of secondary compounds and nutrients exist in the *Cuminum cyminum*.

Using antimicrobials from traditional medicine is one of the suggested strategies for dealing with germs that are resistant to medication. By generating a variety of antimicrobial compounds, including phenolic, terpenoids, flavanones, and quinones, medicinal plant extracts can play a significant part in the fight against bacterial infection (Moradi et al., 2020).

Moreover, mosquito breeding is contaminating water bodies and, hence, the present study investigates the mosquitocidal properties of *C. cyminum* seed extract against malarial, dengue, and filarial vectors. Further we investigate the water purification properties of *C. cyminum* seed extract for three different mosquito-breeding water ecosystems. We also evaluate the phytochemical screening of *C. cyminum* seed extract through GC-MS analysis and finally we investigate the antibacterial effect of *C. cyminum* seed extract against bacterial strains (*Pseudomonas aeruginos* and *Bacillus cereus*).

7.2 MATERIALS AND DISCUSSION

7.2.1 PLANT EXTRACT PREPARATION

Methanolic extracts of *C. cyminum* seed power were extracted with Soxhlet apparatus and the extract was concentrated by Rotary Evaporator. The percentage of yield was also noted. Impurities in the extract were removed by standard fixatives. For preparation of stock solution one g of seed powder was mixed with a hundred milliliter of acetone. From the stock solution, all other experimental concentrations were made.

7.2.2 GC-MS ANALYSIS OF CUMIN

The Clarus Mass Spectrometer, a sophisticated bench-top mass spectrometric, was employed for gas chromatograph analysis of cumin seed extract. The carrier gas was 1 mL/min of pure helium gas (99.99%) flowing continuously. Based on comparisons between the test sample parameters, the phytoconstituents have been identified by following methods and procedures (Kordali et al., 2005).

7.2.3 WATER PURIFICATION PROPERTIES OF *C. CYMINUM* SEED EXTRACT

For water quality parameters, The American Public Health Association's (2005) guidelines were used to analyze water quality metrics as temperature, pH, turbidity, and dissolved oxygen and salinity. Treatment coagulants were prepared using the Schwarz-reported techniques (2000). *C. cyminum* seeds were shade-dried, ground into a powder, and added to with water to make a semisolid paste. The necessary strength was maintained on the basis of turbidity of the raw water and respective strength was kept between 50 and 150, and >150 mg/l (Schwarz, 2000). Quick agitation was done for 30 seconds and insoluble coagulant stuff was filtered by special mesh. Special care was taken and the purified water was undisturbed for 1 to 2 hours.

7.2.4 MOSQUITO REARING

The Mettupalayam field offices of the Mosquito Breeding Centre, Government of India donated eggs of three mosquito vectors (Tamil Nadu, India). In 18×13× 4 cm cups filled with 500 ml of water, with standard light and temperature regimes, mosquito larvae were fed (dog) biscuits, and 500 ml dechlorinated water was used in glass beakers for further experiments (Rajaganesh et al., 2016).

7.2.5 LARVAL/PUPAL TOXICITY ASSAYS UNDER LABORATORY CONDITIONS

Twenty-five pupae or larvae of three different mosquito species at various stages were incubated in a container for 24 hours with the appropriate amount of *C. cyminum* seed methanol extract, in addition to 250 ml of distilled water (100, 200, 300, 400, and 500 parts per million). 0.75 mg food was given for every experimental dose (Kovendan et al., 2012). There were five replications for every concentration. A 24-hour exposure to the solvent at that concentration was given to the control mosquitoes. The percentage larval mortality was noted down with following formula: % mortality = (number of dead individuals / number of treated individuals) × 100.

7.2.6 OVICIDAL ACTIVITY OF *C. CYMINUM* SEED EXTRACT

The ovicidal activity of *C. cyminum* seed powder extracts was studied by Su and Mulla (1998). Eggs were collected from the experimental ovipositional cup from the Laboratory at School of Life Sciences and exposed with different doses. Freshly oviposited eggs of the concerned species were treated with experimental concentrations. There were five replicates. After 20 hours of exposure, all the eggs at all the concentrations were individually transferred into the container which contained double distilled water and hatchability was noted under microscope. The observation period was 48 hours. The percentage of egg hatching used the methods of Chenniappan and Kadarkarai (2008) and Rajaganesh et al. (2016).

7.2.7 BACTERIAL STRAIN EXPERIMENTATIONS

7.2.7.1 Disk Diffusion Method

The antibacterial activity of cumin seed extract was tested against gram-negative and gram-positive bacterial strains and the Kirby-Bauer Disk Diffusion Susceptibility Test technique was used (Bauer et al., 1966) for assessing the activity of plant compounds against pathogens. All the materials were kept sterile. The antimicrobial test analysis, by a sterile blank antimicrobial susceptibility disc, used diverse solution materials of various test regimes, for example the discs were loaded with different solutions containing 25, 50, and 75 µg/ml of *C. cyminum* seed extract (one millimolar concentration). The experimental setup with an agar plate allowed for 24 hour incubation at 37 °C. The zone of inhibition was noted by the following formula with standard time period: % inhibition = O.D (control) – O.D (sample) × 100 O.D (control).

7.2.7.2   Growth Curve Analysis of *C. cyminum* Seed Extract against Bacterial Strains

We analyzed the growth curves of *C. cyminum* treated with various bacterial strains. In order to comprehend their growth pattern, 25, 50, and 75 µg/ml of *C. cyminum* seed extract was used to treat the strains over time. Bacteria were injected into LB broth and incubated at 37°C and 180 rpm for an entire night. After that, a growth medium inoculums containing different concentrations of bacterial populations had *C. cyminum* uniformly spread out. To achieve the requisite dilution, *C. cyminum* seed extract from a stock suspension of 1 mg/ml was diluted. The negative control was designated as no treatment and it was maintained at 37°C and 180 rpm. Absorbance of the sample was performed at 600 nm by adding seed extract of *C. cyminum* for 10 h starting from the 0 hour.

7.2.8   STATISTICAL ANALYSIS

The SPSS (16.0 version) is used for the management of data. Data on bacteria growth inhibition and larval and laboratory results of pupal bioassay information have been used for the lab tests that were converted into arcsine/proportion values two-way ANOVA with two components (i.e., doses and various larval and pupa stages). The means were separated by Tukey's HSD test. Finney's (1971) method was adopted for the estimation of larval and pupal toxicity and its IC_{50}, LC_{50}, and LC_{90} values. For two-factor ANOVA analysis the ovicidal results were converted into arcsine proportion values and implemented (i.e. dose and species). To calculate means (P <0.05), the Tukey's HSD was performed.

7.3   RESULTS AND DISCUSSION

7.3.1   GC-MS EXPERIMENTATION OF *C. CYMINUM* SEED EXTRACT

Methanolic seed extracts of *C. cyminum* were identified as containing 19 distinct volatile compounds by GC-MS chromatogram (Figure 7.1 and Table 7.1). The main components were found to be Butylbenzene, P-Cymene, α-Terpinene, Safranal, Cuminal Cumaldehyde, Styrene Glycol, trans-β-Farnesene, glycosyl compounds, methyl palmitate, Elaidic acid, and margaric acid. Ravi et al. (2013) reported the oil of cumin contained several prominent volatile compounds, with terpinene at 26.36 and 27.73%; samples were collected from different agro-climatic regions of India. Romeilah et al. (2010) isolated 20 compounds and a further GC-MS study conducted by Ramya et al. (2022) reported the presence of 4-isopropylbenzaldehyde as an important compound.

7.3.2   WATER PURIFICATION PROPERTIES OF SEED EXTRACT AT CLEAN/DRINKING, STAGNANT, AND SEWAGE WATER

Water purification properties of *C. cyminum* seed extract were studied in three different ecosystems using five parameters: pH, turbidity, temperature, salinity, and dissolved oxygen (Table 7.2). All five were reduced significantly after treatment with *C. cyminum* seed extract: for drinking/clean water it was 7.2 ± 1.41, 22.4 ± 1.47, 26.6 ± 1.92, 0.6 ± 2.15, and 7.4 ± 2.46; for stagnant water 7.8 ± 1.92, 29.4 ± 1.64, 19.2 ± 1.70, 0.8 ± 1.49, and 8.2 ± 1.51; for sewage water 8.8 ± 1.53, 37.4 ± 1.24, 29.2 ± 1.72, 1.6 ± 2.86, and 4.4 ± 1.68. In Asia and other regions, the experimental plant extract has been used in community medicine and it cures many illnesses. When it comes to their ability to purify water, *C. cyminum* were successful, producing clean water that may be utilized for a variety of things outside drinking. Dishna Schwarz (2000) claimed that vetiveria and *M. oleifera* seeds might be utilized as a natural water treatment component for tap water and also as a public

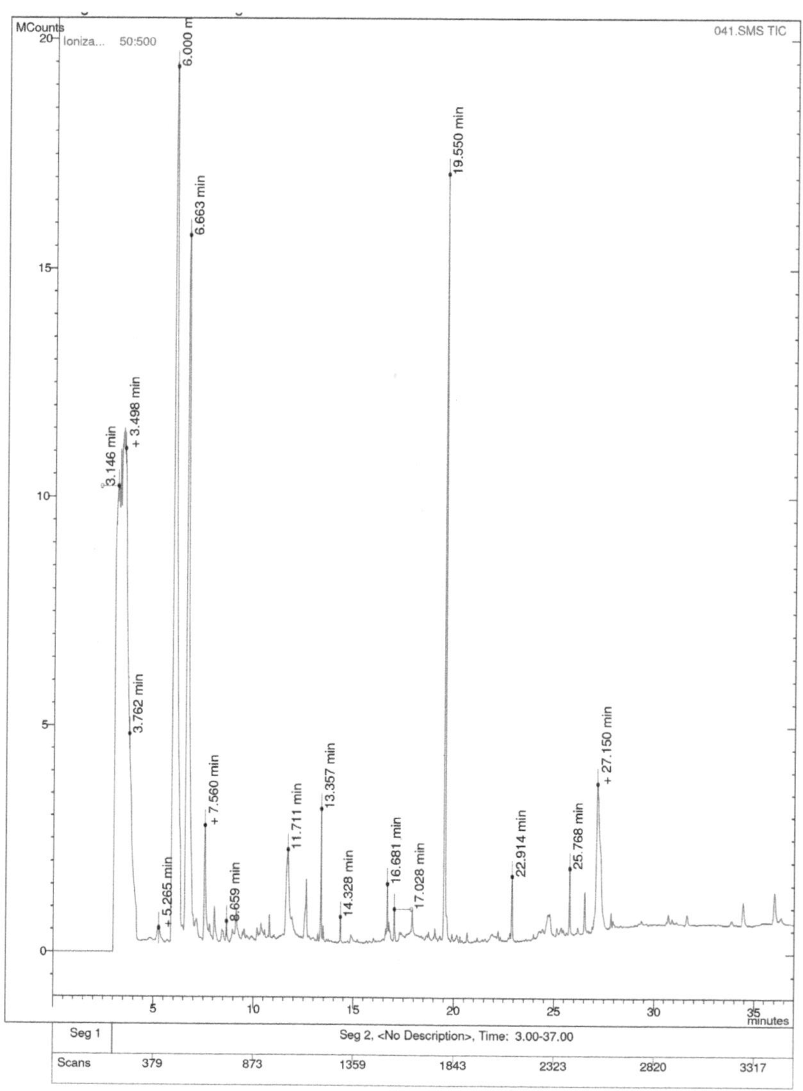

FIGURE 7.1   GC-MS mass spectrum of *C. cyminum* seed extract.

water treatment practice. Further, Folkard et al. (1992) used laboratory tests to discover that six distinct Moringa species have natural coagulant capabilities. Additionally, *C. dactylon* had the best ability to purify water, followed by vetiveria and *Coleus amboinicus*. Water quality metrics like color, pH, and turbidity were altered after treatment with plant extracts, turning the water samples into potable water Arjunan et al. (2012). Pritchard et al. (2009) reported the water purification and related parameters. The findings demonstrated that the quality of fluid can be significantly enhanced by the addition of drum stick pod powder and *Jatropha curcas* and gum of guar plant (*Cyamopsis tetragonoloba*). Efficiency of 90% was noted for all the plant extracts in the well water ecosystem. All extracts resulted in a coliform decrease of roughly 80%. At the prescribed dose level, the level of the water's pH was initially a little higher and remained stable and was suitable as drinking water. These provide compelling proof that phytoconstituents derived from plants have the ability to naturally purify water.

TABLE 7.1

GC-MS Analysis of *C. cyminum* Seed Extract

S. No	Phytochemical Compound	Retention Time	Percentage
1.	tert-Butylbenzene	3.146	8.333
2.	4-Isopropyltoluene	3.196	4.514
3.	1,3,8-p-Menthatriene	3.263	4. 338
4.	1,4-Cyclohexadiene, 1-methyl-4-(1-methylethyl)-	3.435	11.995
5.	1,3-Cyclohexadiene-1-methanol, 4-(1-methyl)	5.265	0.349
6.	Benzaldehyde, 4-(1-methylethyl)-	6.000	26.291
7.	1,2-Ethanediol, 1-phenyl-	6.662	15.368
8.	Cyclohexene, 2-ethenyl-1,3,3-trimet	6.961	0.009
9.	Phthalic acid, 4-bromophenyl methyl ester	7.558	1.335
10.	Benzeneacetic acid, .alpha.-methoxy	8.057	0.390
11.	1,6,10-Dodecatriene, 7,11-dimethyl-	8.659	0.157
12.	Methyl(methyl 4-O-methyl-.alpha.-d-mannopyranoside)uronate	11.711	1.481
13.	4-(3,3-Dimethyl-but-1-ynyl)-4-hydro	13.357	0.733
14.	Hexadecanoic acid, methyl ester	14.328	0.123
15.	9-Octadecenoic acid, methyl ester	16.681	0.264
16.	Heptadecanoic acid, 16-methyl-, methyl	17.028	0.213
17.	4,4,5',5'-Tetramethyl-bicyclohexyl-6-ene-2,3'-dione	19.550	6.135
18.	3-Methyl-but-2-enoic acid, 2,2-dime	26.531	0.315
19.	12-Oleanen-3-yl acetate, (3.alpha.)-	27.146	1.432

TABLE 7.2

Water Purification Properties of *C. cyminum* Seed Extract in Three Different Mosquito-breeding Water Ecosystems

Parameter	Drinking Water		Stagnant Water		Sewage Water	
	Before Treatment	After Treatment	Before Treatment	After Treatment	Before Treatment	After Treatment
pH	8.4 ±2.79	7.2±1.41	9.4±2.63	7.8± 1.92	11.6 ± 1.87	8.8±1.53
Turbidity(mg/L)	32.8±1.56	22.4±1.47	48.6±1.51	29.4±1.64	56.6±1.36	37.4± 1.24
Temperature	24.6± 2.51	26.6± 1.92	18.6 ± 2.64	19.2±1.70	28.2± 2.65	29.2± 1.72
Salinity (ppt)	1.6 ± 2.30	0.6 ± 2.15	1.8 ± 1.65	0.8 ± 1.49	4.8 ± 1.74	1.6 ±2.86
Dissolved Oxygen	6.2 ± 1.58	7.4 ± 2.46	7.6 ± 2.67	8.2 ± 1.51	2.4 ± 2.58	4.4 ± 1.68

7.3.3 Larval/pupal Toxicity Assays under Laboratory Condition against Mosquito Vectors

Under laboratory circumstances, *C. cyminum* seed extract showed toxicity against young larvae of mosquito vectors. In earlier research on other plant extracts, a dose-dependent impact was discovered (Panneerselvam et al. 2013a, b; Benelli et al. 2015; Roni et al. 2015; Suresh et al., 2015). Lethal concentration$_{50}$ (LC_{90}) values of *C. cyminum* seed extract against *An. stephensi* were 212.095 (407.804) ppm (larva I), 243.452 (473.035) ppm (II), 277.607 (561.187) ppm (III), 320.279 (652.175) ppm (IV), and 355.449 (712.117) ppm (pupa) (Table 7.3). LC_{50} (LC_{90}) values of *C. cyminum* seed extract against *Ae. aegypti* were 215.943 (409.495) ppm (larva I), 242.896 (496.171) ppm (II), 280.396

TABLE 7.3

Larval/pupal Toxicity Assays under Laboratory Condition on *An. stephensi* after the Treatment of *C. cyminum* Seed Extract

		95% Confidence Limit LC_{50} (LC_{90})			
Target Instars	LC_{50} (LC_{90})	**LCL**	**UCL**	**Regression Equation**	x^2 (d.f.=4)
1st Instar	212.095 (407.804)	189.438 (378.799)	232.365 (445.749)	$y = -1.389+0.007x$	4.804 n.s
2nd Instar	243.452 (473.035)	219.220 (437.358)	265.668 (520.921)	$y = -1.359+ 0.006x$	2.150 n.s
3rd Instar	277.607 (561.187)	250.139 (511.146)	303.837 (632.582)	$y = -1.255+0.005x$	0.523 n.s
4th Instar	320.279 (652.175)	290.101 (583.552)	352.124 (756.468)	$y = -1.237+0.004x$	0.090 n.s
Pupa	355.449 (712.117)	323.088 (629.993)	393.192 (841.668)	$y = -1.277+0.004x$	0.060 n.s

Notes:

LC_{50} = lethal concentration that kills 50% of the exposed organisms

LC_{90} = lethal concentration that kills 90% of the exposed organisms

LCL = lower confidence limit

UCL = upper confidence limit

χ^2 = chi-square; n.s. = not significant (α = 0.05).

TABLE 7.4

Larval/pupal Toxicity Assays under Laboratory Condition on *Ae.aegypti* after the Treatment of *C. cyminum* Seed Extract

		95% Confidence Limit LC_{50} (LC_{90})			
Target Instars	LC_{50} (LC_{90})	**LCL**	**UCL**	**Regression Equation**	x^2 (d.f.=4)
1st Instar	215.943 (409.495)	193.736 (380.741)	235.943 (447.012)	$y = -1.430+0.007x$	4.932 n.s
2nd Instar	242.896 (496.171)	216.114 (456.117)	267.019 (551.181)	$y = -1.229 + 0.006x$	0.611 n.s
3rd Instar	280.396 (575.365)	252.002 (522.177)	307.562 (652.293)	$y = -1.218+0.004x$	0.888 n.s
4th Instar	331.572 (667.264)	301.181 (595.870)	364.584 (776.543)	$y = -1.266+0.004x$	1.191 n.s
Pupa	377.816 (748.260)	343.480 (657.376)	420.446 (894.902)	$y = -1.307+0.003x$	0.160 n.s

Notes:

LC_{50} = lethal concentration that kills 50% of the exposed organisms

LC_{90} = lethal concentration that kills 90% of the exposed organisms

LCL = lower confidence limit

UCL = upper confidence limit

χ^2 = chi-square; n.s. = not significant (α = 0.05).

(575.365) ppm (III), 331.572 (667.264) ppm (IV), and 377.816 (748.260) ppm (pupa) (Table 7.4). LC_{50} (LC_{90}) values of *C. cyminum* seed extract against *C. quinquefasciatus* were 377.816 (748.260) ppm (larva I), 269.084 (520.277) ppm (II), 305.317 (584.414) ppm (III), 355.598 (687.530) ppm (IV), and 390.471 (728.195) ppm (pupae) (Table 7.5). There was a presence of volatile compounds, which affect the central region of the digestive system of mosquito larvae, and concomitant toxicity, which prevents the functioning of the respiration organ of mosquitoes (David et al. 2000). Similar studies were conducted with methanolic extracts of more than 64 botanicals from the Colombian Caribbean region for the larval toxicity against dengue vector and they have attempted to test the other predatory organisms in the stagnant water system. Again, the lethal effects of ethanolic extracts

TABLE 7.5

Larval/pupal Toxicity Assays under Laboratory Condition on *Culex quinquefasciatus* after the Treatment of *C. cyminum* Seed Extract

| Target Instars | LC_{50} (LC_{90}) | 95% Confidence limit LC_{50} (LC_{90}) | | Regression Equation | x^2 (d.f.=4) |
		LCL	UCL		
1st Instar	377.816 (748.260)	343.480 (657.376)	420.446 (894.902)	y = -1.454+0.006x	1.348 n.s
2nd Instar	269.084 (520.277)	244.068 (478.790)	292.729 (577.244)	y = -1.373 + 0.006x	0.281 n.s
3rd Instar	305.317 (584.414)	279.185 (532.959)	331.703 (657.692)	y = -1.402+0.005x	0.614 n.s
4th Instar	355.598 (687.530)	325.249 (613.790)	390.514 (800.521)	y = -1.373+0.004x	0.984 n.s
Pupa	390.471 (728.195)	358.131 (855.240)	430.843 (855.240)	y = -1.482+0.004x	0.370 n.s

Notes:

LC_{50} = lethal concentration that kills 50% of the exposed organisms

LC_{90} = lethal concentration that kills 90% of the exposed organisms

LCL = lower confidence limit

UCL = upper confidence limit

χ^2 = chi-square; n.s. = not significant (α = 0.05).

TABLE 7.6

Ovicidal Toxicity Assays under Laboratory Condition on *Anopheles stephensi, Aedes aegypti* and *Culex quinquefasciatus* after the Treatment of *C. cyminum* Seed Extract

| Target | Percentage of Egg Hatchability Concentrations (ppm) | | | | | |
	Control	100	200	300	400	500
An. stephensi	99.8 ± 0.3	54.2 ± 1.64	41.2 ± 1.30	18.0 ± 1.87	6.4 ± 1.14	NH
Ae. aegypti	98.2 ± 0.2	58.2 ± 1.48	46.6 ± 1.34	21.8 ± 0.83	10.4 ± 1.81	NH
Cx. quinquefasciatus	100.0 ± 0	60.2 ± 1.78	47.4 ± 1.51	24.6 ± 1.14	15.2 ± 1.92	NH

Notes:

Values were means±SD of five replicates.

NH: no hatchability (100% mortality).

of 16 plants have been studied with young larval instars of mosquitoes (Singha and Chandra, 2011; Oliveros-Díaz et al., 2022).

7.3.4 Ovicidal Activity against Three Species of Mosquitoes

Considerable ovicidal effect was found to exist after the treatment of *C. cyminum* seed extract against eggs of different mosquito vectors; moreover, egg hatchability was severely affected at the higher dose level and 100% egg hatchability was noted with 500 ppm (Table 7.6). So far, not much research has been done to clarify plant extracts' ovicidal characteristics. A. Reegan et al. (2013) reported on similar findings regarding the negative effects of *Cliona spp.* extracts on *filarial vector* and *Aedes* eggs. The eggs of both species were discovered to be extremely vulnerable to methanol extract. Methanol extract demonstrated 100% ovicidal efficacy at 500 ppm in *C. quinquefasciatus* and 72% in *Ae. aegypti* after 120 hours post-treatment. Four herbal plants were recently studied using ethanolic leaf extracts to demonstrate the ovicidal activity of all four plants. At greater doses, all plant extracts caused 70–90% mortality (Torawane et al., 2021).

7.3.5 Antibacterial Activity of *C. cyminum* Seed Extract

Findings of agar well diffusion tests showed that the *C. cyminum* seed extract disc had a clear zone surrounding it, indicating that it had antibacterial activity that could prevent the growth of bacterial populations. The zone of inhibition is presented in Figure 7.2. The exponential growth curves of

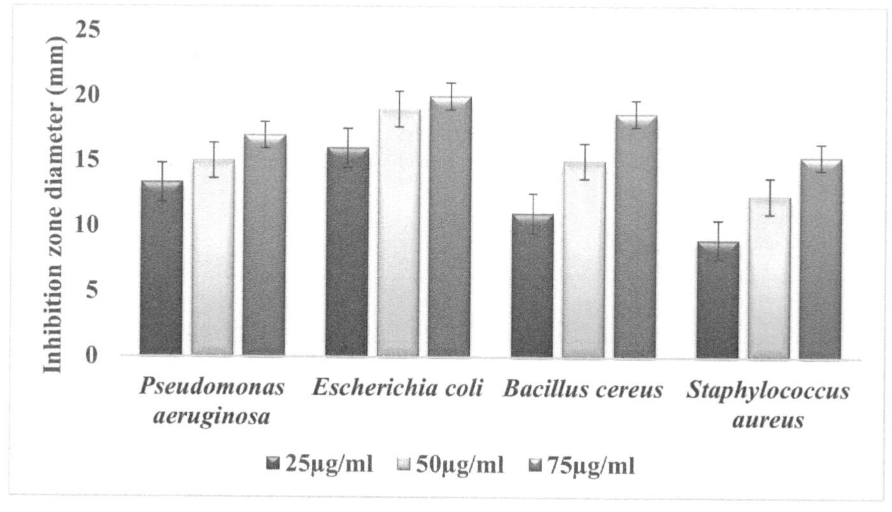

FIGURE 7.2   Zone of inhibition by *C. cyminum* seed extract in gram-positive and gram-negative strains.

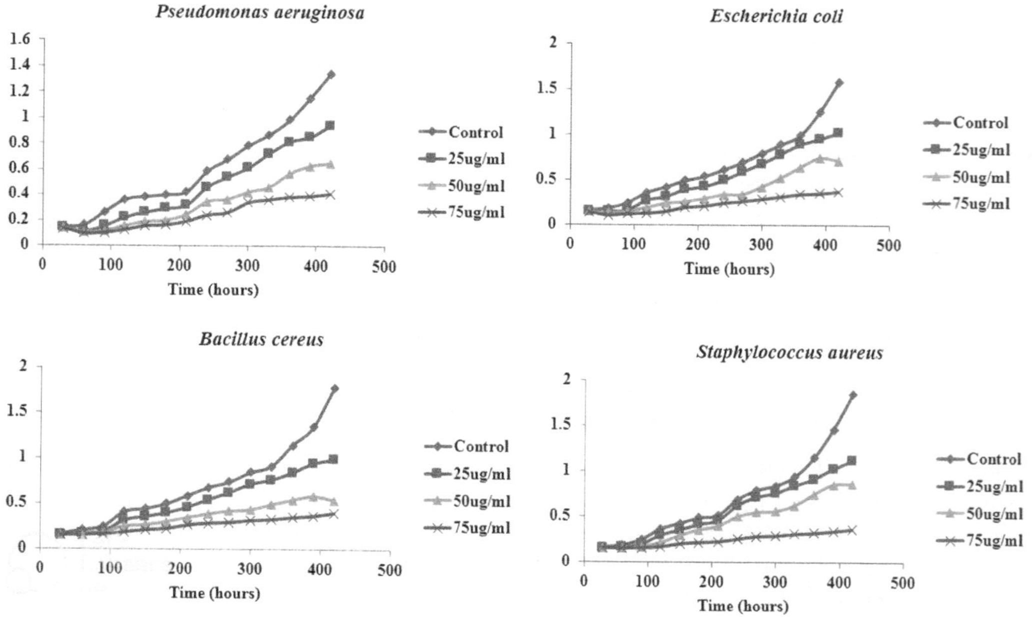

FIGURE 7.3   Growth curve analysis of gram-positive and gram-negative strains.

the strains exposed to various concentrations of *C. cyminum* seed extract (25, 50, and 75 µg/mL) include chosen bacterial strains. The antibacterial activity of *C. cyminum* seed extract or its bacterial inhibition behavior was convincingly demonstrated by the growth curve data (Figure 7.3). At a higher dose level of *C. cyminum* seed extract, the growth curves of all four distinct pathogens showed improved activity. The report of earlier work by Almaary et al. (2021), with marine sea grass, obtained comparable findings. Sakee et al. (2011) revealed that *B. balsamifera* hexane extract at 19.2 mg/mL provided the highest growth inhibition in a dose-dependent manner. Very recently Yogeswara et al. (2022) studied how using *Blumea balsamifera* extract treatment considerably suppressed three bacterial strains, such as B.cereus.

7.4 CONCLUSION

Development of drug resistance in mosquitoes and bacteria is a significant management challenge. The GC-MS analysis in this work has supported the chemical compounds that were used to identify the phytochemical compounds in the *C. cyminum* seed extract. Methanolic extracts of *C. cyminum* seeds contain many active compounds with different compositions, giving them very interesting mosquitocidal properties that have multiple actions rather than a single molecule as with that of chemical insecticides. Therefore, it can be said that these *C. cyminum* seed extracts are the most efficient at keeping the targeted mosquitoes under control. Additionally, the *C. cyminum* seed aqueous methanol extract was efficient against the bacterial populations. Current research points to sequential extraction as a potential improvement in extracting phytochemicals with the highest bioactivity. When utilized in three different mosquito breeding water ecosystems, *C. cyminum* seed extract significantly reduced and improved the temperature, pH, dissolved oxygen, and turbidity. In this study, *C. cyminum* seed extract was chosen to focus more on a green and eco-friendly method that prevents environmental contamination or pollution.

CONFLICT OF INTEREST

The authors declare no conflict.

REFERENCES

Almaary, Khalid S., Naiyf S. Alharbi, Shine Kadaikunnan, Jamal M. Khaled, Govindan Rajivgandhi, Govindan Ramachandran, Chelliah Chenthis Kanisha, Manavalan Murugan, Khalid F. Alanzi, and Natesan Manoharan. "Anti-bacterial effect of marine sea grasses mediated endophytic actinomycetes against K. pneumoniae." *Journal of King Saud University-Science* 33, no. 6 (2021): 101528.

Al-Snafi, Ali Esmail. "The pharmacological activities of Cuminum cyminum-A review." *IOSR Journal of Pharmacy* 6, no. 6 (2016): 46–65.

Anoopkumar, A. N., and Embalil Mathachan Aneesh. "A critical assessment of mosquito control and the influence of climate change on mosquito-borne disease epidemics." *Environment, Development and Sustainability* 24, no. 6 (2022): 8900–8929.

Arjunan, Nareshkumar, Kadarkarai Murugan, Pari Madhiyazhagan, Kalimuthu Kovendan, Kanagarajan Prasannakumar, Sundaram Thangamani, and Donald R. Barnard. "Mosquitocidal and water purification properties of Cynodon dactylon, Aloe vera. Hemidesmus indicus and Coleus amboinicus leaf extracts against the mosquito vectors." *Parasitology Research* 110, no. 4 (2012): 1435–1443.

Barakat, M. A. "New trends in removing heavy metals from industrial wastewater." *Arabian Journal of Chemistry* 4, no. 4 (2011): 361–377.

Bauer, A. W. "Antibiotic susceptibility testing by a standardized single disc method." *American Journal of Clinical Pathology* 45 (1966): 149–158.

Benelli, Giovanni, Kadarkarai Murugan, Chellasamy Panneerselvam, Pari Madhiyazhagan, Barbara Conti, and Marcello Nicoletti. "Old ingredients for a new recipe? Neem cake, a low-cost botanical by-product in the fight against mosquito-borne diseases." *Parasitology Research* 114, no. 2 (2015): 391–397.

Benelli, Giovanni, Roman Pavela, Filippo Maggi, Riccardo Petrelli, and Marcello Nicoletti. "Commentary: Making green pesticides greener? The potential of plant products for nanosynthesis and pest control." *Journal of Cluster Science* 28, no. 1 (2017): 3–10.

Bolisetty, Sreenath, Mohammad Peydayesh, and Raffaele Mezzenga. "Sustainable technologies for water purification from heavy metals: Review and analysis." *Chemical Society Reviews* 48, no. 2 (2019): 463–487.

Bustamante, Jaleesa. "Mosquito deaths & Mosquito borne disease statistics [2020]." *MosquitoReviews*, 18 Dec. 2019, mosquitoreviews.com/learn/disease-death-statistics.

Castillo, Ruth M., Elena Stashenko, and Jonny E. Duque. "Insecticidal and repellent activity of several plant-derived essential oils against Aedes aegypti." *Journal of the American Mosquito Control Association* 33, no. 1 (2017): 25–35.

Chellappandian, Muthiah, Prabhakaran Vasantha-Srinivasan, Sengottayan Senthil-Nathan, Sengodan Karthi, Annamalai Thanigaivel, Athirstam Ponsankar, Kandaswamy Kalaivani, and Wayne B. Hunter. "Botanical essential oils and uses as mosquitocides and repellents against dengue." *Environment International* 113 (2018): 214–230.

Chenniappan, Kuppusamy, and Murugan Kadarkarai. "Oviposition deterrent, ovicidal and gravid mortality effects of ethanolic extract of Andrographis paniculata Nees against the malarial vector Anopheles stephensi Liston (Diptera: Culicidae)." *Entomological Research* 38, no. 2 (2008): 119–125.

David, Jean-Philippe, Delphine Rey, Marie-Paule Pautou, and Jean-Claude Meyran. "Differential toxicity of leaf litter to dipteran larvae of mosquito developmental sites." *Journal of Invertebrate Pathology* 75, no. 1 (2000): 9–18.

Demok, Samuel, Nancy Endersby-Harshman, Rebecca Vinit, Lincoln Timinao, Leanne J. Robinson, Melinda Susapu, Leo Makita, Moses Laman, Ary Hoffmann, and Stephan Karl. "Insecticide resistance status of Aedes aegypti and Aedes albopictus mosquitoes in Papua New Guinea." *Parasites & Vectors* 12, no. 1 (2019): 1–8.

Finney, D. J. "Statistical logic in the monitoring of reactions to therapeutic drugs." *Methods of Information in Medicine* 10, no. 04 (1971): 237–245.

Folkard, Geoff K., John P. Sutherland, and W. D. Grant. "Natural coagulants at pilot scale." (1992). *Water, environment and management. Proc. of the 18th WEDC Conference, Kathmandu*, Nepal, 30 Aug.–3 Sept. 1992. Loughborough University Press, pp 51–54

Fu, Fenglian, and Qi Wang. "Removal of heavy metal ions from wastewaters: A review." *Journal of Environmental Management* 92, no. 3 (2011): 407–418.

Gohari, Ahmad Reza, and Soodabeh Saeidnia. "A review on phytochemistry of Cuminum cyminum seeds and its standards from field to market." *Pharmacognosy Journal* 3, no. 25 (2011): 1–5.

Griffin, Jamie T., Samir Bhatt, Marianne E. Sinka, Peter W. Gething, Michael Lynch, Edith Patouillard, Erin Shutes et al. "Potential for reduction of burden and local elimination of malaria by reducing Plasmodium falciparum malaria transmission: A mathematical modelling study." *The Lancet Infectious Diseases* 16, no. 4 (2016): 465–472.

Hamama, Heba M., Ola H. Zyaan, Ola A. Abu Ali, Dalia I. Saleh, Hend A. Elakkad, Mohamed T. El-Saadony, and Shaimaa M. Farag. "Virulence of entomopathogenic fungi against Culex pipiens: Impact on biomolecules availability and life table parameters." *Saudi Journal of Biological Sciences* 29, no. 1 (2022): 385–393.

Hemingway, Janet, Hilary Ranson, Alan Magill, Jan Kolaczinski, Christen Fornadel, John Gimnig, Maureen Coetzee et al. "Averting a malaria disaster: Will insecticide resistance derail malaria control?." *The Lancet* 387, no. 10029 (2016): 1785–1788.

Iwamura, Takuya, Adriana Guzman-Holst, and Kris A. Murray. "Accelerating invasion potential of disease vector Aedes aegypti under climate change." *Nature Communications* 11, no. 1 (2020): 1–10.

Kordali, S., Kotan, R., Mavi, A., Cakir, A., Ala, A. and Yildirim, A., 2005. "Determination of the chemical composition and antioxidant activity of the essential oil of Artemisia dracunculus and of the antifungal and antibacterial activities of Turkish Artemisia absinthium, A. dracunculus, Artemisia santonicum, and Artemisia spicigera essential oils." *Journal of Agricultural and Food Chemistry*, 53(24), pp.9452–9458.

Kovendan, Kalimuthu, Kadarkarai Murugan, Savariar Vincent, and Donald R. Barnard. "Studies on larvicidal and pupicidal activity of Leucas aspera Willd.(Lamiaceae) and bacterial insecticide, Bacillus sphaericus, against malarial vector, Anopheles stephensi Liston.(Diptera: Culicidae)." *Parasitology Research* 110, no. 1 (2012): 195–203.

Kumar, C. P. "Water resources issues and management in India." *Journal of Scientific and Engineering Research* 5, no. 9 (2018): 137–147.

Leta, Samson, Tariku Jibat Beyene, Eva M. De Clercq, Kebede Amenu, Moritz UG Kraemer, and Crawford W. Revie. "Global risk mapping for major diseases transmitted by Aedes aegypti and Aedes albopictus." *International Journal of Infectious Diseases* 67 (2018): 25–35.

Moradi, F., N. Hadi, and A. Bazargani. "Evaluation of quorum-sensing inhibitory effects of extracts of three traditional medicine plants with known antibacterial properties." *New Microbes and New Infections* 38 (2020): 100769.

Oliveros-Díaz, Andrés Felipe, Yina Pájaro-González, Julian Cabrera-Barraza, Catherine Hill, Wiston Quiñones-Fletcher, Jesús Olivero-Verbel, and Díaz-Castillo Fredyc. "Larvicidal activity of plant extracts from Colombian North Coast against Aedes aegypti L. mosquito larvae." *Arabian Journal of Chemistry* (2022): 104365.

Panneerselvam, Chellasamy, and Kadarkarai Murugan. "Adulticidal, repellent, and ovicidal properties of indigenous plant extracts against the malarial vector, Anopheles stephensi (Diptera: Culicidae)." *Parasitology Research* 112, no. 2 (2013a): 679–692.

Panneerselvam, C., K. Murugan, K. Kovendan, P. Mahesh Kumar, and J. Subramaniam. "Mosquito larvicidal and pupicidal activity of Euphorbia hirta Linn.(Family: Euphorbiaceae) and Bacillus sphaericus against Anopheles stephensi Liston.(Diptera: Culicidae)." *Asian Pacific Journal of Tropical Medicine* 6, no. 2 (2013b): 102–109.

Pratheeba, T., V. Taranath, DVR Sai Gopal, and D. Natarajan. "Antidengue potential of leaf extracts of Pavetta tomentosa and Tarenna asiatica (Rubiaceae) against dengue virus and its vector Aedes aegypti (Diptera: Culicidae)." *Heliyon* 5, no. 11 (2019): e02732.

Pritchard, M., T. Mkandawire, A. Edmondson, J. G. O'neill, and G. Kululanga. "Potential of using plant extracts for purification of shallow well water in Malawi." *Physics and Chemistry of the Earth, Parts A/B/C* 34, no. 13–16 (2009): 799–805.

Rahman, Mohammad Shadiqur, SM Majharul Islam, Anamul Haque, and Md Shahjahan. "Toxicity of the organophosphate insecticide sumithion to embryo and larvae of zebrafish." *Toxicology Reports* 7 (2020): 317–323.

Rajaganesh, Rajapandian, Kadarkarai Murugan, Chellasamy Panneerselvam, Sudalaimani Jayashanthini, Mathath Roni, Udaiyan Suresh, Subrata Trivedi et al. "Fern-synthesized silver nanocrystals: Towards a new class of mosquito oviposition deterrents?." *Research in Veterinary Science* 109 (2016): 40–51.

Ramya, S., T. Loganathan, M. Chandran, R. Priyanka, K. Kavipriya, G. Grace Lydial Pushpalatha, Devaraj Aruna, L. Ramanathan, R. Jayakumararaj, and Vikrant Saluja. "Phytochemical Screening, GCMS, FTIR profile of Bioactive Natural Products in the methanolic extracts of Cuminum cyminum seeds and oil." *Journal of Drug Delivery and Therapeutics* 12, no. 2-S (2022): 110–118.

Ravi, Ramasamy, Maya Prakash, and Kodangala Keshava Bhat. "Characterization of aroma active compounds of cumin (Cuminum cyminum L.) by GC-MS, e-Nose, and sensory techniques." *International Journal of Food Properties* 16, no. 5 (2013): 1048–1058.

Reegan, Appadurai Daniel, Arokia Valan Kinsalin, Michael Gabriel Paulraj, and Savarimuthu Ignacimuthu. "Larvicidal, ovicidal, and repellent activities of marine sponge Cliona celata (Grant) extracts against Culex quinquefasciatus Say and Aedes aegypti L.(Diptera: Culicidae)." *International Scholarly Research Notices* 2013 (2013).

Romeilah, Ramy M., Sayed A. Fayed, and Ghada I. Mahmoud. "Chemical compositions, antiviral and anti-oxidant activities of seven essential oils." *Journal of Applied Sciences Research* 6, no. 1 (2010): 50–62.

Roni, Mathath, Kadarkarai Murugan, Chellasamy Panneerselvam, Jayapal Subramaniam, Marcello Nicoletti, Pari Madhiyazhagan, Devakumar Dinesh et al. "Characterization and biotoxicity of Hypnea musciformis-synthesized silver nanoparticles as potential eco-friendly control tool against Aedes aegypti and Plutella xylostella." *Ecotoxicology and Environmental Safety* 121 (2015): 31–38.

Sakee, Uthai, Sujira Maneerat, TP Tim Cushnie, and Wanchai De-Eknamkul. "Antimicrobial activity of Blumea balsamifera (Lin.) DC. extracts and essential oil." *Natural Product Research* 25, no. 19 (2011): 1849–1856.

Schwarz, D. "Water clarification using Moringa oleifera. Gate Technical Information." 67726 (2000): 1–7.

Shearer, Freya M., Joshua Longbottom, Annie J. Browne, David M. Pigott, Oliver J. Brady, Moritz UG Kraemer, Fatima Marinho et al. "Existing and potential infection risk zones of yellow fever worldwide: A modelling analysis." *The Lancet Global Health* 6, no. 3 (2018): e270–e278.

Singha, Someshwar, and Goutam Chandra. "Mosquito larvicidal activity of some common spices and vegetable waste on Culex quinquefasciatus and Anopheles stephensi." *Asian Pacific Journal of Tropical Medicine* 4, no. 4 (2011): 288–293.

Smith, Letícia B., Colin Sears, Haina Sun, Robert W. Mertz, Shinji Kasai, and Jeffrey G. Scott. "CYP-mediated resistance and cross-resistance to pyrethroids and organophosphates in Aedes aegypti in the presence and absence of kdr." *Pesticide Biochemistry and Physiology* 160 (2019): 119–126.

Soon, Wei Long, Mohammad Peydayesh, Raffaele Mezzenga, and Ali Miserez. "Plant-based amyloids from food waste for removal of heavy metals from contaminated water." *Chemical Engineering Journal* 445 (2022): 136513.

Su, Tianyun, and M. S. Mulla. "Ovicidal activity of neem products (azadirachtin) against Culex tarsalis and Culex quinquefasciatus (Diptera: Culicidae)." *Journal of the American Mosquito Control Association* 14, no. 2 (1998): 204–209.

Suresh, Udaiyan, Kadarkarai Murugan, Giovanni Benelli, Marcello Nicoletti, Donald R. Barnard, Chellasamy Panneerselvam, Palanisamy Mahesh Kumar, Jayapal Subramaniam, Devakumar Dinesh, and Balamurugan Chandramohan. "Tackling the growing threat of dengue: *Phyllanthus niruri*-mediated synthesis of silver nanoparticles and their mosquitocidal properties against the dengue vector Aedes aegypti (Diptera: Culicidae)." *Parasitology Research* 114, no. 4 (2015): 1551–1562.

Tjaden, Nils Benjamin, Cyril Caminade, Carl Beierkuhnlein, and Stephanie Margarete Thomas. "Mosquito-borne diseases: Advances in modelling climate-change impacts." *Trends in Parasitology* 34, no. 3 (2018): 227–245.

Torawane, Sarika, Ramnath Andhale, Radhakrishna Pandit, Digambar Mokat, and Samadhan Phuge. "Screening of some weed extracts for ovicidal and larvicidal activities against dengue vector Aedes aegypti." *The Journal of Basic and Applied Zoology* 82, no. 1 (2021): 1–9.

Uraki, Ryuta, Andrew K. Hastings, Doug E. Brackney, Philip M. Armstrong, and Erol Fikrig. "AgBR1 antibodies delay lethal Aedes aegypti-borne West Nile virus infection in mice." *NPJ Vaccines* 4, no. 1 (2019): 1–4.

WHO, 2020. Vectore borne diseases. Retrieved from www.who.int/news-room/fact-sheets/detail/vector-borne-diseases.

Yamashita, Mason M., Laura Wesson, George Eisenman, and David Eisenberg. "Where metal ions bind in proteins." *Proceedings of the National Academy of Sciences* 87, no. 15 (1990): 5648–5652.

Yogeswara, Ida Bagus Agung, I. Gusti Ayu Wita Kusumawati, and Ni Wayan Nursini. "Antibacterial activity and cytotoxicity of sequentially extracted medicinal plant Blumea balsamifera Lin. (DC)." *Biocatalysis and Agricultural Biotechnology* 43 (2022): 102395.

Zhang, Chao, Xiaohui Yi, Chen Chen, Di Tian, Hongbin Liu, Lingtian Xie, Xiuping Zhu, Mingzhi Huang, and Guang-Guo Ying. "Contamination of neonicotinoid insecticides in soil-water-sediment systems of the urban and rural areas in a rapidly developing region: Guangzhou, South China." *Environment International* 139 (2020): 105719.

8 Advanced Molecular and Metagenomic Approaches for Microbial Diversity Analysis in Wastewater Treatment Systems

Reshmi Sasi and T.V. Suchithra

8.1 INTRODUCTION

Wastewater treatment is a necessary process to recycle polluted water, which in turn helps to maintain a hygienic environment. Secondary treatment is considered one of the crucial steps in wastewater treatment. The application of microbial flora is an extensively studied approach in wastewater treatment processes. Hence, these waste-degrading communities of microorganisms are the core components of biological wastewater treatment plants (BWWTPs). Microbial diversity studies are intended to aid biological research by providing insight into the relationship between the structure and diversity of microbial communities and their actual function during wastewater processing (Jena et al. 2016). This chapter describes the most advanced molecular and metagenomic approaches that shed light on the phylogenetic characteristics and functional activities of microbial communities in BWWTPs.

8.2 CONVENTIONAL METHODS AT A GLANCE

Old methods used to identify the microflora in crude wastewater samples mainly involve isolating pure cultures and characterization at different levels. Standard conventional microbial diversity analysis methods in BWWTPs are community-level physiological profiling, plate counts, and fatty acid analysis (Edet et al. 2017).

8.2.1 COMMUNITY-LEVEL PHYSIOLOGICAL PROFILING (CLPP)

CLPP is also known as the sole carbon source utilization method. The basic principle behind this method is the ability of microorganisms to use different compounds as the sole carbon and energy source. CLPP is commonly used to detect microbial functional diversities.

8.2.2 PLATE COUNT METHOD

The plate count method allows the study of microbial diversity using appropriate media. It mainly involves serial dilution, plating on media, and obtaining the total viable or heterotrophic or aerobic

DOI: 10.1201/9781003408352-9

counts. This is followed by subculturing the isolates to get them purified. Once purified, they are subjected to identification and characterization experiments.

8.2.3 Fatty Acid Analysis

In fatty acid analysis, bacterial cell wall lipids are analyzed. The bacterial cell wall is made of proteins and lipids. It could be exploited for the identification of community-level diversity in bacteria. Fatty acids are more stable than nucleic acids, and those with two to twenty-four carbons (C2 – C24) are remarkably preserved in microbes. These conserved fatty acids are helpful in community diversity profiling.

All of these methods work mainly on the presumption that culture-dependent techniques aid in isolating most organisms present in the sample, leaving the fact that the extent of microbial diversity is much beyond expectations. This is one of the significant drawbacks of the conventional methods, making them inappropriate for diversity studies in natural and engineered ecosystems. Other major drawbacks of conventional methods are listed below (Carraro et al. 2011).

- Pure culture isolation of microbes from their natural habitats is often restricted by limited knowledge about their nutritional and physio-chemical needs
- The use of selective media and differential media lead to the oversight of other important microbes present in the sample
- It is tough to get the complete picture of an ecosystem's microbial diversity

8.3 MOLECULAR APPROACHES FOR WASTEWATER MICROBIOLOGY STUDIES

The invention of molecular techniques has revolutionized microbial diversity studies previously constrained by the inability to grow most of the prokaryotes in an ecosystem by culture-dependent techniques. Molecular techniques, especially nucleic acid amplification procedures (i.e., Polymerase Chain Reaction (PCR)), provide rapid and sensitive analytical tools for detecting microbial diversity. The conserved ribosomal rRNA genes often help to explain the phylogenetic relations among microbes. The 16s rRNA gene sequencing is one of the accurate and reliable methods for the genotypic identification of prokaryotes (Kumar et al. 2020; 2021). Amplified ribosomal DNA restriction analysis, 16s-restriction fragment length polymorphism, and ribosomal intergenic spacer analysis are some of the rRNA sequencing-based methods (Wagner and Loy 2002).

PCR-based methods helped uncover key players' diversity in biological wastewater treatment plants and how genome diversity is structured in connection with environmental parameters (Wells et al. 2011). However, difficulties with primers and PCR amplifications can limit the quantitative data obtained from these procedures. Perhaps the most critical aspect of these methodologies, which can directly impact the identification of a broad collection of functional genes, is primer designing. Approaches like fluorescent in situ hybridization, which uses target-specific oligonucleotide probes to eliminate PCR-related biases, have been used to identify the dominant group of bacteria engaged in different stages of waste removal during wastewater treatment (Joss et al. 2011; Natuscka et al. 1999).

Apart from the initial applications of molecular approaches, the commencement of the *omics era* has become a turning point in studying functional and phylogenetic diversity in wastewater treatment systems. It leads to the development of high-throughput sequencing technologies that allow the production of millions of copies of sequences at low cost and time. Whole genome sequencing of isolated pollutant-degrading microbes was the first attempt to explain the functional potential of microorganisms relevant to wastewater treatment. Similarly, metatranscriptomics and metagenomics have also been applied to uncover the microbes' genomic potential without the need for individual species isolation and PCR amplification (Yu and Zhang 2012). However, the information derived from these methodologies represents only the most abundant members of the community. In order to study the actual extent of microbial diversity and gene expression patterns in wastewater treatment plants, a considerable sequencing investment and a combination of metatranscriptomics and metagenomics are required, but such studies are rare (Yu and Zhang 2012). Great effort is needed to analyze the functional capabilities of microbes in BWWTPs. Also, it is essential to understand the gene expression pattern and the circumstances of gene expression, which will help to unveil the structural and functional relationship between these microbial-mediated processes. Some significant molecular techniques used in analyzing wastewater microbial diversity are discussed below.

8.3.1 NUCLEIC ACID HYBRIDIZATION TECHNIQUES FOR MICROBIAL DETECTION

Hybridization approaches are based on the interaction between labeled single-stranded nucleic acid molecules (probes) and their complementary targets in the sample DNA to evaluate the relative abundance of genes and their transcriptional products.

8.3.1.1 Microarray Analysis

The microarray technique follows complementarity, where nucleic acids bind only with complementary sequences. In this method, probes (short-length nucleic acid sequences) targeting functional or taxonomic genes are fixed on a chemically treated glass slide (Spotting). The samples were incubated with probes under conditions that favor hybridization. Since the probes are previously labeled with fluorescent materials or radioactive compounds, the emission and intensity of fluorescence or radiation indicate a specific group in the sample. This method allows the identification of a particular species from the rest of the microbes from wastewater samples. It can also be used to detect gene expression (whether the gene is turned on or off in the given sample). This is one of the best methods for characterizing uncultured microorganisms in WWTPs (Figure 8.1.A) (Sebat et al. 2003).

8.3.1.2 Fluorescent in situ Hybridization (FISH)

The FISH method uses fluorescent-labeled probe-based detection technology to enable phylogenetic identification and quantification of individual microbial cells. The methodology involves target-specific hybridization of the 16SrRNA gene with fluorescent-labeled probes. Initially, the test microbes are treated with ethanol (sample fixation), and by using fluorescent-labeled rRNA-targeted probes, the targeted rRNA genes are hybridized. Epifluorescence microscopy may be used to identify the probes attached to rRNA since they are labeled at the 5 end with a fluorescent dye (Figure 8.1.B). The strength of the signal is determined by the amount of cellular rRNA and the pace of growth. The metabolic dynamics of the community under investigation can be calculated using these variables. The following are some of the benefits of this method: (i) it is a quick method if the probes are available; (ii) it helps in the identification and differentiation of active microbes; (iii) professionals are not required to perform this; (iv) it avoids artifacts and bias caused by DNA extraction, PCR, and cloning (Andreas, L., and M. 1998).

8.3.1.3 Reverse Sample Genome Probing

This technique rapidly detects cultivable species in natural and engineered ecosystems, including WWTPs. This blotting-based technique involves the isolation of DNA from known individual strains, blotting the genetic material onto the nylon membrane, and hybridizing with unknown DNA extracted from environmental samples. It enables highly specific hybridization. DNA from environmental samples is isolated, radiolabeled, and hybridized with the DNA probes of known microbes blotted on a nylon membrane (Figure 8.2.A). This method is not applicable for identifying the

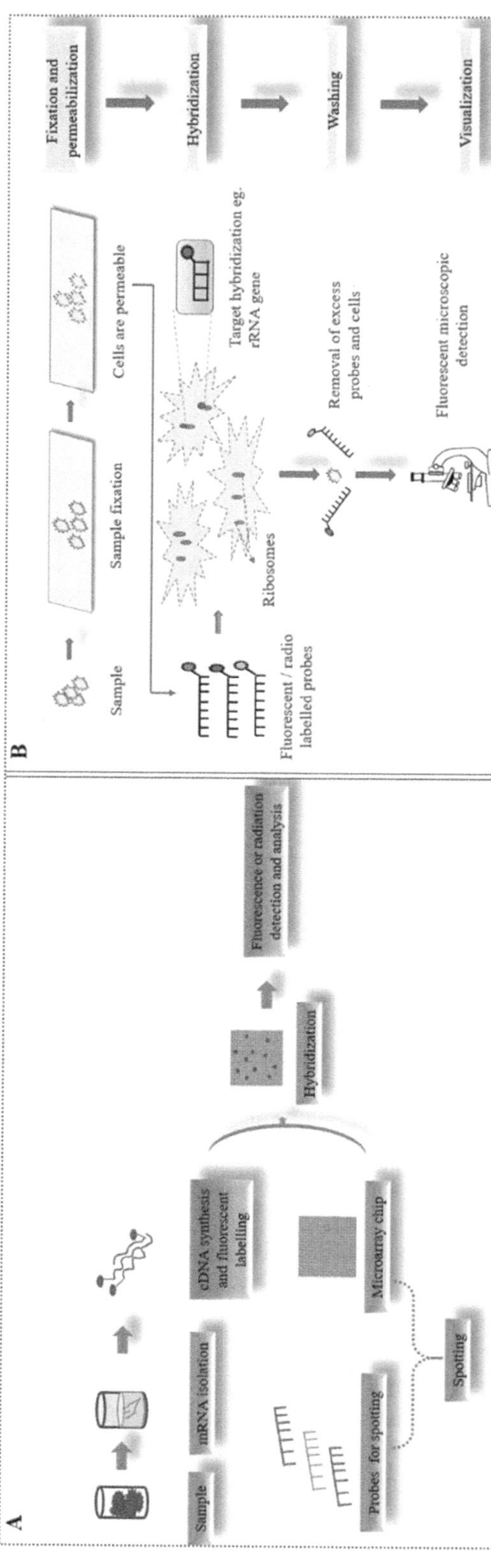

FIGURE 8.1 A. General procedure of microarray analysis showing sample preparation, spotting, and hybridization during microbial species identification. B. Stages of the fluorescent in situ hybridization procedure for species identification, including sample fixation and permeabilization, target DNA hybridization using fluorescent labeled probes, removal of excessive probes by washing, and visualization of hybridized targeted DNA fragments using fluorescent microscope.

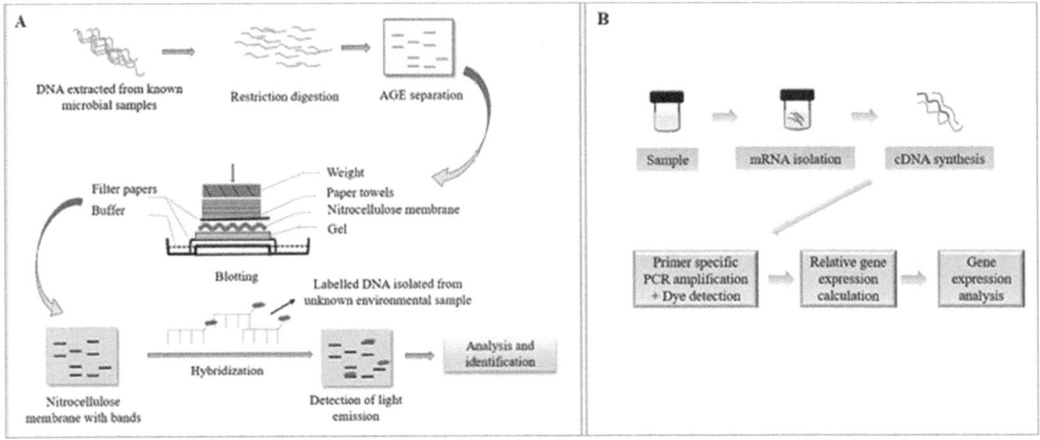

FIGURE 8.2 A. Identification of microbial species via reverse sampling genome probing. B. qRT-PCR methodology for microbial diversity analysis in wastewater treatment plants.

uncultured microbes in WWTPs or other ecosystems. It helps determine the microbial diversity of a selected environment by following the targeted cultivable community members. Reverse genome sampling is gaining more attention as one of the most promising methods for microbial diversity analysis. In 2003, Greene and Voordouw (Greene and Voordouw 2003) analyzed different environmental microbial communities by reverse sample genome probing, and they found that this method is more reliable than the 16SrRNA gene cloning method.

8.3.1.4 Quantitative Real-time PCR

The qRT-PCR method is a quantitative technique that provides better means for biodiversity analysis in different ecosystems, including WWTPs. It is a sensitive and dynamic method that accurately quantifies genes in wastewater samples. As the amplification progresses, it provides information on the real-time amplification and quantification of targeted genes. In the case of total wastewater bacteria estimation, specific primers can quantify the total number of 16S rRNA genes. In order to monitor the accumulation of PCR amplicons in real-time as the amplification progresses, the qRT-PCR employs either SYBR Green (intercalating fluorescent dye) or fluorescent probes (Figure 8.2.B). qRT-PCR was used to explore the dynamics and metabolic activities of the *Pseudomonas* population from pulp mill effluent microbial communities in a study by Muttray et al. (Muttray, Yu, and Mohn 2001).

8.4 DNA FINGERPRINTING OR COMMUNITY PROFILING

In the DNA fingerprinting approach, a region on the genome from all members of the wastewater sample is amplified using PCR and used for microbial identification. DNA fingerprints are the DNA profiles obtained using this approach.

8.4.1 Denaturing and Temperature Gradient Gel Electrophoresis (DGGE and TGGE)

DGGE follows the separation of DNA fragments based on their mobility under different denaturing conditions. Nucleic acid fragments are loaded into a denaturing gel incorporated with a denaturant (urea or formamide), resulting in the denaturation of the nucleic acids at various stages. This is followed by electrophoresis allowing the denatured fragments to move through the gel and are analyzed individually. Muyzer et al. (1993) developed this technique and applied it to microbial

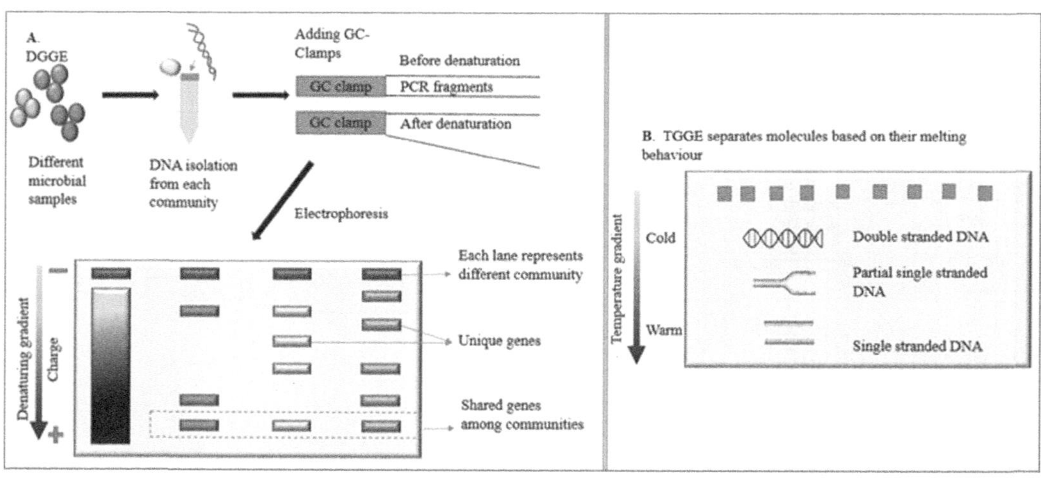

FIGURE 8.3   A. Step by step procedure of DGGE. B. Basic principle of TGGE.

ecology. This method is widely used for studying the composition and changes in the microbial population. Microbial communities can be differentiated by comparing the banding patterns obtained from various samples, and also it is possible to monitor the community dynamics in a sample (Mühling et al. 2008). The procedure involves recovering genetic materials from environmental samples, amplifying specific primers, and electrophoretic separation. It relies on the differences in GC pairing stability. G-C-rich DNA fragments are more stable and intact; until they reach a higher denaturant concentration and migrate faster relative to denatured single-stranded fragments. This difference in the GC pairing stability allows the separation of DNA fragments based on their mobility in an acrylamide gel. The gel is usually stained with DNA-binding fluorescent dyes (for better visualization) and is observed under UV light. Each band on the gel represents a single species; hence the total number of bands gives an idea about the species diversity. The gene fragments excised from the gel can be eluted and amplified for sequencing (Figure 8.3.A) (Duan et al. 2013).

TGGE is also a gradient-based gel electrophoresis method that uses a temperature gradient instead of a chemical gradient. Melting behavior is determined by the sequences of the different amplicons. A "GC clamp" is used during amplification to avoid the complete denaturation of DNA strands. For the phylogenetic studies, the gel must be excised, eluted, amplified, and sequenced (Figure 8.3.B).

Both of these techniques allow for handling many samples at a time and are reliable, reproducible, and fast. Both DGGE and TGGE are used to study mixed cultures. However, PCR bias, poor resolution of non-dominant species, and issues of co-migration are the common problems encountered during these procedures (Fakruddin 2015).

8.4.2 Amplified Ribosomal DNA Restriction Analysis (ARDRA)

The ARDRA technique is widely used to analyze microbial diversity in activated sludge (Cardinali-Rezende et al. 2012). It involves three significant steps: PCR amplification of 16s ribosomal gene, digestion using tetra cutter restriction endonucleases (Hae III and Alu I), and band separation by agarose gel electrophoresis (Figure 8.4.A). Whole community fingerprints can be generated via PCR amplification using universal primers, while species-specific primers are used for finding subgroups. The major challenge in this method is the generation and analysis of restriction profiles from multiple microbial communities using agarose gel electrophoresis (AGE)/polyacrylamide gel electrophoresis (PAGE), which is laborious and time-consuming.

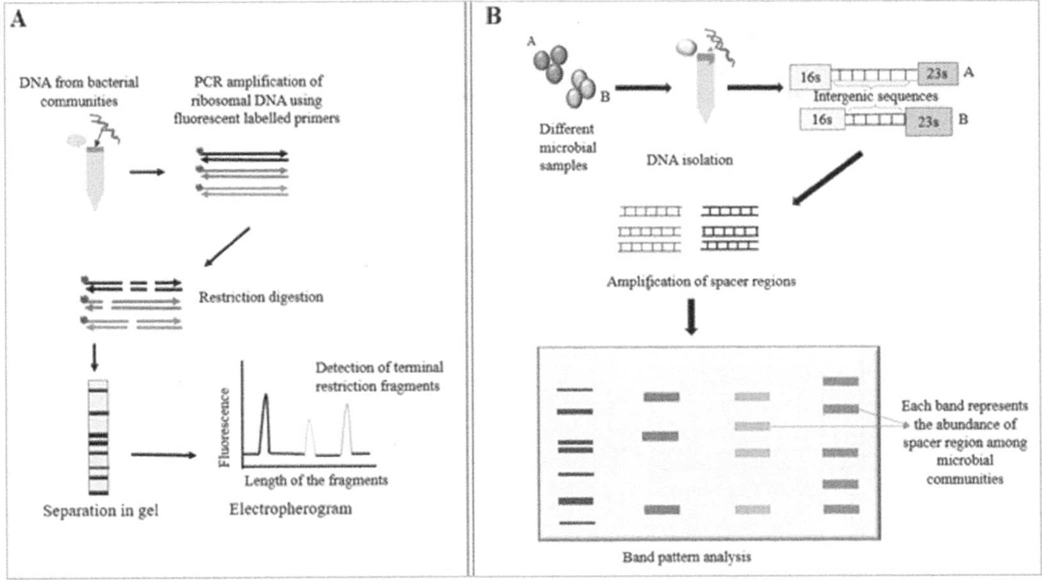

FIGURE 8.4 A. Stages involved in ARDRA analysis. B. Various stages of RISA for microbial community analysis.

8.4.3 RIBOSOMAL INTERGENIC SPACER ANALYSIS (RISA)

The RISA technique is considered one of the most potent molecular tools for microbial community analysis. It involves the amplification of the Intergenic Spacer Region (ISR- rRNA gene operon) between the large (23S) and small (16S) subunits of the 30s rRNA subunit followed by gel electrophoresis and band pattern analysis (Figure 8.4.B). The intergenic spacer region shows considerable heterogeneity in both nucleotide sequence and length. It provides a better tool for comparing environments. This method is more effective than 16S-RFLP and allows the comparison of different environmental samples without bias. It also helps to recognize variations between samples. It enables the generation of community-specific profiles where each band corresponds to at least one organism from the original microbial community (Ciesielski et al. 2013; Zhongtang and W. 2001).

8.4.4 TERMINAL RESTRICTION FRAGMENT LENGTH POLYMORPHISM (T-RFLP)

The T-RFLP technique is a PCR-based technique that allows the molecular profiling of microbial communities based on the restriction site of an amplified gene at its labeled end. The process involves restriction digestion of an amplified variant of a single gene using fluorescent-labeled specific primers with one or more restriction enconucleases followed by separating the fragments by capillary electrophoresis or PAGE coupled to a DNA sequencer (Figure 8.5). The size of various terminal fragments is measured using a fluorescence detector. As the amplicons are analyzed using a sequencer, only the terminal fragments are read, while all others are ignored. This helps to simplify the banding pattern and allows detailed analysis.

8.5 SEQUENCING-BASED TECHNIQUES

Environmental DNA research has been aided by recent advances in molecular biology and the availability of high-throughput sequencing tools (Garlapati et al. 2019). The know-how on microbial ecology is essential for developing and improving biological processes such as WT (Widder

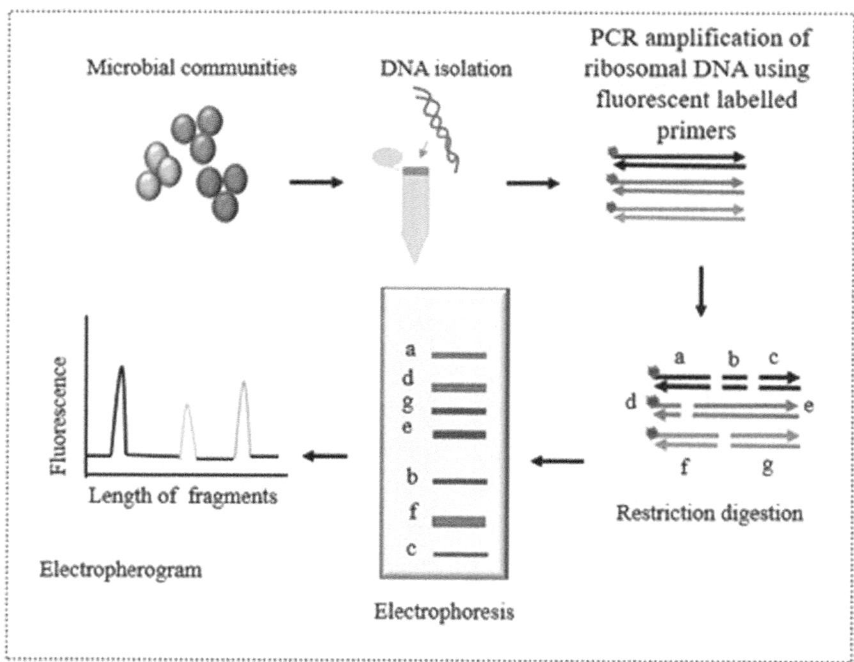

FIGURE 8.5   Stages involved in T-RFLP analysis.

et al. 2016). Sequencing-based techniques mainly allow the taxonomic identification of wastewater microorganisms (Urrea-valencia 2021).

8.5.1   16s rDNA Cloning and Sequencing

16S rRNA gene codes for the RNA component of the 30S ribosomal subunit in prokaryotes. Sequencing the 16S rRNA gene is one of the most widely used techniques in microbial ecology for identifying and characterizing prokaryotes. This technique is based on detecting sequence variation (polymorphisms) in the hypervariable regions of the 16S rRNA gene. This helps in species-level identification of bacteria and assists with differentiating between closely related bacterial species. The methodology involves nucleic acid extraction, amplification, cloning, and sequencing of 16s rRNA genes. Later, the sequence data is analyzed and identified with the help of phylogenetic tools (Figure 8.1). Genomic DNA extracted from microbial communities needs to be cloned before sequencing, but this could be avoided for amplicons generated from pure cultures (M. Kim and Chun 2014).

8.5.2   454 Pyrosequencing Method

The pyrosequencing technology is based on the sequencing-by-synthesis principle. Pyrosequencing relies upon detecting light emitted from a chain reaction caused by the release of pyrophosphate when nucleotide bases are attached to the template DNA. It involves binding DNA onto a bead with an adapter, followed by cloning. The DNA strands are then incubated with DNA polymerase, luciferase, ATP sulfurylase, and apyrase, where ammonium persulfate (APS) and luciferin act as substrates. This will initiate the reaction, and nucleotides will be added to the template DNA one by one. The addition of each nucleotide will cause the release of a pyrophosphate, which is then converted to ATP by the enzyme sulfurylase in the presence of APS. ATP associates with luciferase and converts luciferin to oxyluciferin, causing light generation. Apyrase enzyme degrades unused nucleotide and ATP, which initiates the subsequent reaction (Liao et al. 2013). A light detection system detects the emitted light and will be converted into a pyrogram proportional to the number of nucleotides incorporated. Later, a computer interprets the data to align smaller sequences into a complete genome sequence.

8.5.3 Massively Parallel Signature Sequencing (MPSS)

The MPSS technique is commonly used in gene expression studies by quantifying the mRNA level present in the samples. With the help of the reverse transcriptase enzyme, the mRNA isolated from the sample is converted to complementary DNA (cDNA). The cDNA is amplified using labeled primers, and the resulting amplicons are coupled to microbeads. This is followed by sequencing using fluorescently labeled probes. A fluorescent imaging system captures the signal produced by each bead, and on average, sequencing information of 1620 bp (base pair) can be determined from each bead (Read et al. 2011; Tan et al. 2015).

8.5.4 Illumine Sequencing

The Illumina technology is also based on the sequencing-by-synthesis principle. For sequencing millions of clusters of DNA on the flow surface, all four fluorescent-labeled nucleotides and DNA polymerases are progressively introduced to the flow cell channels. The wastewater sample's DNA is fragmented (or PCR amplifies the 16s rRNA gene) and joined to the adapters' ends. In order to link DNA strands and produce bridges between double-stranded DNA (dsDNA), unlabeled nucleotides and DNA polymerase are added. dsDNA is denatured into single-stranded DNA by heating. In each flow channel, the denaturation stage produces millions of condensed DNA clusters. The emitted fluorescence from each cluster is collected using laser excitation, and the bases are identified. It is a highly accurate method, and the sequencing is done by base analysis. This technique was adopted to analyze microbial diversity studies in many types of research, such as microbial diversity studies in wastewater from the leather industry (Wang et al. 2014).

8.5.5 Iron Torrent Sequencing

The Ion torrent method also follows the sequencing-by-synthesis principle, similar to pyrosequencing. Rather than fluorescence, it monitors H^+ ion release during the addition of bases in the newly synthesized strand. The first stage in the Ion torrent method involves target amplification or DNA splitting, followed by binding single-stranded DNA fragments into Ion torrent adapters. These adapter-linked DNA fragments are further attached to sequencing beads and are amplified by emulsion PCR. Then, the coated beads will be placed in chip wells. The loaded chips are finally taken to the Ion torrent sequencer, where the sequencing process occurs. DNA polymerase enzyme will add new nucleotides to the template strands. As the synthesis starts, the addiction of each nucleotide will generate a proton due to pH change. A sensor will detect these proton releases in the system, transforming the chemical changes into digital data. Ion torrent is a versatile technique used in many wastewater studies, such as microbial analysis during food-waste wastewater processing (E. Kim et al. 2017).

8.5.6 Single-cell Genome Sequencing

It involves sequence information analysis of individual cells using next-generation sequencing (NGS) technology, providing insights into the cellular and molecular functions within the microenvironment. The major steps involved are the isolation of cells, PCR amplification, analysis of whole-genome-amplification products, identifying single-cell variants, and establishing genetic relationships between single cells (Nikolaki and Tsiamis 2013).

The advantages and disadvantages of the above-mentioned techniques are summarized in Table 8.1.

TABLE 8.1

Advantages and Disadvantages of Molecular Techniques Used for Molecular Diversity Analysis in Wastewater Treatment Plants

Techniques	Advantages	Disadvantages
Microarray technique	It enables the detection of many species at the same time It is swift and precise	Incorrect probe-DNA interactions might cause a low signal intensity
FISH	Simple detection and analysis Experimenting does not necessitate the use of expert trainers	Finding targets with few DNA copies is tough Time-consuming and laborious
qRT-PCR	Simultaneously amplifies and measures the DNA sequence of interest It enables real-time amplification monitoring	It counts the number of copies of the tagged gene rather than the number of cells
DGGE/TGGE	Sensitivity is high, and bands from gels may be removed for amplification and sequencing	Due to comparable GC contents, dissimilar DNA sequences from different bacterial species may display the same pattern of separation
ARDRA	It is a quick and precise procedure	It has a lower discriminating power when compared to other fingerprinting methods
RISA	It has a high level of discrimination and is less likely to produce inconsistent results	Only ISR (Intergenic Space Region) fragment alterations can be detected Bacteria having the same ISR cannot be differentiated
T-RFLP	Sensitive and dependable The fluorochrome-based detecting technique is precise	Restriction enzymes are distinct and may differ amongst microorganisms
454 Pyrosequencing	The average read length with paired-end sequencing can exceed 1000 bp A throughput of up to 450 mb is feasible	High-priced technology It can only process short nucleotide sequences
MPSS	The number of transcripts present per million molecules represents the amount of unique gene expression	A lack of restriction enzyme recognition sites and uncertainty in tag annotation can result in the loss of many transcripts
Illumine Sequencing method	Clonal amplification using bridge amplification The average read length with paired-end sequencing can exceed 300 bp	High-priced technology
Iron torrent method	Emulsion PCR for clonal amplification With paired-end sequencing, the average read length can reach 400 bp	High-priced technology
Single-cell genome sequencing	It aids in the detection of cell heterogeneity	Single-cell isolation and culture are necessary

Source: Silveira et al. 2021.

8.6 METAGENOMIC APPROACHES IN MICROBIAL DIVERSITY ANALYSIS

Metagenomics is the application of modern genomic techniques to study communities of microorganisms directly in their natural environment, bypassing the need for isolation and laboratory cultivation of individual species (Sleator, Shortall, and Hill 2008). It is also known as community, environmental, or microbial eco-genomics. It allows one to mainly decipher the interaction of uncultured microbes with other biotic and abiotic factors in their natural habitat (Figure 8.6). In contrast, 16S rRNA gene-based metagenomics approaches do not differentiate between viable and dead microorganisms. The significant events in wastewater metagenome analysis involve the following procedures.

8.6.1 EXPERIMENTAL DESIGNING AND SAMPLING

The significant events in metagenomic studies start with experimental designing and sampling. This stage involves asking and answering questions about environmental sample collection, DNA extraction, purification, amplification, and metagenome analysis before the actual study begins. Sampling is one of the most crucial stages in metagenomic studies since it can directly affect both the quantity and quality of the final result. This stage deals with collecting samples, proper storage, safe transportation, and finally, analysis of the samples in the laboratory. The collected samples could be environmental (seas, ponds, streams, lagoons, rivers, ponds, and oceans) or non-environmental (guts, skin, blood, and feces). Environmental samples can be heterologous or homogenous. A sampling methodology that will eliminate exogenous microbes and possible contaminants is considered the best (Thomas, Gilbert, and Meyer 2012).

8.6.2 DNA EXTRACTION AND AMPLIFICATION

An efficient DNA extraction procedure is needed for metagenome analysis of the environmental samples. The selection of extraction procedure may vary depending on the sample source and type.

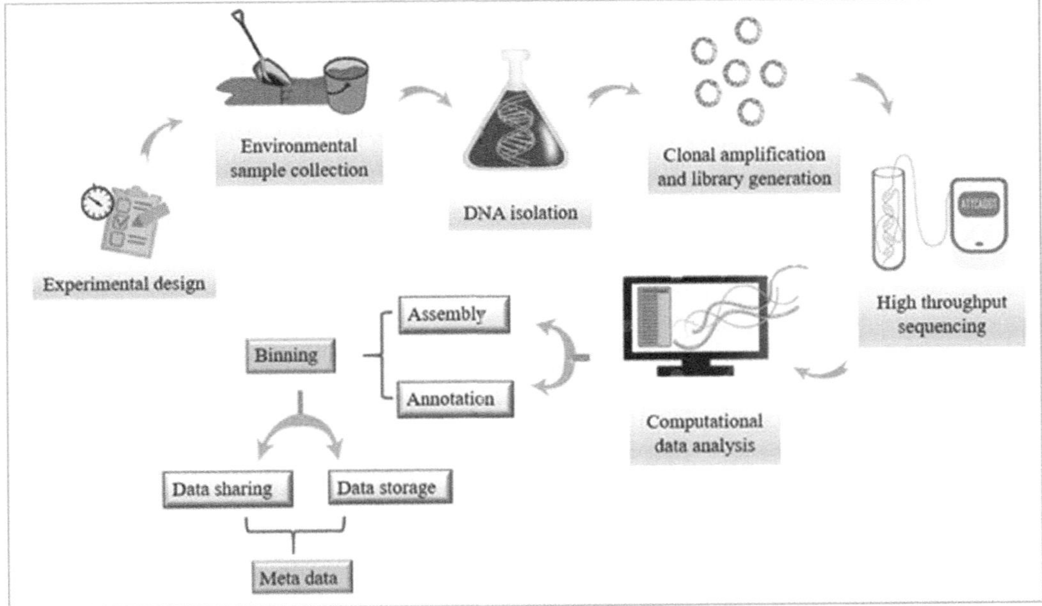

FIGURE 8.6 Metagenome analysis for species identification.

The extracted DNA should be of good quality and quantity for the subsequent library generation and sequencing.

8.6.3 Polymerase Chain Reaction (PCR)

Metagenomics and PCR are the key events invented in the last century that have revolutionized and shaped microbial ecology, molecular biology, microbiology, and biotechnology. PCR involves *in-vitro* amplification of target DNA sequence with the help of specific primers. Although the PCR steps are the same for metagenome analysis, DNA extraction and PCR are likely to be the possible sources of errors in next-generation sequencing, affecting the final results (Brooks et al. 2015).

8.6.4 Sequencing

Two major sequencing technologies for metagenome analysis are whole metagenome shotgun sequencing (WMSS) and high-throughput sequencing (Thomas, Gilbert, and Meyer 2012).

8.6.4.1 Whole Metagenome Shotgun Sequencing (WMSS)

WMSS combines shotgun sequencing with metagenomics and provides information about both the organism and the metabolic process. It reveals genes present in environmental samples. Clone library preparation was a necessary step in sequencing in earlier stages, but with the advancements in high-throughput sequencing technologies, cloning steps became avoidable. The latest trend in this area is a combination of shotgun sequencing and Hi-C (Chromosome conformation capture). This helps to track microbial genome assembly by measuring the closeness of any two genomic sequences within a cell. Some other choices to get long shotgun sequencing reads are long read sequencing technologies, including PacBio Sequel, PacBio RSII, Grigion, PrometION, and Nanopore MinION. All these techniques aid in a smooth assembling process (Segata et al. 2013).

8.6.4.2 High-throughput Sequencing

High-throughput sequencing is a broad term for describing technologies that help sequence DNA or RNA quickly and cost-effectively. The most attractive feature of high-throughput sequencing is that it does not require cloning before sequencing. The most widely used high-throughput sequencing platforms for metagenome analysis are pyrosequencing, the Illumina MiSeq or HiSeq, Ion torrent personal genome, and the Applied Biosystems SOLiD system. The first high-throughput metagenomic study used massively parallel 454 pyrosequencing platforms. All these high-throughput platforms produce reads shorter than that of Sanger sequencing (Mohan et al. 2021).

8.6.5 Assembly, Binning, and Annotation

Genome assembly involves the arrangement of short scattered DNA sequences obtained after sequencing to get the complete genome of an organism. This primarily helps to identify uncultured microbes, and without genome assembly, it is hard to analyze the complex genetic elements and repeated classes. Reference-based and de-novo assemblies are the two practical approaches to genome assembly in metagenomics. Binning is the process of sorting short DNA sequences into different groups, which helps to describe a genome or an individual organism. Many algorithms are available for the binning process. They work based on two facts: first, targeted genes are used to bin the DNA sequence by referring to such genes in reference databases; second, sequences that have conserved G-C (guanine-cytosine) regions will be reflected in the fragments of the genome. Annotation is the process of identification of gene locations and assigning functions to them. It can be done either by functional annotation or feature prediction. Significant advantages of using metagenomic approaches in wastewater treatment include process optimization, identifying uncultured microbes, and providing a complete picture of microbial diversity. Also, it can reveal the information of functional genes of an organism that help to utilize different pollutants (Edet et al. 2017).

Besides nucleic acid-based techniques, protein-based approaches are also useful for microbial diversity analysis. Metaproteomics is an advanced approach for studying the proteins involved in major processes and unknown metabolic pathways and identifying the functional roles of protein families whose functions are unknown. Sometimes, simple gene modifications can lead to changes in the specificity and activity of a particular protein or enzyme. Puttker et al. (2015) used metaproteomics to identify the proteins involved in the citrate cycle, glycolysis, and nitrogen removal during wastewater processing in a biological treatment plant. Metabolomics (an approach of the molecular *omics* used for analyzing the cellular metabolites) is one of the least practiced approaches due to its complex analytical procedures, but it is a critical approach for studying the interaction between microbe to microbiome and microbe to molecules.

Although molecular omics methods are beneficial in studying the functional and genetic diversity within an ecosystem, we cannot neglect the power of culture-dependent species identification and diversity analysis methods. Culture-dependent methods have formed the base of our understanding of microbial environments and are necessary for studying the physiological aspects of simple and managed environments. Culture-dependent methods are commonly used for the generation of reference genomes. In contrast, identifying the genome of uncultured microbes is now possible because of advances in single-cell sorting, manipulation, and whole-genome amplification (Blainey 2013; Dichosa et al. 2021). Despite significant gains in understanding wastewater microbes' identification and function, we are still far from fully comprehending these immensely complex microbial communities. The key to successfully characterizing these microbial consortia will be combining these approaches with systematic measurements and experimental validations.

8.7 CONCLUSION

This book chapter highlights the current status of techniques used for biodiversity studies in various WWTPs, which house dynamic and diverse microbial populations with a wide range of critical metabolic processes. Molecular techniques quickly become the trusted method for studying microbial communities, laying the groundwork for various applications, especially while connecting metabolic potential with gene function and regulation know-how. Integrating existing molecular techniques and the expected development of new procedures will create novel ecological methods that can aid in resolving structural and functional correlations in wastewater microbiology.

REFERENCES

Andreas, Felske, Akkermans Antoon D L., and De Vos Willem M. 1998. "In Situ Detection of an Uncultured Predominant Bacillus in Dutch Grassland Soils." *Applied and Environmental Microbiology* 64 (11). American Society for Microbiology: 4588–90. doi:10.1128/AEM.64.11.4588-4590.1998.

Blainey, Paul C. 2013. "The Future Is Now: Single-Cell Genomics of Bacteria and Archaea." *FEMS Microbiology Reviews* 37 (3): 407–27. doi:10.1111/1574-6976.12015.

Brooks, J Paul, David J Edwards, Michael D Harwich, Maria C Rivera, Jennifer M Fettweis, Myrna G Serrano, Robert A Reris, et al. 2015. "The Truth about Metagenomics: Quantifying and Counteracting Bias in 16S RRNA Studies." *BMC Microbiology* 15 (1): 66. doi:10.1186/s12866-015-0351-6.

Cardinali-Rezende, Juliana, Luís F D B Colturato, Thiago D B Colturato, Edmar Chartone-Souza, Andréa M A Nascimento, and José L Sanz. 2012. "Prokaryotic Diversity and Dynamics in a Full-Scale Municipal Solid Waste Anaerobic Reactor from Start-up to Steady-State Conditions." *Bioresource Technology* 119: 373–83. doi:https://doi.org/10.1016/j.biortech.2012.05.136.

Carraro, Lisa, Michela Maifreni, Ingrid Bartolomeoli, Maria Elena Martino, Enrico Novelli, Francesca Frigo, Marilena Marino, and Barbara Cardazzo. 2011. "Comparison of Culture-Dependent and -Independent

Methods for Bacterial Community Monitoring during Montasio Cheese Manufacturing." *Research in Microbiology* 162 (3): 231–39. doi:https://doi.org/10.1016/j.resmic.2011.01.002.

Ciesielski, Slawomir, Katarzyna Bułkowska, Dorota Dabrowska, Dariusz Kaczmarczyk, Przemyslaw Kowal, and Justyna Możejko. 2013. "Ribosomal Intergenic Spacer Analysis as a Tool for Monitoring Methanogenic Archaea Changes in an Anaerobic Digester." *Current Microbiology* 67 (2): 240–48. doi:10.1007/s00284-013-0353-2.

Dichosa, E. K. Armand, Karen W. Davenport, Po-E Li, Sanaa A. Ahmed, Hajnalka Daligault, Cheryl D. Gleasner, Yuliya Kunde, et al. 2021. "Draft Genome Sequence of Thauera Sp. Strain SWB20, Isolated from a Singapore Wastewater Treatment Facility Using Gel Microdroplets." *Genome Announcements* 3 (2). American Society for Microbiology: e00132–15. doi:10.1128/genomeA.00132-15.

Duan, Zheng Hua, Liu Ming Pan, Hua Wang, and Ning Tao Li. 2013. "Analysis of Bacterial Communities in A2O Membrane Bioreactor Treating Oily Wastewater." *Advanced Materials Research* 641–642. Trans Tech Publications Ltd: 87–91. doi:10.4028/www.scientific.net/AMR.641-642.87.

Edet, U O, S P Antai, A A Brooks, A D Asitok, and O Enya. 2017. "An Overview of Cultural, Molecular and Metagenomic Techniques in Description of Microbial Diversity" 7 (2): 1–19. doi:10.9734/JAMB/2017/37951.

Fakruddin, M. 2015. "Methods for Analyzing Diversity of Microbial Communities in Natural Environments." *Ceylon Journal of Science (Biological Sciences)*, 42 (1): 19–33. https://doi.org/10.4038/cjsbs.v42i1.5896.

Garlapati, Deviram, B Charankumar, K Ramu, P Madeswaran, and M V Ramana Murthy. 2019. "A Review on the Applications and Recent Advances in Environmental DNA (EDNA) Metagenomics." *Reviews in Environmental Science and Bio/Technology* 18 (3): 389–411. doi:10.1007/s11157-019-09501-4.

Greene, E. Anne, and Gerrit Voordouw. 2003. "Analysis of Environmental Microbial Communities by Reverse Sample Genome Probing." *Journal of Microbiological Methods* 53 (2): 211–19. doi:https://doi.org/10.1016/S0167-7012(03)00024-1.

Jena, Jyotsnarani, Ravindra Kumar, Md Saifuddin, Anshuman Dixit, and Trupti Das. 2016. "Anoxic–Aerobic SBR System for Nitrate, Phosphate and COD Removal from High-Strength Wastewater and Diversity Study of Microbial Communities." *Biochemical Engineering Journal* 105: 80–89. doi:https://doi.org/10.1016/j.bej.2015.09.007.

Joss, Adriano, Nicolas Derlon, Clementine Cyprien, Sabine Burger, Ilona Szivak, Jacqueline Traber, Hansruedi Siegrist, and Eberhard Morgenroth. 2011. "Combined Nitritation-Anammox: Advances in Understanding Process Stability." *Environmental Science & Technology* 45 (November): 9735–42. doi:10.1021/es202181v.

Kim, Eunji, Seung Gu Shin, Md Abu Hanifa Jannat, Jovale Vincent Tongco, and Seokhwan Hwang. 2017. "Use of Food Waste-Recycling Wastewater as an Alternative Carbon Source for Denitrification Process: A Full-Scale Study." *Bioresource Technology* 245: 1016–21. doi:https://doi.org/10.1016/j.biortech.2017.08.168.

Kim, Mincheol, and Jongsik Chun. 2014. "Chapter 4 – 16S RRNA Gene-Based Identification of Bacteria and Archaea Using the EzTaxon Server." In *New Approaches to Prokaryotic Systematics*, edited by Michael Goodfellow, Iain Sutcliffe, and Jongsik B T – Methods in Microbiology Chun, 41:61–74. Academic Press. doi:https://doi.org/10.1016/bs.mim.2014.08.001.

Kumar, V., Singh, K., Shah, M.P., Singh, A.K, Kumar, A., Kumar, Y. 2021. Application of omics technologies for microbial community structure and function analysis in contaminated environment. In Shah, M.P., Sarkar, A., Mandal, S. (Ed.), *Wastewater treatment: Cutting edge molecular tools, techniques & applied aspects in waste water treatment*. Elsevier. https://doi.org/10.1016/B978-0-12-821925-6.00013-7.

Kumar, V., Thakur, I.S., Singh, A.K., Shah, M.P. 2020. Application of metagenomics in remediation of contaminated sites and environmental restoration. In: Shah M, Rodriguez-Couto S, Sengor SS (Eds), *Emerging technologies in environmental bioremediation*. Elsevier. https://doi.org/10.1016/B978-0-12-819860-5.00008-0.

Liao, Runhua, Ke Shen, Ai-Min Li, Peng Shi, Yan Li, Qianqian Shi, and Zhu Wang. 2013. "High-Nitrate Wastewater Treatment in an Expanded Granular Sludge Bed Reactor and Microbial Diversity Using 454 Pyrosequencing Analysis." *Bioresource Technology* 134: 190–97. doi:https://doi.org/10.1016/j.biortech.2012.12.057.

Mohan, Shruthi, Premchand Subhash Chigadannavar, Zeba Quadri, Priyadarshini Dey, Deepak Gola, Nitin Chauhan, and Randhir K Bharti. 2021. "16 – Microbial Community Diversity in a Wastewater Treatment

Plant." In, edited by Maulin P Shah and Susana B T – *Wastewater treatment reactors rodriguez-couto*, 373–85. Elsevier. doi:https://doi.org/10.1016/B978-0-12-823991-9.00013-7.

Mühling, Martin, John Woolven-Allen, J Colin Murrell, and Ian Joint. 2008. "Improved Group-Specific PCR Primers for Denaturing Gradient Gel Electrophoresis Analysis of the Genetic Diversity of Complex Microbial Communities." *The ISME Journal* 2 (4): 379–92. doi:10.1038/ismej.2007.97.

Muttray, Annette F, Zhontang Yu, and William W Mohn. 2001. "Population Dynamics and Metabolic Activity of Pseudomonas Abietaniphila BKME-9 within Pulp Mill Wastewater Microbial Communities Assayed by Competitive PCR and RT-PCR." *FEMS Microbiology Ecology* 38 (1): 21–31. doi:10.1111/j.1574-6941.2001.tb00878.x.

Muyzer, G, de Waal E C, and A G Uitterlinden. 1993. "Profiling of Complex Microbial Populations by Denaturing Gradient Gel Electrophoresis Analysis of Polymerase Chain Reaction-Amplified Genes Coding for 16S RRNA." *Applied and Environmental Microbiology* 59 (3). American Society for Microbiology: 695–700. doi:10.1128/aem.59.3.695-700.1993.

Natuscka, Lee, Nielsen Per Halkjær, Andreasen Kjær Holm, Juretschko Stefan, Nielsen Jeppe Lund, Schleifer Karl-Heinz, and Wagner Michael. 1999. "Combination of Fluorescent In Situ Hybridization and Microautoradiography – a New Tool for Structure-Function Analyses in Microbial Ecology." *Applied and Environmental Microbiology* 65 (3). American Society for Microbiology: 1289–97. doi:10.1128/AEM.65.3.1289-1297.1999.

Nikolaki, Sofia, and George Tsiamis. 2013. "Microbial Diversity in the Era of Omic Technologies." Edited by Dimitrios Karpouzas. *BioMed Research International 2013*. Hindawi Publishing Corporation: 958719. doi:10.1155/2013/958719.

Puttker, Sebastian, Fabian Kohrs, Dirk Benndorf, Robert Heyer, Erdmann Rapp, and Udo Reichl. 2015. "Metaproteomics of Activated Sludge from a Wastewater Treatment Plant – A Pilot Study." *PROTEOMICS* 15 (20). John Wiley & Sons, Ltd: 3596–3601. doi:https://doi.org/10.1002/pmic.201400559.

Read, Suzanne, Massimo Marzorati, Beatriz C M Guimarães, and Nico Boon. 2011. "Microbial Resource Management Revisited: Successful Parameters and New Concepts." *Applied Microbiology and Biotechnology* 90 (3): 861–71. doi:10.1007/s00253-011-3223-5.

Sebat, Jonathan L., Frederick S. Colwell, and Ronald L. Crawford. 2003. "Metagenomic Profiling: Microarray Analysis of an Environmental Genomic Library." *Applied and Environmental Microbiology* 69 (8). American Society for Microbiology: 4927–34. doi:10.1128/AEM.69.8.4927-4934.2003.

Segata, Nicola, Daniela Boernigen, Timothy L Tickle, Xochitl C Morgan, Wendy S Garrett, and Curtis Huttenhower. 2013. "Computational Meta'omics for Microbial Community Studies." *Molecular Systems Biology* 9 (May). European Molecular Biology Organization: 666. doi:10.1038/msb.2013.22.

Silveira, D D, P Belli Filho, L S Philippi, M E Cantão, A Foulquier, S Bayle, T P Delforno, and P Molle. 2021. "In-Depth Assessment of Microbial Communities in the Full-Scale Vertical Flow Treatment Wetlands Fed with Raw Domestic Wastewater." *Environmental Technology* 42 (20). Taylor & Francis: 3106–21. doi:10.1080/09593330.2020.1723709.

Sleator, R D, C Shortall, and C Hill. 2008. "Metagenomics." *Letters in Applied Microbiology* 47 (5). John Wiley & Sons, Ltd: 361–66. doi:https://doi.org/10.1111/j.1472-765X.2008.02444.x.

Tan, BoonFei, Charmaine Ng, Jean Nshimyimana, Lay-Leng Loh, Karina Gin, and Janelle Thompson. 2015. "Next-Generation Sequencing (NGS) for Assessment of Microbial Water Quality: Current Progress, Challenges, and Future Opportunities." *Frontiers in Microbiology*. www.frontiersin.org/article/10.3389/fmicb.2015.01027.

Thomas, Torsten, Jack Gilbert, and Folker Meyer. 2012. "Metagenomics – a Guide from Sampling to Data Analysis." *Microbial Informatics and Experimentation* 2 (1). BioMed Central: 3. doi:10.1186/2042-5783-2-3.

Urrea-valencia, Salomé. 2021. "Molecular Techniques to Study Microbial Wastewater Communities." *Brazilian Archives of Biology and Technology* 64. https://doi.org/10.1590/1678-4324-2021200193.

Wagner, Michael, and Alexander Loy. 2002. "Bacterial Community Composition and Function in Sewage Treatment Systems." *Current Opinion in Biotechnology* 13 (3): 218–27. doi:https://doi.org/10.1016/S0958-1669(02)00315-4.

Wang, Zhu, Xu-Xiang Zhang, Xin Lu, Bo Liu, Yan Li, Chao Long, and Aimin Li. 2014. "Abundance and Diversity of Bacterial Nitrifiers and Denitrifiers and Their Functional Genes in Tannery Wastewater

Treatment Plants Revealed by High-Throughput Sequencing." *PLOS ONE* 9 (11). Public Library of Science: e113603. https://doi.org/10.1371/journal.pone.0113603.

Wells, George F, Hee-Deung Park, Brad Eggleston, Christopher A Francis, and Craig S Criddle. 2011. "Fine-Scale Bacterial Community Dynamics and the Taxa–Time Relationship within a Full-Scale Activated Sludge Bioreactor." *Water Research* 45 (17): 5476–88. doi:https://doi.org/10.1016/j.watres.2011.08.006.

Widder, Stefanie, Rosalind J Allen, Thomas Pfeiffer, Thomas P Curtis, Carsten Wiuf, William T Sloan, Otto X Cordero, et al. 2016. "Challenges in Microbial Ecology: Building Predictive Understanding of Community Function and Dynamics." *The ISME Journal* 10 (11): 2557–68. doi:10.1038/ismej.2016.45.

Yu, Ke, and Tong Zhang. 2012. "Metagenomic and Metatranscriptomic Analysis of Microbial Community Structure and Gene Expression of Activated Sludge." *PLOS ONE* 7 (5). Public Library of Science: e38183. https://doi.org/10.1371/journal.pone.0038183.

Zhongtang, Yu, and Mohn William W. 2001. "Bacterial Diversity and Community Structure in an Aerated Lagoon Revealed by Ribosomal Intergenic Spacer Analyses and 16S Ribosomal DNA Sequencing." *Applied and Environmental Microbiology* 67 (4). American Society for Microbiology: 1565–74. doi:10.1128/AEM.67.4.1565-1574.2001.

Section II

Environmental Nexus in Waste Management

9 Trends in the Generation, Behaviour, and Fate of Microplastics from Municipal Solid Waste Management Systems

*Ana Laura De la Colina Martínez and
David Joaquín Delgado-Hernández*

9.1 INTRODUCTION

Plastics have become an integral part of the world economy, but the municipal waste management systems that have been put in place globally are insufficient in the face of the large amount of plastic waste generated annually, which causes extensive damage to the climate, wildlife, ecosystems, and human beings. In 2019, 353 million tonnes (Mt) of plastic waste were produced. Almost two thirds were packaging (40%), consumer goods (12%), and clothing and textiles (11%). By the end of 2022, 380 Mt more will have been produced. Of this, 83 Mt (22%) will be emitted into the environment (OECD.Stat, 2022). By 2060, this waste, driven by 'single-use plastics', is expected to reach 1,014 Mt (OECD.Stat, 2022), of which 153 Mt will be released into the natural environment (OECD, 2022a, 2022c; OECD.Stat, 2022).

For the last 70 years, all the millions of tonnes of plastic waste dispersed in the environment have been at risk of fragmenting, giving rise to microplastics (<5 mm) (Thompson et al., 2004) and nanoplastics (1–100 nm) (Ryan, 2020; Gigault et al., 2021) due to weathering, light, and abrasion. Microplastics and nanoplastics have been found floating in the Pacific, Atlantic, Arctic, and Antarctic oceans. They have also been found in more than 170 marine species. These taxa consist of zooplankton, coral reefs, marine invertebrates (including molluscs and crustaceans), fish (including sharks), seabirds (including penguins), reptiles (including turtles), and marine mammals (including seals, dolphins, and whales) (Miller, Santillo and Johnston, 2016; Nelms et al., 2018; Soares et al., 2020). Moreover, they have also been found in freshwater species such as the temporary frog (from a high-mountain ecosystem) (Pastorino et al., 2022) and tadpoles (Hu et al., 2022), and in terrestrial species such as terrestrial microarthropods (Zhu et al., 2018), *Lumbricus terrestris* (Baeza et al., 2020), *Eisenia fetida* (Liu et al., 2022), and macroinvertebrate species (Ferreira-Filipe et al., 2022).

These microplastics are highly bioavailable to aquatic and terrestrial organisms, either through direct ingestion or indirectly. They are transferred within the food chain when predators ingest prey that has consumed plastics. They can even accumulate in prey and predators, altering their behaviour. However, what are their effects on human health? Recent evidence indicates that humans constantly inhale and ingest microplastics. Indeed, these microplastics have already been found in the placenta and in human blood (Miller, Santillo and Johnston, 2016; Nelms et al., 2018; Fragão et al., 2021; Ragusa et al., 2021; Vethaak and Legler, 2021; Leslie et al., 2022).

DOI: 10.1201/9781003408352-11

In this context, is of concern that primary microplastics (Cao et al., 2022) are still being produced industrially and that, after seven decades, plastics are still being inadequately managed, despite the existence of multiple technologies to address this. Even though all plastics that saturate and inhabit the planet are a resource, currently only 10% of the world's discarded plastics are recycled; the rest (90%) are landfilled (49%), incinerated (19%), and, most alarmingly, disposed of in the natural environment (22%) (UNEP, 2015; OECD, 2022a, 2022c; OECD.Stat, 2022). On the other hand, the COVID-19 pandemic generated a massive amount of plastic waste, such as disposable gloves and masks, which poses a challenge for solid waste management worldwide due to its environmental and public health impact (Dey et al., 2023). Furthermore, inadequate management of solid plastic waste contributes to the presence of microplastics in wastewater treatment plants, where activated sludge becomes a major reservoir for microplastics (Bhat et al., 2022).

Some techniques are envisioned to remove microplastics in wastewater treatment plants and activated sludge (Chen et al., 2020; Bhat et al., 2022); and biological treatment of microplastics by certain fungi and bacteria could help mitigate microplastic waste generated by COVID-19 (Dey et al., 2023). However, globally, the main problem persists, and that is the mismanagement of plastic waste that deeply pollutes the planet (Lake Ontario, 2016).

9.2 GENERATION OF MICROPLASTICS

The productive activities conducted by consumer societies generate diverse types of solid waste. Among them is municipal solid waste (MSW), which comes from the cleaning of public and private spaces, which are distributed throughout the urban area. These spaces are usually: houses, condominiums, flats, shops, offices, doctors' offices, laboratories, clinics, hospitals, schools, squares, streets, public gardens, pavements, and avenues (SEMARNAT, 2005).

At present, MSW has a global production of 1,300 Mt per year, with forecasts of 2,200 Mt by 2025. MSW consists of organic and inorganic waste. Within the latter type are plastic solid wastes (PSW), which comprise an important part of MSW (OECD, 2022b). Collection and disposal of solid wastes is one of the major problems in both urban and rural areas in numerous developed and developing countries. Particularly, MSW management solutions must be viable, sustainable, in accordance with society and within the law, always seeking the benefit of natural ecosystems (Abdel-Shafy and Mansour, 2018).

PSW includes packaging, consumer/institutional products, electrical/electronic appliances and a variety of textiles (Abdel-Shafy and Mansour, 2018). However, PSW may also contain plastic pieces larger than 5 mm, primary and secondary microplastics (<5 mm in size) (Thompson et al., 2004), as well as nanoplastics (size, 1–100 nm) (Ryan, 2020). PSW generated today is not properly managed (collected, recovered, or disposed of), contributing to the increasing accumulation of plastic wastes in soils, rivers, and oceans. Besides, when plastic waste is landfilled, space availability decreases and waste management costs tend to increase. Plastic waste must be effectively treated through different valorisation methods, including in situ recycling, mechanical recycling, thermolysis, chemolysis, and energy recovery (Idumah and Nwuzor, 2019).

It is estimated that in the last six decades, 8,300 Mt of plastics (primary production) have been produced. Of them, 2,500 Mt are immobilised in construction (or latent). Of the remaining 5,800 Mt, 500 Mt (9%) were recycled, 700 Mt (12%) were incinerated, and 4,600 Mt (79%) were discarded. Of these discarded plastics 79% (4,600 Mt), one part was dumped in landfills, probably 50% (2,900 Mt), and the other part was emitted into the environment (Idumah and Nwuzor, 2019; Ritchie, Samborska and Roser, 2023), probably 29% (1,700 Mt). This can be estimated if it is taken into account that globally, according to (OECD.Stat, 2022), 50% of the plastic waste currently generated is dumped in landfills and 22% is emitted into the natural environment.

Considering that 1,700 Mt of plastic waste has been released into the environment from 1950 to 2015, it is expected that a significant amount of microplastics could be actively polluting every corner of the water, soil, and atmosphere.

By the end of 2022, 380 Mt of PSW will have been generated. Of them, only 38 Mt (10%) will be recycled, and 83 Mt (22%) will be emitted into the environment (OECD.Stat, 2022). By 2060, this plastic waste, driven by 'single-use products', is expected to reach 1,014 Mt (OECD.Stat, 2022). Of this, 176 Mt will be recycled, and 153 Mt will be dispersed in the environment (OECD, 2022a; OECD.Stat, 2022), which will increase environmental and health impacts unprecedentedly.

The generation of microplastics in the environment can be approached from their classification into primary and secondary. This is detailed below.

Primary microplastics are manufactured at the micro scale and are designed to be used in some specific applications (OECD, 2022b). For instance, microfibers are in microfiber clothing (sportswear), nappies, and cigarette butts. Although one of the most common ways for microfibers from clothing to reach the environment is through the water used in the laundry (Lake Ontario, 2016), microfibers from nappies and cigarette butts can reach the environment through the mismanage of MSW and leakage. Microbeads are made principally for personal care and cosmetic products (PCCPs), such as deodorants, shampoos, toothpastes, and conditioners (Arpia et al., 2021), and cleaning products such as detergents (Dimerc Office, 2013). Pellets are small pieces of plastic used to manufacture plastic goods. Microfibers, microbeads and pellets are recurrently rejected into the environment during processing, transportation, and managing. Due to their size, they may be washed into the drainage system by rainfall, or remain trapped in soils and rivers (Lake Ontario, 2016; OECD, 2022b).

Secondary microplastics are the product of the fragmentation of larger plastics (packaging, textiles, toys) (van Wijnen, Ragas and Kroeze, 2019; Meijer et al., 2021). Despite their chemical stability (Thompson et al., 2004), fragmentation can occur during production, use or after disposal (van Wijnen, Ragas and Kroeze, 2019). Various processes such as weathering, solar irradiation, and abrasion cause ageing and fragmentation (Cheng et al., 2020). These microplastics are particularly evident on coasts, where wave action makes plastic articles fragile, increasing their fragmentation rate (van Wijnen, Ragas and Kroeze, 2019).

The most ubiquitous types of microplastics are PS, PE, PP, PA, and PVC. The impact of the first three has attracted additional attention because they can be transported more easily (Bamshad and Cho, 2021). Although microplastics (primary and secondary) are smaller than 5 mm in size, some researchers make explicit reference to nanoplastics size (1–100 nm) (Ryan, 2020). Therefore, *the microplastics addressed in this study are plastic particles in the range of 5 mm to 0.0001 mm (100 nm).*

According to the OECD, the quantities of macro- and microplastics entering the terrestrial and aquatic environment have many types and sources, as described below (OECD, 2022a).

In 2022, 20 Mt of macroplastics were leaked. The main cause was the mismanagement of municipal and non-municipal plastic waste, as well as the dumping of plastic objects that have reached the end of their useful life. Fishing and some other sea-related activities were similarly relevant; due to the loss and abandonment of nets at sea, as well as wear and tear (abrasion) of dolly ropes and other waste not directly related to fishing nets (OECD, 2022a). Macroplastics can also be seen in the remains of household paint (Turner, 2021). See Figure 9.1.

In the same way, 3 Mt of microplastics came from leakage in 2022. This leakage comes mainly from road transport: tyre abrasion, brake wear, and eroded road markings. Another important source is the dust generated by the wear and tear of the soles of shoes worn by millions of pedestrians every day. Even household dust created by clothing is included in these losses, as well as the losses from construction and demolition (OECD, 2022a). However, recent studies report that microplastics from architectural paint, found in house dust (interior paint) and on house facades (exterior paint that wears and peels over time), can leach into the environment, circumventing MSW management

FIGURE 9.1   White paint macroplastics (brush, roller, and dried paint residues) exposed to the environment.

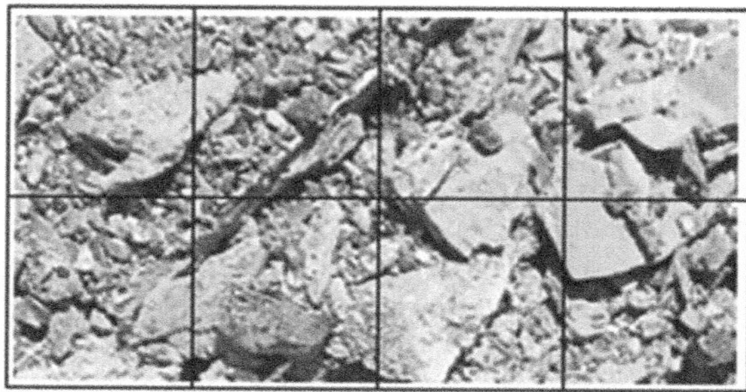

FIGURE 9.2   Microplastics of paint from a facade left in the environment (the size of each square is 5 x 5 mm).

(Paruta, Pucino and Boucher, 2021). See Figure 9.2. In addition, Suzuki et al. (2022) reveal that losses of microplastics also occur in the mechanical recycling of plastic waste.

Since 2010 it has been known that the sources contributing to global leakage of microplastics come from poorly managed waste (80%). Primary microplastics contribute 15% and the remaining 5% is due to the loss of fishing nets at sea. Some studies considered the contribution of paint to total microplastic leakage ranged between 9.6% and 21%; however, using Monte Carlo analysis, a new study estimates that it is 58% (Paruta, Pucino and Boucher, 2021).

By 2060, the amount of macroplastic leakages into nature is expected to increase to 38 Mt and microplastics to 6 Mt (OECD.Stat, 2022). There remains an urgent need to improve the understanding of macro/microplastics leakage and to align existing assessment methodologies (OECD, 2022a; OECD.Stat, 2022).

Once microplastics flow into the environment, it can go global with environmental impacts that can last hundreds of years (UNEP, 2022). Plastic pollution is growing relentlessly as waste management and recycling fall short, says OECD (OECD, 2022c). World production has increased to such an extent that some authors do not hesitate to call the 20th century the age of plastics or the Plasticene Era (Arpia et al., 2021).

9.3 BEHAVIOUR OF MICROPLASTICS

9.3.1 WEATHERING OF MICROPLASTICS

Degradation of plastics refers to a series of reactions that break down their structure, which reduces its molecular weight. When this occurs on the Earth's surface (mainly due to water, solar radiation, air, and temperature changes, as well as the presence of animals, plants, salts, or acids), it is namely weathering of plastics. This weathering can involve a series of joint degradations (physical, chemical, and biological) that do not stop because plastics are constantly exposed to weather phenomena. This is a dynamic situation, where macro, micro, and nanoplastics are continuously changing (Tanaka et al., 2023).

Weathering of microplastics, shown in Figure 9.3, changes the physical characteristics (size, colour, morphology, density, and crystallinity) and the chemical properties of microplastics, due to the breakdown and modification of chains and functional groups (Cao et al., 2022). The degradation of microplastics in the environment can be mechanical, photo-oxidative, and biological. Some examples reported in the literature are described below.

9.3.1.1 Photo-oxidative Degradation

Plastics undergo photo-oxidation when exposed to UV radiation and oxygen. It takes place on the surface (outer layer) because oxygen cannot penetrate to the layers that are located on the inside of them. Effects can be, e.g., yellowing of microplastics, ageing, fragmentation, and degradation.

- *Yellowing of microplastics*
  By analysing primary plastic pellets found on some beaches on the island of Malta, the abundance of which was greater than 1,000 microplastics/m^2 on the surface, they were observed to

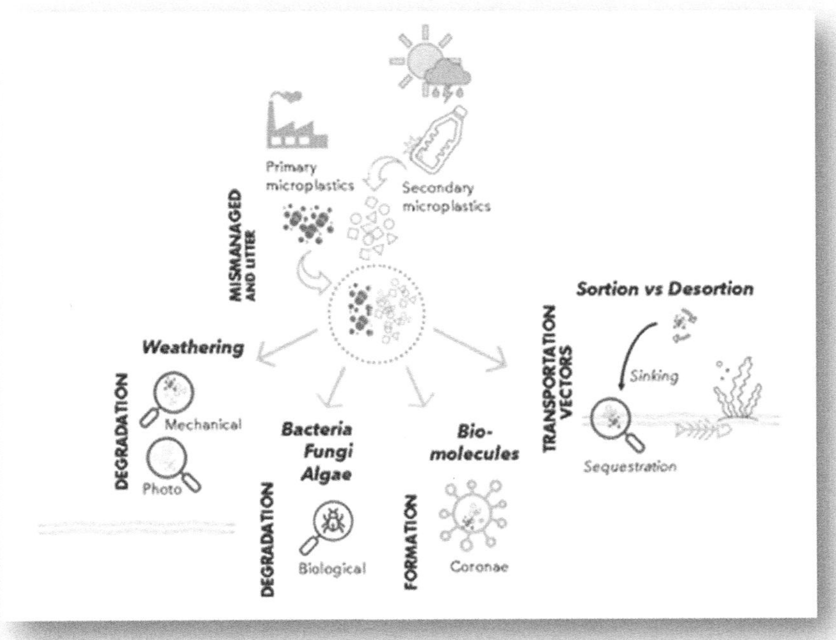

FIGURE 9.3 The behaviour (degradation, coronae formation, and transportation vectors) of primary and secondary microplastics present in the environment, generated due to mismanaged plastic waste and litter.

be yellow, amber, and brown in colour. The degree of yellowing (or darkening) was found to be associated with an increase in the carbonyl index, due to photo-oxidation or ageing. These pellets are like those seen in other areas of the Mediterranean, indicating their persistence in the region for more than three decades (Rummel et al., 2022). The rate of yellowing of microplastics is higher on land than in marine environments, due to the higher oxygen concentration and temperature. In addition, microplastics on land are more exposed to the elements, are less prone to developing biofilms, and are generally not covered by water (Guo and Wang, 2019). See Figure 9.4.

- *Degradation of products*
  A solar-box system was employed to determine the effect of solar radiation on microplastics. Some micro-powders (857–509 µm) of polymers such as PP, PS, PE, and PET were aged (see Figure 9.5) artificially (4 weeks) under intense conditions: temperature 40 °C, irradiance 750 W/m^2, and relative humidity around 60%; equivalent to a 6-month exposure in central Italy. The results, at the end of the 4 weeks of induced ageing, were as follows (Biale et al., 2021):

FIGURE 9.4   Yellowing of microplastics exposed to sunlight and weathering for months (scale is in cm).

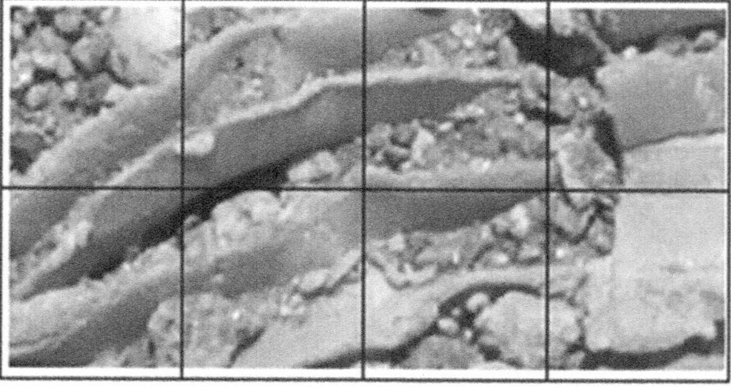

FIGURE 9.5   Natural ageing observed in an electrical polyduct, exposed to sunlight and weathering for months (the size of each square is 5 x 5 mm).

PP. It degraded the most, due to the reactivity of its tertiary bonds. The photo-oxidative degradation products had a low molecular weight. Short-chain carboxylic and dicarboxylic acids were identified, as well as oligomers with different chain lengths. PP has a maximum thermal degradation temperature (TD max) at 453 °C. In addition, a decrease in the onset temperature of the thermal degradation peak from 421 to 350 °C is observed, i.e., a difference of (ΔT = 71 °C).

PS. Its degradation products were styrene, benzoic acid, its dimer, and trimer. Different acids and alcohols derived from benzoic acid, as well as carboxylic and dicarboxylic acids, were also detected. Fractions with molecular weights ranging from 10,200 to 500 Da and even some of low molecular weight (<500 Da) were also observed. The TD max was recorded at 406 °C. There was a slight decrease of the onset temperature (ΔT = 8 °C) that can be associated with the formation of photo-oxidised products, which coincides with the processes of thermal degradation of PS, characterised by depolymerisation.

LDPE & HDPE. Both have a similar response to induced ageing, but LDPE shows a higher number of oxidative degradation products. On the other hand, LDPE has a TD max at 454 °C, while the TD max of HDPE is higher (474 °C). On ageing, both observe a slight decrease in the starting temperature of (ΔT = 7 °C) and (ΔT = 6 °C) respectively, which may be related to the formation of photo-oxidation products.

PET. It shows a clear photo-oxidative stability, as no significant amount of degradation products were produced, highlighting its stability to induced ageing. The TDmax was recorded at 416 °C. Only a slight decrease of the onset temperature (ΔT = 7 °C) is observed, which may be related to the formation of photo-oxidation products.

In another similar study, PS microplastics were exposed to controlled accelerated weathering conditions in the Q-SUN XE-3 industrial test chamber. In the test, the PS microplastics were constantly agitated in a vessel containing quartz glass and water at 55 °C. The total irradiance was 594 W/m² and the relative humidity was 50%. The results showed that microplastics with a size of 0.16 mm, after 3200 hours, acquired a size of 0.022 mm (Strohriegl et al., 2021).

On the other hand, tyre particles combined with pavement particles (from the roads) were exposed to UV rays. Exposure for 1.5 years produced ketone intermediates. Exposure for 3 years produced carboxylic acids in addition to ketones. UV radiation produced more degradation products than wet/dry and freeze/thaw exposure (Thomas et al., 2022).

9.3.1.2 Mechanical Degradation

This type of microplastics degradation leads to fragmentation, particle size reduction, and morphological changes. See Figure 9.6. Some examples reported in the literature are described below.

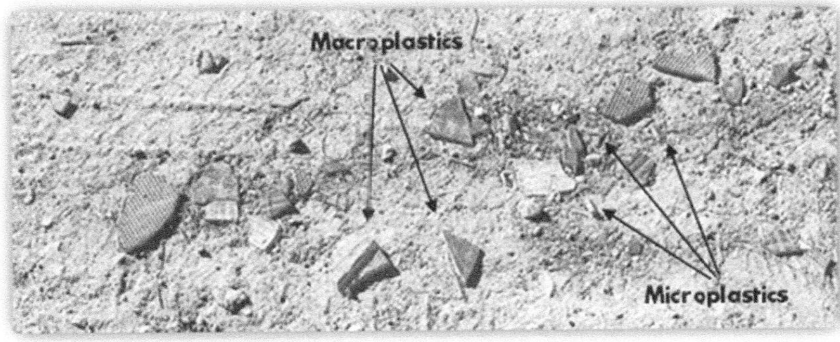

FIGURE 9.6 Fragmentation of macro- and microplastics by mechanical impact (due to the passage of people, bicycles, and motorbikes on the dirt road).

The mass and number of fragments produced after 24 hours in a mixer containing pebbles were analysed for each plastic. Each type of plastic was cut into squares (2 x 2 cm). The plastics used came from different sources, e.g., PP-solid from the side walls of single-use beverage glasses, PS-foam from the standard 2-cm thick building heat insulation plates, LDPE-film from the garbage bags (thickness–0.005 mm) from the local supermarket, and PS-solid from the 'corrugated' bottoms of the usual disposable plates.

After 24 h of active abrasion, the total number of microplastics (5–0.5 mm) generated was: 550 PP-solid particles (exponential ratio); 20,000 PS-foam particles (quadratic and linear ratios); 36,000 LDPE-film particles (exponential ratio); and 1,100,000 PS-solid particles (quadratic ratio). On the other hand, the corresponding mass for each plastic was: 97.6% of the PP-solid remained unchanged (>5 mm). However, the 2 x 2 cm squares were softer and more flexible, had a smoother and matt surface; 72.8% of the PS-foam remained 2 x 2 cm and was generally cubic in shape; 31% of the total LDPE-film was transformed into microplastics, while the number of nanoparticles was negligible. Finally, 99.8% of the microplastics obtained from PS-solid decreased during the experiment to 82%, as 18% were transferred to an uncountable number of nanoparticles (<0.5 mm) (Efimova et al., 2018).

A curious fact about the thermal stability of microplastics obtained by mixing machines is that they probably become more resistant compared to the source material, e.g., LDPE beads (primary microplastics) with a diameter of 3 mm were mechanically fragmented to obtain smaller particles. The TGA obtained (Astner et al., 2019) for the LDPE samples shows that the microplastics (250 μm) have a higher thermal resistance compared to the original beads (3 mm) and nanoplastics (375 nm) obtained.

9.3.2 BIODEGRADATION

Biodegradation (see Figure 9.3) transforms microplastics through enzymatic depolymerisation of microorganisms such as bacteria, fungi, and algae (Debroy, George and Mukherjee, 2021; Dey et al., 2023). Some examples reported in the literature are described below.

Bacteria, fungi, and algae colonise the plastic polymers, generating a biofilm. Colonies of these microbes produce *proteinases* to catalyse the hydrolysis of the plastic polymers, resulting in the formation of monomers and metabolic intermediates that, eventually, mineralise into carbon dioxide and water (Habib et al., 2020).

For example, no microorganisms have yet been reported that, in addition to growing at extreme temperatures, can utilise microplastics as their unique source of energy. Besides, certain structural features of PP make it difficult for microbes to adhere to its surface. Despite this, the biodegradation of PP microplastics by the Antarctic soil bacteria *Pseudomonas* sp. ADL15 and *Rhodococcus* sp. ADL36 resulted, after 40 days, in a 17.3% and 7.3% weight loss of the microplastics, respectively. In this case, the ability of ADL15 to synthesise biosurfactants may have been one of the important factors that facilitated the adhesion of bacterial cells to the microplastic surface. Because of this, the resulting hydrophilic PP is more exposed to microbial attack and further degradation (Habib et al., 2020).

On the other hand, sewage sludge, which comes from wastewater treatment plants (industrial, domestic, and stormwater), is high in organic matter and nutrients and is widely used in agricultural soils. However, it is one of the main sources of soil pollution by microplastics. A new hyperthermophilic composting technology (at 70 °C) allows in situ biodegradation of microplastics in sewage sludge. In 45 days, it is possible to eliminate almost 44% of plastics. This far exceeds conventional thermophilic composting (at 40 °C). *Thermus, Bacillus*, and *Geobacillus* bacteria accelerated the biodegradation via bio-oxidation (Chen et al., 2020). This technology could, possibly, also be applied to the large amount of microplastics contained in landfills used for the disposal of plastic waste from MSW.

Biodegradation may be the most lucrative route since there are no secondary pollutants generated. Sustainable degradation is expected to make significant advances in a few years, as research related to microplastics and their biodegradation is capturing the attention of scientists (Arpia et al., 2021).

9.3.3 Transportation Vectors

Traditionally, the toxicity of microplastics is attributed to the chemicals adsorbed on them (Hongwei et al., 2022). Microplastics in the environment may show certain relevant changes, such as marked surface oxidation and small cracks. These types of affectations have been shown to increase the sorption capacity of microplastics compared to pristine microplastics. Factors dominating this change include physical interactions, partition coefficient, electrostatic interactions, and intermolecular hydrogen bonds (Vieira et al., 2021).

Among the substances that can sorb and desorb microplastics (biodegradable and non-degradable) are petroleum hydrocarbons. It was found that the capacity to sorb these hydrocarbons increases in the following order PVC < PLA < PET < PS < PS < PE < PA; with pseudo-second-order sorption kinetics (Song et al., 2021).

It is known that many plastics can be transported geographically through air, water, and soil. Thus, microplastics and the substances they carry can be transported and transferred (desorbed). Microplastics, carrying sorbed substances, are known to settle on riverbanks or shorelines at high tide. A study on the sorption of microplastics (225 μm) present in seawater showed that PE sorbed more phenanthrene than PP and PVC (PE > PP > PVC) and even more than the natural sediments analysed. This is relevant because it shows that phenanthrene can be transported into the sediments, increasing its concentration in them (Teuten et al., 2007). It could also happen that the microplastics residing in the sediments release the phenanthrene into the surrounding water. This would initiate an almost endless cycle of sorption, transport, and desorption.

It should be noted that the addition of very small amounts of microplastics that have sorbed phenanthrene to sediments inhabited by *Arenicola marina* shows the accumulation of phenanthrene in *Arenicola marina*. Therefore, microplastics may be important transport vectors for hydrophobic contaminants into the sediments and the organisms living there (Teuten et al., 2007).

On the other hand, the sinking of microplastics that have adsorbed pollutants from the sea surface can lead to the transport (sequestration) of pollutants into deep-sea sediments (see Figure 9.3). Due to the strong alliance between microplastics and pollutants, the latter are no longer bioavailable and can hardly be degraded by bacteria (Teuten et al., 2007). These results suggest that the pollutant is masked, which may 'not justify' the need for remediation (Guo and Wang, 2019).

Finally, the transportation and release of additives, such as flame retardants, incorporated during the manufacture of plastics could have a very negative impact on the environment and ecosystems. Even the ageing of microplastics is generally considered as a factor that accelerates the release of these additives (Cheng et al., 2020).

9.3.4 Coronae Formation

When microplastics reach the planet's aquatic environments (marine and freshwater), they attach to biomolecules, forming what is described as an eco-corona (see Figure 9.3), which can facilitate organisms ingesting the microplastics (Ramsperger et al., 2020). Some examples described in the literature are illustrated below.

In this study, microplastics were incubated to create an eco-corona under conditions like those experienced by microplastics in nature. It was observed that the treated microplastics were internalised into macrophages significantly more frequently than pristine microplastics. It was also able to identify biomolecules that form an eco-corona on the outer part of the microplastic. The environmental characteristics create the conditions for the cellular internalisation of the microplastics.

This indicates that cellular internalisation is a key route by which microplastic particles translocate into tissues, with implications for the environment and human health (Ramsperger et al., 2020).

Microplastics can interact with secretory proteins of aquatic organisms, and form protein-coated microplastic corona complexes. The protein-coated PS and the corona-microplastic complex inhibit digestion in adult zebrafish (*Danio rerio*). Therefore, the ecological risk of microplastics may be underestimated (Hongwei et al., 2022).

9.4 FATE OF MICROPLASTICS

The fate of microplastics is related to their small size (<5 mm) and their shape (granules, foams, fragments, flakes, films, fibres, and sponges) (Cao et al., 2022). They can be transported and hosted in the three fundamental sources of the biosphere: soil, water, and air. When present in these environments they become bioavailable, so, they can bioaccumulate (be absorbed and concentrate) in many organisms (plants, animals, fungi, microbes) (Madani, 2020). This can lead to the death of many of these organisms.

Today, there is a method designed to geographically regionalise macroplastic and microplastic emissions to soil, water, and air, according to plastic type. Geographical and statistical data, including records of plastic emissions in Switzerland, were applied to develop this method. Although these emissions can be found in areas with high human activity, it is expected that the emission maps will be used to obtain and predict the fate patterns of macro- and microplastics, globally (Kawecki and Nowack, 2020).

9.4.1 RIVERS

Rivers are one of the main routes by which microplastics reach the oceans (Liu et al., 2021). These plastic particles, when transported downstream, accumulate in riverbed sediments. To predict the residence times of microplastics in rivers, a model was devised that considers the particle exchange processes observed between surface water and riverbed sediments. The simulation results, after validation (Roter Main River in Germany), show that the residence times from the headwaters to the main channel are longer in the headwaters. Values can range from 5 hours/km (high-flow = stormflow conditions) to 7 years/km (low-flow) (Drummond et al., 2022).

A study carried out in the Dafeng River shows that, although microplastics (>1 mm) have a high abundance in the river water (80%), they accumulate in a lower proportion (40%) in the sediments (the top 3 cm of the surface layer). Smaller microplastics (<1 mm), on the other hand, sediment in a higher proportion (60%), although their abundance in the water is much lower (20%) (Liu et al., 2021).

The above results may indicate that one part of the larger microplastics (>1 mm) are carried away and fragmented by the current. On the other hand, these smaller fractions (<1 mm), as they are formed, are filtered, and stored in the inner layers of the sediments. Thus, although the common perception is to view rivers and streams as only transport pathways for microplastics, recent studies indicate that river systems are also areas where microplastic accumulation occurs. For instance, the abundance (~30,000 particles/kg dry weight) of microplastics (0.02–0.05 mm) was observed in the sediments (60 cm deep frozen cores) (Frei et al., 2019).

These results suggest the accumulation of microplastics in the hyporheic zone of the river, which is of concern as it is the habitat of invertebrates that are the source of river food webs. Furthermore, when looking at the exchange of microplastics, in the sediment zone and the surface water flow of the river, there is concern that microplastics may be transported to local aquifers and be present during drinking water abstraction (Frei et al., 2019).

The quantity of microplastics that rivers can concentrate, while linked to residential activities, is also related to seasonal changes. For example, the flow of the Dafeng River in dry seasons can have up to three times more microplastics than in rainy seasons. In fact, it is estimated that the microplastic load of the Dafeng River can reach 830 million plastic particles annually (Liu et al., 2021).

On the other hand, the characteristics of the microplastics present in the water, in the sediments, and ingested by the fish in the Dafeng River, are discussed below (Liu et al., 2021):

Colour: although blue microplastics have a significantly lower abundance (8%), both in water and sediments, fish are preferentially inclined to ingest them (56%), even though white microplastics are present in higher proportions, both in water (80%) and sediments (75%), i.e., white microplastics are ingested in just 35%. It can also be observed that the preference of fish for other colours is as follows: red (8%), yellow (4%), black (3%), and green (2%).

Polymer: fish ingest considerably more PET (55%) than Rayon (16%), even though the abundance of both is very similar, both in the water (22%) and in the sediments (26%). At the same time, PS (8%) and Nylon (3%), which are present in the water and in the sediments, seem not to be ingested by fish. In contrast, PU (2%) and PS (1%), present in smaller proportions in both water and sediment, are ingested, 3% and 2% respectively.

Form: fish ingest far more fibres (90%) than films (8%) or particles (2%). This is almost consistent with the abundance of fibres (78%) in both water and sediments. However, it seems that among all forms, foams present in the water (13%) and in sediments (7%) do not seem to be ingested by fish.

Size: the microplastics most frequently ingested by fish (74%) are in the range of 5 to 0.5 mm, while smaller microplastics (<0.5 mm) are ingested by just 26%. This pattern of abundances is like that of the sediments. However, in river water, the abundance of larger microplastics (5–0.5 mm) is 90%, while the abundance of smaller ones (<0.5 mm) is 10%.

9.4.2 ESTUARIES

Estuaries as rivers also are a key pathway for microplastics entering the oceans. In proximity to populated or urban areas, the diversity and abundance of microplastics has been found to increase in rivers and coastal marine environments. Many persons live in coastal ecosystems that are estuaries. However, microplastic contamination is inadequately studied contrasted to river and sea. River basins can provide a pathway for plastic waste to enter estuaries, such as poorly managed plastic waste and waste from sewage treatment plants. However, secondary microplastics are difficult to associate to particular causes, and it is probable that some ways and origins remain unknown (Adam, Jessica L and Malcolm D, 2022).

Regarding the abundance of microplastics in the water column of three estuaries that are subject to different scales of human impact, the following is reported. The Clyde estuary, which has few human interventions, had the lowest amount of microplastics (98 particles/m^3). The Bega estuary, which has a small municipality and a single sewage treatment plant, had the second highest amount of microplastics (246 particles/m^3). Finally, the Hunter Estuary, which has multiple municipalities, multiple sewage treatment plants, and heavy industry, had the highest amount of microplastics (1,032 particles/m^3). These particles were mainly in the form of fragments rather than fibres. (< 0.2 mm). In the Hunter Estuary, the abundance of microplastics was sometimes as high as zooplankton. This high level has also been observed in other systems around the world. The investigations show that higher microplastic contamination leads to greater levels of anthropogenic impact (Hitchcock and Mitrovic, 2019).

9.4.3 SEAS AND OCEANS

The accumulation of microplastics in the marine environment has been well documented. It is estimated that 80% of the waste that is in the sea is plastic, and that plastic has already invaded the most remote marine environments (Cheng et al., 2020). Microplastic fibres contribute 91% of all plastics collected in global seawater samples (Walkinshaw et al., 2020).

An estimated 8 Mt of plastic waste is discarded into the oceans every year. Of this, 80% comes from land-based sources. At the same time, 50% of the plastic products frequently found on beaches are single-use (bags, food packaging, bottles, and utensils) (Pon, 2019) .

The first reports of plastic particles in the sea date from 1970–80. Subsequently, the widespread presence of microplastics in the oceans was reported. The plastics found were PVC, PS, PE, PA, PET, and PP (Vieira et al., 2021).

Some 80% of the world's ocean plastics enter the ocean via 1,656 rivers (Meijer et al., 2021), while the other 20% come from marine sources such as fishing nets, ropes, and fleets (Data, 2022).

Microplastics have been found floating in the Pacific, Atlantic, Arctic, and Antarctic oceans. They have also been found in more than 170 marine species. These taxa consist of zooplankton, coral reefs, marine invertebrates (including molluscs and crustaceans), fish (including sharks), seabirds (including penguins), reptiles (including turtles), and marine mammals (including seals, dolphins, and whales) (Miller, Santillo and Johnston, 2016; Nelms et al., 2018; Soares et al., 2020). See Figure 9.7.

These microplastics are highly bioavailable to aquatic organisms, either through direct ingestion or indirectly. They are transferred within the food chain when predators ingest prey that has consumed plastics. They can even accumulate in prey and predators, altering their behaviour (Nelms et al., 2018).

Today, research conducted in the marine environment on microplastics that are present in the sea focuses only on their presence and abundance. On the other hand, research carried out in the laboratory tends to focus on the ingestion of microplastics by zooplankton, and its effect on feeding, reproduction, and growth. However, it has also been observed that the microplastics used in these experiments are far from those found in the marine environment, especially in terms of age, shape, type, and concentration (Botterell et al., 2019).

9.4.4 SOIL

Less attention has been paid to the soil. It is a larger reservoir of microplastics than the ocean. It is estimated that microplastic contamination on land could be up to 23 times higher than in the oceans. Soils may be contaminated due to the billions of tonnes of plastic waste that have been buried underground for decades, as well as from sewage irrigation, plastic mulching, sludge use, and atmospheric

FIGURE 9.7 Dead fish on the beach, probably due to ingestion of microplastics which have been found in more than 170 marine species, including fish.

deposition. In fact, they have been found in plains, beaches, and shallow and deep agricultural soils. Also, accidental loss and improper handling of plastic waste in landfills and in urban and industrial centres can lead to soil input and contamination of terrestrial soils (Cao et al., 2022).

Although not much is known about the cascade of events at the fundamental levels of terrestrial ecosystems, microplastics are known to alter the biophysical properties of soil. For example, they can cause significant changes in soil microbial activities and plant biomass. It was observed in this study that microplastics, similar in shape to other natural soil particles, cause few changes in soil structure and water dynamics. Polyester fibres and polyamide beads caused the most pronounced impacts on plant traits and function. This implies that widespread microplastic contamination in soil may affect plant performance, terrestrial biodiversity, and agroecosystems (De Souza Machado et al., 2019).

On the other hand, some studies show that animal movement is an important contributor to the transport of microplastic particles. For example, the predator–prey relationship significantly promotes the movement of microplastic particles, more so than if there were only one species. It is probably predation pressure that increases the dispersal of microplastics in the soil. Soil microarthropods, due to their small body size, can enter through soil pores. This can expose other biota to microplastics, changing the physical properties of soils. The pores through which microarthropods enter the soil can serve as a pathway for other organisms to be exposed to the microplastics, which can also alter the physico-chemical properties of the soil (Zhu et al., 2018).

9.4.5 Air

Atmospheric microplastics have been of great concern because the wind can carry those long distances. The microplastics that can be displaced by the wind come from synthetic fibres (clothing) as well as from construction materials, incineration of plastics, and road dust. Most atmospheric microplastics are microfibers, and their concentration in indoor environments is much higher than in outdoor areas. The reported size of atmospheric microplastics ranges from 0.002 to 9.55 mm, but it is not known how long they can be suspended in the air (Cao et al., 2022).

9.4.6 Humans

Humans are exposed daily to microplastics by inhalation (Danopoulos et al., 2022). Atmospheric microplastics are particles so fine that they can reach the lower airways and cross the air–blood barrier in mammals, leading to severe lung diseases such as pulmonary fibrosis and pleural granuloma formation (Cao et al., 2022).

Humans are also constantly ingesting microplastics (Danopoulos et al., 2022). Unfortunately, microplastics are ingested by many animals, and by this route they can also enter the human diet. The abundance of microplastics in molluscs, crustaceans, fish, and echinoderms indicates that each human may consume up to 55,000 marine microplastics per year (Cao et al., 2022).

Although most ingested microplastics are excreted, existing data suggest that accumulation of these particles in human tissues would have long-term adverse effects (Danopoulos et al., 2022). Indeed, it has been shown that some microplastics can cross the intestinal barrier and reach the liver and spleen. Microplastic excretion rates can be closely related to the composition, size, and shape of these particles; for example, particles with a round, smooth shape may be excreted more rapidly than particles with a pointed shape or fibres (Cao et al., 2022).

The abundance of microplastics in molluscs, crustaceans, fish, and echinoderms means that each human can ingest up to 55,000 marine microplastics per year (Cao et al., 2022). Prevailing data suggest that accumulation of these particles in human tissues would have long-term adverse effects (Danopoulos et al., 2022).

Today, four different types of microplastics (0.514 mm to 700 nm) have been identified and quantified for the first time in the blood of 22 donors: PET, PE, PS, and PMMA. The average blood concentration of microplastics was 1.6 µg/ml. This pioneering study observed that plastic particles can be absorbed through the membranes of the human body and demonstrated that microplastics are bioavailable in the bloodstream. It remains to be seen whether the existence of these particles in the blood constitutes a public health risk (Leslie et al., 2022).

In addition, microplastics have been detected in many food products, such as vegetables, fruit, honey, desserts, soups, water, beverages, and salt. Also it is known that packaging can release significant amounts of microplastics into the food it contains. Furthermore, numerous microplastics have also been detected in the form of flakes or films (size, 0.6–332 µm) in the sterilisation of plastic baby bottles, and it is known that babies can consume up to 660,000 microplastics if they drink from a plastic bottle for a year (Cao et al., 2022).

9.5 MAIN GLOBAL CHALLENGES

The analysis in the previous sections shows that it is essential to solve the problem of microplastic generation and pollution. The three most relevant aspects, which cannot be avoided, are the production, use, and disposal of plastics. From this perspective, the main challenges to be addressed at the global level could be divided along 4 lines: business, governments, urban and rural authorities, and citizens.

9.5.1 BUSINESS

- Reduce the production and use of primary plastics, prioritising their rationalised use and the invention, production, and appropriate use of biodegradable plastics.
- Reduce the generation of plastic waste in both the goods and services sectors.
- Take responsibility and clean up the air, water, and soil that is polluted by plastics.

9.5.2 GOVERNMENTS

- Promote an efficient solid-waste resource management system, in which plastics and all other waste are treated as resources and not as waste.
- Ban the use of non-biodegradable plastics in single-use items.
- Create laws that encourage companies to clean up the air, water, and soil contaminated by plastics.

9.5.3 URBAN AND RURAL AUTHORITIES

- Establish an efficient solid-waste resource management system, in which plastics and all other waste are treated as resources and not as waste. Prioritise the separation of plastics before collection, as well as paper, cardboard, metals, glass, and organic matter.
- Prevent plastics from being sent to illegal landfills, buried, or incinerated.
- Support the reuse and recycling of plastics. Promote plastics as a resource and not as waste. Make it easier for citizens to compost organic matter in their homes and businesses.

9.5.4 CITIZENS

- Minimise the consumption of plastic products. Reduce, reuse, and recycle plastic waste and do not burn it.

- Consider plastics as a resource and not as waste. Separate plastic waste as well as paper, cardboard, metals, and glass. Do not mix them with organic matter, so that they can be used again or recycled.
- Compost organic matter without mixing it with other waste.

9.6 INNOVATIONS AND FUTURE RESEARCH

9.6.1 MICROROBOTS

Microrobots (on-the-fly) capture nanoplastics in three-dimensional (3D) space. These microrobots, derived from MXene, have a powerful 3D motion, and a programmable surface charge. In water, the microrobots can rapidly trap the nanoplastics with an efficiency of 97%, and they can be magnetically removed from water. *Could microbots also trap microplastics?*

9.6.2 HYDROGENOLYSIS

Hydrogenolysis could be the most viable catalytic recycling technology of the future. For example, under mild conditions, it can transform HDPE into lubricants and jet fuel. Besides, hydrogenolysis of LDPE produces oils and waxes. *Could hydrogenolysis transforms microplastics into lubricants, jet fuel, oils, and waxes?*

9.6.3 CATALYTIC DEPOLYMERISATION

Catalytic depolymerisation of certain polyester and PC wastes can transform them into diols, taking their recycling to a higher level, i.e., upcycling (when recycling generates value-added products). *Could catalytic depolymerisation achieve upcycling of microplastics?*

9.6.4 SUPER/SUBCRITICAL WATER TECHNOLOGY

Hydrolysis of ABS in super/subcritical water gave thirteen important products, including toluene, benzene, styrene, naphthalene, biphenyl, and terphenyl. Super/subcritical water will play a key role in future waste recycling. *Could hydrolysis in super/subcritical water give key products such as toluene or benzene from microplastics?*

9.7 CONCLUSIONS

The productive activities carried out by consumer societies generate MSW, the global production of which is expected to reach 2,200 Mt by 2025. MSW is a very important part of PSW. Primary microplastics may be contained in PSW. Secondary microplastics can be generated from PSW. Both types of microplastics (size <5 mm) pollute the environment (rivers, lakes, soil, seas, and oceans) when MSW is mismanaged. By the end of 2022, 380 Mt of PSW will have been generated. Of this, 83 Mt (22%) will be released into the environment. In addition, according to the OECD, 20 Mt of macroplastics (naturally occurring plastic waste >5 mm in size, which is an active source of microplastics) and 3 Mt of microplastics will have leaked into the environment. An estimated 1,700 Mt of PSW have been released into the environment in the last six decades. Thus, it is expected that an extensive amount of secondary microplastics is actively polluting every corner of the planet. While the quantities of macroplastics and microplastics entering the terrestrial and aquatic environment come from many sources, the most ubiquitous types of microplastics are PS, PE, PP, PA, and PVC. It is imperative that PSW is effectively treated through different recovery methods, including in situ recycling, mechanical recycling, thermolysis, chemolysis, biological treatments, and energy recovery so that it does not generate microplastics that pollute the environment.

When present in the environment, due to mismanagement of MSW and litter, microplastics (primary and secondary) may go through mechanical, photo-oxidative, and biological degradation. Also, coronae formation (protein-coated microplastic corona complexes) and microplastics may be important transport vectors for hydrophobic contaminants, with the subsequent environmental and health implications.

The ubiquity of microplastics is expressed in an ever-changing and extensive fate. That is, its fate may be in zooplankton, fish, and even whales. In the same way, it may be destined for the air, the ground, fresh water, and even in the seas and oceans. But the fate of microplastics can also be in the blood and organs of human beings.

In summary, it is essential to solve the problem of microplastic. The three most relevant aspects, which cannot be avoided, are the production, use, and disposal of plastics. From this perspective, the main challenges to be addressed at the global level could be divided along 4 lines: business, governments, urban and rural authorities, and citizens. In general, the objectives to reduce the generation of microplastics on the planet are a) minimising the consumption of plastic products; b) reducing, reusing, and recycling plastic waste and not burning it; c) considering plastics as a resource and not as waste; d) separating plastic waste, as well as paper, cardboard, metals, glass, and organic matter; and e) composting organic matter without mixing it with other waste.

It is expected that new technologies (focusing on microrobots, hydrogenolysis, catalytic depolymerisation, and super/subcritical water) could reduce the amount of microplastics in the environment, minimising the pollution they produce, with the advantage of being able to transform microplastics into key products such as lubricants, jet fuels, oils, waxes, diols, toluene, or benzene, taking their recycling to a higher level (upcycling), when recycling generates value-added products. Could these innovations minimise microplastics and their pollution?

ABBREVIATIONS AND ACRONYMS

HDPE	High-density polyethylene
LDPE	Low-density polyethylene
PA	Polyamide
PC	Polycarbonate
PE	Polyethylene
PET	Polyethylene terephthalate
PLA	Polylactic acid (biodegradable)
PMMA	Polymethylmethacrylate
PP	Polypropylene
PS	Polystyrene

GLOSSARY

This glossary considers the terms and definitions presented by the OECD, in the book *Global Plastics Outlook: Policy Scenarios to 2060* (OECD, 2022a).

DEGRADATION

The partial or total decomposition of a polymer due to its vulnerability to oxygen, UV radiation, or biological factors. This implies modification of the properties, such as yellowing, cracking, and fragmentation.

MACROPLASTICS

Littered plastic items that are still recognisable, such as bottles and packaging. In this chapter, the term covers plastics larger than 5 mm in diameter.

MICROPLASTICS

Solid synthetic polymers less than 5 mm in diameter.

MISMANAGED WASTE

Waste that is burned in open holes, abandoned in seas and fresh water, or placed in illegal or contaminated landfills, i.e., waste that is not collected or treated with advanced technology.

PLASTIC LEAKAGE

Plastic leakage describes plastics that go into terrestrial and aquatic environments.

PLASTIC LITTER

Waste that results from littering by individuals in the environment and from fly-tipping. Littered waste is distinct from mismanaged waste. Littered waste generally remains uncollected and leaks into the environment or can be collected for further disposal.

PLASTIC POLLUTION

In general, all risks and emissions related to plastics at any stage of their production, use, or waste treatment, as well as those resulting from mismanagement, leaks, and litter.

PRIMARY MICROPLASTICS

Plastics that have a diameter of less than 5 mm and that have been designed and manufactured for some purpose, such as microfibers, cosmetic cleaning agents, and plastic pellets.

PRIMARY PLASTICS

Plastics made from fossil (crude oil) or bio-based (corn, sugar cane, wheat) raw materials. In other words, they have never been utilised or processed before.

SANITARY LANDFILL

The final deposition of household waste above- or underground, in a controlled manner and in accordance with state-of-the-art health, safety, and ecological requirements.

SECONDARY PLASTICS

Plastic polymers made from recycled material.

Secondary Microplastics

Microplastics formed from the fragmentation of larger plastics. These can be microplastics derived from tyre abrasion, microfibers released from synthetic textile products, and microplastics generated from macroplastics found abandoned in the environment.

Subcritical Water Zone

It ranges between 100 and 374 °C and between 1 and 22.1 MPa. As the temperature of subcritical water increases, its dielectric constant decreases, allowing it to dissolve non-polar substances.

Supercritical Water

It is found at pressure and temperature conditions above those that define its critical point (22.1 MPa and 374 °C). The dielectric constant of water makes it an effective solvent for organic compounds. Moreover, once the critical point is exceeded, the ionic product decreases drastically. Thus, free radical reactions in supercritical water are favoured.

Waste Incineration

Incineration in a state-of-the-art industrial facility.

REFERENCES

Abdel-Shafy, H.I. and Mansour, M.S.M. (2018) 'Solid waste issue: Sources, composition, disposal, recycling, and valorization', *Egyptian Journal of Petroleum*. doi:10.1016/j.ejpe.2018.07.003.

Adam, B.-W., Jessica L, S. and Malcolm D, H. (2022) 'The estuarine plastics budget: A conceptual model and meta-analysis of microplastic abundance in estuarine systems', *Estuarine, Coastal and Shelf Science [Preprint]*.

Arpia, A.A. et al. (2021) 'Microplastic degradation as a sustainable concurrent approach for producing bio-fuel and obliterating hazardous environmental effects: A state-of-the-art review', *Journal of Hazardous Materials*, 418. doi:10.1016/j.jhazmat.2021.126381.

Astner, A.F. et al. (2019) 'Mechanical formation of micro- and nano-plastic materials for environmental studies in agricultural ecosystems', *Science of the Total Environment*, 685. doi:10.1016/j.scitotenv.2019.06.241.

Baeza, C. et al. (2020) 'Experimental Exposure of Lumbricus terrestris to Microplastics', *Water, Air, and Soil Pollution*, 231(6). doi:10.1007/s11270-020-04673-0.

Bamshad, A. and Cho, H.J. (2021) 'A novel print-and-release method to prepare microplastics using an office-grade laserjet printer; a low-cost solution for preliminary studies', *Marine Pollution Bulletin*, 170. doi:10.1016/j.marpolbul.2021.112601.

Bhat, S.A. et al. (2022) 'Removal Potential of Microplastics in Organic Solid Wastes via Biological Treatment Approaches', in *Microbial Biotechnology: Role in Ecological Sustainability and Research*.

Biale, G. et al. (2021) 'A systematic study on the degradation products generated from artificially aged microplastics', *Polymers*, 13(12). doi:10.3390/polym13121997.

Botterell, Z.L.R. et al. (2019) 'Bioavailability and effects of microplastics on marine zooplankton: A review', *Environmental Pollution*. doi:10.1016/j.envpol.2018.10.065.

Cao, J. et al. (2022) 'Coronas of micro/nano plastics: a key determinant in their risk assessments', *Particle and Fibre Toxicology* [Preprint]. doi:10.1186/s12989-022-00492-9.

Chen, Z. et al. (2020) 'Enhanced in situ biodegradation of microplastics in sewage sludge using hyperthermophilic composting technology', *Journal of Hazardous Materials*, 384. doi:10.1016/j.jhazmat.2019.121271.

Cheng, H. et al. (2020) 'Release kinetics as a key linkage between the occurrence of flame retardants in microplastics and their risk to the environment and ecosystem: A critical review', *Water Research*. doi:10.1016/j.watres.2020.116253.

Danopoulos, E. et al. (2022) 'A rapid review and meta-regression analyses of the toxicological impacts of microplastic exposure in human cells', *Journal of Hazardous Materials*. doi:10.1016/j.jhazmat.2021.127861.

Data, O.W. in (2022) *Where does the plastic in our oceans come from?*

De Souza Machado, A.A. et al. (2019) 'Microplastics can change soil properties and affect plant performance', *Environmental Science and Technology*, 53(10). doi:10.1021/acs.est.9b01339.

Debroy, A., George, N. and Mukherjee, G. (2021) 'Role of biofilms in the degradation of microplastics in aquatic environments', *Journal of Chemical Technology and Biotechnology* [Preprint]. doi:10.1002/jctb.6978.

Dey, S., Anand, U., Kumar, V., Kumar, S., Ghoraj, M., Ghosh, A., Kant, N., Suresh, S., Bhattacharya, S., Bontempi, E., Bhat, S.A., Dey, A., 2023. Microbial strategies for degradation of microplastic generated from COVID-19 healthcare waste. *Environmental Research* 216, 114438. https://doi.org/10.1016/j.env res.2022.114438.

Dimerc Office (2013) *Technical data sheet: Detergente Regular Ariel.*

Drummond, J.D. et al. (2022) 'Microplastic accumulation in riverbed sediment via hyporheic exchange from headwaters to mainstems', *Science Advances*, 8(2). doi:10.1126/sciadv.abi9305.

Efimova, I. et al. (2018) 'Secondary microplastics generation in the sea swash zone with coarse bottom sediments: Laboratory experiments', *Frontiers in Marine Science*, 5(SEP). doi:10.3389/fmars.2018.00313.

Ferreira-Filipe, D.A. et al. (2022) 'Are mulch biofilms used in agriculture an environmentally friendly solution? An insight into their biodegradability and ecotoxicity using key organisms in soil ecosystems', *Science of the Total Environment*, 828. doi:10.1016/j.scitotenv.2022.154269.

Fragão, J. et al. (2021) 'Microplastics and other anthropogenic particles in Antarctica: Using penguins as biological samplers', *Science of the Total Environment*, 788. doi:10.1016/j.scitotenv.2021.147698.

Frei, S. et al. (2019) 'Occurrence of microplastics in the hyporheic zone of rivers', *Scientific Reports*, 9(1). doi:10.1038/s41598-019-51741-5.

Gigault, J. et al. (2021) 'Nanoplastics are neither microplastics nor engineered nanoparticles', *Nature Nanotechnology*. doi:10.1038/s41565-021-00886-4.

Guo, X. and Wang, J. (2019) 'The chemical behaviors of microplastics in marine environment: A review', *Marine Pollution Bulletin*. doi:10.1016/j.marpolbul.2019.03.019.

Habib, S. et al. (2020) 'Biodeterioration of untreated polypropylene microplastic particles by antarctic bacteria', *Polymers*, 12(11). doi:10.3390/polym12112616.

Hitchcock, J.N. and Mitrovic, S.M. (2019) 'Microplastic pollution in estuaries across a gradient of human impact', *Environmental Pollution*, 247. doi 10.1016/j.envpol.2019.01.069.

Hongwei, L. et al. (2022) 'Protein-coated microplastics corona complex: An underestimated risk of microplastics', *Science of The Total Environment*, 851.

Hu, L. et al. (2022) 'Accumulation of microplastics in tadpoles from different functional zones in Hangzhou Great Bay Area, China: Relation to growth stage and feeding habits', *Journal of Hazardous Materials*, 424. doi:10.1016/j.jhazmat.2021.127665.

Idumah, C.I. and Nwuzor, I.C. (2019) 'Novel trends in plastic waste management', *SN Applied Sciences*. doi:10.1007/s42452-019-1468-2.

Kawecki, D. and Nowack, B. (2020) 'A proxy-based approach to predict spatially resolved emissions of macro- and microplastic to the environment', *Science of the Total Environment*, 748. doi:10.1016/j.scitotenv.2020.141137.

Lake Ontario (2016) 'Zooming in on the Five Types of Microplastics', *Swim Drink Fish.* www.swimdrinkf ish.ca/.

Leslie, H.A. et al. (2022) 'Discovery and quantification of plastic particle pollution in human blood', *Environment International*, 163. doi:10.1016/j.envint.2022.107199.

Liu, J. et al. (2022) 'The protective layer formed by soil particles on plastics decreases the toxicity of polystyrene microplastics to earthworms (Eisenia fetida)', *Environment International*, 162. doi:10.1016/j.envint.2022.107158.

Liu, S. et al. (2021) 'The distribution of microplastics in water, sediment, and fish of the Dafeng River, a remote river in China', *Ecotoxicology and Environmental Safety*, 228. doi:10.1016/j.ecoenv.2021.113009.

Madani, S. (2020) 'Analyse et protection de l'environnement', *Polycopié pour la matière Analyse et Protection de l'Environnement*. https://elearning.univ-msila.dz/moodle/pluginfile.php/232145/mod_resource/cont ent/1/Chapitre%203%20%28APE%29%203LEE.pdf.

Meijer, L.J.J. et al. (2021) 'More than 1000 rivers account for 80% of global riverine plastic emissions into the ocean', *Science Advances*, 7(18). doi:10.1126/sciadv.aaz5803.

Miller, K., Santillo, D. and Johnston, P. (2016) 'Plastics in Seafood – full technical review of the occurrence, fate and effects of microplastics in fish and shellfish, Greenpeace', *Greenpeace Research Laboratories Technical Report*. www.greenpeace.to/greenpeace/wp-content/uploads/2016/07/plastics-in-seafood-technical-review.pdf.

Nelms, S.E. et al. (2018) 'Investigating microplastic trophic transfer in marine top predators', *Environmental Pollution*, 238. doi:10.1016/j.envpol.2018.02.016.

OECD (2022a) '*Global Plastics Outlook: Policy Scenarios to 2060*', Organisation for Economic Cooperation and Development. https://doi.org/10.1787/aa1edf33-en.

OECD (2022b) '*Global Plastics Outlook Economic Drivers, Environmental Impacts and Policy Options*', Organisation for Economic Cooperation and Development. https://rds.org.co/apc-aa-files/205ec78c9 cca6d1850bdca24e20e50bf/document.pdf.

OECD (2022c) '*Plastic pollution is growing relentlessly as waste management and recycling fall short, says OECD*', Organisation for Economic Cooperation and Development. www.oecd.org/environment/plas tic-pollution-is-growing-relentlessly-as-waste-management-and-recycling-fall-short.htm.

OECD.Stat (2022) *Organisation for Economic Cooperation and Development*. www.oecd-ilibrary.org/econom ics/data/oecd-stat_data-00285-en.

Paruta, P., Pucino, M. and Boucher, J. (2021) '*Plastic Paints the Environment*', www.e-a.earth/wp-content/uplo ads/2023/07/plastic-paint-the-environment.pdf.

Pastorino, P. et al. (2022) 'Microplastics Occurrence in the European Common Frog (Rana temporaria) from Cottian Alps (Northwest Italy)', *Diversity*, 14(2). doi:10.3390/d14020066.

Pon, J. (2019) 'Instrumentos para la implementación efectiva y coherente de la dimensión ambiental de la agenda de desarrollo. Caso 4 Residuos', *Comisión Económica para América Latina y el Caribe* [Preprint].

Ragusa, A. et al. (2021) 'Plasticenta: First evidence of microplastics in human placenta', *Environment International*. doi:10.1016/j.envint.2020.106274.

Ramsperger, A.F.R.M. et al. (2020) 'Environmental exposure enhances the internalization of microplastic particles into cells', *Science Advances*, 6(50). doi:10.1126/sciadv.abd1211.

Ritchie, H. and Roser, M. (2022) '*Plastic Pollution*', *Our World In Data*. https://ourworldindata.org/plastic-pollution.

Rummel, C.D. et al. (2022) 'Effects of leachates from UV-weathered microplastic on the microalgae Scenedesmus vacuolatus', *Analytical and Bioanalytical Chemistry*, 414(4). doi:10.1007/s00216-021-03798-3.

Ryan, L. (2020) *AQA GCSE Chemistry*. Third edit. Oxford University Press.

SEMARNAT (2005) 'Generación de residuos sólidos municipales', *Secretaría de Medio Ambiente y Recursos Naturales*. https://apps1.semarnat.gob.mx:8443/dgeia/informe_resumen/08_residuos/cap8.html#2.

Soares, M. de O. et al. (2020) 'Microplastics in corals: An emergent threat', *Marine Pollution Bulletin*, 161. doi:10.1016/j.marpolbul.2020.111810.

Song, Xiaowei et al. (2021) 'Sorption and desorption of petroleum hydrocarbons on biodegradable and nondegradable microplastics', *Chemosphere*, 273. doi:10.1016/j.chemosphere.2020.128553.

Strohriegl, P. et al. (2021) 'Reconstructing the environmental degradation of polystyrene by accelerated weathering', *Environmental Science and Technology*, 55(12). doi:10.1021/acs.est.0c07718.

Suzuki, G. et al. (2022) 'Mechanical recycling of plastic waste as a point source of microplastic pollution', *Environmental Pollution*, 303. doi:10.1016/j.envpol.2022.119114.

Tanaka, S. et al. (2023) *The Handbook of Environmental Chemistry*. Springer. Electronic ISSN 1616-864X. www.springer.com/series/0698.

Teuten, E.L. et al. (2007) 'Potential for plastics to transport hydrophobic contaminants', *Environmental Science and Technology*, 41(22). doi:10.1021/es071737s.

Thomas, J. et al. (2022) 'Investigation of abiotic degradation of tire cryogrinds', *Polymer Degradation and Stability*, 195. doi:10.1016/j.polymdegradstab.2021.109814.

Thompson, R.C. et al. (2004) 'Lost at sea: Where is all the plastic?', *Science*, 304(5672), p. 838. doi:10.1126/ science.1094559.

Turner, A. (2021) 'Paint particles in the marine environment: An overlooked component of microplastics', *Water Research X*, 12(100110). doi:10.1016/j.wroa.2021.100110.

UNEP (2015) *Plastics in cosmetics, are we polluting the environment through our personal care?, The Global Programme of Action for the Protection of the Marine Environment from Land-based Activities (GPA)*. United Nations Environment Programme. https://wedocs.unep.org/bitstream/handle/20.500.11822/9664/-Plastic_in_cosmetics_Are_we_polluting_the_environment_through_our_personal_care_-2015Plas.pdf?sequence=3&%3BisAllowed=.

UNEP (2022) *World leaders set sights on plastic pollution*. United Nations Environment Programme. www.unep.org/news-and-stories/story/world-leaders-set-sights-plastic-pollution.

van Wijnen, J., Ragas, A.M.J. and Kroeze, C. (2019) 'Modelling global river export of microplastics to the marine environment: Sources and future trends', *Science of the Total Environment*, 673. doi:10.1016/j.scitotenv.2019.04.078.

Vethaak, A.D. and Legler, J. (2021) 'Microplastics and human health', *Science*, 371(6530). doi:10.1126/science.abe5041.

Vieira, Y. et al. (2021) 'Microplastics physicochemical properties, specific adsorption modeling and their interaction with pharmaceuticals and other emerging contaminants', *Science of the Total Environment*, 753(141981). doi:10.1016/j.scitotenv.2020.141981.

Walkinshaw, C. et al. (2020) 'Microplastics and seafood: Lower trophic organisms at highest risk of contamination', *Ecotoxicology and Environmental Safety*, 190, p. 110066. doi:10.1016/J.ECOENV.2019.110066.

Zhu, D. et al. (2018) 'Trophic predator-prey relationships promote transport of microplastics compared with the single Hypoaspis aculeifer and Folsomia candida', *Environmental Pollution*, 235. doi:10.1016/j.envpol.2017.12.058.

10 Antibiotic Pollution
Toxicity Assessment and Bioremediation Strategies

Gangar Tarun, Hazarika Risha, and Patra Sanjukta

10.1 INTRODUCTION

The use of antibiotics in human health and medicine has changed the track of the evolution of antibiotic resistance, especially in disease-causing pathogens. This has arisen because of undue selective pressure on the human and animal microbiota. This phenomenon also occurs in the antibiotic-polluted environment. The selection pressure mobilized the antibiotic resistance genes (ARGs) horizontally among many bacterial species, especially disease-causing pathogens. The culmination of the above events has led to consequences we are dealing with nowadays, such as increased antibiotic-resistant infections, increased mortality, and difficult-to-treat infections. Sometimes, the bacterium containing ARGs is directly transferred to the host and causes resistant infections. This phenomenon is called ecological connectivity.

Antibiotics make their way into the environment via various anthropogenic routes, primarily through urine and feces, improper and overproduced products, waste discharge from wastewater treatment plants (WWTPs), and direct contamination via aquaculture (Li et al. 2021).

When in the environment, antibiotics go through a range of events which include:

1. Degradation: some antibiotic molecules are naturally susceptible to degradation by existing environmental factors (photolysis, hydrolysis, etc.). They do not pose much threat to the environment in their native form.
2. Interaction with bacteria: upon their interaction with bacteria, antibiotics put selective pressure on the existing bacterial population to acquire resistance or develop certain mechanisms to escape the effect of antibiotics on them.
3. Interaction with anthropogenic factors: the harmful impact of antibiotics is more pronounced when they work in conjunction with some existing pollutants; examples include pharmaceutically active ingredients, endocrine disruptors, and microplastics. Antibiotics are also shown to stabilize when they are adsorbed onto soil particles.

Anthropogenic activities are the leading cause of the influx of antibiotics into the environment. Our treatment processes are not developed to treat/degrade molecules like antibiotics, which let them to escape in their native form. When antibiotics escape wastewater treatment plants, they encounter native and naïve organisms, which confer changes in the microbial community structure and function.

For the treatment of infectious diseases in both humans and animals, antibiotics are frequently administered. They are administered to cattle to boost meat production by reducing the risk of infections or disease outbreaks and fostering worldwide growth. Antibiotic manufacturing is still

DOI: 10.1201/9781003408352-12

TABLE 10.1
Average Concentration of Antibiotics Found in Surface Water of Different Nations

Country	Antibiotic	Average concentration (μgL^{-1})	References
South Africa (Streams/Rivers)	Nadilic acid	23	(Agunbiade and Moodley 2016)
	Ciprofloxacin	14	
Vietnam	Ofloxacin	17.7	(aus der Beek et al. 2016)
	Sulfamethoxazole	14.3	
Mozambique	Sulfamethoxazole	53.8	(Segura et al. 2015)
Kenya	Sulfamethoxazole	38.9	(Segura et al. 2015)
Ghana	Amoxicillin	0.0003	(Azanu et al. 2018)
	Cefalexin	0.868	
Australia	Cefalexin	0.027	(Mirzaei et al. 2018)
Croatia	Sulfamerzine	11	(Bielen et al. 2017)

growing, and global annual usage has increased from 100,000 to 200,000 tonnes during the first decade of the 21st century.

Antibiotic use is expected to increase globally by 200% by 2030 compared to 2015, with low- and middle-income nations witnessing the most significant increases. The global defined daily dose (DDD) consumption of antibiotics increased by 65% between 2000 and 2015, reaching 42 billion DDDs in 76 different nations. The United States, France, and Italy were the top three high-income antibiotic consumers in 2015, whereas India, China, and Pakistan were the top three consumers of antibiotics among low- and middle-income nations (Klein et al. 2018). The trends in antibiotic consumption in European nations differ significantly. According to the European Center for Disease Prevention and Control's (ECDC) *Antimicrobial Consumption in the EU/EEA: Annual Epidemiological Report for 2020*, Bulgaria, Croatia, Estonia, Greece, Hungary, Ireland, Italy, Latvia, Lithuania, Poland, and Slovakia showed increased usage of antibiotics from 2011 to 2020, while Belgium, Finland, Norway, Portugal, and Slovenia showed a statistically significant trend of decreasing antibiotic usage during the same time period (ECDC 2022).

The concentration of antibiotics inside the human body is approximately a million times more than that of the environment. If there is continuous drift in the microbial genetic pool in the environment, then imagine what could be the condition of human gut microbiota that encounters such a heavy dose of antibiotics.

From Table 10.1, it can be deduced that there is great variability in the concentration of antibiotics found in the streams and rivers of different nations. One thing which can be brought out here is that low- and middle-income countries (LMICs) have a greater concentration of antibiotics in their environmental streams even though the consumption of antibiotics is higher in high-income countries (HICs).

10.2 ANTIBIOTICS: USE AND ABUSE

The use of antibiotics and their consequential effect on the day-to-day lives of living organisms is manifold. They are commonly used for infection prevention and treatment purposes in humans, animals, or plants. Because antibiotics alter the course of bacterial infections, they have also contributed to longer predicted lifespans. Around the world, antibiotics reduce the morbidity and mortality brought on by food-borne and other poverty-related diseases in developing nations with poor sanitation. In animals, antibiotics are used as growth promoters and feed enhancers, leading to higher meat and milk production for poultry and cattle. This is also true for aquaculture produce. Apart from this, antibiotics are ubiquitously present in feed and water, leading to metaphylaxis or mass medication. As a result, animals receive antibiotics in substantial dosages even when it is not required. This administration of antibiotics in animals by the farmers and livestock rearers also provides sure economic benefits/profits, leading to their widespread use or excessive dosages. This ultimately results in the misutilization of many over-the-counter antibiotics.

10.3 SIDE EFFECTS OF ANTIBIOTICS

10.3.1 Effect of Antibiotics on Humans

Although necessary to treat infections, there are several adverse effects of antibiotics. Allergic reactions, photosensitivity, cross-reactivity with other drugs, digestive problems, nausea, vomiting, pain, diarrhea, and loss of appetite are some of the symptoms experienced. Since antibiotics target bacteria, a severe disadvantage of using them is the removal of the "good microflora" or the body's natural bacteria also. This imbalance may further cause a fungal infection of the vagina, digestive tract, or mouth (candidiasis). Vaginal yeast infections can also rise while treating urinary tract infections with antibiotics.

Among these good bacteria, the human gut microbiota is the most common non-pathogenic microorganisms that are affected by the administration of antibiotics. The perturbations of the gut bacteria due to surplus antibiotic treatment often lead to a persistent condition known as dysbiosis. It is characterized as an alteration in the microbial composition and function brought on by changes in the host-related environment that outweigh the microbial ecosystem's capacity for resistance. Gut dysbiosis is strongly linked with conditions like rheumatoid arthritis, inflammatory bowel disease, type II diabetes, and atopy. Long-term effects of antibiotic usage in humans include antibiotic-associated diarrhea (AAD) (which occurs in around 5–35% of patients due to low release of antimicrobial peptides), *Clostridium difficile*-associated diarrhea, *Helicobacter pylori* infection, increased susceptibility to *Salmonella typhimurium*, *Citrobacter rodentium*, and development of chronic conditions like cardiovascular disease, cancer, and even premature death.

In newborn and young children, consequences of antibiotic exposure are seen in terms of delayed gut microbiota development, asthma, obesity, adiposity, and modulations to the innate and adaptive immunity.

10.3.2 Effect of Antibiotics on Animals and Plants

Adding non-therapeutic antibiotics to the livestock feed adversely affects the farm animals' normal microbiota, leading to dysbiosis and the emergence of antibacterial resistance. It also causes several severe consequences for human health whereby antibiotic-resistant strains, like *Campylobacter coli/ jejuni*, *Salmonella enterica*, *Yersinia enterocolitica*, and MRSA CC398, under selective pressure propagate and spread, either via direct contact with veterinary slaughterhouses and farm animal handlers/workers or indirectly through the food chain, water, soil, or fertilizers contaminated with the animal excreta or genetic transfer between resistant bacteria in animals and pathogenic microorganisms infecting humans.

Antibiotics can potentially cure plant diseases; also primarily those brought on by bacteria. Their ability to be effective at low doses and minimal toxicity to plants sets antibiotics apart from metal-based bactericides. The most common disease in plants is the fire blight of apples and pears caused by the bacterium *Erwina amylovora*. The bacteria infect the flower through natural apertures. It multiplies and migrates to branches killing the plant quickly and causing gradual necrosis and wilting, rendering any chemical treatment ineffective. Therefore, antibiotics like streptomycin oxy-tetracycline, gentamicin, and oxalinic acid are applied on the flower surfaces to prevent the growth of pathogens. The frequency of antibiotics used in plant agriculture is low as they are relatively expensive. Therefore, they are mainly administered on high-value food and vegetable plants and plants of ornamental value where their worth might be recovered. Apart from this, antibiotics are also utilized in agriculture to improve crop productivity. However, several side effects of antibiotics have been observed, exemplified by the emergence of streptomycin-resistant plant pathogens like *E. amylovora Pectobacterium, Pectobacterium carotovora, Pseudomonas chichorii, Pseudomonas lachrymans, Pseudomonas syringae* pv. *Papulans*, and *Pseudomonas syringae* pv. *syringae Xanthomonas dieffenbachiae* (Mann et al. 2021).

Antibiotics released into the environment, like agricultural lands or soil ecosystems, via animal grazing, manure, or excreta can also indirectly affect plant growth and development, delay germination, disrupt the photosynthetic machinery, and lead to phytotoxicity (Carballo et al. 2021).

10.3.3　Effect of Antibiotics on Microbes

More than 90% of the bacterial population residing in the gut belongs to five significant phyla: *Firmicutes, Bacteroidetes, Actinobacteria, Proteobacteria,* and *Verrucomicrobia*. The remaining bacteria belong to less prevalent phyla: *Fusobacteria* and *Fibriobacteria*. The human gut microbiota mainly aids in performing metabolic functions like a breakdown of complex carbohydrates, synthesis of vitamins and amino acids, and furnishing nutrients to the host. These immunomodulatory functions produce bacteriocins to stop the growth of other pathogenic and competitive microorganisms and other regulatory processes. Excessive use of antibiotics diminishes the gut's microbial diversity, thus disrupting the existing balance between the various species. With the administration of antibiotics like meropenem, gentamicin, and vancomycin, the incidence of *Enterobacteriaceae* and other pathobionts increases while *Bifidobacterium* and other butyrate-producing species decrease. This leads to several alterations in the metabolite levels depending on the prevalent species in the gut. Metabolites produced by gut microbes facilitate the growth and multiplication of many other members, which is impaired due to these perturbations. Antibiotics like clindamycin target anaerobic gut bacteria, which results in the proliferation/colonization of *C. difficile* and increases the risk of pseudomembranous colitis, compared to the gut of healthy individuals. At the same time, the use of claritomycin and amoxicillin causes significant changes in the gut microbial compositions with the reduction of *Actinobacteria* and *Bifidobacterium* and dominance of *Enterobacteriaceae, Fermicutes,* and *Proteobacteria*. The impact of antibiotics on the resident/commensal bacteria of the gut also transcends to disrupted bacterial signaling, reduced secretion of antimicrobial peptides, disruption of the gut immune cells, and impairment of host-systemic immunity. The appearance of novel strains with resistant markers and point mutations has also been observed with the continuous misuse of antibiotics.

Therefore, the underlying effect of antibiotic overexploitation can be observed in both gram-positive and gram-negative bacteria, causing dysbiosis and affecting the immunomodulatory metabolic responses. The status of the microbiota, range of antibiotic administration food habits, duration, route, and post-antibiotic exposure environment also plays an essential role on how these changes will influence the individual's overall health.

10.3.4　Effect of Antibiotics on Environmental Microbiota

Antibiotics have been found in various ecosystems because of their massive manufacturing and heavy consumption in the health care, veterinary, aquaculture, etc. sectors, raising concerns on a global scale. Through excretion of human, animal, and domestic effluents in urine, feces, and manure, antibiotic drugs and their by-products are regularly dumped into natural ecosystems and linger around through bioaccumulation, leaching into groundwater, aquatic surfaces, and sediments. Their prevalence in the aquatic setting can cause significant alterations in the makeup and organization of bacterial communities, such as the capacity to create and spread resistance genes. In natural ecosystems, the maintenance of biogeochemical cycles is mediated primarily by the commensal microbial community. The influx of antibiotics into the environment alters the structure of the non-target organism and even causes their suppression (bactericidal or bacteriostatic) or extinction, which affects the growth and enzymatic (metabolic) activities and eventually influences ecological processes, like biomass generation and nutrient transformation, and results in the loss of functional stability. Changes in nitrogen transformation, methanogenesis, sulfate reduction, nutrient cycling, and organic matter degradation are some of the other activities affected due to such influx. Be it a broad-spectrum or narrow-spectrum antibiotic, it can obstruct the various interactions between species. Sulphonamides, for instance, have been reported to alter microbial diversity by lowering both biomass and interactions between bacteria and fungi. In contrast, fluoroquinolones and sulphonamides both have been demonstrated to impede the denitrification activity of the nitrogen cycle. The application of tylosine-harboring swine manure to the soil also has been observed to amend the nitrogen transformation processes by the resident bacterial communities.

Soil microorganisms also carry out numerous essential functions and regulate the soil's health and purity. They are necessary for the turnover of organic matter, release of the nutrients, maintenance of soil fertility, and soil structural stabilization. Overuse of antibiotics leads to the disruption of soil's homeostasis and differentially impedes the development of soil microbes affecting the makeup of the microbial population in the ecosystem. Impacts of antibiotics on the soil are seen in terms of decreased soil respiration, changes in the nitrification/denitrification processes, and altered enzyme activities.

Another rising topic of concern related to antibiotic pollution is the release of hospital wastewater containing numerous hazardous and emerging micropollutants like pharmaceutically active chemicals (PhACs), heavy metals, adsorbable organic halogens, radioisotopes, and antibiotics. Antibiotics are given top priority among emerging pollutants because of their negative impact on human health and the natural habitat of different beings. Because of the lack of metabolism of antibiotics in human bodies, most of their polar and soluble by-products are flushed into sewers either in an active or inactive form.

10.4   RESISTANCE AGAINST ANTIBIOTICS: DEVELOPMENT

The penultimate effect of the misuse and overuse of antibiotics is the emergence of antibiotic resistance. Be it the impact on humans, soil agriculture, animals, or environment, the ultimate repercussion is seen in terms of development of ARGs harboring bacteria. Any decrease in the vulnerability/response of a bacteria relative to the normal wild-type is referred to as antibiotic resistance (AbR). Because most antibiotics are not fully metabolized in humans and animals, a large amount of administered drugs are released into water and soil through municipal wastewater, animal manure, sewage sludge, and biosolids (nutrient-rich organic materials left over after sewage treatment), which are frequently used to irrigate and fertilize agricultural lands (Cui et al. 2018; 2019). Although various wastewater treatment approaches can break down antibiotics, their removal rates vary noticeably. This can be attributable to variations in treatment procedures, such as the influent, the capacity of the treatment facility, and the technology employed. Tetracyclines, erythromycin, and lincomycin, respectively, have been shown to be eliminated in urine and feces at rates of 75–80%, 50–90%, and 60% of the doses taken. When such a huge influx of antibiotics occurs in conventional wastewater treatment plants, mostly 90% of the influx of antibiotics escape the treatment process and are fed into the receiving waters, where they pose major threats, such as the development of antibiotic resistance by the inhabiting microbial communities.

The primary concern regarding exposure and release of antibiotics via the various processes explained previously is the spread of bacterial resistance genes and their concomitant effects like chronic diseases due to failure of possible treatment leading to further utilization of antibiotics and their negative consequences on microbiota and hence directing towards antibiotic resistance development. Increases in fatal cases, relapse, patient expenditures, and extended hospital stays are also some of the other consequences of AbR. The development of antibiotic resistance occurs as a result of continuous evolution and spontaneous mutation among microorganisms, which has given us an insight into a significant emerging problem. Bacteria have adapted various mechanisms to counter the effect of antibiotics, resulting in resistance development as described in the following sections.

10.4.1 INTRINSIC RESISTANCE/NATURAL RESISTANCE

In inherent resistance, the resistance genes occur naturally and are expressed only after antibiotic exposure. The two most frequent bacterial mechanisms implicated in intrinsic resistance are the decreased permeability of the outer membrane (LPS) to limit drug uptake and the normal activation of efflux pumps. Another typical mechanism of induced resistance is multi-drug efflux pumps. This mechanism is used mainly by gram-negative bacteria only, and therefore, they are inherently resistant to cell wall-targeting drugs like β-lactams and glycopeptides.

10.4.2 ACQUIRED RESISTANCE

On the other hand, acquired resistance develops by the acquisition of genetic material or via the horizontal change transfer methods – transformation, conjugation, and transduction – or the chromosomal DNA of the bacteria may undergo mutations. This mechanism adopted by the bacteria is also known as the genetic basis of resistance. It results from mutations that alter the antimicrobial targets, reduce drug absorption, increase the influx and extrusion of molecules, and modify the metabolic pathways.

Genomic mutations are the alterations in the bacterial genome that arise as a result of selective pressure put in by the antibiotics on the microbiome administered. This points toward the fact that the external environment is rarely a factor involved in the mutation-based selection of resistant pathogens, although there might be the case where various environmental factors aid such a process where such changes are more likely to occur.

There is a diverse plethora of antibiotic-resistant genes and other factors which the environment harbors. This number exceeds the combined number of humans and animals. An astonishing fact is that there exists some or other kind of resistance that the antibiotic has encountered. This is true for natural, semi-synthetic, and synthetic molecules approved as antibiotics to date. Hence, it can be rightly said that the environment harbors nearly all the factors responsible for the resistance against all antibiotics.

Process of genomic integration of extracellular ARG.

1. The ARGs get associated with the insertion sequences (Razavi et al. 2020) and move within the genome or they form cassettes and get incorporated in the integrons.
2. The gene is relocated to either a plasmid or integrative conjugative element for its autonomous movements.
3. The correctly integrated (mobilized) gene is transferred horizontally.

The other mechanism taken up by the bacteria to evade antibiotic effect and develop AbR is the mechanistic basis of resistance. This occurs mainly because of changes in antimicrobial molecules, omission or replacement of target sites, blockage of chemicals from the targets, and global cell adaptive mechanisms. These workings adopted by the bacteria are further explained as follows.

10.4.3 Alterations in the Antimicrobial Target Cells

A few targets may be altered by the bacterium to confirm drug resistance to those drugs via strategies like mutations that change the structure of the target site, guarding the target site so that drug binding may become entirely or partially hindered. Changes like point mutations in the target site-encoding genes, and enzymatic modification to the binding site exemplified by methylation by *erm* genes, may occur. Occasionally the bacteria utilize another approach whereby they discard the original target entirely and yield new targets that are similar in function but are immune/ resistant to the antibiotic or overproduce the antibiotic target to escape the action of the antibiotic, which is also termed "bypassing the metabolic pathway."

10.4.4 Changes in Antimicrobial Molecule

Antibiotics can also be acquired by the action of enzymes that induce chemical alteration in the drug molecule. These enzymes catalyze biochemical reactions like acetylation, phosphorylation, and adenylation, ultimately resulting in steric hindrance lowering the drugs' affinity for its target and creating higher MIC. Destruction of the compounds by enzymes is another AbR mechanism acquired by bacteria. β-lactamases are the most prominent examples where the amide link is broken, rendering the antibiotic useless.

10.4.5 Extrusion of Antibiotics by Efflux Pumps

Genes for efflux pumps are chromosomally encoded in bacteria. Apart from permanent expression, mutations also cause induction or overexpression of the efflux pumps, which generally occurs in the presence of a stimulus (substrate or environment), leading to the expulsion of the antibiotic, resulting in AbR.

10.4.6 Global Cell Adaptive Processes

Over time, bacteria have evolved sophisticated coping strategies and regulated systems to survive in the most adverse environments, including the human body. To escape the immune system's attacks and prevent disruption of crucial physiological processes like homeostasis and cell wall formation, they must adapt and deal with these circumstances. These strategies, with time, give rise to antibiotic resistance.

One question might arise why naturally produced antibiotics do not cause the problem of resistance? The simple and straightforward answer to this is the natural production of antibiotics by organisms occurs at the micro level, where the concentration of the antibiotic drops to a very low level once in the environment, which is not enough to develop resistance. Contrary to this, the production of antibiotics by synthetic means occurs on a macro level and hence provides enough selection pressure so as to confer resistance across microbial communities (Larsson and Flach 2021). Even a very low concentration of antibiotics of the order 10–100 times less than the clinically obtained MIC is sufficient to cause the increase in the resistant bacterial population, and this directly impacts the adaptive evolution rate of the community thriving in that very environment. Understandably, the majority of research has concentrated on the problem of illuminating the genetic causes of secondary resistance or mutations that impart higher tolerance to antibiotics following exposure. This is distinct from primary or intrinsic resistance, which can be caused by a variety of factors, such as gram-negative bacteria's inability to bind to the target site of the majority of penicillin.

There is a limited understanding of factors governing the widespread dissemination of resistance across bacterial communities. Regardless, the presence of a sublethal concentration of antibiotics has positively affected with regards to the phenotypic as well as physiological changes that render antibiotics non-functional. This phenomenon is further enhanced by the favorable environmental changes governing this change.

There is a study supporting the fact that the presence of microplastics can be an added benefit for the spread of resistance genes and organisms in the aquatic environment. It is crucial to understand the circumstances that can lead to a rise in ARGs as well as dynamics that promote and restrict the spread of resistance among bacterial populations in the natural environment. In addition to antibiotics, other substances that contribute to antibiotic resistance include biocides like disinfectants, metals, and phytochemicals. Unlike other chemical pollutants, bacteria and their ARGs can stay and propagate in the environment, eventually passing on their resistance to human infections. Antibiotics are formed and broken down by enzymes, which is key in the antibacterial resistance mechanism. Antibiotic-resistant bacteria, ARG expression, and the generation of antibiotic-degrading enzymes are all connected. Additionally, several researches revealed that ARGs involving the development of aminoglycoside modification and lactamase enzymes have existed in the environment for a very long time prior to using synthetic antibiotics for therapeutic purposes.

This points towards the fact that ARGs and ARBs existed before the time when antibiotics were synthesized and used in the treatment of various pathogenic infections. But still, the problem of antibiotic-resistant infections did not exist before. Hence it is only the mismanagement and misuse of antibiotics which has led us to the situation where we are short of drugs to treat certain infections. Complex regimes are being followed for infections that otherwise were treatable with simpler drugs.

10.5 CASE STUDIES

10.5.1 Drug Resistance in *Mycobacterium tuberculosis*

One of the significant problems associated with tuberculosis (TB) is the emergence of resistance to the first-line and second-line anti-TB drugs. The first anti-TB drug, streptomycin, discovered in 1944, was initially effective in treating patients. However, the patient's health started declining with the treatment after a few months, mainly due to the development of resistant *Mtb* strains making streptomycin ineffective. To prevent and control the emergence of resistance, para-aminosalicylic acid and streptomycin were used in combination to treat pulmonary TB. New anti-tuberculous drugs were introduced in the ensuing years with the discovery of rifampicin in 1965. Its consequent application in the treatment of TB allowed for a remarkable reduction in treatment time from 18 months to 9 months. In the 1990s, the DOTS therapy (Directly Observed Therapy Short Course) was created where the four first-line anti-TB drugs – isoniazid, rifampicin, ethambutol, and pyrazinamide – were prescribed for a drug-susceptible TB treatment regimen. Despite this, various drug-resistant forms continued to evolve. This is due to non-compliance with lengthy medication regimens, inadequate healthcare facilities, incorrect treatment recommendations, low-quality medications, lack of medication availability, and improper administration of anti-TB drugs. Low socioeconomic status, substance misuse, immigration status, and co-infection with other diseases are additional risk factors for drug-resistant TB. The drug-resistant TB can be multi-drug-resistant (MDR), extensively drug-resistant (XDR), and extremely drug-resistant (XXDR) as shown in Figure 10.1.

MDR-TB necessitates a lengthier course of therapy than drug-sensitive TB. To achieve a cure for MDR-TB, treatment using previous regimens may continue up to two years and needs the use of drugs with severe toxicities and injectable treatments. People who cannot accept or adhere to therapy due to its length or toxicity have a significant failure rate, which frequently results in the emergence of XDR-TB.

The genes undergoing mutations to develop resistance against anti-TB drugs are the *rpoB* gene for rifampicin, *embB* gene for ethambutol, and *pncA* gene for pyrazinamide. In 1992, it was reported that mutation in the *katG* gene was responsible for isoniazid (INH) resistance in *Mycobacterium tuberculosis*. These are mostly point mutations where one amino acid changes and is replaced by

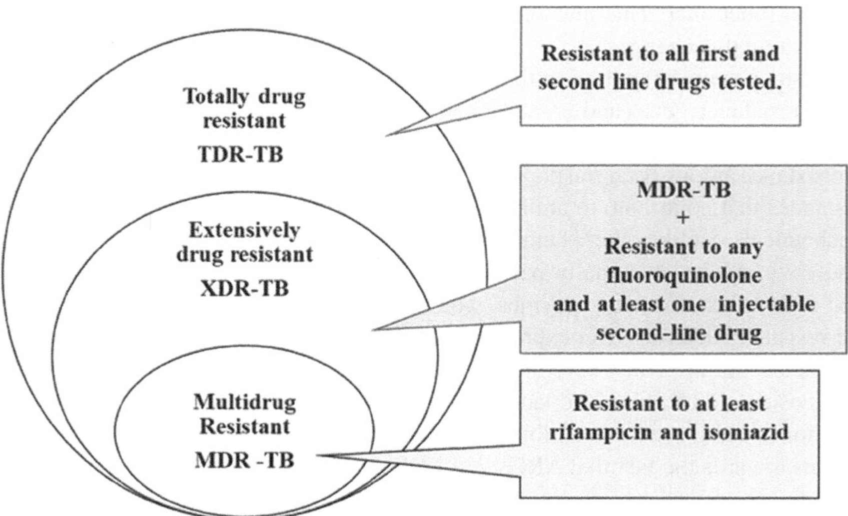

FIGURE 10.1   Types of drug-resistant TB.

another. Among the various *Mtb* strains, the Beijing strain (lineage 2) is associated with the spread of MDR-TB and TB outbreaks mainly due to its increased virulence, transmissibility, mutability, capacity to evade BCG vaccine-induced immunity and ability to acquire resistance (Gupta et al. 2020).

In the past ten years, significant attempts have been made to develop MDR-TB and XDR-TB treatment plans that do not require injectables, are less expensive, and expose patients to fewer toxic drugs. TB PRACTECAL is a clinical trial of three 6-month all-oral regimens consisting of a 9–24-month treatment course of oral and injectable treatments. Bedaquiline, pretomanid, and linezolid were the main components of the experimental therapy in TB PRACTECAL (BPaL). After 16 weeks of treatment, the linezolid dose was decreased from 600 mg to 300 mg to reduce toxicity. Bedaquiline, pretomanid, and linezolid were administered to one research group, while moxifloxacin 400 mg or clofazimine 100 mg were given to the other experimental groups. Linezolid exposure was observed to enhance operability without reducing effectiveness.

Currently, three treatment regimes are being followed for MDR/RR TB (multi-drug resistant/ Rifampicin-resistant tuberculosis) as per WHO guidelines (Migliori, Tiberi, and Migliori 2022):

1. The BPaL (bedaquiline [BDQ], pretomanid [Pa], linezolid [LZD] 600 mg and moxifloxacin [MFX]) regimen for 6 months or only BPaL without MFX if any fluoroquinolones resistance is recognized.
2. All-oral, BDQ-containing regimes for 9 months in adults and children.
3. The previous specific lengthier treatment of more than 18 months based on 2020 WHO recommendations are still considered for XDR-TB patients, patients not qualified for the regimes mentioned above, and those who have failed to respond to shorter treatment plans.

The administration of antibiotics to treat TB patients raises the question of side effects or toxicity. The probability of toxicity depends on various factors like intake, length of treatment, simultaneous use of other drugs, the intensity of the TB disease, and associated co-morbidities. The potential toxic effects of first-line anti-Tb drugs are neurological, hepatotoxic, allergic reactions, hemoglobinuria, hemolytic anemia and kidney damage with acute renal failure, GI intolerance, multiple joint pains, inflammatory arthritis, and inflammation of optic nerves. In children, adverse reactions like liver failure and hepatitis due to INH have been observed in some instances while in elderly patients, hepatotoxicity is one of the significant risks associated with the TB treatment.

10.5.2 Drug Resistance in *Staphylococcus aureus*

Staphylococcus aureus is a major human pathogen associated with healthcare-transmitted infections like minor skin or soft tissue infections and even fatal cases. The two resistant strains – Methicillin-resistant *S. aureus* (MRSA) and vancomycin-resistant *S. aureus* (VRSA) – continue to pose a significant global public health concern. In the absence of adequate containment and treatment options, MRSA and VRSA have the potential to cause significantly severe worldwide mortality. Vancomycin, a glycopeptide, is administered against MRSA strains as the last therapeutic option. With VRSA being multi-drug-resistant against available antibiotics, drug resistance in *S. aureus* is a grave concern. Heterogeneous vancomycin-intermediate *S. aureus* (hVISA), vancomycin-intermediate *S. aureus* (VISA), and high-level vancomycin-resistant *S. aureus* (VRSA) are three groups of vancomycin-resistant *S. aureus* that have different vancomycin susceptibilities.

Initially, methicillin was used against Penicillin-resistant *Staphylococcus aureus* (PRSA) strains, which were first detected in the early 1940s. Soon after, methicillin resistance emerged in *S. aureus* in 1961 when the drug was introduced for therapeutic use. Apart from methicillin, MRSA strains are resistant to oxacillin, flucloxacillin, and other β-lactam antibiotics. Vancomycin became the first drug of choice for treating MRSA in the 1980s and was alleged to be immune to developing any resistance. However, this antibiotic was also used in hospital patients to treat other bacterial infections. Consequently, in 1996, in Japan, the first (VISA) vancomycin-intermediate-resistant strain, and its predecessor the hetero-VISA (hVISA) strain, was isolated. Eventually, the first VRSA strain was reported in Michigan, USA, and to date, a total of 52 VRSA strains have been reported all around the world (Wu et al. 2021).

The current treatment options for MRSA include administering antibiotics – vancomycin, linezolid, daptomycin, ceftaroline, teicoplanin, mupirocin, oritavancin, tigecycline, iclaprim, dalbavancin, delafloxacin, and cethromycin are some of the other semi-synthetic drugs used against MRSA infections. Combination or multi-drug therapy with synergistic effects has also become a novel approach for treating such resistant organisms. There is no treatment recommendation due to the low number of VRSA infection patients. Only a small number of the reported VRSA cases included comprehensive clinical information. These treatment experiences and CDC recommendations suggest that antimicrobial therapy with trimethoprim/sulfamethoxazole, daptomycin, ceftaroline, linezolid etc., and preventive measures should be followed to treat VRSA infections generally (Cong, Yang, and Rao 2020; Nandhini et al. 2022; "VISA/VRSA Infections – Epidemiology" 2022).

The antibiotics used also have their related side effects and toxicity levels. In adults and children, prolonged use of vancomycin causes renal toxicity or nephrotoxicity, which may lead to an extended hospital stay, dialysis, and more expenses. Another common side-effect of vancomycin seen in pediatric patients is vancomycin-associated acute kidney injury (vAKI) (Moffett et al. 2018). Studies showed that linezolid caused thrombocytopenia, leucopenia, anemia, increased liver enzymes, and skin reactions. Treatment with daptomycin is associated with cardiovascular issues, insomnia, a neuromuscular disorder like myopathy, anaphylactic reactions, and eosinophilic pneumonia ("Daptomycin – Wikipedia" 2022; "Daptomycin – Infectious Diseases – MSD Manual Professional Edition" 2022; "Daptomycin – StatPearls – NCBI Bookshelf" 2022).

10.6 DEGRADATION OF ANTIBIOTICS

The process of degradation of antibiotics occurs by both abiotic and biotic means. Abiotic factors include oxidation, hydrolysis, photolysis, sorption, and reduction, and biotic factors largely include microorganisms.

The degradability of antibiotics depends largely on the environment in which it thrives. Also, the chemical structure, physical form, etc. are the deciding factors that decide the degradability of antibiotics. Depending upon this, more than 70% of the antibiotics administered are found unchanged or as degradation products that may or may not be recalcitrant. There are reports which support the fact that there are some antibiotics which pose a greater risk when in the environment than the disease they treat.

Different classes of antibiotics are degraded by different abiotic stresses: for example, sulfamethoxazole and oxytetracycline are oxidized by oxygen and ozone treatment. Tetracyclines are more sensitive to UV and photolysis. On the other hand, sulphonamides, tylosin, and nitrofuran antibiotics are UV sensitive. The presence of metal ions (copper, cadmium, zinc, cobalt, mercury) also results in the cleavage of the β-lactam ring structure of β-lactam antibiotics (Kumar et al. 2019).

Depending on the various biotic and abiotic influences an antibiotic molecule encounters, they may transform or degrade via the following mechanisms:

1. Sorption/desorption
2. Uptake by plants
3. Runoff and transport into groundwater
4. Hydrolysis
5. Photodegradation
6. Reductive and oxidative transformation

10.6.1 BIOREMEDIATION

The process of employing living organisms to extract and/or detoxify toxins is known as bioremediation. Both in-situ and ex-situ techniques are used in bioremediation. Ex-situ bioremediation refers to removing contamination that needs to be treated elsewhere, whereas in-situ bioremediation involves removing contaminants on-site (Ezzariai et al. 2018).

Qualities of bacteria for bioremediation of antibiotics:

1. Should be able to survive extreme environmental conditions
2. Should be resistant to the antibiotics
3. Should be resistant to uptake of foreign ARGs

The organism which can degrade most of the antibiotics listed in Table 10.2 is *Trametes versicolor*, and the enzyme mainly involved is laccase which may work with or without a suitable mediator.

10.6.2 BIOLOGICAL TREATMENT

The varying oxygen requirements of biological treatments can be divided into three categories: aerobic, anaerobic, and combined aerobic and anaerobic approaches. The biological aerated filter (BAF) system is the primary aerobic technique. Anaerobic digestion (AD), up-flow anaerobic sludge blanket (UASB), anaerobic filter (AF), and anaerobic baffled reactor (ABR) technologies are the basic anaerobic techniques. The sequencing batch reactor (SBR) and membrane bioreactor (MBR) procedures are the most often used combined aerobic and anaerobic techniques. The BAF, AD, SBR, and MBR methods are frequently used to remove antibiotics from breeding wastewater.

10.6.2.1 The Sequencing Batch Reactor

As the name suggests, the sewage is introduced into the system in batch mode. The sequential reactors are influent feeding, anoxic phase, aerobic phase, sludge settling, and effluent discharge. The control and administration of such a reactor system are difficult, although it is versatile. The reason for this is that it involves a number of reactor systems. The removal rate of tetracycline was 88%, with an HRT (hydrolic retention time) of 3–5 days and a minimal COD (chemical oxygen

TABLE 10.2
Microorganisms Involved in the Bioremediation of Antibiotics

Sr. no.	Antibiotic	Microorganism	Reference
1.	Sulfamethoxazole	*Phoma sp.*	(Hofmann and Schlosser 2016)
		Acinetobacter sp.	(S. Wang, Hu, and Wang 2018)
		Ochrobacter sp.	(Larcher and Yargeau 2011)
		Rhodococcus equi	
		Rhodococcus rhodocrous	
		Pseudomonas aeruginosa	
		Rhodococcus erythropolis	
		Bacillus subtilis	
		Rhodococcus zopfi	
		Trametes versicolor	(Rahmani et al. 2015)
2	Tetracycline	*Klebsiella sp.*	(Shao et al. 2018)
		Raoltella sp.	
		Burkholderia cepacia	(Hong et al. 2020)
		Cerrena sp.	(Yang et al. 2017)
		Trametes versicolor	(de Cazes et al. 2015)
		Stenotrophomonas	(Leng et al. 2016)
3	Ciprofloxacin	*Thermus thermophilus*	(Pan et al. 2018)
		Bradyrhizobium	(Nguyen, Nghiem, and Oh 2018)
		Lactobacillus gasseri	(Liyanage and Manage 2022)
		Aspergillus oryzce	(Sutar and Rathod 2015)
		Trametes versicolor	(Ding et al. 2016)
		Labrys	(Amorim et al. 2014)
4	Oxytetracycline	*Cerrena sp.*	(Yang et al. 2017)
		Pleurotus ostreatus	(Migliore et al. 2012)
		Trametes versicolor	(Ding et al. 2016)
5	Ofloxacin	*Rhodococcus sp*	(Maia, Tiritan, and Castro 2018)
		Trametes versicolor	(Ding et al. 2016)
6	Norfloxacin	*Staphylococcus caprae*	
		Trametes versicolor	(Ding et al. 2016)
		Bacillus sp.	(Erickson et al. 2014)
7	Sulfamethazine	Bacterial consortia (*Bacillus licheniformis, Pseudomonas putida, Alcaligenes sp., Aquamicrobium defluvium*)	(Islas-Espinoza et al. 2012)
		Echinodontium taxodii	(Shi et al. 2014)
		Trametes versicolor	(Rodríguez-Rodríguez et al. 2012)
		Microbacterium	(Topp et al. 2013)
8	Gentamicin	Bacterial consoria (*Providencia vermicola, Brevundimonas diminuta, Alcaligenes sp., Acinetobacter*)	(Liu et al. 2017)
		Aspergillus terreus	(Liu et al. 2016)
9	Cefdinir	*Ustilago*	(Selvi, Salam, and Das 2014)
10	Cefuroxime axetil	*Imleria badio*	(Dąbrowska et al. 2018)
		Lentinula edodes	
11	Sulfapyridine	*Trametes versicolor*	(Rodríguez-Rodríguez et al. 2012)
12	Sulphathiozole		
13	Sulphadimethoxine	*Perenniporia* strain	(Weng, Liu, and Lai 2013)
14	Sulfamonomethoxine		
15	Penicillin G	*Burkholderia*	(Zhang and Dick 2014)
16	Erythromycin	*Labrys*	(Mulla et al. 2018)
17	Deoxycyclin	*Ochrobactrum*	(Mulla et al. 2018)
18	Chloamphenicol	*Klebsiella*	(Xin et al. 2012)

demand) load, with sludge adsorption and biological degradation rate of 30% and 58%, respectively. The rate of biodegradation of Sulfamethoxazole was roughly 96% (Zheng et al. 2018).

10.6.2.2 Biological Aerated Filters

The BAF system combines biological contact oxidation and filtration. It has three phases: a solid phase to enhance microbial growth, a liquid phase to immerse the solid material, and a gas phase to add air into the system. This system is advantageous because it automatically has a low carbon footprint and, above all, the running cost is very low. Since the basic mechanism behind its functioning is the strong adhesion of biofilms and the presence of a large surface area, the better absorption of sludge promotes the efficient removal of antibiotics. With the BAF process, the total antibiotic removal can be up to 91% with a hydraulic retention time of 48 hours and a hydraulic load rate of 2.8 cm/h (Chen et al. 2017).

10.6.2.3 Membrane Bioreactors

An MBR is a biological wastewater treatment system that integrates biological technology and contemporary membrane separation technology. While the MBR has several benefits, like flexible operation, minimal sludge production, and strong nitrification performance, it also has some drawbacks, including high energy consumption and operational expenses. In a study by Xu et al. 2019, MBRs are highly effective (>90%) at removing sulfamethazine and tetracycline from swine wastewater but only around 70% as effective at removing quinolones with hydraulic retention time between 33 and 51 hours. HRT, sludge concentration, membrane type, water quality (pH value, temperature), and antibiotic properties are the key variables impacting biodegradation efficiency (Luo et al. 2014; Tiwari et al. 2017; J. Wang and Wang 2016). Longer sludge retention times and higher biomass concentrations, which lengthen the time that microorganisms spend in contact with antibiotics, can increase the effectiveness of antibiotic elimination.

10.6.3 CONSTRUCTED WETLAND

In addition to broad-spectrum decontamination capabilities, constructed wetlands are affordable, effective, and simple to maintain. As per Huang et al. 2017, artificial wetlands remove sulfamethazine, tetracycline, and quinolone from sewage used for breeding at high rates ranging from 49.43% to 85%, 69% to 100%, and 82% to 100%, respectively. Additionally, it has been demonstrated that artificial wetlands effectively remove the antibiotics enrofloxacin, oxytetracycline, and antibiotic-resistant bacteria from marine aquaculture wastewater.

The following factors mostly influence the performance of constructed wetlands:

1. Plant species
   Plants aid in the degradation of antibiotics by directly absorbing them and by providing an attachment surface to microorganisms, causing an increase in microbial activity. For example, *Eichhornia crassipes* have a TC removal rate of more than 70% (initial concentration, 3.0–15.0 mg/L) (Lu et al. 2014). *Azolla* can remove sulfadimethoxine up to 86% with an initial concentration range of 50–450 mg/L. On the other hand, *Lolium sp.* can use sulfamethazine and sulfamethoxazole up to a maximum of 99% with an initial concentration of 100µg/L.

2. Material of the fillers used
   The fill material has adsorption and interception properties that can speed up the breakdown of antibiotics by bacteria. The effectiveness of antibiotic elimination is probably influenced by the pore size, specific surface area, and chemical makeup of the fill material. Examples of filler material include zeolite and volcanic rocks, of which zeolite results in a slightly higher removal rate of sulfamethazine, oxytetracycline, and ciprofloxacin.

3. Hydraulic loading rate (HLR)

   The HLR is the daily capacity of a constructed wetland to treat a specific volume or area of wastewater. Multiplying the HLR and the initial antibiotic concentration can determine the daily antibiotic intake in a constructed wetland. The sulfamethoxazole removal rates in integrated vertical flow constructed wetlands and horizontal subsurface flow constructed wetlands respectively increased from 4% and 3% to 59% and 55%. However, it has been demonstrated that the HRT has no appreciable impact on the elimination of enrofloxacin and florfenicol. This is due to the fact that each of these three antibiotics has a unique structure that necessitates a unique clearance method (Xiangfeng et al. 2022).

10.6.4 Membrane Technology

Membrane technology helps to trap the pollutants as wastewater passes through small membranous pores. The basic mechanism works on the principle of micro-, ultra-, reverse filtration and reverse osmosis. Although membrane technology has rarely been used to remove antibiotics from other types of wastewater, this technique has potential applications in wastewater treatment facilities. Furthermore, additional toxins in breeding wastewater might be removed using membrane technology. In order to successfully remove different ARGs, nitrogen, phosphorus, and other contaminants from swine wastewater, Lan et al. discovered that reverse osmosis and nanofiltration methods might be applied (Lan et al. 2019).

10.7 LATEST DEVELOPMENTS IN THE TREATMENT OF ANTIBIOTICS

10.7.1 CRISPR Approach

To improve degradation efficiency, CRISPR-Cas9 can be employed to delete, add, or mutate a gene in a particular microbial strain. The CRISPR-Cas system exhibits antibacterial activity that may disable ARGs or cleave crucial genome regions, resulting in toxicity in bacteria when administered as an alternative to or in addition to traditional antibiotics. According to Citorik, Mimee, and Lu (2014), programmed gyrA phagemid was only lethal to quinolone-resistant *E. coli* strains with altered chromosomal gyrA and not to wild-type, demonstrating that the CRISPR-Cas system can differentiate between resistant and susceptible strains. The CRISPR-Cas system, in which Cas9 was directed toward a gene producing the virulence component intimin, was used by to successfully cure a *Galleria mellonella* infection model infected with an *E. coli* strain resistant to chloramphenicol. In enteropathogenic and enterohemorrhagic *E. coli* strains, intimin, an adhesin, plays a significant role in virulence. MRSA strains have been successfully eliminated from mixed bacterial populations using Cas9 phagemids. Contrary to antibiotics, which may also impact nearby bacteria, the CRISPR-Cas system exclusively targets pathogenic bacteria, and this specificity is another benefit of this technique.

10.7.2 Microbial Fuel Cell

Antibiotic degradation has recently been achieved using microbial fuel cells, and bioelectrochemical systems that combine microbial and electrochemical processes (S. Wang, Hu, and Wang 2018). *Geobacter, Shewanella, Propionibacterium, Pseudomonas,* and *Rhodoferax* are preferred choices in microbial fuel cell systems as they belong to electrochemically active groups. Microbial electron mediating cells (MERCs) are used by Quejigo et al. (2019) for the mineralization of ^{14}C Sulfamethazine in compost manure. After an incubation period of one month, the native microbial population showed only 17% mineralization of the ^{14}C Sulfamethazine in the absence of electrodes. On the other hand, the degradation efficiency increased ten-fold in the first 14 days of incubation in the presence of negative anode potential.

In microbial fuel cell (MFC) reactors, sulfamethoxazole and its by-product 3-amino-5-methylisoxazole (3A5MI) were successfully broken down. About 85% of the 20 ppm of sulfamethoxazole was broken down within 12 hours. Additionally, 3A5MI that is produced during the degradation process may be further mineralized. Sulfamethoxazole's biotoxicity toward *Shewanella oneidensis* MR-1 and *E. coli* DH5α significantly decreased following MFC treatment, as demonstrated by an antibacterial activity test. After a 40-hour treatment, the sediment microbial fuel cell electro-Fenton (SMFC-E-Fenton) system displayed a 97.4% rate of sulfamethoxazole breakdown (Y. Wang et al. 2019).

10.7.3 Metal Nanoparticles

The sensitivity of ESKAPE pathogens concerning metal nanoparticles is adequately known (Lee, Ko, and Hsueh 2019). This property of metal nanoparticles (MNP) against the pathogens can be exploited to combat resistant pathogens. Antimicrobial effects of titanium dioxide photocatalytic nanoparticles are present against methicillin-resistant *S. aureus* and multidrug-resistant *E. coli*. Metal nanoparticles can also work in conjunction with various currently used antibiotics to produce beneficial effects. Using antibiotics and nanoparticles together is advantageous because it reduces the likelihood of resistance developing. The combination use of antibiotics along with transition metal nanoparticles proved beneficial for the treatment of first-line drug-resistant *Mycobacterium tuberculosis* (Montelongo-Peralta et al. 2019).

Baaloudj et al. 2021 showed 89% removal of cefixime via solar light-assisted $ZnBi_2O_4$ nanoparticles in 30 mins. Similar effects were demonstrated by nano $N\text{-}TiO_2$/graphene oxide/titan grid sheets and $ZnO/\alpha\text{-}Fe_2O_3$, which showed 99.1% and 80% degradation of cefixime, respectively. In the photocatalytic method, electron-hole separation occurs via photon absorption. To speed up the chemical reaction from photoelectrons, the prolonged electron-hole charge separation enhances reaction effectiveness, photocatalytic activity, and the mobility of charge carriers. As a result, photocatalytically produced nanoparticles are employed to remove antibiotics from the environment effectively. Although nanoparticles have many advantages, their toxicity and environmental risk have recently drawn much attention.

10.8 CONCLUSION

Antibiotic pollution is ubiquitous citing its presence in two major domains of the environment i.e. soil and water. A disturbing amount of antibiotics are being found in nearly all the domains of the environment. This can be attributed to the fact that antibiotics are being misused and industrial production is also happening by non-compliance with the rules and regulations as laid down by the government. As a result of all these malpractices, there is an emerging problem of antibiotic resistance, which has affected the human healthcare sector to a much greater extent. Various disease-causing pathogens have gained resistance against conventionally used antibiotics. Notable ones are MDR, MRSA, VISA, VRSA, etc. These pathogens have taken a heavy toll on general public health, and the healthcare sector is struggling to deal with them. Not only this, but antibiotic pollution has also affected various domains of the environment, including plants, animals, and microbes.

This issue of antibiotic pollution needs critical thinking as to what can be done to curb this problem. Various remediation techniques are there which can be modified for the treatment of specific antibiotics. Several newer technologies, especially CRISPR and nanoparticle mediated, can be effectively integrated with conventional methodologies to bring out the change and contain this ever-expanding problem.

CONFLICT OF INTEREST

The authors declare no conflict of interest regarding the publication of this manuscript.

ACKNOWLEDGMENT

The authors acknowledge the Indo-EU Horizon 2020 project (BT/IN/EU-WR/60/SP/2018), funded by the Department of Biotechnology, New Delhi. Further, the authors acknowledge the Indian Institute of Technology Guwahati, Assam, India, for providing research fellowship and infrastructure resources for carrying out the present work.

CONTRIBUTION

Gangar Tarun and Hazarika Risha contributed equally to the preparation of this manuscript. Patra Sanjukta conceptualized the overall manuscript and directed the flow.

REFERENCES

Agunbiade, Foluso O., and Brenda Moodley. 2016. "Occurrence and Distribution Pattern of Acidic Pharmaceuticals in Surface Water, Wastewater, and Sediment of the Msunduzi River, Kwazulu-Natal, South Africa." *Environmental Toxicology and Chemistry* 35 (1). John Wiley & Sons, Ltd: 36–46. doi:10.1002/ETC.3144.

Amorim, Catarina L., Irina S. Moreira, Alexandra S. Maia, Maria E. Tiritan, and Paula M.L. Castro. 2014. "Biodegradation of Ofloxacin, Norfloxacin, and Ciprofloxacin as Single and Mixed Substrates by Labrys Portucalensis F11." *Applied Microbiology and Biotechnology* 98 (7). Springer Verlag: 3181–90. doi:10.1007/S00253-013-5333-8/FIGURES/4.

aus der Beek, Tim, Frank Andreas Weber, Axel Bergmann, Silke Hickmann, Ina Ebert, Arne Hein, and Anette Küster. 2016. "Pharmaceuticals in the Environment – Global Occurrences and Perspectives." *Environmental Toxicology and Chemistry* 35 (4). Wiley Blackwell: 823–35. doi:10.1002/ETC.3339.

Azanu, David, Bjarne Styrishave, Godfred Darko, Johan Juhl Weisser, and Robert Clement Abaidoo. 2018. "Occurrence and Risk Assessment of Antibiotics in Water and Lettuce in Ghana." *Science of The Total Environment* 622–623 (May). Elsevier: 293–305. doi:10.1016/J.SCITOTENV.2017.11.287.

Baaloudj, Oussama, Achraf Amir Assadi, Mohamed Azizi, Hamza Kenfoud, Mohamed Trari, Abdeltif Amrane, Aymen Amine Assadi, and Noureddine Nasrallah. 2021. "Synthesis and Characterization of ZnBi2O4 Nanoparticles: Photocatalytic Performance for Antibiotic Removal under Different Light Sources." *Applied Sciences* 11 (9). Multidisciplinary Digital Publishing Institute: 3975. doi:10.3390/APP11093975.

Bielen, Ana, Ana Šimatović, Josipa Kosić-Vukšić, Ivan Senta, Marijan Ahel, Sanja Babić, Tamara Jurina, Juan José González Plaza, Milena Milaković, and Nikolina Udiković-Kolić. 2017. "Negative Environmental Impacts of Antibiotic-Contaminated Effluents from Pharmaceutical Industries." *Water Research* 126 (December). Pergamon: 79–87. doi:10.1016/J.WATRES.2017.09.019.

Carballo, Matilde, Antonio Rodríguez, and Ana de La Torre. 2021. "Phytotoxic Effects of Antibiotics on Terrestrial Crop Plants and Wild Plants: A Systematic Review." *Archives of Environmental Contamination and Toxicology* 82: 48–61. doi:10.1007/s00244-021-00893-5.

Cazes, M. de, M. P. Belleville, M. Mougel, H. Kellner, and J. Sanchez-Marcano. 2015. "Characterization of Laccase-Grafted Ceramic Membranes for Pharmaceuticals Degradation." *Journal of Membrane Science* 476 (February). Elsevier: 384–93. doi:10.1016/J.MEMSCI.2014.11.044.

Chen, Jun, You Sheng Liu, Jin Na Zhang, Yong Qiang Yang, Li Xin Hu, Yuan Yuan Yang, Jian Liang Zhao, Fan Rong Chen, and Guang Guo Ying. 2017. "Removal of Antibiotics from Piggery Wastewater by Biological Aerated Filter System: Treatment Efficiency and Biodegradation Kinetics." *Bioresource Technology* 238 (August). Elsevier: 70–77. doi:10.1016/J.BIORTECH.2017.04.023.

Citorik, Robert J., Mark Mimee, and Timothy K. Lu. 2014. "Sequence-Specific Antimicrobials Using Efficiently Delivered RNA-Guided Nucleases." *Nature Biotechnology* 32 (11). Nature Publishing Group: 1141–45. doi:10.1038/nbt.3011.

Cong, Yanguang, Sijin Yang, and Xiancai Rao. 2020. "Vancomycin Resistant Staphylococcus Aureus Infections: A Review of Case Updating and Clinical Features." *Journal of Advanced Research* 21 (January). Elsevier: 169. doi:10.1016/J.JARE.2019.10.005.

Cui, G., Li, F., Li, S., Bhat, S.A., Ishiguro, Y., Wei, Y., Yamada, T., Fu, X. and Huang, K. 2018. Changes of Quinolone Resistance Genes and their Relations with Microbial Profiles During Vermicomposting of Municipal Excess Sludge. *Science of the total environment*, 644, 494–502.

Cui, G., Bhat, S.A., Li, W., Wei, Y., Kui, H., Fu, X., Gui, H., Wei, C. and Li, F. 2019. Gut Digestion of Earthworms Significantly Attenuates Cell-Free and-Associated Antibiotic Resistance Genes In Excess Activated Sludge By Affecting Bacterial Profiles. *Science of the Total Environment*, 691, 644–653.

Dąbrowska, M., B. Muszyńska, M. Starek, P. Żmudzki, and W. Opoka. 2018. "Degradation Pathway of Cephalosporin Antibiotics by in Vitro Cultures of Lentinula Edodes and Imleria Badia." *International Biodeterioration and Biodegradation* 127 (February). Elsevier Ltd: 104–12. doi:10.1016/J.IBIOD.2017.11.014

"Daptomycin – Infectious Diseases – MSD Manual Professional Edition." 2022. Accessed October 31. www.msdmanuals.com/en-in/professional/infectious-diseases/bacteria-and-antibacterial-drugs/daptomycin.

"Daptomycin – StatPearls – NCBI Bookshelf." 2022. Accessed October 31. www.ncbi.nlm.nih.gov/books/NBK470407/.

"Daptomycin – Wikipedia." 2022. Accessed October 31. https://en.wikipedia.org/wiki/Daptomycin.

Ding, Huijun, Yixiao Wu, Binchun Zou, Qian Lou, Weihao Zhang, Jiayou Zhong, Lei Lu, and Guofei Dai. 2016. "Simultaneous Removal and Degradation Characteristics of Sulfonamide, Tetracycline, and Quinolone Antibiotics by Laccase-Mediated Oxidation Coupled with Soil Adsorption." *Journal of Hazardous Materials* 307 (April). Elsevier: 350–58. doi:10.1016/J.JHAZMAT.2015.12.062.

ECDC. 2022. "ESAC-NET AER 2020 – Antimicrobial Consumption in the EU/EEA." Accessed October 29.

Erickson, Bruce D., Christopher A. Elkins, Lisa B. Mullis, Thomas M. Heinze, R. Doug Wagner, and Carl E. Cerniglia. 2014. "A Metallo-β-Lactamase Is Responsible for the Degradation of Ceftiofur by the Bovine Intestinal Bacterium Bacillus Cereus P41." *Veterinary Microbiology* 172 (3–4). Elsevier: 499–504. doi:10.1016/J.VETMIC.2014.05.032.

Ezzariai, Amine, Mohamed Hafidi, Ahmed Khadra, Quentin Aemig, Loubna el Fels, Maialen Barret, Georges Merlina, Dominique Patureau, and Eric Pinelli. 2018. "Human and Veterinary Antibiotics during Composting of Sludge or Manure: Global Perspectives on Persistence, Degradation, and Resistance Genes." *Journal of Hazardous Materials* 359 (October). Elsevier: 465–81. doi:10.1016/J.JHAZMAT.2018.07.092.

Gupta, Anamika, Pallavi Sinha, Vijay Nema, Pramod K. Gupta, Pampi Chakraborty, Savita Kulkarni, Nalin Rastogi, and Shampa Anupurba. 2020. "Detection of Beijing Strains of MDR M. Tuberculosis and Their Association with Drug Resistance Mutations in KatG, RpoB, and EmbB Genes." *BMC Infectious Diseases* 20 (1). BioMed Central Ltd: 1–7. doi:10.1186/S12879-020-05479-5/TABLES/4.

Hofmann, Ulrike, and Dietmar Schlosser. 2016. "Biochemical and Physicochemical Processes Contributing to the Removal of Endocrine-Disrupting Chemicals and Pharmaceuticals by the Aquatic Ascomycete Phoma Sp. UHH 5-1-03." *Applied Microbiology and Biotechnology* 100 (5). Springer Verlag: 2381–99. doi:10.1007/S00253-015-7113-0/FIGURES/6.

Hong, Xiaxiao, Yuechun Zhao, Rudong Zhuang, Jiaying Liu, Guantian Guo, Jinman Chen, and Yingming Yao. 2020. "Bioremediation of Tetracycline Antibiotics-Contaminated Soil by Bioaugmentation." *RSC Advances* 10 (55). The Royal Society of Chemistry: 33086–102. doi:10.1039/D0RA04705H.

Huang, Xu, Jialun Zheng, Chaoxiang Liu, Lin Liu, Yuhong Liu, Hongyong Fan, and Tingfeng Zhang. 2017. "Performance and Bacterial Community Dynamics of Vertical Flow Constructed Wetlands during the Treatment of Antibiotics-Enriched Swine Wastewater." *Chemical Engineering Journal* 316 (May). Elsevier: 727–35. doi:10.1016/J.CEJ.2017.02.029.

Islas-Espinoza, Marina, Brian J. Reid, Margaret Wexler, and Philip L. Bond. 2012. "Soil Bacterial Consortia and Previous Exposure Enhance the Biodegradation of Sulfonamides from Pig Manure." *Microbial Ecology* 64 (1). Springer: 140–51. doi:10.1007/S00248-012-0010-5/FIGURES/6.

Klein, Eili Y., Thomas P. van Boeckel, Elena M. Martinez, Suraj Pant, Sumanth Gandra, Simon A. Levin, Herman Goossens, and Ramanan Laxminarayan. 2018. "Global Increase and Geographic Convergence in Antibiotic Consumption between 2000 and 2015." *Proceedings of the National Academy of Sciences of the United States of America* 115 (15) National Academy of Sciences: E3463–70. doi:10.1073/PNAS.1717295115/SUPPL_FILE/PNAS.1717295115.SAPP.PDF.

Kumar, Mohit, Shweta Jaiswal, Kushneet Kaur Sodhi, Pallee Shree, Dileep Kumar Singh, Pawan Kumar Agrawal, and Pratyoosh Shukla. 2019. "Antibiotics Bioremediation: Perspectives on Its Ecotoxicity and Resistance." *Environment International* 124 (March). Pergamon: 448–61. doi:10.1016/J.ENVINT.2018.12.065.

Lan, Lihua, Xianwang Kong, Haoxiang Sun, Changwei Li, and Dezhao Liu. 2019. "High Removal Efficiency of Antibiotic Resistance Genes in Swine Wastewater via Nanofiltration and Reverse Osmosis Processes." *Journal of Environmental Management* 231 (February). Academic Press: 439–45. doi:10.1016/J.JENVMAN.2018.10.073.

Larcher, Simone, and Viviane Yargeau. 2011. "Biodegradation of Sulfamethoxazole by Individual and Mixed Bacteria." *Applied Microbiology and Biotechnology* 91 (1). Springer: 211–18. doi:10.1007/S00253-011-3257-8/FIGURES/2.

Larsson, D., G. Joakim, and Carl Fredrik Flach. 2021. "Antibiotic Resistance in the Environment." *Nature Reviews Microbiology* 2021 20:5 20 (5). Nature Publishing Group: 257–69. doi:10.1038/s41579-021-00649-x.

Lee, Nan Yao, Wen Chien Ko, and Po Ren Hsueh. 2019. "Nanoparticles in the Treatment of Infections Caused by Multidrug-Resistant Organisms." *Frontiers in Pharmacology* 10. Frontiers Media S.A.: 1153. doi:10.3389/FPHAR.2019.01153/BIBTEX

Leng, Yifei, Jianguo Bao, Gaofeng Chang, Han Zheng, Xingxing Li, Jiangkun Du, Daniel Snow, and Xu Li. 2016. "Biotransformation of Tetracycline by a Novel Bacterial Strain Stenotrophomonas Maltophilia DT1." *Journal of Hazardous Materials* 318 (November). Elsevier: 125–33. doi:10.1016/J.JHAZMAT.2016.06.053.

Li, W., Su, H., Li, J., Bhat, S.A., Cui, G., Han, Z.M., Nadya, D.S., Wei, Y. and Li, F. 2021. Distribution of extracellular and intracellular antibiotic resistance genes in sludge fractionated in terms of settleability. *Science of the Total Environment*, 760, 143317.

Liu, Yuanwang, Huiqing Chang, Zhaojun Li, Yao Feng, Dengmiao Cheng, and Jianming Xue. 2017. "Biodegradation of Gentamicin by Bacterial Consortia AMQD4 in Synthetic Medium and Raw Gentamicin Sewage." *Scientific Reports* 2017 7:1 7 (1). Nature Publishing Group: 1–11. doi:10.1038/s41598-017-11529-x.

Liu, Yuanwang, Huiqing Chang, Zhaojun Li, Cheng Zhang, Yao Feng, and Dengmiao Cheng. 2016. "Gentamicin Removal in Submerged Fermentation Using the Novel Fungal Strain Aspergillus Terreus FZC3." *Scientific Reports* 2016 6:1 6 (1). Nature Publishing Group: 1–10. doi:10.1038/srep35856.

Liyanage, G Y, and Pathmalal M Manage. 2022. "Removal of Ciprofloxacin (CIP) by Bacteria Isolated from Hospital Effluent Water and Identification of Degradation Pathways." *International Journal of Medical, Pharmacy and Drug Research (IJMPD)* 2 (3). Accessed October 27. doi:10.22161/ijmpd.2.3.1.

Luo, Yunlong, Wenshan Guo, Huu Hao Ngo, Long Duc Nghiem, Faisal Ibney Hai, Jian Zhang, Shuang Liang, and Xiaochang C. Wang. 2014. "A Review on the Occurrence of Micropollutants in the Aquatic Environment and Their Fate and Removal during Wastewater Treatment." *Science of The Total Environment* 473–474 (March). Elsevier: 619–41. doi:10.1016/J.SCITOTENV.2013.12.065.

Lu, Xin, Yan Gao, Jia Luo, Shaohua Yan, Zed Rengel, and Zhenhua Zhang. 2014. "Interaction of Veterinary Antibiotic Tetracyclines and Copper on Their Fates in Water and Water Hyacinth (Eichhornia Crassipes)." *Journal of Hazardous Materials* 280 (September). Elsevier: 389–98. doi:10.1016/J.JHAZMAT.2014.08.022.

Maia, Alexandra S., Maria Elizabeth Tiritan, and Paula M.L. Castro. 2018. "Enantioselective Degradation of Ofloxacin and Levofloxacin by the Bacterial Strains Labrys Portucalensis F11 and Rhodococcus Sp. FP1." *Ecotoxicology and Environmental Safety* 155 (July). Academic Press: 144–51. doi:10.1016/J.ECOENV.2018.02.067.

Mann, Avantika, Kiran Nehra, J. S. Rana, and Twinkle Dahiya. 2021. "Antibiotic Resistance in Agriculture: Perspectives on Upcoming Strategies to Overcome Upsurge in Resistance." *Current Research in Microbial Sciences* 2 (December). Elsevier: 100030. doi:10.1016/J.CRMICR.2021.100030.

Migliore, Luciana, Maurizio Fiori, Anna Spadoni, and Emanuela Galli. 2012. "Biodegradation of Oxytetracycline by Pleurotus Ostreatus Mycelium: A Mycoremediation Technique." *Journal of Hazardous Materials* 215–216 (May). Elsevier: 227–32. doi:10.1016/J.JHAZMAT.2012.02.056.

Migliori, G B, S Tiberi, and Giovanni Battista Migliori. 2022. "WHO Drug-Resistant TB Guidelines 2022: What Is New?" Accessed October 31. doi:10.5588/ijtld.22.0263.

Mirzaei, Roya, Masud Yunesian, Simin Nasseri, Mitra Gholami, Esfandiyar Jalilzadeh, Shahram Shoeibi, and Alireza Mesdaghinia. 2018. "Occurrence and Fate of Most Prescribed Antibiotics in Different Water Environments of Tehran, Iran." *Science of The Total Environment* 619–620 (April). Elsevier: 446–59. doi:10.1016/J.SCITOTENV.2017.07.272.

Moffett, Brady S., Jennifer Morris, Charissa Kam, Marianne Galati, Ankhi Dutta, and Ayse Akcan-Arikan. 2018. "Vancomycin Associated Acute Kidney Injury in Pediatric Patients." *PloS One* 13 (10). PLoS One. doi:10.1371/JOURNAL.PONE.0202439.

Montelongo-Peralta, L. Z., A. León-Buitimea, J. P. Palma-Nicolás, J. Gonzalez-Christen, and J. R. Morones-Ramírez. 2019. "Antibacterial Activity of Combinatorial Treatments Composed of Transition-Metal/Antibiotics against Mycobacterium Tuberculosis." *Scientific Reports 2019 9:1* 9 (1). Nature Publishing Group: 1–6. doi:10.1038/s41598-019-42049-5.

Mulla, Sikandar I., Anyi Hu, Qian Sun, Jiagwei Li, Fidèle Suanon, Muhammad Ashfaq, and Chang Ping Yu. 2018. "Biodegradation of Sulfamethoxazole in Bacteria from Three Different Origins." *Journal of Environmental Management* 206 (January). Academic Press: 93–102. doi:10.1016/J.JENVMAN.2017.10.029.

Nandhini, Palanichamy, Pradeep Kumar, Suresh Mickymaray, Abdulaziz S. Alothaim, Jayaprakash Somasundaram, and Mariappan Rajan. 2022. "Recent Developments in Methicillin-Resistant Staphylococcus Aureus (MRSA) Treatment: A Review." *Antibiotics* 11 (5). Multidisciplinary Digital Publishing Institute (MDPI). doi:10.3390/ANTIBIOTICS11050606.

Nguyen, Luong N., Long D. Nghiem, and Seungdae Oh. 2018. "Aerobic Biotransformation of the Antibiotic Ciprofloxacin by Bradyrhizobium Sp. Isolated from Activated Sludge." *Chemosphere* 211 (November). Pergamon: 600–607. doi:10.1016/J.CHEMOSPHERE.2018.08.004.

Pan, Lan jia, Jie Li, Chun xing Li, Xiao da Tang, Guang wei Yu, and Yin Wang. 2018. "Study of Ciprofloxacin Biodegradation by a Thermus Sp. Isolated from Pharmaceutical Sludge." *Journal of Hazardous Materials* 343 (February). Elsevier: 59–67. doi:10.1016/J.JHAZMAT.2017.09.009.

Rahmani, Kourosh, Mohammad Ali Faramarzi, Amir Hossain Mahvi, Mitra Gholami, Ali Esrafili, Hamid Forootanfar, and Mahdi Farzadkia. 2015. "Elimination and Detoxification of Sulfathiazole and Sulfamethoxazole Assisted by Laccase Immobilized on Porous Silica Beads." *International Biodeterioration & Biodegradation* 97 (January). Elsevier: 107–14. doi:10.1016/J.IBIOD.2014.10.018.

Razavi, Mohammad, Erik Kristiansson, Carl-Fredrik Flach, and D. G. Joakim Larsson. 2020. "The Association between Insertion Sequences and Antibiotic Resistance Genes." *MSphere* 5 (5). American Society for Microbiology. doi:10.1128/MSPHERE.00418-20/SUPPL_FILE/MSPHERE.00418-20-ST002.DOCX.

Rodrigo Quejigo, Jose, Sara Tejedor-Sanz, Reiner Schroll, and Abraham Esteve-Núñez. 2019. "Electrodes Boost Microbial Metabolism to Mineralize Antibiotics in Manure." *Bioelectrochemistry* 128 (August). Elsevier: 283–90. doi:10.1016/J.BIOELECHEM.2019.04.008.

Rodríguez-Rodríguez, Carlos E., Ma Jesús García-Galán, Paqui Blánquez, M. Silvia Díaz-Cruz, Damià Barceló, Glòria Caminal, and Teresa Vicent. 2012. "Continuous Degradation of a Mixture of Sulfonamides by Trametes Versicolor and Identification of Metabolites from Sulfapyridine and Sulfathiazole." *Journal of Hazardous Materials* 213–214 (April). Elsevier: 347–54. doi:10.1016/J.JHAZMAT.2012.02.008.

Segura, Pedro A , Hideshige Takada, José A. Correa, Karim el Saadi, Tatsuya Koike, Siaw Onwona-Agyeman, John Ofosu-Anim, et al. 2015. "Global Occurrence of Anti-Infectives in Contaminated Surface Waters: Impact of Income Inequality between Countries." *Environment International* 80 (July). Pergamon: 89–97. doi:10.1016/J.ENVINT.2015.04.001.

Selvi, A., Jaseetha Abdul Salam, and Nilanjana Das. 2014. "Biodegradation of Cefdinir by a Novel Yeast Strain, Ustilago Sp. SMN03 Isolated from Pharmaceutical Wastewater." *World Journal of Microbiology and Biotechnology* 30 (11). Kluwer Academic Publishers: 2839–50. doi:10.1007/S11274-014-1710-4/FIGURES/9.

Shao, Sicheng, Yongyou Hu, Jianhua Cheng, and Yuancai Chen. 2018. "Degradation of Oxytetracycline (OTC) and Nitrogen Conversion Characteristics Using a Novel Strain." *Chemical Engineering Journal* 354 (December). Elsevier: 758–66. doi:10.1016/J.CEJ.2018.08.032.

Shi, Lili, Fuying Ma, Yuling Han, Xiaoyu Zhang, and Hongbo Yu. 2014. "Removal of Sulfonamide Antibiotics by Oriented Immobilized Laccase on Fe3O4 Nanoparticles with Natural Mediators." *Journal of Hazardous Materials* 279 (August). Elsevier: 203–11. doi:10.1016/J.JHAZMAT.2014.06.070.

Sutar, Rahul S., and Virendra K. Rathod. 2015. "Ultrasound Assisted Laccase Catalyzed Degradation of Ciprofloxacin Hydrochloride." *Journal of Industrial and Engineering Chemistry* 31 (November). Elsevier: 276–82. doi:10.1016/J.JIEC.2015.06.037.

Tiwari, Bhagyashree, Balasubramanian Sellamuthu, Yassine Ouarda, Patrick Drogui, Rajeshwar D. Tyagi, and Gerardo Buelna. 2017. "Review on Fate and Mechanism of Removal of Pharmaceutical Pollutants from Wastewater Using Biological Approach." *Bioresource Technology* 224 (January). Elsevier: 1–12. doi:10.1016/J.BIORTECH.2016.11.042.

Topp, Edward, Ralph Chapman, Marion Devers-Lamrani, Alain Hartmann, Romain Marti, Fabrice Martin-Laurent, Lyne Sabourin, Andrew Scott, and Mark Sumarah. 2013. "Accelerated Biodegradation of Veterinary Antibiotics in Agricultural Soil Following Long-Term Exposure, and Isolation of a Sulfamethazine-Degrading Microbacterium Sp." *Journal of Environmental Quality* 42 (1). John Wiley & Sons, Ltd: 173–78. doi:10.2134/JEQ2012 0162

"VISA/VRSA Infections – Epidemiology." 2022. Accessed October 31. www.vdh.virginia.gov/epidemiology/epidemiology-fact-sheets/vancomycin-intermediate-staphylococcus-aureus-visa-and-vancomycin-resistant-staphylococcus-aureus-vrsa-infections/.

Wang, Jianlong, and Shizong Wang. 2016. "Removal of Pharmaceuticals and Personal Care Products (PPCPs) from Wastewater: A Review." *Journal of Environmental Management* 182 (November). Academic Press: 620–40. doi:10.1016/J.JENVMAN.2016.07.049.

Wang, Shizong, Yuming Hu, and Jianlong Wang. 2018. "Biodegradation of Typical Pharmaceutical Compounds by a Novel Strain Acinetobacter Sp." *Journal of Environmental Management* 217 (July). Academic Press: 240–46. doi:10.1016/J.JENVMAN.2018.03.096.

Wang, Yuezhu, Hanmin Zhang, Yujie Feng, Baikun Li, Mingchuan Yu, Xiaotong Xu, and Lu Cai. 2019. "Bio-Electron-Fenton (BEF) Process Driven by Sediment Microbial Fuel Cells (SMFCs) for Antibiotics Desorption and Degradation." *Biosensors and Bioelectronics* 136 (July). Elsevier: 8–15. doi:10.1016/J.BIOS.2019.04.009.

Weng, Shin Sian, Shiu Mei Liu, and Hong Thih Lai. 2013. "Application Parameters of Laccase–Mediator Systems for Treatment of Sulfonamide Antibiotics." *Bioresource Technology* 141 (August). Elsevier: 152–59. doi:10.1016/J.BIORTECH.2013.02.093.

Wu, Qianxing, Niloofar Sabokroo, Yujie Wang, Marzieh Hashemian, Somayeh Karamollahi, and Ebrahim Kouhsari. 2021. "Systematic Review and Meta-Analysis of the Epidemiology of Vancomycin-Resistance Staphylococcus Aureus Isolates." *Antimicrobial Resistance and Infection Control* 10 (1). BioMed Central Ltd: 1–13. doi:10.1186/S13756-021-00967-Y/TABLES/3.

Xiangfeng, Huang, Wang Shen, Chen Guoxin, Lu Lijun, Liu Jia, Huang Xiangfeng, Wang Shen, Chen Guoxin, Lu Lijun, and Liu Jia. 2022. "Typical Pollutants Removal Efficiency from Aquaculture Wastewater by Using Constructed Wetlands." *Chinese Journal of Environmental Engineering* 10 (1). Chinese Journal of Environmental Engineering: 12–20. Accessed October 31. doi:10.12030/J.CJEE.20160102.

Xin, Zhao, Tian Fengwei, Wang Gang, Liu Xiaoming, Zhang Qiuxiang, Zhang Hao, and Chen Wei. 2012. "Isolation, Identification and Characterization of Human Intestinal Bacteria with the Ability to Utilize Chloramphenicol as the Sole Source of Carbon and Energy." *FEMS Microbiology Ecology* 82 (3). Oxford Academic: 703–12. doi:10.1111/J.1574-6941.2012.01440.X.

Xu, Zhicheng, Xiaoye Song, Yun Li, Guoxue Li, and Wenhai Luo. 2019. "Removal of Antibiotics by Sequencing-Batch Membrane Bioreactor for Swine Wastewater Treatment." *Science of the Total Environment* 684 (September). Elsevier: 23–30. doi:10.1016/J SCITOTENV.2019.05.241.

Yang, Jie, Yonghui Lin, Xiaodan Yang, Tzi Bun Ng, Xiuyun Ye, and Juan Lin. 2017. "Degradation of Tetracycline by Immobilized Laccase and the Proposed Transformation Pathway." *Journal of Hazardous Materials* 322 (January). Elsevier: 525–31. doi:10.1016/J.JHAZMAT.2016.10.019.

Zhang, Qichun, and Warren A. Dick. 2014. "Growth of Soil Bacteria, on Penicillin and Neomycin, Not Previously Exposed to These Antibiotics." *Science of The Total Environment* 493 (September). Elsevier: 445–53. doi:10.1016/J.SCITOTENV.2014.05.114.

Zheng, Wei, Zhenya Zhang, Rui Liu, and Zhongfang Lei. 2018. "Removal of Veterinary Antibiotics from Anaerobically Digested Swine Wastewater Using an Intermittently Aerated Sequencing Batch Reactor." *Journal of Environmental Sciences* 65 (March). Elsevier: 1–17. doi:10.1016/J.JES.2017.04.011.

11 Antibiotics and Antibiotic Resistance Genes and Their Management in Organic Wastes

Shokufeh Rafieyan, Zahra Etemadifar, Reihaneh Moridshahi, Farzaneh Dianatdar, and Sharareh Harirchi

11.1 INTRODUCTION

Many toxic substances, such as various drugs and antibiotics, toxic heavy metals, pathogens (with antibiotic resistance genes (ARGs)), and microplastics, inevitably enter domestic wastewater (DWW) and municipal solid waste (MSW) as a result of anthropogenic activities (Yu et al., 2020, Anand et al., 2021; Kumar et al., 2022). Microplastics and antibiotics are discharged into surface and ground water after the wastewater treatment process and cause contamination of drinking or agricultural water sources, ultimately harming human and animal bodies (Bhat et al., 2022, Li et al., 2021b). Antibiotics and ARGs are of great importance due to the spread of resistance genes between bacterial strains, and it is necessary to ensure their decomposition before the discharge of treated DWW and MSW into the environment (Cui et al., 2018, 2019; Huang et al., 2021).

Since Alexander Fleming discovered penicillin in 1929, antibiotics have been extensively utilized. In 1996, 10,200 tons of antibiotics were utilized in Europe, with 50% going toward veterinary medical growth stimulation. In 2008, America consumed 16,200 tons of antibiotics, with 70% of those being used in livestock farming (Kümmerer, 2003). However, it is believed that between 100,000 and 200,000 tons of antibiotics are consumed globally each year, and 50 million tons of antibiotics have been produced in the past 50 years. Diverse countries have different patterns of antibiotic usage. China is the world's biggest producer and consumer of antibiotics (Goel, 2015). According to research, animal husbandry uses antibiotics significantly more than human treatment. The first global map of antibiotic use in animal husbandry in 2010 (63,151 tons; 228 countries) estimated that by 2030, antibiotic use would increase by 67% and possibly double in a number of countries (Brazil, India, China, Russia, and South Africa) (Polianciuc et al., 2020).

Excessive consumption of antibiotic residues and bioproducts results in their release into the environment, which poses a risk to human health. Antibiotics enter the environment in many ways, including: excretion of urine and feces from humans and animals, disposal of excess drugs in sewage or domestic waste; and direct contamination of the environment through aquaculture, plant cultivation, livestock and poultry farming, and the pharmaceutical industry. Antibiotic concentrations of 1 microgram per liter have been detected in various media, including sea water, and up to 10,000 in polluted surface water (Larsson and Flach, 2022). High concentrations of several different antibiotics can also cause significant concern because of the emergence and spreading of ARGs and antibiotic-resistant bacteria (ARB) (Lu et al., 2020; Cui et al., 2022). One of the most important sources for dispersion of ARGs in the environment is wastewater treatment systems, along with sludge settling

DOI: 10.1201/9781003408352-13

as intra- or extracellular ARGs (Li et al., 2021b). Most ARGs are capable of horizontal and vertical gene transfer from pathogenic bacteria to environmental microorganisms. As a result, it's critical to monitor the microbes in the environment to predict the development of antibiotic resistance there (Iwu et al., 2020). There have been several studies on detecting and evaluating ARGs and ARB in the environment, as well as numerous strategies to remove antibiotics and ARGs from the environment that are categorized into three main categories: physical, chemical, and biological methods. ARGs can be detected in the environment using a variety of polymerase chain reaction (PCR) techniques and metagenomics. In this chapter, recent advances in antibiotic and ARGs pollution and techniques for detecting and removing these pollutants are discussed comprehensively.

11.2 SOURCE OF ANTIBIOTICS IN THE ENVIRONMENT

Antibiotics are used to treat infections that affect humans, animals, plants, and aquatic animals. They are discharged into the wastewater through urine and feces and can concentrate in manure because they are not completely metabolized by these organisms. Since manure is used as fertilizer in agriculture, antibiotics are released into groundwater and accumulate in the soil, where they can be consumed by plants and animals (Figure 11.1) (Kümmerer, 2003).

11.2.1 Antibiotics Residues in Surface Water

The pharmaceutical business and the use of antibiotics in the livestock and aquaculture industries are strongly connected with the geographical and temporal distribution of antibiotics in aquatic ecosystems. For instance, antibiotic pollution in China is concentrated mainly in aquatic areas that are close to these sectors, and sources like drinking water, surface water, and rivers have been reported to include antibiotics such as chloramphenicol, sulfonamide, tetracycline, and quinolone (Liu et al., 2021). Rodriguez-Mozaz et al. (2020a) examined the water and wastewater of seven European countries and found that the highest amounts of antibiotics in these countries were

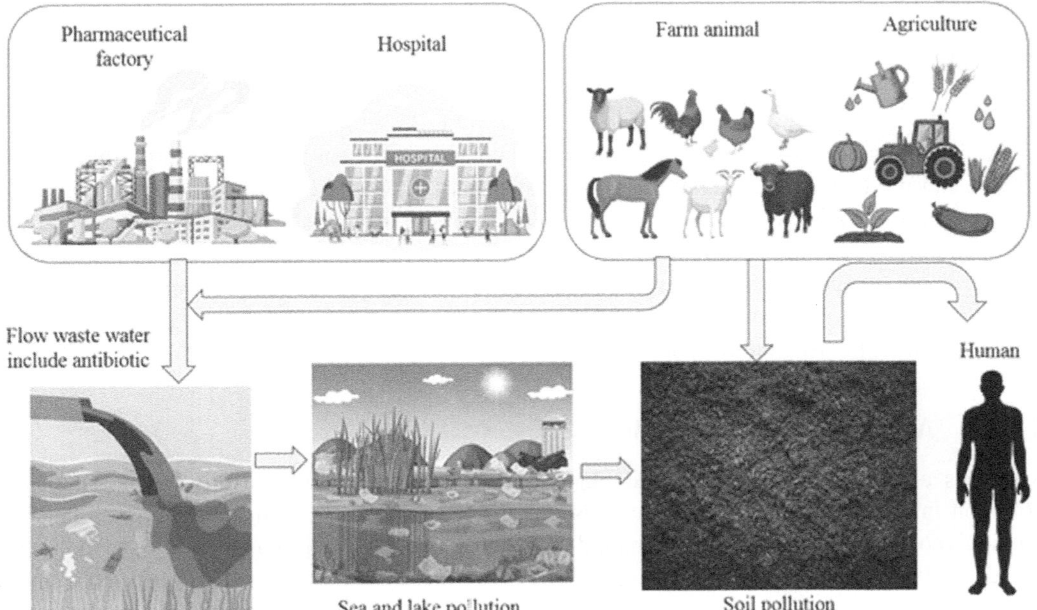

FIGURE 11.1   Antibiotic circulation cycle in the environment.

macrolides and fluoroquinolones. Gray et al. (2020a) examined 16 samples from wells and streams and identified 15 types of antibiotics from the penicillin, lincosamide, sulfonamide, macrolide, quinolone, and aminoglycoside groups. In this research, the antibiotics sulfamethoxazole, sulfamerazine, danofloxacin, and erythromycin were more abundant.

11.2.2 Antibiotics Residues in Sediment

Antibiotics used by humans, livestock, agriculture, and aquaculture, including tetracyclines, sulfonamides, and macrolides, have been detected in sediments. Antibiotics are poured into lakes and seas by urban, livestock, and aquaculture sewage, and then the antibiotics settle in the sediments. Björklund et al. (1991) investigated the antibiotics in sediments under fish farms; however, only a few pieces of research have been done on antibiotic concentration in sediments. As a result, further investigation into this topic is required.

Antibiotics can be readily absorbed by sediment particles. In contrast to surface waters, the concentration of antibiotics in sediments is less sensitive to environmental conditions such as dilution, particle absorption, and photodegradation. Antibiotics are therefore more stable in sediments than in surface water. Since sediments absorb antibiotics differently depending on the water conditions, there is a regionally and geographically diverse distribution of antibiotics in sediments. However, additional research is still required to understand how antibiotics travel through sediments and how they settle in sediments (Liu et al., 2021, Kümmerer, 2009).

11.2.3 Antibiotics Residues in Aquaculture

In aquaculture, antibiotics are used to control infectious diseases. Examples of antibiotics used in aquaculture include oxytetracycline, florfenicol, sarafloxacin, and erythromycin. The majority of these antibiotics—between 70% and 80%—are discharged into the water, where they alter the microbial communities by allowing the resistant strains of the environmental microbiota to thrive (Iwu et al., 2020). According to research by Grave et al. (2008), over the years 2000–2005, Norwegian aquaculture for Atlantic fish farming has increased the use of antimicrobial medicines in farmed fish. China consumed 57.9% of total global antibiotic consumption in aquaculture in 2017 (Schar et al., 2020).

11.2.4 Antibiotics Residues in Poultry Tissues

In the poultry industry, antibiotics are utilized to cure various diseases, promote growth, and boost weight and productivity. Sometimes non-specialist personnel in the poultry sector use excessive doses of antibiotics, which causes antibiotic levels in the environment to rise and accumulate. The use of antibiotics near the time of slaughter and without a period of cleaning antibiotics from poultry tissue creates many risks for consumers. The highest amounts of antibiotics were found in liver samples. The most common drugs in muscles and organs include enrofloxacin, chloramphenicol, oxytetracycline, fluoroquinolone, and tetracycline (Mohammadzadeh et al., 2022).

11.2.5 Antibiotics Residues in Milk

Milk has a special place in the food plan, especially for children. Mastitis, one of the most common infectious disorders in animals, is treated by injecting antibiotics such as beta-lactams, aminoglycosides, tetracycline, sulfonamides, macrolides, and quinolones into the mammary glands. Milk should not be consumed after using antibiotics until the antibiotic is removed from the animal's body (Bahramian et al., 2022). According to worldwide research, the beta-lactam group, tetracyclines, fluoroquinolones, sulfonamides, and aminoglycosides are the antibiotics found in milk, with proportions of 36.54%, 14.01%, 13.46%, 12.64%, and 10.44%, respectively. Consumption of milk containing antibiotics causes the spreading of ARGs and changes intestinal microflora, which increases the risk of many other diseases (Sachi et al., 2019).

11.2.6 ANTIBIOTIC RESIDUES IN HUMAN MEDICINE

The per capita and individual shares of each antibiotic vary in different countries. The number of antibiotics used without a prescription varies significantly between countries. Levels of antibiotic usage vary based on the country; for example, vancomycin is used more widely in the United States than in Germany. Around 50–70% of the antibiotics consumed worldwide are beta-lactams, which include penicillins, carbapenems, cephalosporins, and others. In contrast to conceptions, the origin of antibiotics in wastewater is not from hospitals; however, the major source of antibiotics is wastewater from antibiotics used in the community and homes. In America and Britain, 10% of antibiotics in waste are from hospital. The main source of third- and fourth-generation cephalosporin antibiotics is hospital wastewater. Most antibiotics are metabolized in the liver. Information on the quantity of antibiotics consumed in Germany indicates that 70% of the total amount is discharged both unaltered and as active compounds. The majority of antibiotics are water soluble and excreted through the urine. The metabolism of some antibiotics leads to the production of chloramphenicol, and some antibiotics are transformed into more toxic substances than the original antibiotics, such as the acetylation of sulfamethoxazole (Kümmerer, 2009). According to a survey from 2016, Iran's use of antibiotics was approximately three times greater than the OECD[1] average. The drug utilization 90% (DU90%) of antibiotics in Iran include amoxicillin (32.4%), co-amoxiclav (12.7%), azithromycin (12.3%), ciprofloxacin (5.3%), doxycycline (5.2%), cephalexin (4.3%), cotrimoxazole (1.3%), and clarithromycin (1.3%) (Abbasian et al., 2019).

11.2.7 ANTIBIOTIC RESIDUES IN THE COVID-19 PANDEMIC

During the COVID-19 pandemic, the viral illness gave rise to an infectious illness that was treated with antibiotics in 95% of cases. However, a significant portion of COVID-19 patients receive unnecessary antibiotic treatment, and some even take antibiotics at home to fight the disease. Doxycycline and amoxicillin concentrations in British sewage increased during the COVID-19 epidemic. Macrolides, including azithromycin, are antibiotics used against gram-positive bacteria and respiratory tract infections that have demonstrated immunomodulatory and anti-inflammatory effects and are consequently beneficial in treating viral respiratory infections (Sosa-Hernández et al., 2021). Additionally, following the emergence of COVID-19, beta-lactam antibiotic use—including amoxicillin and cephalexin—increased in Isfahan, Iran's urban wastewater treatment plants (Samandari et al., 2022).

11.2.8 ANTIBIOTICS RESIDUES IN PLANT AGRICULTURE

Antibiotics have been utilized to combat bacterial infections in fruits, vegetables, and ornamental plants since the 1950s. Streptomycin and oxytetracycline are now the main antibiotics for plants that are most frequently utilized. For instance, the use of proper antibiotics is employed to combat the fire blight brought on by *Erwinia amylovora* in apples, pears, and ornamental trees. Less than 0.5% of all antibiotics are used in the US for plant-related purposes. Streptomycin and oxytetracycline are the most often used antibiotics for the treatment of bacterial illnesses in fruit trees. Being active outside or inside the plant and being resistant to oxidation, UV radiation, precipitation, and high temperatures are all characteristics of antibiotics that are suitable for agricultural application. These characteristics are the ones that cause issues in the environment (Kümmerer, 2009). Different sources of antibiotic concentrations in some countries are shown in Table 11.1.

TABLE 11.1
Antibiotics Residues from Different Sources

Source	Regions	Antibiotic	Concentration	Reference
Wetland water	Weining, China	Sulfonamides	50.5 (ng/L)	(Liu et al., 2021)
		Fluoroquinolones	43.2 (ng/L)	
		Macrolides	22.6 (ng/L)	
		Chloramphenicol	15.9 (ng/L)	
Surface water in aquaculture areas	Yancheng, China	Sulfonamides	1 ~ 2.35 (ng/L)	(Liu et al., 2021)
		Fluoroquinolones	3.77 ~ 9.76 (ng/L)	
		Macrolides	0.15 ~ 2.05 (ng/L)	
		Chloramphenicol	2.91 ~ 3.44 (ng/L)	
Drinking water	Dongguan, China	Sulfonamides	24.63 ~ 38.7 (ng/L)	(Liu et al., 2021)
		Fluoroquinolones	11.11 ~ 82.24 (ng/L)	
		Macrolides	27.25 ~ 206.65 (ng/L)	
		Tetracyclines	20.08 ~ 43.02 (ng/L)	
River	Kan rivers, Iran	Amoxicillin	0.03183 (µg/L)	(Mirzaei et al., 2019)
		Penicillin	0.06344 (µg/L)	
		Ciprofloxacin	0.02227 (µg/L)	
		Cefixime	0.13699 (µg/L)	
		Cephalexin	0.1844 (µg/L)	
		Azithromycin	0.05276 (µg/L)	
		Erythromycin	0.01802 (µg/L)	
Wastewater	Tehran, Iran	Amoxicillin	0.09457 (µg/L)	(Mirzaei et al., 2019)
		Penicillin	0.01713 (µg/L)	
		Ciprofloxacin	0.2487 (µg/L)	
		Cefixime	0.1244 (µg/L)	
		Cephalexin	0.02893 (µg/L)	
		Azithromycin	0.01445 (µg/L)	
		Erythromycin	0.05137 (µg/L)	
Surface water in animal farms areas	Tehran, Iran	Tetracyclines	5.4 ~ 8.1 (ng/L)	(Javid et al., 2016)
Wastewater	Norway	Cephalosporins	60.7 (ng/L)	(Rodriguez-Mozaz et al., 2020b)
		Fluoroquinolones	159.2 ~ 60.7 (ng/L)	
		Lincosamides	97.1 (ng/L)	
		Macrolides	149.7 (ng/L)	
		Nitroimidazole	93.2 (ng/L)	
		Quinolones	7.5 (ng/L)	
		Tetracyclines	179.2 (ng/L)	
Wastewater	Portugal	Cephalosporins	38.4 (ng/L)	(Rodriguez-Mozaz et al., 2020b)
		Fluoroquinolones	457.3 ~ 89.7 (ng/L)	
		Lincosamides	8.5 (ng/L)	
		Macrolides	361.8 ~ 74.2 (ng/L)	
		Quinolones	15.0 (ng/L)	
		Tetracyclines	231.2 (ng/L)	
Surface water	United States America, North Carolina	Danofloxacin	8.31 ~ 299.62 (ng/L)	(Gray et al., 2020b)
		Erythromycin	2.50 ~ 8.16 (ng/L)	
		Sulfamerazine	0.35 ~ 16.41 (ng/L)	
		Sulfamethoxazole	5.97 ~ 14.54 (ng/L)	
Bivalves	Spain, Ebro Delta	Ronidazole	1 ~ 1.8 (ng/g)	(Baralla et al., 2021)
		Sulfamethoxazole	<0.02 ~ 2.9 (ng/g)	
		Azithromycin	1.3 ~ 3 (ng/g)	

TABLE 11.1 (Continued)
Antibiotics Residues from Different Sources

Source	Regions	Antibiotic	Concentration	Reference
Bivalves	China, Hailing Island	Erythromycin	0.9 (ng/g)	(Baralla et al., 2021)
		Salinomycin	14.5 (ng/g)	
		Narasin	7.5 (ng/g)	
Bivalves	California, United States America	Lomefloxacin	29 (ng/g)	(Baralla et al., 2021)
		Sulfamethazine	24 (ng/g)	
		Enrofloxacin	1.3 (ng/g)	
		Erythromycin	0.14 (ng/g)	
		Ofloxacin	1.2 (ng/g)	
Bivalves	Brazil, Sepetiba Bay and Parnaiba Delta River	Chloramphenicol	0.5 (ng/g)	(Baralla et al., 2021)
Sediment	Xiangjiang river, China	Sulfonamides	0.98 (ng/L)	(Liu et al., 2021)
		Fluoroquinolones	3.77 ~ 487 (ng/L)	
		Macrolides	0.33 ~ 9.44 (ng/L)	
		Chloramphenicol	9.03 (ng/L)	
		Tetracyclines	1.56–1115 (ng/L)	
Sediment	Mangrove nature reserves, China	Sulfonamides	39.1 (ng/L)	(Liu et al., 2021)
		Fluoroquinolones	28 (ng/L)	
		Macrolides	126.6 (ng/L)	
		Chloramphenicol	7.7 (ng/L)	
		Tetracyclines	95.6 (ng/L)	
Sediment	Baltic Sea, Poland	Sulfathiazole	1.77 (ng/g)	(Siedlewicz et al., 2016)
		Sulfamethazine	1.76 (ng/g)	
		Sulfamethiazole	12.85 ~ 20.84 (ng/g)	
		Sulfachloropyridazine	0.47 ~ 1.07 (ng/g)	
Poultry muscle	Tehran, Iran	Fluoroquinolone	72.59 (ng/g)	(Mohammadzadeh et al., 2022)
		Tetracycline	15.35 (ng/g)	
		Sulfonamide	36.52 (ng/g)	
Poultry liver	Iran	Enrofloxacin	18.35 (ng/g)	(Mohammadzadeh et al., 2022)
		Oxytetracycline	576.65 (ng/g)	
		Chloramphenicol	47.10 (ng/g)	
Milk	Bangladesh	Oxytetracycline	61.29 µg/l	(Rahman et al., 2021)
		Amoxicillin	124 µg/l	
WWTP A during the prevalence of covid-19	Isfahan, Iran	Amoxicillin	509.64 (µg/L)	(Samandari et al., 2022)
		Cephalexin	189.42 (µg/L)	

11.3 ANTIBIOTICS AS MICRO-CONTAMINANTS

Micropollutants, also known as micro-contaminants, are biological or chemical pollutants that enter groundwater and surface water in low (nanograms per liter) to extremely low (picograms per liter) quantities as a result of human activity. Natural and synthetic organic compounds that are chemical micropollutants include drugs, personal care products (PCP), industrial chemicals (e.g., perfluoroalkyl and per- and polyfluoroalkyl substances (PFAS)), detergents, steroid hormones, agricultural pesticides, and mineral micropollutants, which include metals, metalloids, and elements such as lead (Pb), cadmium (Cd), mercury (Hg), arsenic (As), antimony (Sb), radon (Rn), uranium (U),

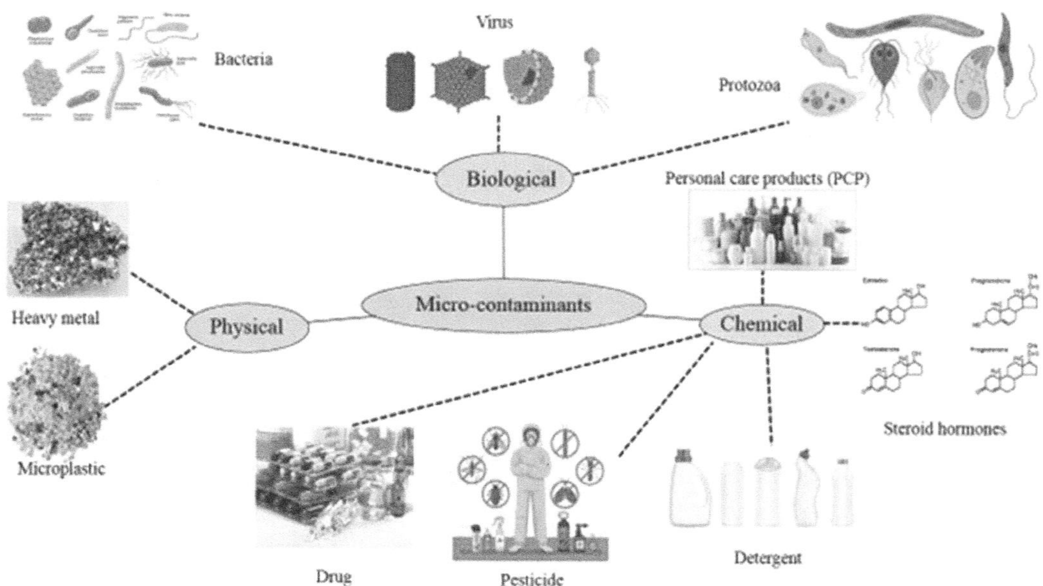

FIGURE 11.2 Various types of micro-contaminants.

and chromium (Cr). Biological micropollutants include a variety of viruses, protozoa, and bacteria (Bruchet and Janex-Habibi, 2007).

There are many different sources of environmental micropollutants (Figure 11.2). Domestic sewage, however, is the primary source of micropollutants in aquatic environments. Micropollutant drugs are often found in aquatic environments and obtained over the counter from hospitals, pharmacies, and stores. Medicines are made to treat humans and animals. However, the body does not entirely metabolize them. Drug metabolites and residues are discharged into wastewater. Drugs that have passed their expiration dates and production-related waste are further sources of pharmaceutical contamination in aquatic environments. The direct discharge of untreated wastewater into rivers, lakes, and reservoirs causes the release of micropollutants. The natural deposition of micropollutants on sediments or their transit to other locations are also possible. The chemical and biological destruction of these compounds in surface waters can also convert them into byproducts. However, some micropollutants are stable and accumulate in surface water. Among the micropollutants, drugs have biological accumulation within the food chain in the ecosystem and cause unknown and chronic diseases in humans (Monteiro and Boxall, 2010). Chemicals known as contaminants of emerging concern (CEC) have recently attracted attention because of their ability to harm humans, aquatic life, and the environment. They have also been found in numerous environments in greater than predicted amounts. Antibiotics are known as the primary group of emerging micro-contaminants, and they make up the largest class of CECs, along with other pharmaceuticals, including anti-inflammatories and pain relievers (Goel, 2015).

11.4 ARGS AS AN EMERGING POLLUTANT

Unnecessary, incorrect, and excessive usage of antibiotics has caused microbial resistance and accelerated the development of ARB and ARGs among bacterial species (García et al., 2020). Unlike antibiotics, the release of ARGs does not depend on the local and continuous release of antibiotics. These genes can spread between various species because they are present in the environment. Although there is no direct evidence that ARGs may be transferred from the environment to the human body, some studies suggest that they can be distributed and exchanged between various genera of environmental microbes and even in organisms from a completely different phylum, which is the basis for the emergence of acute diseases that are really hazardous to the public's health (Kumar et al., 2019a). Due to their widespread release in various environmental sectors and the public health problems caused by this release, ARGs can be considered as emerging environmental pollutants (Iwu et al., 2020). ARGs, as pollutants, with a tendency to continuously accumulate in the environment, are easily created but are resistant to elimination. Therefore, ARGs' acquisition by clinically relevant microorganisms has led to the formation of "superbugs" that seriously affect health and threaten human well-being. Consequently, one of the main issues of the 21st century, according to WHO,[2] is the rise in antimicrobial resistance (AMR) genes and ARGs (Huang et al., 2021). According to studies, AMR accounts for 25,000 deaths each year in the European Union (EU) and 700,000 worldwide. As a result, it is expected that by 2050, AMR will be more lethal than either cancer or diabetes (Zalewska et al., 2021). Antibiotic resistance is defined as an increase in the minimum inhibitory concentration (MIC) of antibiotics toward bacteria. Prolonged exposure to low antibiotic concentrations is the main cause of antibiotic resistance (Kumar et al., 2019a).

Through horizontal gene transfer (HGT), vertical gene transfer (VGT), and de novo mutation, antibiotic resistance can develop between various bacterial hosts (Haffiez et al., 2022). HGT relies more on the movement of ARGs by mobile genomic elements (MGEs), accelerates the transfer rate. Bacterial cells can exchange intercellular MGEs such as conjugative plasmids, bacteriophages, and gene transfer agents (phage-like particles). Contrarily, intracellular MGEs like integrons, transposons, and insertion sequences can only be transported within the same bacterial cell. However, there may be interactions between intracellular and extracellular MGEs that could increase the stability and subsequent release of ARGs (Xie et al., 2018, Koch et al., 2021). Plasmids are regarded as crucial MGE elements in the transmission of ARGs since they are able to multiply apart from the bacterial chromosome and can thus cross phylogenetic barriers (Nguyen et al., 2021). Transposons and integrons are mobile elements that are particularly important for the horizontal transfer of environmental ARGs because of their potential to acquire and spread gene cassettes (Xie et al., 2018).

Various factors play different roles in the horizontal transfer and spread of ARGs and ARB in diverse environments, and their fate in an ecosystem are dynamic. The selective pressure caused by the increasing production and consumption of antibiotics is the most important factor in the expansion of ARGs because it facilitates the acquisition of ARGs, increases the fitness of certain bacteria, and enables rapid emergence and spread on a global scale (Kumar et al., 2019a). On the other hand, exposure to stressful conditions such as antimicrobial substances, heavy metals, nanoparticles, and disinfectants significantly facilitates the development and release of ARGs. In fact, these factors cause a series of changes in the expression of different bacterial genes; for example, as a result of being exposed to stressful factors, the expression of genes related to the conjugation process is changed, leading to further stimulation of sexual pili at the bacterial surface, and promoting the conjugation process. An increase in the expression of the SOS response system may raise genetic instability and the rate of DNA mutations (Cycoń et al., 2019). Similarly, when the cell is exposed to stressful factors, it produces reactive oxygen species that disrupt the order of the bacterial cell membrane, increase its permeability, and stimulate HGT processes. The evidence suggests that heavy metals and antibiotics can have the same regulatory responses; therefore, the co-presence of heavy metals and antibiotics may speed up the selection of ARGs through co-selection and cross-resistance mechanisms. Also, metal stress causes a thousand-fold increase in the permissibility of the microbial community plasmid, which has a significant effect on the ARGs profile (Zalewska et al., 2021). Spatial distribution of bacteria, oxygen levels, and even low doses of chlorine disinfectant in wastewater treatment systems have also been introduced as factors affecting the transfer rate of ARGs between bacterial communities. For example, the findings indicate that there is a higher proportion of proteobacteria (27%) in aerobic sludge than in anaerobic sludge (21%), so the abundance of the plasmid is twice as high, and the transfer rate increases accordingly (Nguyen et al., 2021).

Soil is recognized as being a significant source of ARGs due to the presence of a wide variety of bacteria that can create natural antibiotics. In addition, the widespread use of antibiotics in the livestock production industry to increase growth results in the release of large amounts of veterinary antibiotics and high levels of antibiotic resistance in agricultural lands. Overall, antibiotic resistance has significantly changed the soil ecosystem and turned it into one of the most important reservoirs of pollution (Zalewska et al., 2021). The presence of ARGs in soil-dwelling bacteria makes them easily transmissible to new hosts via HGT, which promotes the spread and abundance of resistance genes in the soil. Additionally, ARB found in manure may stick to plants grown in fertilized soil, proliferate quickly, and dominate the local bacterial population (Cycoń et al., 2019, Xie et al., 2018).

ARGs have been detected in surface and subterranean water near hospitals, fish breeding ponds, and pig farms as a direct result of the usage of antibiotics in these facilities. Eventually, this antibiotic resistance spreads to bacteria in drinking water or the food chain (Polianciuc et al., 2020). ARB and ARGs can be transferred to food crops by surface streams' irrigation water that contains ARGs. Additionally, human activities may also contaminate groundwater used to irrigate crops. Until the source of contamination is eliminated, this contamination will remain for a long time. Irrigation water is the most common way antibiotic-resistant pathogens infect vegetables in the pre-harvest stage (Xie et al., 2018). As effluents from wastewater treatment plants (WWTPs) are treated during the wastewater treatment process, non-target hosts may be exposed to ARGs. The typical biological wastewater treatment concept is that the majority of ARB and ARGs are eliminated from the water phase and go into the sludge phase; as a result, sludge and biosolids will include a significant amount of ARGs. Many ARGs remain in the treated effluent and cannot be completely degraded, only changing from a liquid to a solid phase. As a result, some microorganisms may develop antibiotic resistance under the pressure of antibiotic selection. Since the microorganisms in the aeration tank have a high reproduction rate, it increases the possibility of ARGs release by HGT (Cui et al., 2020). The number of ARGs, transposons, and integrons detected in the receiving river water and downstream sediments is significantly influenced by the effluent from WWTPs because the final effluents are subsequently released back into the environment. Upon coming into contact with people, ARGs hosts can transmit ARGs to commensal bacteria and pathogens through HGT (Nguyen et al., 2021). It's important to keep in mind that not all ARGs pose the same risk to human health. ARGs with a high transfer probability, i.e., those associated with MGE factors, can transmit to human pathogens and develop resistance to commonly used antibiotics. In contrast, ARGs with low risk are ones where the possibility of their transmission from environmental bacteria to human pathogens is rare (Nguyen et al., 2021).

11.5 ADVERSE EFFECTS OF ANTIBIOTICS ON THE ENVIRONMENT AND ASSOCIATED MICROORGANISMS

Antibiotics are continuously released from large reservoirs, including hospital wastewater, industrial wastewater, animal dung, and municipal WWTPs; as a result, environmental organisms are constantly exposed to these contaminants. In this way, the effect on the structure of microbial populations and their metabolic activity remains uniform (Riaz et al., 2020). Risk coefficients are indicators that researchers use to assess the environmental risks related to the use of antibiotics in various ecosystems. In a number of studies, general toxicity indices such as LC50, EC50, and IC50 have been used to investigate how antibiotics affect non-target sensitive organisms such as cyanobacteria, *Daphnia*, algae, oysters, and other aquatic organisms. Undeniably, it should be noted that antibiotic residues in the environment are mixed and rarely exist individually, so synergistic or antagonistic effects due to the mixture of antibiotics should be expected (Polianciuc et al., 2020). HGT in the human intestine accelerates, and it causes the appearance of multidrug resistance genes in pathogens such as *Staphylococcus aureus* resistant to methicillin (He et al.,

2020). In a study, Shang et al. (2015) examined how different antibiotics, including aureomycin, oxytetracycline, and tetracycline, negatively affected the proliferation of *Microcystis aeruginosa*. These findings show that the tetracycline antibiotic, in similar environmental concentrations, affects the pigment and soluble protein content and induces oxidative stress; therefore, cell growth and protein synthesis are affected in *M. aeruginosa*. Among the 226 antibiotics examined, 45 antibiotics induced highly toxic effects on algae and 100 antibiotics on daphnids. Lethal effects on *Daphnia* may cause disturbances in other aquatic populations and reduce the health of fish and other aquatic plankton predators (Sanderson et al., 2004). Moreover, the negative and threatening effects of sulfonamides, tetracyclines, and macrolides at low environmental concentrations on *Synechocystis* sp., *Lemna minor*, and *Selenastrum capricornutum* have also been proven (Gul et al., 2022).

The bacterial reaction to various antibiotics is concentration-dependent. High doses of antibiotics have antimicrobial effects on sensitive cells, whereas sub-inhibitory amounts cause a range of biological reactions in bacteria. Antibiotics may operate as extracellular molecules that bacteria perceive in non-lethal quantities to cause a variety of cellular reactions, some of which may result in a changed profile of antibiotic resistance or tolerance (Bernier and Surette, 2013). In soils containing sulfamethoxazole, sulfamethazine, sulfadiazine, and trimethoprim, a significant decrease in soil respiration (SR) has also been noted. However, this benefit was temporary and was based on how quickly these antibiotics disappeared (Cycoń et al., 2019).

High environmental concentrations of fluoroquinolones and sulfonamides have shown an inhibitory effect on denitrification. The use of pig manure containing the antibiotic tylosin in the soil caused changes in the nitrogen cycle mediated by microbial communities. Antibiotics have a selective effect on some microbial populations, leading to the emergence of resistance among them. This results in genetic and phenotypic diversity, which in turn influences various physiological processes and interactions between various species. It should be noted that the effect of antibiotics on microbial communities is not the same and is influenced by various factors, including antibiotic concentration, duration of exposure, microbial community, soil texture, water content, temperature, and even natural weather processes. Moreover, freezing, drying, and rewetting can change the physicochemical properties of antibiotics. Low concentrations of antibiotics (ng/liter/kg or sub-inhibitory, non-lethal, and sub-MIC concentrations) indirectly affect microbial species or consortia that are not directly exposed to them. In these concentrations, one of the important effects that occur in bacterial communities is the creation of microbial resistance, which is either created through the enrichment of existing resistant bacteria or through the formation of new resistance. In these concentrations, antibiotics influence the virulence, biofilm formation, and gene expression of microbes by acting as signaling molecules. The emergence of genetic and phenotypic variety through adaptive evolution can be considered one of the consequences of antibiotics in low concentrations. Furthermore, the physiology and dynamics of microbial populations are impacted by the existence of resistance genes in natural ecosystems. For instance, in gram-positive bacteria, peptidoglycan structure is changed by beta-lactam resistance (Grenni et al., 2018).

11.6 ELIMINATION AND MANAGEMENT OF ANTIBIOTICS

The demand for food and medical services is increasing as a result of the world population's exponential growth. Therefore, a large number of antibiotics are produced globally to treat a variety of diseases in humans and animals as well as to stimulate growth in livestock and aquaculture. As a result, antibiotic residue has heavily contaminated the biosphere, posing a serious threat to both human and environmental health. Thus, it is crucial to remove antibiotic residues from the environment (Riaz et al., 2020). Antibiotics and antimicrobial genes can be treated using a variety of physical, chemical, and biological techniques.

11.6.1 Physical Methods

In physical processes, the targeted contaminant is removed from the relevant environment without the use of any chemicals. The chemical characteristics of the environment and pollutants are unaffected by this procedure (Riaz et al., 2020). The physical processes, which include membrane filtration, adsorption, photolysis, and UV radiation, are briefly discussed in the coming sections (Table 11.2).

11.6.1.1 Membrane Filtration

Chemical substances are separated using a membrane filtration technique. This technique is regarded as the fundamental technology for removing contaminants from polluted sources. Reverse osmosis (RO), microfiltration (MF), ultrafiltration (UF), and nanofiltration (NF) are the four types of pressure-driven membrane filtration for separating pollutants from the liquid phase (Phoon et al., 2020). Low-pressure membrane filtration methods include MF (0.1–1 m pore size) and UF (0.01 to 0.1 m pore size), whereas high-pressure membrane filtration methods include NF (0.3–2 kDa molecular weight cut-off) and RO (b100 Da molecular weight cut-off). Studies reveal that the membranes have a strong potential for acting as an ARGs barrier. The molecular weight cut-off or pore size of the MF, UF, and NF membranes has a significant impact on ARGs reduction. For example, in the case of the *bla*TEM gene reduction, at pore sizes of 0.45 m and 0.1m up to 1 log unit; at 100 kDa, 1.7 log units; at 10 kDa, 4.7 log units; and at 1 kDa, > 5.7 log units of gene copies/100 ml removal were reported (Breazeal et al., 2013, Hiller et al., 2019). There have been some reports of antibiotics being eliminated using membrane filtration. For removing antibiotics from liquid waste, NF and RO are thought to be the most effective membrane filtration techniques (Riaz et al., 2020). According to Koynuncu et al., tetracyclines have a high adsorptive affinity. In this investigation, it was observed that over 80% of the chlortetracycline was adsorbed onto the NF

TABLE 11.2
Various Physical Methods for Antibiotics Removal

	Physical Methods			
Treatment Method	Targeted Antibiotic	Initial Concentration	Removal Efficiency (%)	References
Membrane filtration				
Nanofiltration	Sulfamethoxazole,	20 µg/L	40%	(Li et al., 2021a)
	Azithromycin		99%	
	Ciprofloxacin	24 ng/L	>99%	(Cristóvão et al., 2020)
	Levofloxacin	10 mg/L		
Reverse Osmosis	Ciprofloxacin	50, 200 and 500 µg/L	>90%	(Alonso et al., 2018)
Ultrafiltration	Ciprofloxacin	40 mg/L	70%–80%	(Palacio et al., 2018)
Adsorbents				
Chitosan Film	Cefotaxime	5 mg/L	60%	(Li et al., 2021c)
Lanthanum Modified Zeolite (La-Z)	Chlortetracycline	5 mg/L	98.4%	(Awasthi et al., 2020)
3D graphene/MnO$_2$	Tetracycline	5 mm	94.0%	(Song et al., 2019)
Core–shell-structure activated carbon	Sulfamethoxazole	5 mg/L	99.0%	(Pamphile et al., 2019)
		10 mg/L	97.9%	
Photolysis				
UV/H$_2$O$_2$,	Enrofloxacin	1 g/L	100%	(Qiu et al., 2019)
UV/Fe (II), UV/H$_2$O$_2$/Fe (II)	Pefloxacin	1 g/L	79.1%	
	Sulfaquinoxaline	4 mg/L	100%	

membrane. Additionally, they found 50% adsorption for doxycycline and 11% to 20% for the sulfonamide antibiotic group (Koyuncu et al., 2008).

11.6.1.2 Adsorption

Among the various methods for treating water, adsorption stands out as one of the most remarkable techniques for removing and treating pollutants. Chemicals are accumulated or adsorbed from the liquid phase into the solid phase during adsorption, which is a mass transfer process (Phoon et al., 2020).

Due to its strong adsorption power for some organic compounds and removal capacity, this method offers a simple, efficient, eco-friendly, and economical approach for treating wastewater. For the removal of antibiotics, a variety of nano-adsorbent materials, including biomass and nanomaterial biochar, have been used successfully (Riaz et al., 2020). By pyrolyzing different kinds of biomass, such as wood, leaves, manure, agricultural leftovers, garbage, etc. under controlled anaerobic circumstances, biochar is created, which is a material rich in carbon. For high-efficiency antibiotic adsorption, biochar can be utilized as an adsorbent (Lu et al., 2020). The removal of the antibiotic ciprofloxacin was conducted using magnetic biochar (M-BC) made from plant medicinal waste (*Astragalus membranaceus*) (Chaturvedi et al., 2021). Méndez-Díaz et al. (2010) conducted another experiment on adsorbent material for antibiotic removal that demonstrated the impact of batch adsorption of imidazole and sulfonamide on activated carbon. According to the results, 90% of both antibiotics were extracted from the aqueous phase.

11.6.1.3 Photolysis

Antibiotics are more water soluble than some other pharmaceuticals; therefore, oxidative or transformational antibiotic treatments, like photolysis, are more effective than sorption-based ones. Antibiotics can be photolyzed either directly under sunlight absorption or indirectly through interactions with highly reactive species that are produced under exposure to light (Riaz et al., 2020). The self-sensitization process is another method of photolysis in which organic materials absorb light energy and create reactive oxygen species, which then degrade the ground-state molecules of the organic compounds. Tetracycline and fluoroquinolone are two antibiotics that can be removed by photolysis, and studies have discovered that the main method for oxytetracycline removal from surface water is direct photolysis. The problem of photolysis is that it only functions with antibiotics that are photosensitive, like tetracyclines (Shi et al., 2020).

11.6.1.4 UV Irradiation

UV irradiation is a treatment technique that uses ultraviolet (UV) light to destroy pathogens' DNA nucleotides (Hiller et al., 2019). With an increase in UV light intensity, ARGs inactivation increases. This is due to the cell's RNA and DNA absorbing UV light through ultraviolet-transparent structures. Constant UV exposure does not harm the cell membrane, but it directly impacts plasmids carrying ARGs, which can result in the loss of either the donor or the recipient (Riaz et al., 2020). According to studies, the total abundances of sulfamethoxazole- and tetracycline-resistant fecal coliforms (FC) and tetracycline-resistant heterotrophic bacteria (HB) significantly decreased at typical UV disinfection Fluences (doses). Despite the efficient eradication of tetracycline-resistant FCs, some studies have revealed that ARB are more UV-tolerant than non-resistant bacteria. It's been suggested that some proteins that result in tetracycline resistance may also absorb UV light to protect bacteria. At the typical UV Fluences used for disinfection (200 mJ/m^2), only a small amount of ARGs were reported to be eliminated. Therefore, substantial Fluences are required to eliminate ARGs in comparison to the fast inactivation of ARB (Hiller et al., 2019).

11.6.2 Chemical Methods

The chemical treatment strategy is based on the chemical reaction between pollutants and the reactive oxides or chemical oxidizing agents generated during the reaction process. Subsequently, pollutants' chemical molecular structure is destroyed, and they are further transformed into non-toxic, small molecules, or they are completely mineralized and removed (Huang et al., 2022). Advanced oxidation techniques, including Fenton, photo-Fenton, ozonation, and chlorination, are the main methods of chemical procedures (Riaz et al., 2020) (Table 11.3).

11.6.2.1 Advanced Oxidation Processes (AOPs)

Advanced oxidation is a process that uses the hydroxyl radical to break down large, hazardous organic chemicals into smaller, substantially less dangerous inorganic molecules (OH). AOPs are a new and efficient technology for the treatment of antibiotics in wastewater since they can degrade all

TABLE 11.3

Various Chemical Methods for Antibiotics Removal

	Chemical Methods			
Treatment Method	**Targeted Antibiotic**	**Initial Concentration**	**Removal Efficiency (%)**	**References**
Photocatalysis				
UV/NiO	Amoxicilin	(25–200 mg/L)	96%	(Balarak and Mostafapour, 2019)
$BiVO_4$ plate with Fe and Ni oxyhydroxide cocatalysts	Sulfadimethoxine	20 mg/L	95.43%	(Ma et al., 2020)
Isotype heterojunction g-C_3N_4-MU	Tetracycline	10 mg/L	85.6%	(Solehudin et al., 2020)
Sonocatalytic Irradiation				
Zinc oxide nanostructures	Tetracycline	50 mg/L	87.6%	(Akhil et al., 2021)
$Y_3Al_5O_{12}$ –AuBiVO$_4$ composite	Sulfanilamide	10 mg/L	95.64%	
Ozonation				
Heteroatoms doped graphene: reductive graphene oxides N doped graphene oxides P doped graphene oxides	Sulfamethoxazole	50 mg/L	83.0% 95.0% 99.0%	(Yin et al., 2017)
Fe_3O_4–Multiwalled carbon nanotubes	Sulfamethazine	20 mg/L	>90%	(Bai et al., 2018)
Fenton				
Catalyst: FeCl$_3$-activated biochar	Sulfamethoxazole	100 mg/L	99.94%	(Zeng and Kan, 2022)
Discharge plasma system based on graphene-Fe$_3$O$_4$ nanocomposites	Ofloxacin	-	99.9%	(Guo et al., 2021)
$[H_2O_2]:[Fe^{2+}]=3.97$	Ciprofloxacin	56 mM	85.62%	(Basturk et al., 2021)
Electro-Fenton				
Iron electrodes H_2O_2/COD ratio: 1.09	Cephalexin, Ciprofloxacin, Clarithromycin	100 µg/L	99.12%, 98.65%, 99.38%	(Basturk et al., 2021)
modified graphite felt cathode	Oxytetracycline	22, 44, 66 µM	100%	(Lai et al., 2020)
Steel-graphite felt electrodes	Ciprofloxacin	-	95.1%	(Basturk et al., 2021)
Photo-Fenton				
Catalyst: 1 g/L H_2O_2	Clarithromycin	-	100%	(Basturk et al., 2021)
Immobilized FA/FS/TiO$_2$	Cephalexin	-	89%	

compounds that traditional oxidants such as chlorine, ozone, and oxygen are unable to oxidize. The effectiveness of AOP treatment for removing antibiotics from sewage effluent has been the subject of several studies (Godoy and Sánchez, 2020).

11.6.2.2 Fenton Oxidation and Photo-Fenton

Due to the presence of Fenton's reagent, a powerful oxidizing agent, the Fenton oxidation and photo-Fenton processes are capable of completely decomposing resistant contaminants as well as converting harmful chemicals into compounds of lower toxicity. Fenton oxidation is a particularly efficient method for destroying antibiotics and eliminating different ARB/ARGs. The Fenton oxidative reaction uses iron salt (Fe (II)) as the photocatalyst and H_2O_2 as the oxidizing agent to produce highly active oxidative species. Photo-Fenton is a slightly modified form of Fenton oxidation that produces more hydroxyl radicals. UV light is used along with iron salts (Fe (II) or Fe (III)) and H_2O_2 in this process (Riaz et al., 2020, Chaturvedi et al., 2021). According to Karaolia et al. (2014) numerous antibiotics, including clarithromycin and sulfamethoxazole, can be successfully hydrolyzed by using Fenton's oxidation method. They also demonstrated that this method is effective at disinfection and inactivating ARB and enterococci resistant to sulfamethoxazole and clarithromycin, respectively.

11.6.2.3 Ozonation

Ozone, the primary ingredient in the ozonation process, is a potent oxidizing agent. The vast majority of applications have involved using ozonation to treat wastewater. There are two mechanisms involved in pollutant breakdown during ozonation. The first is a low-rate reaction called highly selective ozone direct oxidation. The majority of molecular ozone's reactions take place at lower pH levels, where organic molecules are attacked by electrophiles and broken down into fatty acids. The second method, an indirect one, produces hydroxyl radicals at high pH levels (>8) (Phoon et al., 2020).

Strong oxidizing agents produced by ozonation have the potential to interact with antibiotics and, as a consequence, can increase their removal. It has been discovered that various antibiotics, including b-lactams, trimethoprim, sulfonamides, macrolides, quinolones, and tetracyclines, undergo direct oxidative distortion during wastewater ozonation. The major purpose of ozonating wastewater is to make antibiotics inactive by disrupting their functional sites. The amount of ozone used during the treatment process and the duration of their contact time both have an impact on antibiotic removal from wastewater. According to research, using ozone at a dosage of 2 mg/l may remove 80% of sulfonamides, trimethoprim, and macrolides, and increasing the ozone concentration to 7.1 mg/l can improve the rate of antibiotic elimination up to 95% (Riaz et al., 2020).

11.6.2.4 Chlorination

Chlorination is the most popular and cost-effective method of disinfection for wastewater treatment. There have been studies published on the removal of antibiotics and ARBs/ARGs from wastewater using the chlorination approach. Wastewater chlorination requires a significant amount of chlorine and time. The ability of different chlorines to eliminate antibiotics from wastewater differs. Hypochlorite, for instance, exhibited the highest standard oxidation potential, whereas chlorine dioxide and chlorine gas displayed lower oxidation potentials, and chloramine was associated with the least effect. Studies have shown that this process tends to deactivate and decrease the number of ARB, but the chlorination process is unable to effectively destroy ARGs (Riaz et al., 2020). According to studies, chlorination can be used to remove ciprofloxacin and norfloxacin as well as promote the degradation of the b-lactam antibiotic group (penicillin, amoxicillin, and cefadroxil). Also, researchers discovered that the rate of deterioration increased as the pH increased (Chaturvedi et al., 2021). Despite all of the benefits of chlorination, the chlorination procedure has drawbacks for removing antibiotic contamination, including the issue that chlorine is an extremely toxic gas with numerous possible dangers when used, and the byproducts that are created during the disinfection procedure could also be hazardous.

11.6.3 BIOLOGICAL METHODS

There are certain drawbacks to using physicochemical methods for antibiotic remediation, such as high operating costs, weak function in sewage containing high concentrations of antibiotics, the need for a significant amount of labor, the expense of the equipment, and the production of harmful byproducts. One example of a poisonous byproduct is the photolysis of the cephalosporine byproduct, which was found to be more dangerous when examined using the Microtox assay (Warsito, 2022). Consequently, biological methods that are reliable, efficient, and eco-friendly ways to remove contaminants like antibiotics have gained attention. These approaches are particularly helpful when the antibiotic molecule has unique properties, including temperature and pH stability, which make it difficult to break down using traditional chemical or physical approaches (Warsito, 2022, Dangi et al., 2019).

Plants, microbes, and biocatalysts can all be used for antibiotic biodegradation. Microbial cells have the ability to produce or use the enzyme in reaction to their environment. ARB also plays a significant role in the biodegradation of antibiotics by creating enzymes that can change or hydrolyze antibiotic structure (Chaturvedi et al., 2021). The mechanisms for microbial metabolism or microbial co-metabolism can be used by microorganisms to biodegrade substances. In co-metabolism, the microbial community can secrete matching enzymes that can break down antibiotics. Additionally, antibiotics can be used by microbes in their metabolic processes as carbon and energy sources for growth. As a result, microorganisms can effectively break down and transform antibiotics (Huang et al., 2022).

Aerobic or anaerobic treatment methods as well as constructed wetlands can be used for the bioremediation of antibiotics and ARGs. In the aerobic treatment approach, waste carrying antibiotic residues and AMR/ARGs is processed in an oxygen-rich condition. This method uses high-temperature aerobic digestion to rapidly eliminate harmful bacteria and bacteria carrying ARGs. But some ARGs may persist after aerobic digestion. The two main techniques for aerobic treatment are the activated sludge process and the aerobic granular sludge method. Anaerobic bacteria are used in closed systems with no oxygen to perform anaerobic treatment. The organic contaminants are transformed into biogas and bioproducts during this process. Anaerobic membrane bioreactors, anaerobic sequencing reactors, up-flow anaerobic sludge blankets, and other closed-system reactors are available for anaerobic treatment. In a constructed wetland approach, pharmaceuticals are degraded naturally by mechanisms that use plants, sunlight, and microorganisms. The type of bacterial community that exists in the artificial wetland determines its ability to remove antibiotics from pharmaceutical waste (Riaz et al., 2020).

Despite all the advantages of antibiotic bioremediation, there are several disadvantages as well. For example, horizontal and vertical gene transfer between microorganisms can lead to harmful pathogenicity and antibiotic resistance. Besides this, the degradation process can result in more lethal intermediates than the parent compound. There are methods for addressing these issues, such as using a closed treatment system and sterilization after degradation to prevent gene transfer. Utilizing a microbial enzyme with antibiotic-degrading capabilities is the alternative method. Optimizing the fermentation process and studying the substrates can help improve the metabolic process in the case of antibiotics with complicated structures. The identification of the metabolic route and, if necessary, its manipulation can enhance the effectiveness of degradative processes and even prevent the production of toxic substances (Warsito, 2022). Bacteria, fungi, and algae are the main types of microorganisms used in antibiotic bioremediation; some studies on this process are presented in the sections that follow.

11.6.3.1 Bacterial Degradation of Antibiotics

The most common types of microorganisms used in the bioremediation of contaminants, including antibiotics, are bacteria. The microbial breakdown of antibiotics is largely influenced by ARB. These bacteria have the ability to create matching degrading enzymes, which alter or hydrolyze antibiotic molecules to render them inactive. There are four main categories of antibiotic-degrading enzymes: β-Lactamase, chloramphenicol-inactivating enzymes, macrolide passivase, and aminoglycoside-modifying enzymes. Cephalosporin and penicillin chemical bonds can be degraded by the enzyme b-lactamase. Additionally, the structures of the relevant antibiotics can be broken by ring-opening oxidases linked to fosfomycin resistance as well as esterases linked to macrolide resistance (Yang et al., 2019). Numerous studies describe the use of consortia and bacteria in the bioremediation of various antibiotic classes. One example of finding a bacterial consortium capable of degrading antibiotics is the work of (Feng et al., 2019), who used the enrichment approach to find the consortium known as XG, which had up to 63% biodegradation rate for the antibiotic ciprofloxacin. They demonstrated that these bacteria can use ciprofloxacin as a sole carbon source, but in the presence of additional carbon sources, the rate of biodegradation will increase. They showed that this consortium primarily consisted of *Achromobacter*, *Ochrobactrum*, *Bacillus*, *Lactococcus*, and *Enterococcus*, plus at least an additional five minor genera. Another recent study on the bacterial biodegradation of antibiotics was conducted (Zhang et al., 2021). The bacterium *Arthrobacter nicotianae* strain OTC-16, which has the potential to biodegrade oxytetracycline (OTC) and tetracycline (Shi et al., 2021), was used in this study to examine the pattern of antibiotic resistance. They revealed that eight of the 3,561 genes identified as the outcome of the whole genome investigation were connected to antibiotic resistance. And using the qRT-PCR technique, they discovered that several genes—including *ant2ia*, *sul1*, *tet33*, and *cml_e8* in the plasmid and *tetV* in the chromosome—were expressed during OTC breakdown. Table 11.4 displays further recent studies.

11.6.3.2 Fungal Degradation of Antibiotics

Fungal strains have been effectively used to mineralize and break down resistant antibiotics because of their strong enzymatic activity and extensive metabolic abilities (Chaturvedi et al., 2021). In a study by Guo et al. (2014), it was discovered that the white rot fungus *Phanerochaete chrysosporium* biodegrades the antibiotic sulfamethoxazole (SMX). The degradation percentage of the antibiotic was 53% after 24 hours and 74% after ten days in the presence of 10 mg/L antibiotics. It was revealed that the degradation of sulfamethoxazole occurred due to the activity of crude laccase produced by *P. chrysosporium*. Moreover, Becker et al. (2016) used a fungal immobilized laccase (produced by *Trametes versicolor*) with the mediator syringaldehyde (SYR) for the degradation of antibiotics in wastewater. The obtained results showed the removal of 32 of 38 antibiotics in an enzymatic membrane reactor (EMR). Similarly, Singh et al. (2017) reported the ability of a white rot fungus (*Pleurotus ostreatus*) to biodegrade ciprofloxacin at a rate of 95.07%. Fascinatingly, three tetracyclines (TCs), including tetracycline, oxytetracycline, and chlortetracycline, were demonstrated to be aerobically degraded in river sediment. It was revealed that the addition of enzyme extract from spent *Pleurotus eryngii* mushroom compost enhanced the breakdown of TCs in sediment (Chang and Ren, 2015). Table 11.4 lists further recent studies.

11.6.3.3 Algal Degradation of Antibiotics

Algae have been recognized as a powerful substitute for the breakdown of antibiotics because of their higher biomass and susceptibility to pollutants. Compared to other algal species, green algae have demonstrated greater antibiotic resistance and removal effectiveness (Chaturvedi et al., 2021). In a study, researchers used *Chlorella pyrenoidosa* for the breakdown of sulfamethazine. They employed sulfamethazine in doses of 2, 4, and 8 mg/L, and the percentages of biotic breakdown on day 13 were 48.45%, 60.21%, and 69.93%, respectively. The results demonstrated the algae's

TABLE 11.4
Recent Studies on Antibiotic Biodegradation

Microorganism	Antibiotic	Initial Concentration	Degradation	Reference
Bacteria				
Klebsiella sp. SQY5	Tetracycline	61.27 mg/L	A maximum tetracycline reduction rate of 0.113 mg L^{-1}·h^{-1}	(Shao et al., 2018)
Anaerobic sulfate-reducing bacteria (SRB)	Ciprofloxacin	5000 μg/L	Nearly 28.0%	(Jia et al., 2018)
Serratia sp. R1	Penicillin G	10 mg/L	Up to 84% degradation	(Kumar et al., 2019b)
Proteobacteria	Enrofloxacin (ENR) Oxytetracycline (OXY)	1 mg/L	98% 95%	(Harrabi et al., 2019)
Bacillus strain LM-1 and LM-2	Penicillin V Potassium	100 μg/mL	(68% for LM-1, 66% for LM-2 in 48 h)	(Yang et al., 2019)
bacterial consortium XG (dominant genera: *Achromobacter, Bacillus, Lactococcus, Ochrobactrum,* and *Enterococcus*)	Ciprofloxacin	5 mg/L	Up to 63%	(Feng et al., 2019)
Bacillus clausii T and *Bacillus amyloliquefaciens* HM618 (Co culture)	Tetracycline Oxytetracycline Chlortetracycline	10 mg/L	51.9%, 76.6% 88.9%	(Liu et al., 2020)
Alcaligenes sp. MMA	Amoxicillin	5 mg/L	Up to 84% of amoxicillin in 14 days	(Sodhi et al., 2020)
Serratia marcescens strain WW1	Tetracycline	20 mg/L	89.5% of the Tetracycline within 48 h.	(Bhatt et al., 2022)
Aeromonas hydrophila HS01	Thiamphenicol	-	More than 90.0%	(Yang et al., 2022)
Paenarthrobacter sp. P27 *and Norcardiodes* sp. N27	Sulfamethoxazole	5.88 mg/L	100%	(Qi et al., 2022)
Stenotrophomonas maltophilia	Amoxicillin	43 mg/L	88.7%	(Dianatdar et al. 2024)
Fungi				
Trametes hirsute (laccase)	Chloramphenicol	10 mg/L	100% within 48 h in presence of mediators	(Navada and Kulal, 2019)
Phanerochaete chrysosporium	Sulfadiazine	10 mg/L	100% within 6 days	(Zhang et al., 2019)
Pycnoporus sp. SYBC-L10 (laccase)	Tetracycline Oxytetracycline	50 mg/L	100% within 5 min	(Tian et al., 2020)
Mycelium of *Penicillium commune, Epicoccum nigrum Trichoderma harzianum, Aspergillus terreus Beauveria bassiana*	Oxytetracycline	250 mg/L	68.15% 75.8% 77.3% 74.1% 78.3%	(Ahumada-Rudolph et al., 2021)
Trichoderma harzianum BGP115	Sulfamethoxazole	-	71% after 7 days	(Piyaviriyakul et al., 2021)

TABLE 11.4 (Continued)
Recent Studies on Antibiotic Biodegradation

Microorganism	Antibiotic	Initial Concentration	Degradation	Reference
Algae				
Microalgae-bacteria consortium (Dominant microalgae: *Chlorella sorokiniana*)	Sulfamethoxazole	50 μg/L	54.34%	(da Silva Rodrigues et al., 2020)
Chaetoceros muelleri	Sulfamethoxazole (SMX) Ofloxacin (OFX)	0.5 mg/L	39.8% 42.5%	(Mojiri et al., 2021)
Microalgae-bacteria consortium (*Cenedesmus obliquus* FACHB-12)	Sulfonamides Sulfadiazine (SDZ) Sulfamethoxazole (SMX)	0–160 mg/L	5.85% 40.84%	(Wang et al., 2022)

remarkable ability to remove sulfamethazine (Sun et al., 2017). *Scenedesmus obliquus* is another example of a microalga being used for antibiotic biodegradation. Researchers looked at whether or not adding sodium chloride affected the ability of *S. obliquus* to break down the fluoroquinolone antibiotic levofloxacin (1 mg/L) in synthetic wastewater (NaCl). Results indicated that antibiotic removal considerably increased from 4.5% to 93.4% with an increase in salinity from 0 to 171 mM NaCl (Xiong et al., 2017). Table 11.4 displays some current research studies.

11.7 ANALYTICAL TECHNIQUES FOR DETECTION OF ARGS

11.7.1 PCR Methods

The polymerase chain reaction is the most widely used molecular diagnostic technique. This technique allows the detection of pathogens, their resistance, and virulence genes. DNA extraction, amplification, and detection of amplified DNA materials are the three steps in the conventional PCR method. This process takes around 12 hours to complete. Therefore, real-time PCR was developed to reduce this time by simultaneously amplifying and detecting amplified molecules. Quantitative PCR (qPCR) is an alternative PCR technique in which ARGs are independently detected. This method requires the DNA extraction of several microorganisms from various environmental samples. The qPCR allows for the simultaneous quantification of hundreds of ARGs in a single run. However, the related drawback of this method is that it requires prior knowledge of the unknown gene or genes that share similarity with known genes. High-throughput qPCR can be used to overcome this restriction. Droplet digital PCR (ddPCR) is another technique for measuring the amount of nucleic acids in samples. With this method, the PCR reaction is split into hundreds of tiny droplets before being amplified, making it exceedingly sensitive and noticeable. This method permits the measurement of nucleic acids in samples without standard curves (Iwu et al., 2020).

11.7.2 Metagenomic Studies

A metagenomics technique is one that directly examines microbial populations in their natural habitats without first culturing them. This procedure includes extracting whole DNA from the environment, sequencing the DNA with high-throughput sequencing techniques, and applying bioinformatic tools to analyze the results. Analyses based on sequences or functions are used to screen the metagenomics libraries (Harirchi et al., 2022). When compared to culture-based methods, the sequence-based method may be used quickly and is useful for simultaneously checking large numbers of data. However, sequence-based methods have the disadvantage of relying on known sequences and prior knowledge for gene discovery, whereas functional metagenomic can be used to mine functionalities of interest in an objective, culture-independent manner, making it useful for identifying completely novel mechanisms (Sukhum et al., 2019). Metagenomics has been utilized in a variety of environments for long-term and quick monitoring of reservoirs in the case of the emergence and abundance of antibiotic resistance organisms (AROs), ARGs, and the entire resistome (Iwu et al., 2020, Sukhum et al., 2019). A metagenomic investigation of the Chaobai River, a city river in Beijing, China, revealed higher ARGs abundance in this river compared to other areas in the region. In this work, 442 ARGs subtypes from 22 ARGs types, including the two most common types, bacitracin and sulfonamide ARGs, were discovered in the river sediment. The predominant ARGs discovered in this study were linked to human use and were under selective pressure as a result of antibiotic use. The findings of this study demonstrated that metagenomic analysis of ARGs in water sources can be useful for assessing drinking water quality and limiting the usage and release of antibiotics into the environment (Chen et al., 2019). WWTPs have been identified as key hotspots for the spread of ARB and ARGs into other environments (Han and Yoo, 2020). An investigation on the bacterial resistome in untreated sewage from 79 sites in 60 countries was conducted in 2019 for the purpose of monitoring antibiotic resistance on a global scale. A massive number of genes were evaluated simultaneously using metagenomics. The results of this study showed that there are regional differences in the variety and abundance of AMR genes. The results of this study revealed a total of 1546 genera (942–1367 genera per sample), several of which were fecal, including the genera *Faecalibacterium*, *Bacteroides*, *Escherichia*, *Streptococcus*, and *Bifidobacterium*. They also found a few more extremely common taxa with environmental origins, including *Acidovorax* and *Acinetobacter*. These findings demonstrated the complexity of sewage's bacterial composition. The highest level of these genes was found in African countries, although Brazil had the highest abundance of all the countries evaluated in this study's analysis of gene abundance for AMR. A total of 1625 distinct AMR genes from 408 gene families were discovered in this investigation. The AMR genes that code for resistance to aminoglycosides, beta-lactams, macrolides, tetracyclines, and sulfonamides were the most prevalent. Resistance to sulfonamides and phenicols was most prevalent in Asian and African samples, while macrolide resistance genes were more prevalent in European and North American samples. According to the study's findings, improving sanitation, health, and possibly education are useful ways for reducing the global burden of AMR (Hendriksen et al., 2019). Numerous other studies using metagenomics to assess antibiotic resistance at WWTPs have been published. In a metagenomic study, Han and Yoo (2020) analyzed the occurrences and prevalence of ARGs, MGEs, and the bacterial species in bioaerosols (BA), activated sludge (AS), and dewatered sludge (DS). This study identified 153 ARGs subtypes from 19 distinct ARGs types. The results showed that sulfonamide resistance genes were prominent in effluent, dewatered sludge, and surface water. The overall number of these genes in DS was higher than AS. Despite the finding that ARGs frequencies and quantities were much lower in the BA samples than in the AS and DS samples, they noted that the *sul*I was relatively abundant in these samples and that the predominant ARGs in BA were the *erm*, *sul*, and *tet* genes. Additionally, the BA sample had two to five times fewer plasmid and integron-linked genes than the AS and DS samples, but distinct types of MGEs were discovered there. While Firmicutes (25%) and Actinobacteria (20%) dominated in BA samples, Proteobacteria (58%) and Bacteroidetes (18%) dominated in AS and DS samples, similar to many other full-scale WWTPs. Another study by Majeed et al. (2021) examined antibiotic resistance in a standard WWTP at various stages of treatment. They showed a 50% decrease in total ARGs from influent to effluent. Additionally, they found that 90% of the ARGs found in the effluent were also detected in the influent. Virome research from various contexts to explore resistome has been published in multiple studies of viral metagenomics (virome), which also developed quickly over the past decade. Evaluation of an urban surface water viral metagenome by Moon et al. (2020) led to the discovery of a variety of ARGs, including polymyxin resistance genes, multidrug efflux proteins, and beta-lactamases. Additionally, two novel beta-lactamases with distinct sequences, bla_{HRV-1} and bla_{HRVM-1}, were identified in this study by functional assessment of ARGs isolated from viral contigs. These studies were just a small sample of the numerous metagenomic analyses of resistomes conducted in various environments, which suggested that metagenomics might be a potent method for assessing the ARBs and ARGs in various habitats. Metagenomics may be helpful in estimating the spread of ARGs and reducing the emergence of antimicrobial resistance, which has an impact on both human and animal health.

11.8 CONCLUSION

Antibiotics significantly improve human life and have a positive impact on health. Despite all of the benefits, overusing these medications leaves behind residues in the environment that seriously endanger the lives of people and other living organisms. Antibiotics, ARB, and ARGs can all be eliminated using a variety of physical, chemical, and biological approaches, which can also be used in combination. Physical and chemical processes face challenges, such as the generation of hazardous intermediates that must be further degraded. Microorganism-based bioremediation is becoming more attractive because it is an effective and environmentally friendly technique. Metagenomics is a potent tool for identifying ARB and the genes involved in their resistance, as well as for studying the microbial population in various environments. Additionally, by analyzing their genomes and degradative enzymes, it can aid in the detection of potential bacteria with degradation abilities.

NOTES

1   Organisation for Economic Co-operation and Development
2   World Health Organization

REFERENCES

Abbasian, H., Hajimolaali, M., Yektadoost, A. & Zartab, S. 2019. Antibiotic utilization in Iran 2000–2016: Pattern analysis and benchmarking with organization for economic co-operation and development countries. *Journal of Research in Pharmacy Practice,* 8, 162.

Ahumada-Rudolph, R., Novoa, V., Becerra, J., Cespedes, C. & CABRERA-PARDO, J. R. 2021. Mycoremediation of oxytetracycline by marine fungi mycelium isolated from salmon farming areas in the south of Chile. *Food and Chemical Toxicology,* 152, 112198.

Akhil, D., Lakshmi, D., Senthil Kumar, P., Vo, D.-V. N. & Kartik, A. 2021. Occurrence and removal of antibiotics from industrial wastewater. *Environmental Chemistry Letters,* 19, 1477–1507.

Alonso, J. J. S., El Kori, N., Melián-Martel, N. & Del Río-Gamero, B. 2018. Removal of ciprofloxacin from seawater by reverse osmosis. *Journal of Environmental Management,* 217, 337–345.

Anand, U., Reddy, B., Singh, V. K., Singh, A. K., Kesari, K. K., Tripathi, P., Kumar, P., Tripathi, V. & Simal-Gandara, J. 2021. Potential environmental and human health risks caused by antibiotic-resistant bacteria (ARB), antibiotic resistance genes (ARGs) and emerging contaminants (ECs) from municipal solid waste (MSW) landfill. *Antibiotics,* 10, 374.

Awasthi, M. K., Sarsaiya, S., Patel, A., Juneja, A., Singh, R. P., Yan, B., Awasthi, S. K., Jain, A., Liu, T. & Duan, Y. 2020. Refining biomass residues for sustainable energy and bio-products: An assessment of technology, its importance, and strategic applications in circular bio-economy. *Renewable and Sustainable Energy Reviews,* 127, 109876.

Bahramian, B., Sani, M. A., Parsa-Kondelaji, M., Hosseini, H., Khaledian, Y. & Rezaie, M. 2022. Antibiotic residues in raw and pasteurized milk in Iran: A systematic review and meta-analysis. *AIMS Agriculture and Food,* 7, 500–519.

Bai, Z., Yang, Q. & Wang, J. 2018. Catalytic ozonation of sulfamethazine antibiotics using Fe3O4/multiwalled carbon nanotubes. *Environmental Progress & Sustainable Energy,* 37, 678–685.

Balarak, D. & Mostafapour, F. K. 2019. Photocatalytic degradation of amoxicillin using UV/Synthesized NiO from pharmaceutical wastewater. *Indonesian Journal of Chemistry,* 19, 211–218.

Baralla, E., Demontis, M. P., Dessì, F. & Varoni, M. V 2021. An Overview of antibiotics as emerging contaminants: Occurrence in bivalves as biomonitoring organisms. *Animals,* 11, 3239.

Basturk, I., Varank, G., Murat-Hocaoglu, S., Yazici-Guvenc, S., Can-Güven, E., Oktem-Olgun, E. E. & Canli, O. 2021. Simultaneous degradation of cephalexin, ciprofloxacin, and clarithromycin from medical laboratory wastewater by electro-Fenton process. *Journal of Environmental Chemical Engineering,* 9, 104666.

Becker, D., Della Giustina, S. V., Rodriguez-Mozaz, S., Schoevaart, R., Barceló, D., De Cazes, M., Belleville, M.-P., Sanchez-Marcano, J., De Gunzburg, J. & Couillerot, O. 2016. Removal of antibiotics in wastewater by enzymatic treatment with fungal laccase–degradation of compounds does not always eliminate toxicity. *Bioresource Technology,* 219, 500–509.

Bernier, S. P. & Surette, M. G. 2013. Concentration-dependent activity of antibiotics in natural environments. *Frontiers in Microbiology,* 4, 20.

Bhat, S. A., Cui, G., Yaseera, N., Lei, X., Ameen, F. & Li, F. 2022. Removal potential of microplastics in organic solid wastes via biological treatment approaches. *Microbial Biotechnology: Role in Ecological Sustainability and Research,* 255–263. Springer.

Bhatt, P., Jeon, C.-H. & Kim, W. 2022. Tetracycline bioremediation using the novel Serratia marcescens strain WW1 isolated from a wastewater treatment plant. *Chemosphere,* 298, 134344.

Björklund, H., Råbergh, C. & Bylund, G. 1991. Residues of oxolinic acid and oxytetracycline in fish and sediments from fish farms. *Aquaculture,* 97, 85–96.

Breazeal, M. V. R., Novak, J. T., Vikesland, P. J. & Pruden, A. 2013. Effect of wastewater colloids on membrane removal of antibiotic resistance genes. *Water Research,* 47, 130–140.

Bruchet, A. & Janex-Habibi, M. 2007. Fate of anthropogenic micropollutants during wastewater treatment and influence on receiving surface water. *WIT Transactions on Ecology and the Environment,* 103, 387–397.

Chang, B.-V. & Ren, Y.-L. 2015. Biodegradation of three tetracyclines in river sediment. *Ecological Engineering,* 75, 272–277.

Chaturvedi, P., Giri, B. S., Shukla, P. & Gupta, P. 2021. Recent advancement in remediation of synthetic organic antibiotics from environmental matrices: Challenges and perspective. *Bioresource Technology,* 319, 124161.

Chen, H., Bai, X., Jing, L., Chen, R. & Teng, Y. 2019. Characterization of antibiotic resistance genes in the sediments of an urban river revealed by comparative metagenomics analysis. *Science of the Total Environment,* 653, 1513–1521.

Cristóvão, M. B., Tela, S., Silva, A. F., Oliveira, M., Bento-Silva, A., Bronze, M. R., Crespo, M. T. B., Crespo, J. G., Nunes, M. & Pereira, V. J. 2020. Occurrence of antibiotics, antibiotic resistance genes and viral genomes in wastewater effluents and their treatment by a pilot scale nanofiltration unit. *Membranes,* 11, 9.

Cui, G., Bhat, S.A., Li, W., Wei, Y., Kui, H., Fu, X., Gui, H., Wei, C. and Li, F., 2019. Gut digestion of earthworms significantly attenuates cell-free and -associated antibiotic resistance genes in excess activated sludge by affecting bacterial profiles. *Science of the Total Environment,* 691, pp.644–653.

Cui, G., Fu, X., Bhat, S.A., Tian, W., Lei, X., Wei, Y. and Li, F., 2022. Temperature impacts fate of antibiotic resistance genes during vermicomposting of domestic excess activated sludge. *Environmental Research,* 207, p.112654.

Cui, G., Li, F., Li, S., Bhat, S.A., Ishiguro, Y., Wei, Y., Yamada, T., Fu, X. and Huang, K., 2018. Changes of quinolone resistance genes and their relations with microbial profiles during vermicomposting of municipal excess sludge. *Science of the Total Environment,* 644, pp.494–502.

Cui, G., Lü, F., Zhang, H., Shao, L. & He, P. 2020. Critical insight into the fate of antibiotic resistance genes during biological treatment of typical biowastes. *Bioresource Technology,* 317, 123974.

Cycoń, M., Mrozik, A. & Piotrowska-Seget, Z. 2019. Antibiotics in the soil environment—degradation and their impact on microbial activity and diversity. *Frontiers in Microbiology,* 10, 338.

Da Silva Rodrigues, D. A., Da Cunha, C. C. R. F., Freitas, M. G., De Barros, A. L. C., E Castro, P. B. N., Pereira, A. R., De Queiroz Silva, S., Da Fonseca Santiago, A. & Afonso, R. J. D. C. F. 2020. Biodegradation of sulfamethoxazole by microalgae-bacteria consortium in wastewater treatment plant effluents. *Science of the Total Environment,* 749, 141441.

Dangi, A. K., Sharma, B., Hill, R. T. & Shukla, P. 2019. Bioremediation through microbes: Systems biology and metabolic engineering approach. *Critical Reviews in Biotechnology,* 39, 79–98.

Dianatdar, F., Etemadifar, Z. & Momenbeik, F., 2024. Removal of amoxicillin and co-amoxiclav by newly isolated Stenotrophomonas maltophilia DF1. *International Journal of Environmental Science and Technology,* 1–14.

Feng, N.-X., Yu, J., Xiang, L., Yu, L.-Y., Zhao, H.-M., Mo, C.-H., Li, Y.-W., Cai, Q.-Y., Wong, M.-H. & Li, Q. X. 2019. Co-metabolic degradation of the antibiotic ciprofloxacin by the enriched bacterial consortium XG and its bacterial community composition. *Science of the Total Environment,* 665, 41–51.

García, J., García-Galán, M. J., Day, J. W., Boopathy, R., White, J. R., Wallace, S. & Hunter, R. G. 2020. A review of emerging organic contaminants (EOCs), antibiotic resistant bacteria (ARB), and antibiotic resistance genes (ARGs) in the environment: Increasing removal with wetlands and reducing environmental impacts. *Bioresource Technology,* 307, 123228.

Godoy, M. & Sánchez, J. 2020. *Antibiotics as emerging pollutants in water and its treatment. Antibiotic Materials in Healthcare.* Elsevier.

Goel, S. 2015. *Antibiotics in the Environment: A Review. Emerging Micro-Pollutants in the Environment: Occurrence, Fate, and Distribution.* American Chemical Society.

Grave, K., Hansen, M. K., Kruse, H., Bangen, M. & Kristoffersen, A. B. 2008. Prescription of antimicrobial drugs in Norwegian aquaculture with an emphasis on "new" fish species. *Preventive Veterinary Medicine,* 83, 156–169.

Gray, A. D., Todd, D. & Hershey, A. E. 2020a. The seasonal distribution and concentration of antibiotics in rural streams and drinking wells in the piedmont of North Carolina. *Science of the Total Environment,* 710, 136286.

Gray, A. D., Todd, D. & Hershey, A. E. 2020b. The seasonal distribution and concentration of antibiotics in rural streams and drinking wells in the piedmont of North Carolina. *Science of the Total Environment,* 710, 136286.

Grenni, P., Ancona, V. & Caracciolo, A. B. 2018 Ecological effects of antibiotics on natural ecosystems: A review. *Microchemical Journal,* 136, 25–39.

Gul, S., Hussain, S., Khan, H., Khan, K. I., Khan, S., Ullah, S. & Clasen, B. 2022. Advances in bioremediation of antibiotic pollution in the environment. *Biological Approaches to Controlling Pollutants.* Elsevier.

Guo, H., Li, Z., Xie, Z., Song, J., Xiang, L., Zhou, L., Li, C., Li, J. & Wang, H. 2021. Accelerated Fenton reaction for antibiotic ofloxacin degradation in discharge plasma system based on graphene-Fe3O4 nanocomposites. *Vacuum,* 185, 110022

Guo, X.-L., Zhu, Z.-W. & Li, H.-L. 2014. Biodegradation of sulfamethoxazole by Phanerochaete chrysosporium. *Journal of Molecular Liquids,* 198, 169–172.

Haffiez, N., Chung, T. H., Zakaria, B. S., Shahidi, M., Mezbahuddin, S., Hai, F. I. & Dhar, B. R. 2022. A critical review of process parameters influencing the fate of antibiotic resistance genes in the anaerobic digestion of organic waste. *Bioresource Technology,* 127189.

Han, I. & Yoo, K. 2020. Metagenomic profiles of antibiotic resistance genes in activated sludge, dewatered sludge and bioaerosols. *Water,* 12, 1516.

Harirchi, S., Rafieyan, S., Nojoumi, S. A. & Etemadifar, Z. 2022. Microbial Biodegradation and Metagenomics in Remediation of Environmental Pollutants: Enzymes and Mechanisms. *Omics Insights in Environmental Bioremediation.* Springer.

Harrabi, M., Alexandrino, D. A., Aloulou, F., Elleuch, B., Liu, B., Jia, Z., Almeida, C. M. R., Mucha, A. P. & Carvalho, M. F. 2019. Biodegradation of oxytetracycline and enrofloxacin by autochthonous microbial communities from estuarine sediments. *Science of the Total Environment,* 648, 962–972.

He, Y., Yuan, Q., Mathieu, J., Stadler, L., Senehi, N., Sun, R. & Alvarez, P. J. 2020. Antibiotic resistance genes from livestock waste: Occurrence, dissemination, and treatment. *NPJ Clean Water,* 3, 1–11.

Hendriksen, R. S., Munk, P., Njage, P., Van Bunnik. B., Mcnally, L., Lukjancenko, O., Röder, T., Nieuwenhuijse, D., Pedersen, S. K. & Kjeldgaard, J. 2019. Global monitoring of antimicrobial resistance based on metagenomics analyses of urban sewage. *Nature Communications,* 10, 1–12.

Hiller, C., Hübner, U., Fajnorova, S., Schwartz, T. & Drewes, J. 2019. Antibiotic microbial resistance (AMR) removal efficiencies by conventional and advanced wastewater treatment processes: A review. *Science of the Total Environment,* 685, 596–608.

Huang, C., Tang, Z., Xi, B., Tan, W., Guo, W., Wu, W. & Ma, C. 2021. Environmental effects and risk control of antibiotic resistance genes in the organic solid waste aerobic composting system: A review. *Frontiers of Environmental Science & Engineering,* 15, 1–12.

Huang, S., Yu, J., Li, C., Zhu, Q., Zhang, Y., Lichtfouse, E. & Marmier, N. 2022. The effect review of various biological, physical and chemical methods on the removal of antibiotics. *Water,* 14, 3138.

Iwu, C. D., Korsten, L. & Okoh, A. I. 2020. The incidence of antibiotic resistance within and beyond the agricultural ecosystem: A concern for public health. *Microbiologyopen,* 9, e1035.

Javid, A., Mesdaghinia, A., Nasseri, S., Mahvi, A. H., Alimohammadi, M. & Gharibi, H. 2016. Assessment of tetracycline contamination in surface and groundwater resources proximal to animal farming houses in Tehran, Iran. *Journal of Environmental Health Science and Engineering,* 14, 4.

Jia, Y., Khanal, S. K., Shu, H., Zhang, H., Chen, G.-H. & Lu, H. 2018. Ciprofloxacin degradation in anaerobic sulfate-reducing bacteria (SRB) sludge system: Mechanism and pathways. *Water Research,* 136, 64–74.

Karaolia, P., Michael, I., García-Fernández, I., Agüera, A., Malato, S., Fernández-Ibáñez, P. & Fatta-Kassinos, D. 2014. Reduction of clarithromycin and sulfamethoxazole-resistant Enterococcus by pilot-scale solar-driven Fenton oxidation. *Science of the Total Environment,* 468–469, 19–27.

Koch, N., Islam, N. F., Sonowal, S., Prasad, R. & Sarma, H. 2021. Environmental antibiotics and resistance genes as emerging contaminants: Methods of detection and bioremediation. *Current Research in Microbial Sciences,* 2, 100027.

Koyuncu, I., Arikan, O. A., Wiesner, M. R. & Rice, C. 2008. Removal of hormones and antibiotics by nanofiltration membranes. *Journal of Membrane Science,* 309, 94–101.

Kumar, M., Jaiswal, S., Sodhi, K. K., Shree, P., Singh, D. K., Agrawal, P. K. & Shukla, P. 2019a. Antibiotics bioremediation: Perspectives on its ecotoxicity and resistance. *Environment International,* 124, 448–461.

Kumar, M., Sodhi, K. K. & Singh, D. K. 2019b. Bioremediation of Penicillin G by Serratia sp. R1, and enzymatic study through molecular docking. *Environmental Nanotechnology, Monitoring & Management,* 12, 100246.

Kumar, V., Agrawal, S., Bhat, S.A., Américo-Pinheiro, J.H.P., Shahi, S.K., Kumar, S., 2022. Environmental impact, health hazards, and plant-microbes synergism in remediation of emerging contaminants. *Cleaner Chemical Engineering* 2, 100030. https://doi.org/10.1016/j.clce.2022.100030

Kümmerer, K. 2003. Significance of antibiotics in the environment. *Journal of Antimicrobial Chemotherapy,* 52, 5–7.

Kümmerer, K. 2009. Antibiotics in the aquatic environment–a review–part I. *Chemosphere,* 75, 417–434.

Lai, W., Xie, G., Dai, R., Kuang, C., Xu, Y., Pan, Z., Zheng, L., Yu, L., Ye, S. & Chen, Z. 2020. Kinetics and mechanisms of oxytetracycline degradation in an electro-Fenton system with a modified graphite felt cathode. *Journal of Environmental Management,* 257, 109968.

Larsson, D. & Flach, C.-F. 2022. Antibiotic resistance in the environment. *Nature Reviews Microbiology,* 20, 257–269.

Li, D., Zhou, H., Huang, L., Zhang, J., Cui, J. & Li, X. 2021a. Role of adsorption during nanofiltration of sulfamethoxazole and azithromycin solution. *Separation Science and Technology,* 56, 1996–2010.

Li, W., Su, H., Li, J., Bhat, S. A., Cui, G., Han, Z. M., Nadya, D. S., Wei, Y. & Li, F. 2021b. Distribution of extracellular and intracellular antibiotic resistance genes in sludge fractionated in terms of settleability. *Science of the Total Environment,* 760, 143317.

Li, Z., Wang, X., Zhang, X., Yang, Y. & Duan, J. 2021c. A high-efficiency and plane-enhanced chitosan film for cefotaxime adsorption compared with chitosan particles in water. *Chemical Engineering Journal,* 413, 127494.

Liu, C., Tan, L., Zhang, L., Tian, W. & Ma, L. 2021. A review of the distribution of antibiotics in water in different regions of China and current antibiotic degradation pathways. *Frontiers in Environmental Science,* 9, 692298.

Liu, C.-X., Xu, Q.-M., Yu, S.-C., Cheng, J.-S. & Yuan, Y.-J. 2020. Bio-removal of tetracycline antibiotics under the consortium with probiotics Bacillus clausii T and Bacillus amyloliquefaciens producing biosurfactants. *Science of the Total Environment,* 710, 136329.

Llu, Z.-Y., Ma, Y.-L., Zhang, J.-T., Fan, N.-S., Huang, B.-C. & Jin, R.-C. 2020. A critical review of antibiotic removal strategies: Performance and mechanisms. *Journal of Water Process Engineering,* 38, 101681.

Ma, N., Xu, J., Bian, Z., Yang, Y., Zhang, L. & Wang, H. 2020. BiVO4 plate with Fe and Ni oxyhydroxide cocatalysts for the photodegradation of sulfadimethoxine antibiotics under visible-light irradiation. *Chemical Engineering Journal,* 389, 123426.

Majeed, H. J., Riquelme, M. V., Davis, B. C., Gupta, S., Angeles, L., Aga, D. S., Garner, E., Pruden, A. & Vikesland, P. J. 2021. Evaluation of metagenomic-enabled antibiotic resistance surveillance at a conventional wastewater treatment plant. *Frontiers in Microbiology,* 12, 657954.

Méndez-Díaz, J. D., Prados-Joya, G., Rivera-Utrilla, J., Leyva-Ramos, R., Sánchez-Polo, M., Ferro-García, M. A. & Medellín-Castillo, N. A. 2010. Kinetic study of the adsorption of nitroimidazole antibiotics on activated carbons in aqueous phase. *Journal of Colloid and Interface Science,* 345, 481–490.

Mirzaei, R., Mesdaghinia, A., Hoseini, S. S. & Yuresian, M. 2019. Antibiotics in urban wastewater and rivers of Tehran, Iran: Consumption, mass load, occurrence, and ecological risk. *Chemosphere,* 221, 55–66.

Mohammadzadeh, M., Montaseri, M., Hosseinzadeh, S., Majlesi, M., Berizi, E., Zare, M., Derakhshan, Z., Ferrante, M. & Conti, G. O. 2022. Antibiotic residues in poultry tissues in Iran: A systematic review and meta-analysis. *Environmental Research,* 204, 112038.

Mojiri, A., Baharlooeian, M. & Zahed, M. A. 2021. The potential of Chaetoceros muelleri in bioremediation of antibiotics: Performance and optimization. *International Journal of Environmental Research and Public Health,* 18, 977.

Monteiro, S. C. & Boxall, A. B. 2010. Occurrence and fate of human pharmaceuticals in the environment. *Reviews of Environmental Contamination and Toxicology,* 202, 53–154.

Moon, K., Jeon, J. H., Kang, I., Park, K. S., Lee, K., Cha, C.-J., Lee, S. H. & Cho, J.-C. 2020. Freshwater viral metagenome reveals novel and functional phage-borne antibiotic resistance genes. *Microbiome,* 8, 1–15.

Navada, K. K. & Kulal, A. 2019. Enzymatic degradation of chloramphenicol by laccase from Trametes hirsuta and comparison among mediators. *International Biodeterioration & Biodegradation,* 138, 63–69.

Nguyen, A. Q., Vu, H. P., Nguyen, L. N., Wang, Q., Djordjevic, S. P., Donner, E., Yin, H. & Nghiem, L. D. 2021. Monitoring antibiotic resistance genes in wastewater treatment: Current strategies and future challenges. *Science of the Total Environment,* 783, 146964.

Palacio, D. A., Rivas, B. L. & Urbano, B. F. 2018 Ultrafiltration membranes with three water-soluble polyelectrolyte copolymers to remove ciprofloxacin from aqueous systems. *Chemical Engineering Journal,* 351, 85–93.

Pamphile, N., Xuejiao, L., Guangwei, Y. & Yin, W. 2019. Synthesis of a novel core-shell-structure activated carbon material and its application in sulfamethoxazole adsorption. *Journal of Hazardous Materials,* 368, 602–612.

Phoon, B. L., Ong, C. C., Saheed, M. S. M., Show, P.-L., Chang, J.-S., Ling, T. C., Lam, S. S. & Juan, J. C. 2020. Conventional and emerging technologies for removal of antibiotics from wastewater. *Journal of Hazardous Materials,* 400, 122961.

Piyaviriyakul, P., Boontanon, N. & Boontanon, S. K. 2021. Bioremoval and tolerance study of sulfamethoxazole using whole cell Trichoderma harzianum isolated from rotten tree bark. *Journal of Environmental Science and Health, Part A,* 56, 920–927.

Polianciuc, S. I., Gurzău, A. E., Kiss, B., Ştefan, M. G. & Loghin, F. 2020. Antibiotics in the environment: Causes and consequences. *Medicine and Pharmacy Reports,* 93, 231.

Qi, M., Ma, X., Liang, B., Zhang, L., Kong, D., Li, Z. & Wang, A. 2022. Complete genome sequences of the antibiotic sulfamethoxazole-mineralizing bacteria paenarthrobacter sp. P27 and Norcardiodes sp. N27. *Environmental Research,* 204, 112013.

Qiu, W., Zheng, M., Sun, J., Tian, Y., Fang, M., Zheng, Y., Zhang, T. & Zheng, C. 2019. Photolysis of enrofloxacin, pefloxacin and sulfaquinoxaline in aqueous solution by UV/H2O2, UV/Fe (II), and UV/H2O2/Fe (II) and the toxicity of the final reaction solutions on zebrafish embryos. *Science of the Total Environment,* 651, 1457–1468.

Rahman, M. S., Hassan, M. M. & Chowdhury, S. 2021. Determination of antibiotic residues in milk and assessment of human health risk in Bangladesh. *Heliyon,* 7, e07739.

Riaz, L., Anjum, M., Yang, Q., Safeer, R., Sikandar, A., Ullah, H., Shahab, A., Yuan, W. & Wang, Q. 2020. *Treatment technologies and management options of antibiotics and AMR/ARGs. Antibiotics and Antimicrobial Resistance Genes in the Environment.* Elsevier.

Rodriguez-Mozaz, S., Vaz-Moreira, I., Della Giustina, S. V., Llorca, M., Barceló, D., Schubert, S., Berendonk, T. U., Michael-Kordatou, I., Fatta-Kassinos, D. & Martinez, J. L. 2020a. Antibiotic residues in final effluents of European wastewater treatment plants and their impact on the aquatic environment. *Environment international,* 140, 105733.

Rodriguez-Mozaz, S., Vaz-Moreira, I., Varela Della Giustina, S., Llorca, M., Barceló, D., Schubert, S., Berendonk, T. U., Michael-Kordatou, I., Fatta-Kassinos, D., Martinez, J. L., Elpers, C., Henriques, I., Jaeger, T., Schwartz, T., Paulshus, E., O'sullivan, K., Pärnänen, K. M. M., Virta, M., Do, T. T., Walsh, F.

& Manaia, C. M. 2020b. Antibiotic residues in final effluents of European wastewater treatment plants and their impact on the aquatic environment. *Environment International,* 140, 105733.

Sachi, S., Ferdous, J., Sikder, M. H. & Azizul Karim Hussani, S. M. 2019. Antibiotic residues in milk: Past, present, and future. *Journal of Advanced Veterinary and Animal Research,* 6, 315–332.

Samandari, M., Movahedian Attar, H., Ebrahimpour, K. & Mohammadi, F. 2022. Monitoring of Amoxicillin and Cephalexin Antibiotics in Municipal WWTPs During Covid-19 Outbreak: A Case Study in Isfahan, Iran. *Air, Soil and Water Research,* 15, 11786221221103879

Sanderson, H., Brain, R. A., Johnson, D. J., Wilson, C. J. & Solomon, K. R. 2004. Toxicity classification and evaluation of four pharmaceuticals classes: Antibiotics, antineoplastics, cardiovascular, and sex hormones. *Toxicology,* 203, 27–40.

Schar, D., Klein, E. Y., Laxminarayan, R., Gilbert, M. & Van Boeckel, T. P. 2020. Global trends in antimicrobial use in aquaculture. *Scientific Reports,* 10, 1–9.

Shang, A. H., Ye, J., Chen, D. H., Lu, X. X., Lu, H. D., Liu, C. N. & Wang, L. M. 2015. Physiological effects of tetracycline antibiotic pollutants on non-target aquatic Microcystis aeruginosa. *Journal of Environmental Science and Health, Part B,* 50, 809–818.

Shao, S., Hu, Y., Cheng, C., Cheng, J. & Chen, Y. 2018. Simultaneous degradation of tetracycline and denitrification by a novel bacterium, Klebsiella sp. *SQY5. Chemosphere,* 209, 35–43.

Shi, H., Ni, J., Zheng, T., Wang, X., Wu, C. & Wang, Q. 2020. Remediation of wastewater contaminated by antibiotics. A review. *Environmental Chemistry Letters,* 18, 345–360.

Shi, Y., Lin, H., Ma, J., Zhu, R., Sun, W., Lin, X., Zhang, J., Zheng, H. & Zhang, X. 2021. Degradation of tetracycline antibiotics by Arthrobacter nicotianae OTC-16. *Journal of Hazardous Materials,* 403, 123996.

Siedlewicz, G., Borecka, M., Białk-Bielińska, A., Sikora, K., Stepnowski, P. & Pazdro, K. 2016. Determination of antibiotic residues in southern Baltic Sea sediments using tandem solid-phase extraction and liquid chromatography coupled with tandem mass spectrometry. *Oceanologia,* 58, 221–234.

Singh, S. K., Khajuria, R. & Kaur, L. 2017. Biodegradation of ciprofloxacin by white rot fungus Pleurotus ostreatus. *3 Biotech,* 7, 1–8.

Sodhi, K. K., Kumar, M. & Singh, D. K. 2020. Potential application in amoxicillin removal of Alcaligenes sp. MMA and enzymatic studies through molecular docking. *Archives of Microbiology,* 202, 1489–1495.

Solehudin, M., Sirimahachai, U., Ali, G. A., Chong, K. F. & Wongnawa, S. 2020. One-pot synthesis of isotype heterojunction g-C3N4-MU photocatalyst for effective tetracycline hydrochloride antibiotic and reactive orange 16 dye removal. *Advanced Powder Technology,* 31, 1891–1902.

Song, C., Guo, B.-B., Sun, X.-F., Wang, S.-G. & Li, Y.-T. 2019. Enrichment and degradation of tetracycline using three-dimensional graphene/MnO2 composites. *Chemical Engineering Journal,* 358, 1139–1146.

Sosa-Hernández, J. E., Rodas-Zuluaga, L. I., López-Pacheco, I. Y., Melchor-Martínez, E. M., Aghalari, Z., Limón, D. S., Iqbal, H. M. N. & Parra-Saldívar, R. 2021. Sources of antibiotics pollutants in the aquatic environment under SARS-CoV-2 pandemic situation. *Case Studies in Chemical and Environmental Engineering,* 4, 100127.

Sukhum, K. V., Diorio-Toth, L. & Dantas, G. 2019. Genomic and metagenomic approaches for predictive surveillance of emerging pathogens and antibiotic resistance. *Clinical Pharmacology & Therapeutics,* 106, 512–524.

Sun, M., Lin, H., Guo, W., Zhao, F. & Li, J. 2017. Bioaccumulation and biodegradation of sulfamethazine in Chlorella pyrenoidosa. *Journal of Ocean University of China,* 16, 1167–1174.

Tian, Q., Dou, X., Huang, L., Wang, L., Meng, D., Zhai, L., Shen, Y., You, C., Guan, Z. & Liao, X. 2020. Characterization of a robust cold-adapted and thermostable laccase from Pycnoporus sp. SYBC-L10 with a strong ability for the degradation of tetracycline and oxytetracycline by laccase-mediated oxidation. *Journal of Hazardous Materials,* 382, 121084.

Wang, Y., Li, J., Lei, Y., Li, X., Nagarajan, D., Lee, D.-J. & Chang, J.-S. 2022. Bioremediation of sulfonamides by a microalgae-bacteria consortium–Analysis of pollutants removal efficiency, cellular composition, and bacterial community. *Bioresource Technology,* 351, 126964.

Warsito, M. F. *A review on potential microbes for bioremediation of antibiotics contamination in the environment.* IOP Conference Series: Earth and Environmental Science, 2022. IOP Publishing, 012026.

Xie, W. Y., Shen, Q. & Zhao, F. 2018. Antibiotics and antibiotic resistance from animal manures to soil: A review. *European Journal of Soil Science,* 69, 181–195.

Xiong, J.-Q., Kurade, M. B., Patil, D. V., Jang, M., Paeng, K.-J. & Jeon, B.-H. 2017. Biodegradation and metabolic fate of levofloxacin via a freshwater green alga, Scenedesmus obliquus in synthetic saline wastewater. *Algal Research,* 25, 54–61.

Yang, K., Ren, S., Mei, M., Jin, Y., Xiang, W., Shi, Z., Ai, Z., Yi, L. & XIE, B. 2022. Removal of antibiotic thiamphenicol by bacterium Aeromonas hydrophila HS01. *World Journal of Microbiology and Biotechnology,* 38, 1–10.

Yang, X., Li, M., Guo, P., Li, H., Hu, Z., Liu, X. & Zhang, Q. 2019. Isolation, screening, and characterization of antibiotic-degrading bacteria for penicillin V potassium (PVK) from soil on a pig farm. *International Journal of Environmental Research and Public Health,* 16, 2166.

Yin, R., Guo, W., Du, J., Zhou, X., Zheng, H., Wu, Q., Chang, J. & Ren, N. 2017. Heteroatoms doped graphene for catalytic ozonation of sulfamethoxazole by metal-free catalysis: Performances and mechanisms. *Chemical Engineering Journal,* 317, 632–639.

Yu, X., Sui, Q., Lyu, S., Zhao, W., Liu, J., Cai, Z., Yu, G. & Barcelo, D. 2020. Municipal solid waste landfills: An underestimated source of pharmaceutical and personal care products in the water environment. *Environmental Science & Technology,* 54, 9757–9768.

Zalewska, M., Błażejewska, A., Czapko, A. & Popowska, M. 2021. Antibiotics and antibiotic resistance genes in animal manure–consequences of its application in agriculture. *Frontiers in Microbiology,* 640.

Zeng, S. & Kan, E. 2022. FeCl3-activated biochar catalyst for heterogeneous Fenton oxidation of antibiotic sulfamethoxazole in water. *Chemosphere,* 306, 135554.

Zhang, T., Cai, L., Xu, B., Li, X., Qiu, W., Fu, C. & Zheng, C. 2019. Sulfadiazine biodegradation by Phanerochaete chrysosporium: Mechanism and degradation product identification. *Chemosphere,* 237, 124418.

Zhang, X., Zhu, R., Li, W., Ma, J. & Lin, H. 2021. Genomic insights into the antibiotic resistance pattern of the tetracycline-degrading bacterium, Arthrobacter nicotianae OTC-16. *Scientific Reports,* 11, 1–11.

12 Biofiltration Approach for Emerging Contaminants in Organic Wastes

M. Hemalatha, Prerna Kashyap, M. Fathima Sana, V. Mohanasrinivasan, and C. Subathra Devi

12.1 INTRODUCTION

A vast range of organic chemicals is essential to modern society. Pollutants created by the organic waste part of solid waste and their proper disposal are becoming a serious issue on a global scale (Cheng et al., 2016). Due to improper disposal, there is a risk to public health, environmental harm, and a lack of suitable disposal sites. Emerging contaminants (ECs), also known as emerging pollutants, trace organic compounds (TrOCs), or micro pollutants, are one of the most problematic pollutants that are generated from a range of anthropogenic and natural elements and have an enormously detrimental impact on the environment (Conserva et al., 2019; Kumar et al., 2022). They are referred to as emergent because, even though they are not new, the level of concern has escalated. Low concentrations of these pollutants, ranging from nanograms per litre (ng/L) to micrograms per litre (g/L), are usually found in the atmosphere. Even though these chemicals were hardly ever taken into account when the procedures were initially developed, these new pollutants have an impact on all conventional treatment approaches. Therefore, it is essential to emphasize an effective modern method in order to tackle the issue (Devinny et al., 2017).

Anything that was never living can be characterized as a contaminant. A more exact definition would depend on the type and nature of the intended downstream processing for biological processing systems like AD and composting (Dhaka et al., 2019). Some nonliving pollutants, such as small fragments of gypsum drywall in compost, can be thoroughly digested and included in the final product without causing harm. To avoid damaging the system of biological processing, other elements must be removed before processing, such as elements from equipment damage (huge metal pieces, rocks, etc.) or hazardous or harmful chemicals. Glass, metal, plastic, and other contaminants that impair product quality are eliminated both during and following biological processing (Fast et al., 2017). In the majority of organics-recycling separation procedures, one substance must be separated from a complex mixture of constituents. The feedstock is the only input for this process, while the extract (recovered material) and reject (the remaining material) are its two outputs. Each reject stream becomes increasingly homogeneous with each phase of the binary separation process that takes place in materials recovery plants (Ghasemi et al., 2020).

For material separation, the appropriate features must be identified and optimized. For example, a stream of bagged organics and a red bag of medical waste that was mistakenly mixed together could be easily detected and separated based on colour (Kyzas & Kostoglou, 2014). Magnetic attraction is the characteristic that allows for the best separation of nails and screws from crushed pallets. The primary characteristics that divide all recyclables for processing into various groups are density, weight, size, light refraction, hardness, magneticity, and electrical conductivity (Li et al.,2015). Some materials can be distinguished using a variety of traits. To expedite the separation process,

DOI: 10.1201/9781003408352-14

several separation systems use a fluid transfer technique. Air or water is typically used when density is the separating factor, such as in a drum separator or when using a vacuum separator to remove film plastic from a conveyor's discharge belt. Water is used as the separation medium in a float/sink tank to separate floatables, such as light plastics, from sinkables, such as stones and bricks (Mishra et al., 2021). To remove emerging pollutants from organic waste, a variety of techniques have been used, including reverse osmosis, ion exchange, adsorption, coagulation, filtration with coagulation, precipitation, ozonation, and advanced oxidation processes, but none of these are sufficient.

Odours, volatile organic compounds, and other contaminants are eliminated from waste gas or water using the latest pollution control method known as biofiltration. Due to the method's simplicity of use, high removal effectiveness, low energy need, and lack of additional treatment or disposal requirements for the leftover components, it has recently gained in popularity (Sheoran et al., 2022). Utilizing a variety of natural and synthetic substrates, including sand, crushed rock, river gravel, or occasionally plastic or ceramic material shaped into tiny beads and rings, the method uses microorganisms that have been raised on these surfaces (Swanson et al., 1997). Polluted water or air is allowed to pass through the filter media, where bacteria convert the contaminants into safer, simpler molecules that the microorganisms can use for their own growth and development. With a removal effectiveness (RE) of over 90%, this technique has been widely employed to remove toxins from contaminated water and air, and it portends a promising future for the treatment of emerging pollutants found in organic wastes (Setiadi et al., 2022).

12.2 ORGANIC WASTES AND EMERGING CONTAMINANTS (ECS)

Waste is a term used to describe a product or substance that is no longer appropriate for its intended use. Although human-generated garbage is frequently quite robust and takes a very long time to decompose in natural ecosystems, it is typically used as food or a reactant (such as oxygen, carbon dioxide, and dead organic matter) (Gagnon et al.,2008). Waste can be classified based on various factors, such as substance (i.e., the composition), source (i.e., who or what generated the waste), its hazardous properties, and also as a mixture of these concepts. There are two main categories of waste: hazardous and non-hazardous (Lima et al., 2018). Wastes that have been known to cause harm to the environment or humans are known as hazardous waste and require special treatment for their segregation and disposal. Non-hazardous waste is every other kind of waste that has not been categorized as hazardous. It is further divided into subcategories, and organic waste is one of them (Marcoux et al., 2013). Organic waste is any substance that is biodegradable and comes from a plant or an animal. Some types of organic waste are green rubbish, food waste, food-soiled paper, non-hazardous wood waste, food waste, and landscaping debris. These may be biodegradable but can also lead to the production of methane (a greenhouse gas), which is more potent than carbon dioxide (Mishra and Kumar, 2021). In addition to being detrimental on their own, organic wastes can be lethal or at the absolute least exceedingly poisonous to the environment and living things when combined with newly emergent toxins.

12.3 EMERGING CONTAMINANTS (ECS)

ECs are natural or synthetic substances that are defined as substances lacking established health criteria but inflicting actual or perceived harm on the environment or human health. They could potentially be recognized by an unidentified source, a recent human exposure, or a sophisticated detection technique or technology. Pharmaceuticals, perfluorinated compounds (PFCs), personal care products (PPCPs), antimicrobials, hormones from humans and animals, microplastics, water disinfection byproducts, engineered nanoparticles, gasoline additives, and household and industrial detergents are only a few of the many compounds that fall under the broad category of emerging contaminants (Sarma et al., 2022). These chemicals are being used in industry, transportation, agriculture, and

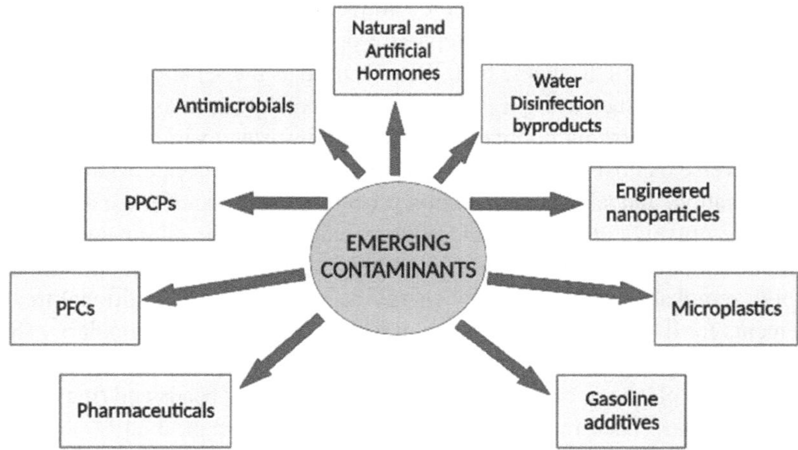

FIGURE 12.1 Emerging contaminants.

urbanization at a rapid rate, and as a result, increasing levels of hazardous waste and nonbiodegradable substances are being discharged into the environment (Lei et al., 2015) (Figure 12.1).

Depending on their usage and application, these substances may infiltrate the environment. They undergo a variety of breakdown processes after being released into the environment. There are still a lot of open problems regarding their modification, mobilization, bioavailability, and retention in soil, sediments, or wastewater.

12.4 EMERGING CONTAMINANTS: TYPES AND TOXICOLOGICAL EFFECTS

ECs, whether man-made, naturally occurring, or created by bacteria that are not frequently seen in the environment, all have the potential to penetrate those environments and have proven or suspected negative consequences on ecosystem health and/or human health (Sarma et al., 2022). The occurrence of ECs in waste, whose effects and composition have not yet been fully understood, can have unexpected effects on health and the environment as well as pose a risk to the treatment processes themselves (Marcoux et al., 2013). The following are some developing pollutants that are causing widespread concern because of their impacts on both human health and the environment.

12.4.1 PERFLUOROCHEMICALS (PFCs)

Perfluorochemicals (PFCs) are a class of man-made compounds that have been used to make goods that are resistant to water, heat, stains, and oil. Perfluorooctanoic acid (PFOA), perfluorooctanesulfonate (PFOS), and their respective salts are the most major PFCs (Thomaidis et al., 2012). Inks, varnishes, coating formulas, cleaning products, firefighting foams, lubricants, metal spray plating, waxes, and water and oil repellents for leather, paper, and textiles are just a few applications where they are frequently used. At this time, PFOA and PFOS have been discovered in human serum, wildlife, surface water, drinking water, and even breast milk. Although the toxicological effects on humans and the environment from such exposure are unknown, they may include reproductive problems, endocrine disruption, and other issues (Lei et al., 2015).

12.4.2 PHARMACEUTICALS

Pharmaceuticals are increasingly being used by both humans and animals, which makes them a significant source of new environmental toxins. The widespread and extensive use of antibiotics has brought attention to this issue on a global scale, increasing worries about the potential effects of pharmaceutical residues in the environment (Thomaidis et al., 2012). In the current scenario, it has been shown that organisms exposed to low dosages of pharmaceuticals over an extended period of time during their lifetimes display significant chronic toxic effects, such as estrogenic effects (for instance, aberrant phenotypic development and lower rates of successful reproduction) (Lei et al., 2015).

12.4.3 HORMONES/STEROIDS

Both organic and synthetic hormones/steroids are excreted by the organism as inactive polar conjugates. However, these compounds are viewed as free, active hormones when they are found in sewage effluents as well as influents. After leaving the body, conjugated estrogens separate chemically or enzymatically, in bacterial sludge, reforming as active estrogens. The presence of estrogenic activity in waste water effluents from the treatment plant (WWTP) have therefore had a detrimental effect on the environment's biota (Thomaidis et al., 2012). It is assumed that endocrine disrupting pollutants are to blame for the physical abnormalities and reproductive abnormalities in fish and amphibians found downstream of polluted water sources.

12.4.4 BYPRODUCTS OF WATER TREATMENT

The pollutant N-nitrosodimethylamine (NDMA), present in drinking water, has become a topic of concern in recent years due to mainly two factors: first, it has the potential to occur during the chlorination of wastewater; and second, it has carcinogenic effects. Additionally, drinking water supplies and surface waterways contain other chlorinated byproducts as well (Naidu et al., 2016).

12.4.5 GASOLINE ADDITIVES

Gasoline is a mixture of more than 500 distinct substances, some of which are recognized or suspected carcinogens, such as methyl tert-butyl ether (MTBE), benzene and 1, 3-butadiene. MTBE spreads to the surrounding area and is easily absorbed by the soil. It is a strong, colourless and insoluble substance in water. It significantly harms the sources of drinking water by contaminating surface and groundwater (Lei et al., 2015). Most developing toxins and their byproducts' chemical characteristics, as well as their level of toxicity for both humans and the environment, are yet unknown. Therefore, effective detection and filtration of these contaminants is necessary for a sustainable future (Lima, 2018).

12.5 REASONS FOR THE INCREASE IN CONTAMINATION

Since 2009, approximately 970 emerging contaminants have been identified. Evidence suggests that EPs in organic wastes when present in the environment degrade water and air quality and have an impact on ecosystems as well as on human health. Hospitals, animal husbandries, domestic waste, pharmaceutical companies, and wastewater and sewage treatment facilities are some of the most significant producers of EPs. Pharmaceutical excretion, inputs from municipal, industrial, and agricultural waste disposal, and accidental spills all have a major influence on the same. According to the global patterns observed in the chemical industry, these emissions will persist and continue to rise.

The use and emissions of specific chemicals and chemical groupings are influenced by various societal and sectoral changes. The following five categories can be used to classify these changes or developments:

(i) Climate change

The occurrence and concentrations of EPs will be majorly impacted by droughts and floods, water scarcity, temperature fluctuations, and the intensity of storms. Moreover, the movement of pollutants across the ecosystem and its components and their transformation are all affected by climate change. Additionally, there are considerable negative effects of climate change on the prevalence of diseases and the usage of drugs to treat them. Increased usage of pharmaceuticals seems inevitable as new disease threats develop and they will be further mixed with organic wastes and pollute the ecosystem (Bunke et al., 2017).

(ii) Changes in demography

There is a foreseen link between demographic shift and pharmaceutical use. Due to demographic changes, there will be an increase in the proportion of people over 65 in the overall population, which will lead to a rise in the incidence of diseases that are common among the elderly, and it is therefore likely that the consumption of medicine will rise in the overall population as a result of these advances. These pharmaceuticals operate as contaminants by entering into both terrestrial and aquatic ecosystems along with other organic wastes being discarded (Bunke et al., 2017).

(iii) Urbanization and population growth worldwide

By 2050, the population of the globe is projected to increase from 7.7 billion to 9.7 billion. In a scenario of population expansion, it is likely that the trend will show that the population of major cities will grow rapidly. This "urbanization" trend will result in a number of environmental issues.

The control of traffic in cities, together with the management of waste and waste water, will become more crucial. Increased ground sealing caused by urban expansion is another major concern. Impervious surfaces, like roads, parking lots, and roofs, are created from natural land surfaces and have detrimental effects on the environment. Snowmelt and rainwater cannot percolate into the soil. Floods may occur more frequently and more intensely as a result of ground sealing. Pesticides, surfactants, medicines, and other new contaminants can be transported by floods to river systems, contributing to local pollution issues.

Some of the compounds used as insulation in home walls are also pollutants of growing concern. Examples include the biocide 2-octyl-2H-isothiazol-3-on and the flame retardant hexabromocyclododecane (HBCDD), which is used as a herbicide (Bunke et al., 2017).

(iv) Technological advancements

There are constant technological advancements occurring in the industrial sector. They exist in all branches and are difficult to predict. New products are created, and new features are added to current ones. These innovations frequently depend on the usage of particular substances.

An example of such a technological advancement is the permanent water resistance feature in outdoor textiles. It has been achieved with the help of per- and polyfluorinated compounds (PFCs). Owing to that, adverse qualities (toxicity, ecotoxicity, persistence, and bioaccumulation) have been established for PFOA, PFOS, and a number of other PFCs (Bunke et al., 2017). The effects of each of these developments on substance emissions and the emergence of new pollutants are still being fully understood.

12.6 TRADITIONAL APPROACHES TO REMOVING EMERGING CONTAMINANTS

The majority of ECs lack uniform regulations but, even in modest concentrations, could have deadly consequences for humans and aquatic life. There is a need for cost-effective tertiary treatment technology because these hazardous pollutants cannot be efficiently removed or degraded by primary or secondary water treatment facilities (Dhangar & Kumar, 2020). The removal or treatment of ECs is expected to be possible using a variety of remediation techniques, including biological therapy, membrane technology, adsorption technology. as well as advanced oxidation technologies (mostly from wastewater) (Ismail et al., 2021).

12.6.1 ADSORPTION

Adsorption is the term for the mass transfer of materials between two phases, such as a liquid–liquid, liquid–solid, gas–liquid, or gas–solid contact (Sharma, 2014). It results from the removal of the component molecules from the bulk phase due to the interaction of physical forces between the porous solids and these molecules. Molecules that get adsorbed are called adsorbates, and the material that adsorbs is called an adsorbent. Physisorption and chemisorption are the two kinds of adsorption. Due to the high operating costs associated with commercial adsorbents, agricultural- and industrial-waste-based adsorbents are being used for removing various emerging contaminants from organic waste, like wastewater and polluted air (Kyzas & Kostoglou, 2014) (Table 12.1). The properties of the adsorbent and the adsorbate determine the most crucial phase in assuring the greatest removal of diverse pollutant kinds. Adsorbent particle size, temperature, pH, selectivity, ionic strength, contact time, and rate of rotation are a few environmental parameters that affect how effectively an adsorption process works (Ismail et al., 2021).

12.6.2 MEMBRANE

One of the potential solutions for effectively removing micropollutants from water is membrane technology. Both biological (membrane bioreactors) and non-biological (reverse osmosis, ultrafiltration, and nanofiltration) processes are used in this technique. Membrane bioreactors (MBRs) are biological reactors for suspended growth combined with membrane-based filtration techniques like microfiltration (MF) or ultrafiltration (UF) (Iorhemen et al., 2016). These days, MBRs are the most well-known and effective methods for separating reasonably clean water from wastewater using a mix of membrane and biological treatments. They are ideal for removing turbidity and microbiological contaminants.

12.6.3 BIOLOGICAL MODE OF TREATMENT

According to numerous research, it has been found that during biological treatment, ECs are primarily removed through biodegradation and adsorption (secondary treatment), which typically include the trickling filter (TF) and the activated sludge process (ASP) (Ismail et al., 2021). ASP is where a massive population of microorganisms mingle with wastewater, while TF has a

TABLE 12.1

List of Emerging Contaminants and the Adsorbent Used to Remove Pollutants

S.No	An adsorbent with a low cost that is used to remove developing pollutants	Emerging contaminants
1	Rice straw, rice husk, bamboo chips, eucalyptus bark, and corn cob	Atrazine, Imidacloprid
2	Rice husk	Heavy metals like Ni, Pb, and Fe; veterinary drug (tetracycline)
3	Coconut husk	Dyes (rhodamine-B (Rh-B) dye)
4	Wood chips and sawdust	Phenol, metals
5	Activated carbon from peach stones	Carbamazepine (psychiatric drug), diclofenac (anti-inflammatory drug) and Caffeine (stimulant)

fixed surface with a large population of microorganisms. The dense microbial biomass is often isolated from water by secondary sedimentation following TF or ASP (Conserva et al., 2019). This treatment is able to remove ofloxacin, diclofenac, and ibuprofen but is unable to remove ampicillin, ciprofloxacin, and naproxen.

12.6.4 METAL–ORGANIC FRAMEWORKS (MOFs)

MOFs have an advantage over traditional adsorbents due to qualities including high surface area, great adsorption capacity, variable porosity, hierarchical structure, and recyclability, although the poor stability of MOFs in water has been a challenge (Dhaka et al., 2019).

12.6.5 ADVANCED OXIDATION METHOD

The most effective method for removing or significantly reducing the levels of certain hormones and PPCPs in wastewater is ozone oxidation (Gagnon et al., 2008). The amounts of bisphenol-A, estriol, and 17-estradiol as well as the estrogenicity of them in secondary effluents were shown to be significantly reduced by ozonation. It has been found that pH, ozone dose, wastewater characteristics, temperature, and contact time are only a few of the numerous factors that could affect the effectiveness of ozonation.

Despite the availability of all these technologies, the emergence of contaminants is still on the rise. These methods have certain limitations too; for example, high operating costs for advanced oxidation. This process also produces hazardous intermediates or byproducts. Furthermore, the aeration tank in the biological treatment requires a lot of energy to run, and the effluent properties fluctuate in highly rigid ways.

12.7 EMERGENCE OF BIOFILTERS

The components of waste gas are broken down by microbiota in biological systems called biofilters. Peat, soil, and compost media are the components of a biofilter's bed. The packing material is adhered to by a moist biofilm that is formed as waste gases are pushed through the bed. The biofilm's microorganisms engage with the contaminants, which in turn cause them to decay. By creating an environment inside the biofilters that enables the microbial cells to grow and utilize the pollutants as a substrate (food and/or energy source), the performance and dependability of biofilters have been improved. Other air pollution management systems are thought to be less efficient, less clean, and less environmentally friendly than biofiltration.

The development of biofiltration as an essential instrument for the management of pungent emissions from composting facilities, wastewater treatment plants, confined animal facilities, the food industry, and other sources of volatile organic compounds (VOCs) and odours has occurred over the last 20 years. It also demonstrates excellent potential for the application of emerging pollutants, especially in the case of organic materials.

The German physicist H. Bach initially proposed the biofilter in 1923. Bach proposed using living creatures to degrade the hazardous gas hydrogen sulphide, produced by waste water treatment facilities. Biofiltration is regarded as one of the least energy-intensive air pollution removal techniques. Although it has been widely used for the past 100 years to cure odours and remove VOCs, it has naturally existed in soil over millions of years (Detchanamurthy and Gostomski, 2012).

In 1941, the Germans became the first to receive a patent for this innovation. There was a significant advancement in the field of biofiltration between 1960 and 1990. In various regions of the world, it has been successfully used to remove organic waste gas, eliminate odours from waste water treatment plants, and more. Additionally, they looked into the RE (removal effectiveness) of various media-filled soil bed biofilters. Up until the late 1980s, the majority of biofiltration research was done in European nations. After the 1980s, a significant number of biofilters were built, and a significant amount of research on the subject was published in journals and conference proceedings (Detchanamurthy & Gostomski, 2012).

During 1990s and the first decade of the 2000s, research was mostly focused on managing pH and moisture, assessing different packing materials, reducing pressure drop, and improving biofilter design to manage massive volumes of contaminated air. These investigations have demonstrated that, when properly optimized, biofilters can reliably manage relatively large loading rates of volatile and odorous air contaminants while also being safe, secure, and reasonably priced. Biofiltration research has changed from a conventional black-box approach to science-based research focused at carefully implementing improved control systems at full scale, thanks to the development of mathematical models and powerful industrial equipment design and simulation software (Rene et al., 2013).

12.8 BIOFILTRATION: PRINCIPLE AND MECHANISM

The biological method used to remediate polluted air and water is called biofiltration. It is a new technology that is used in waste purification in order to manage volatile organic, inorganic, aromatic, poisonous, and odorous components. This is a reliable, environmentally friendly "green" technology because it is based on a process that happens naturally in soil and water. Biofiltration does not produce any dangerous byproducts, and it is also more affordable to use as a method of reducing air pollution (Figure 12.2) (Raju, 2008).

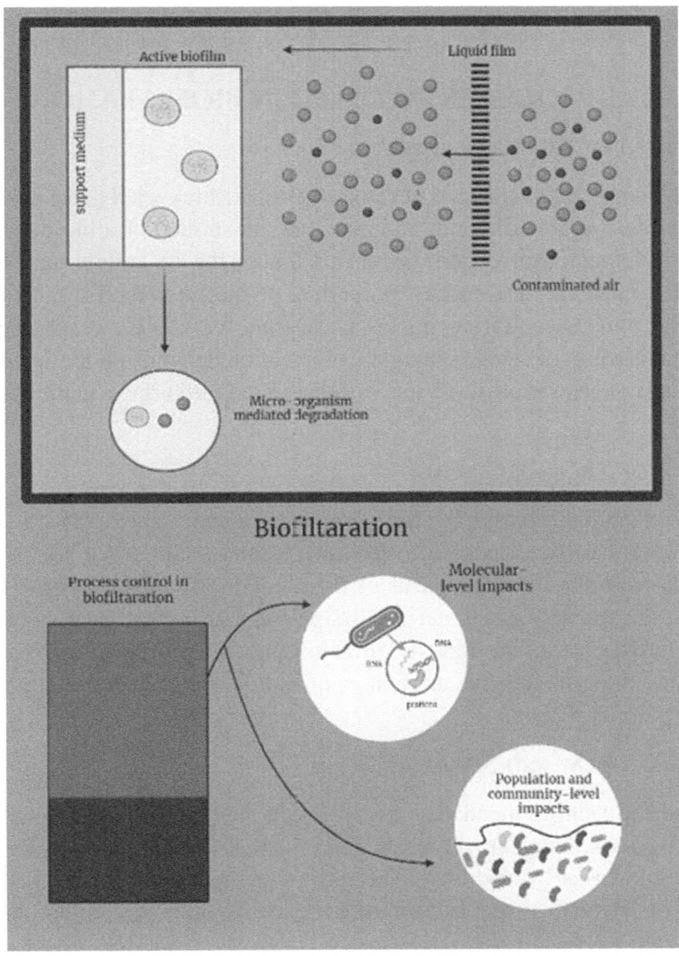

FIGURE 12.2 Biofiltration principle and mechanism.

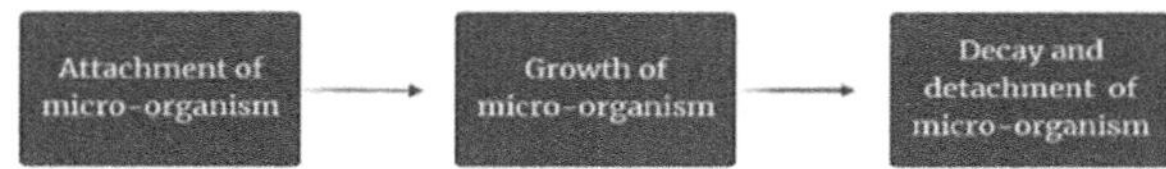

FIGURE 12.3 Process in biofiltration.

It uses in-depth filtration techniques to biotransform organic molecules, nitrogen, phosphorus, and dissolved metal components using autochthonous microbial populations adhering to granular media (Ashraf et al., 2016). For instance, enzymes or microorganisms are introduced to the water purification system in order to break down the pesticides in the water. The first is the pesticide mobility in water resources, where biofiltration design creates a small space through which pesticide-contaminated water can pass, and the second is the easier control of the contaminated water's hydrodynamics, which permits high loading rates of carbon substrates and pesticides despite their modest concentrations (Setiadi et al., 2022).

In conclusion, the removal as well as oxidation of hazardous gases, especially volatile organic compounds derived from contaminated waste, can be characterized as the principle of biofiltration. A horizontal or vertical reactor receives the contaminated air after passing through a compost or soil support. A distribution system with perforated pipes specifically created to allow an even flow of gas through the porous bed is located beneath the bed (Raju, 2008). In order for microorganisms to transform biodegradable pollutant components into innocuous compounds, a gaseous phase must diffuse into an aqueous phase during the biological process of treating the gas (Figure 12.3).

12.9 BIOLOGICAL PROCESSES INVOLVED IN BIOFILTRATION

12.9.1 ATTACHMENT OF MICROORGANISMS

Microorganisms adhere to the media and develop into thick, sticky, jelly-like sheets. There are several ways that microbes might adhere to and colonize the surface of a biofilter's filter media like active microbial movement, convection, sedimentation owing to gravity, and transportation-like diffusion (Brownian motion). The surface properties of the filter media and influent parameters (such as organic type and concentration) are both necessary for a secure attachment. Initial adhesion depends on the total amount of contact energy. Colonization depends on the features of the influent (such as the kind and concentration of organic materials) and the surface qualities of the filter media.

12.9.2 GROWTH OF MICROORGANISMS

Due to concentration gradients within the film, dissolved organics are incorporated into the biofilm. Colloids and suspended particles adhere to the surface and break down into soluble compounds there. In waste and from the vacuum spaces of the medium, dissolved oxygen is used to provide oxygen for aerobic reactions. Waste materials dissipate and are swept away by water currents. The substrate's mass transfer to the biofilm, its diffusion into the biofilm, and its utilization kinetics inside the biofilm are the elements that affect how quickly a substrate is used up within a biofilm.

12.9.3 MICROBIAL DECAY AND SEPARATION

Both anaerobic and endogenous mechanisms manifest at the biofilm-medium surface interaction as the film thickens. Weakened attachment mechanism waste is washed away by the shearing motion. This procedure is known as "sloughing". The rate of organic and hydraulic loading determines sloughing. The diffusivity of oxygen is a limiting factor. The thickness of aerobic zones is capped at 0.1–0.2 mm (Figure 12.3).

12.10 BIOFILTERS AND THEIR TYPES

The several types of biofilters include spinning drums, submerged media, fluidized bed reactors, trickling filters, and rotating biological contactors.

12.10.1 BIOLOGICAL CONTACTOR IN ROTATION

A rotating biological contactor (RBC) is arranged as a cylindrical drum that rotates perpendicular to the water flow in the filter vessel to provide biochemical treatment for wastewater. Drum rotation enables the fixed biofilm and the media for waste treatment to be alternately immersed in and removed from the wastewater, each for oxygenation purposes and wastewater treatment (Cohen, 2001). The required rotating speed is to be maintained to guarantee the preservation of healthy bacterial communities on the medium. If rotation is too quick, it could deprive the biofilm of oxygen. However, rotating too slowly may remove significant chunks of the biofilm from the drum (Cohen, 2001). Either scenario has the potential to reduce the nitrification effectiveness of the filter.

During operation, RBC designs submerge around 40% of the drum at a time. The ideal drum submergence range is 35–50%. Drum rotation generates water turbulence that simultaneously aerates the water column and biofilm (Forero et al., 2018). These forces also make it possible for the RBC to maintain conditions that are relatively clog-resistant and self-cleaning (Rodriguez et al., 2017). Self-cleansing involves the routine elimination of dead microorganisms, enabling the maintenance of an active biofilm (Forero et al., 2018).

Hassard et al. (2015) asserts that biofilm activity regulates filter performance. RBCs have historically been mechanically driven by motor and shaft configurations. These designs have been known to fail while providing adequate filtration performance, mostly because of mechanical issues (Forero et al., 2018). Gear motor breakdown and media removal from the drum shaft were frequent issues. These failures are thought to be a significant issue with RBC filters. As a result, certain RBC filters that have been developed more recently use air to drive drum rotation rather than mechanical functioning.

12.10.2 TRICKLING FILTER

There are many different trickling filter designs, and the media can be anything from rock to different kinds of plastic. Depending on the filter's construction, the medium is either fixed inside the vessel or poured inside. A spray bar distributes the water and allows it to drip naturally when it is pumped to the top of the medium. As water drips downhill through the filter, nitrifying bacteria oxidize nitrogenous waste, stirring the water as they do so. When the water-air interface inside the filter is more actively mixed, oxygen can dissolve into the water and carbon dioxide can be expelled. These characteristics make this filter type suitable for self-aeration and pH buffering.

12.10.3 BEAD FILTER

Bead filters use a submerged bed of tiny plastic beads and are categorized as expandable granular filters (Cohen, 2001). The majority of designs use propeller-washed bead filters from Armant Aquaculture Inc. in Vacherie, Louisiana, which are less dense than water. Some, though, use beads that are a little bit denser than water. Granular filters can be useful because of their huge specific surface area and ability to capture particulates during nitrification (Christen et al., 2002). All oxygen must be provided to the nitrifiers as dissolved oxygen in the culture water because the beaded bed is submerged (Forero et al., 2018). If ambient air is the only source of oxygen, the system's carrying capacity and productivity may be restricted. Productivity is the amount of fish biomass that is actually produced throughout a production cycle, whereas system carrying capacity is the maximum amount of fish biomass that a recirculating system is intended to support. Bead filter accumulation of solids may also reduce system carrying capacity and output. Excessive solids capture can result in medium biofouling when solids break down and hyper-productive biofilm cause the bed to block and divert water flow. Such events should be avoided by periodic bed cleaning, albeit the ensuing biofilm shearing may reduce the filter's nitrification effectiveness (Malakar et al., 2017). The frequency of cleaning could potentially pose issues for maintaining the filter's good upkeep.

12.11 BIOFILTRATION: DESIGN AND DEVELOPMENT

A mixed microflora comprising competitors, predators, and pollutant degraders that is, at least in part, structured as a biofilm is involved in biofiltration. It is a two-step process that involves transferring the chemicals from the gas phase to the water phase (biofilm) and then allowing the microorganisms in the filter bed to oxidize the components that were absorbed (Rene et al., 2013). Biofilters are reactors that pass gaseous organic pollutants through a compost or soil bed where they are broken down by natural microorganisms, or through an inert support where a particular microbe or group of microorganisms is grown. When this occurs, an active biofilm is deposited onto a collection of active and carefully chosen bacteria, which then causes the breakdown of ecologically hazardous substances. On the solid support, the aqueous phase is displayed as humidity.

Therefore, to put it briefly, the main components of a biofilter consist of a solid support surrounded by an active biofilm as well as a bed through which the gas containing hazardous substances flows (Torretta et al., 2015). Processes for treating organic waste use a variety of bioreactor designs. Although the mechanisms of operation of all of these systems are essentially similar, they are distinguished by the liquid phase's movement (constant or nearly stationary) and the location of the microorganisms (freely dispersed or immobilized). The emphasis is on the operational and control requirements essential to provide the appropriate chemical and physical environment for mass transfer and pollutant biodegradation in order to achieve high removal efficiencies.

Biofilters were initially created as simple, porous, open soil beds with suitable air flow mechanisms. After foul air was directed over the porous soil bed in an upflow mode, air pollutants were biodegraded by the local mixed microbial populations. These biofilters were fundamentally changed and reconfigured with better irrigation systems, and novel filter materials were tried in conjunction with the conventional soil beds, despite the fact that they lacked adequate online process control. Since the 1980s, certain biofilters have been developed as closed systems. The majority of research conducted in the 1990s and the first few years of the 2000s was devoted to managing pressure drop, controlling pH and moisture, assessing different packing materials, and creating biofilter designs that could handle enormous volumes of dirty air. All of these investigations have demonstrated that, when properly optimized, biofilters can effectively treat relatively large loading rates of volatile and odorous pollutants in the air. They can also do so safely and affordably. Additionally, the conventional black-box approach to biofiltration research has given way to science-based research that aims to strategically integrate sophisticated control systems at full scale. This is because sophisticated industrial equipment design and simulation software has developed along with mathematical models (Rene et al., 2013).

12.11.1 DESIGN FACTORS

(i) Size of the reactor

When establishing the appropriate size of the biofilter required for a particular application, it is essential to take into account the chemical characteristics of the waste stream, and preliminary study in laboratory- and/or pilot-scale biofilters may be helpful. Once the system's performance has been assessed, modelling work can be done to pinpoint some of the crucial state variables (macro as well as micro-kinetic parameters). The development and sizing of large-scale systems can then be done using this information. In the first scenario (performance evaluation), the biofilter should be put to the test using inlet concentrations of various ranges (minimum to maximum), different rates of mass loading, humidity, and temperatures. However, such information might be site-specific depending on the characteristics of the waste gas evaluated by other authors, the initial inoculum used, the biofilter packing, concentration levels, and other operating parameters. On the other hand, a review of earlier research findings can also give some background data regarding values for removal capacity and efficiency.

(ii) Irrigation method

After deciding on the biofilter's size and packing material, it's critical to consider the watering needs. Microorganisms require a precise amount of moisture to support their metabolic activity. Even if contaminated air is humidified before it enters the biofilter, this is insufficient to sufficiently hydrate the biofilter medium. The moisture content of the bed needs to be regularly monitored because biofilters operating in hot, dry environments can soon dry out. Water losses can be calculated manually by monitoring temperature variations along the filter bed, calculating water loss using psychrometric charts, or measuring the weight loss of the filter medium using load cells.

Moisture sensors can also be installed on the biofilters along the height of the bed. The reactor should have sufficient over-bed sprinkler systems after choosing the operation mode of the biofilter (top or bottom feed). The media can also be kept moist by properly installing soaker hoses, impact irrigation, or drip irrigation. Using the flow control devices on the hose, the amount of water to be given can be chosen. A sprinkler manifold would distribute water from a main water pipe to each pressure-compensated sprinkler head. These heads have the capacity to evenly disperse water across the top of the bed. Additionally, irrigation systems can be installed using a range of medium depths. It is advised to install the water supply (irrigation) at the top and middle of top-feed systems, as well as at the bottom and top of the biofilters on bottom-feed systems.

(iii) Collection and disposal of leachate

Although a mobile or continually flowing liquid phase is not expected for biofilters, a leachate is likely to develop while the device is in use. It should periodically drain off and remove any extra water that was introduced during operation. Condensation from the input air may potentially be the cause of leachate production in biofilters. The leachate collection system can be centralized to meet any industrial setting or is typically situated at the bottom of the biofilter. The leachate contains a variety of nutrients, soluble chemicals, organic and inorganic elements, biodegradation end products, and maybe intermediates, depending on the type of polluted air.

The composition of soluble or suspended material in the leachate is also influenced by the type of pollutants treated by a specific biofilter and the characteristics of the biofilter. Thus, it is essential to constantly assess the leachate's physical, chemical, and biological properties in order to select the most appropriate plan for treatment or disposal. Leachate that doesn't include any potentially hazardous components can be used in a variety of applications as a valuable resource.

The most common method for treating leachates from biofilters is biological. They do a good job of getting rid of the other biodegradable components of the leachate, but they struggle to get rid of inorganic salts. The leachate can then be treated in typical wastewater treatment facilities together with any wastewater the industrial complex produces. Rotating biological contactors, sequencing batch reactors, activated sludge processes, and aerated lagoons are examples of commonly used bioprocesses. A new development in leachate control involves using oxidation substances like hydrogen peroxide or ozone in addition to UV light.

12.12 VARIABLES AFFECTING BIOFILTER PERFORMANCE

The following are some crucial variables that have an impact on a biofilter's operation and microbial development.

12.12.1 MICROORGANISMS

A biological agent serves as the main component and catalyst and initiates the biofiltration process. Biofilm is created by microorganisms such bacteria, protozoa, invertebrates, and fungi (Pachaiappan et al., 2022). The two main heterotrophic microorganisms involved in biodegradation are bacteria and fungus. Microorganisms that are heterotrophic utilize the organic components of off-gases as a source of carbon and energy for growth and metabolic processes. After a period of acclimatization, the population of microorganisms with the greatest resistance is organically chosen, and a microbial hierarchy develops in the filter bed. A smaller population of microorganisms will form at the deeper point of the bed, while a higher density will emerge towards the influent end. A biofilter typically has a biomass density of 106 to 1010 colony forming units (CFU) of bacteria and 103 to 106 CFU of fungi per gramme of bed. Between 1% and 15% of the overall population of microorganisms make up the degrading species.

12.12.2 BIOFILTER BED/ IMMOBILIZING MEDIA

The biofilter bed or packing material is the most important component of the biofiltration system next to microorganisms. The biofilter bed serves as a medium for the growth of a biofilm while immobilizing microorganisms. The packing material for biofilter beds is made up of both organic and inorganic materials. Due to their affordability, accessibility, and compliance with the majority of the necessary bed material requirements, peat, composts, soils, and wood chips are the most commonly used bed materials. These packing materials should have the following qualities.

(1) A large surface area (between 300 and 1000 $m^2\ m^{-3}$) and high porosity to aid in the homogeneous dispersion of influent.
(2) Packed with natural nutrients that encourage the growth of microorganisms and the creation of biofilm.
(3) An abundance of various microorganisms.
(4) A 40–60% increase in water retention capacity to hasten microbial metabolism.
(5) Mechanical and thermal stability to prevent packing of the filter bed.

12.12.3 SUPPLY OF NUTRIENTS

For biofilters to function successfully, the availability of nutrients within the filter bed is a critical component controlling microbial growth and activity. The pollutants supplied into the biofilter are broken down by the microorganisms. These contaminants serve as sources of carbon, which gives the microorganisms energy (Pachaiappan et al., 2022). The materials used as a supporting pack in the biofilter bed provide the needed micronutrients and macronutrients. The main macronutrients are potassium, sulphur, nitrogen, and phosphorus, while the micronutrients include vitamins and metals. Nitrogen is a key component of proteins and nucleic acids and is used by microorganisms to form their cell walls (Rene et al., 2013).

While nutrients may need to be introduced through the liquid phase in biofilters with organic filter beds, they may already be present in bound form in those with inorganic packings. The advantage of using compost is that the nutrients are already in the medium and are used in the right amounts during pollutant degradation. Nutrients can be introduced to biofilters packed with inert packing materials by feeding an aqueous phase made up of inorganic N, inorganic P, and other compounds (Rene et al., 2013). In the biofilter bed, a nutrient solution made primarily of mineral salts is dissolved in water. Common mineral salts include $CaCl_2$, KNO_3, $MgSO_4$, $MnSO_4$, $FeSO_4$, KH_2PO_4, $(NH_4)_2SO_4$, NH_4HCO_3, NH_4Cl and Na_2MoO_4 (Pachaiappan et al., 2022).

12.12.4 Moisture

Low water content in the biofilter pores inhibits microbial activity, while high water content reduces mass transfer (Swanson and Loehr, 1997). It is easy to see how moisture affects biofilms: too little moisture causes the biofilm to dry out, while too much moisture can lead to unwanted or desired organism expansion.

Drying promotes media cracking, which shortens retention duration, while saturated filter media wash away nutrients and produce leachate that needs to be disposed of (Raju, 2008). Furthermore, excessive amounts of water cause an unfavourable high pressure drop and an increase in operating costs when water is introduced to biofilters through the use of humidified air and by misting water on the top of the bed (Van Langenhove et al., 1997).

Each biofilter differs in terms of heat generation from biological oxidation, heat capacity measured in terms of media mass and type, ambient conditions measured in terms of heat loss or absorption from the outside, biofilter housing insulation coefficients, water irrigation nozzle spacing, and irrigation nozzles themselves. Industrial biofilters are irrigated with an over-bed spray (OBS) system that is controlled by a microprocessor, typically a programmable logic controller (PLC), which allows the user to control water application by zone, by time of day, and by day of the week in order to take into account all of these variables.

Applying water on a 24/7 schedule would result in unmanageable reactions that would blind the media and cause an excessive pressure decrease because some processes only run eight hours a day, for five days in a week. The ability of the plant to function is often impacted by the increase in pressure drop, which typically denotes either unbalanced flow over the media or scaled back flow.

12.12.5 Temperature

The biological activity of the bacteria, or how quickly they can migrate, reproduce, and digest any given molecule, is directly influenced by temperature. When the temperature rises over 104 °F, thermophilic microbes that cannot or only partially break down the chemicals begin to proliferate. In terms of the rate of breakdown or molecules per hour degraded by a certain microorganism, increasing biological activity does not always imply increased hunger. In recent years, a lot of time and money have been put into researching and measuring species that can digest the chemicals effectively.

According to a study conducted by the Institute of Fisheries of the National Academy of Agrarian Science of Ukraine that looked at the effects of temperature on the activity of biofilters for the process of nitrification (removal of ammonia from wastewater) (Fast et al., 2017), the nitrification rate is reduced at lower temperature as compared to the normal temperature.

Since temperature can impact physicochemical parameters like solubility, diffusivity, and Henry's Law coefficient of the molecules in the gas stream in addition to influencing biological activity, temperature effects in a biofilter are actually quite difficult to predict.

12.12.6 pH

The effectiveness of biofiltration technologies can be considerably impacted by pH. The ideal pH value makes nutrients and their functions in the efficient breakdown of contaminants by bacteria feasible. Conducting biofiltration operations in a pH range that is below optimal can seriously disrupt microbial activity (Fulazzaky et al., 2014). The biofilters' living habitat has a pH of 7, making it neutral for the neutrophilic heterotrophic microorganisms that thrive there (Pachaiappan et al., 2022). pH also has an impact on how the proton pump system works on the membrane of microbial cells. Such a mechanism can be negatively impacted by abrupt pH changes, which can make the cells toxic. The system's effectiveness can be impacted by an improper pH level, which will also have an impact on how effectively contaminants are removed

12.12.7 Pressure Drop

The characteristics of the packing material, including the particle size and shape, its moisture content, and biomass growth, are the key determinants of pressure drop. In addition to this, the value of P (phosphorus) would be impacted by the surface gas velocity. In a biofilter, irrigation would lead to a rise in P largely because of pore reduction, which is then followed by a decrease in the effective cross-section of the bed. The pressure drop via a biofilter can range from 20 to 100 Pa m^{-1} and even up to 1000 Pa m^{-1} in some cases (Rene et al., 2013). When calculating operational expenses, the pressure drop over the bed is a crucial factor. Power usage increases with increasing pressure drops (Sheoran et al., 2022).

12.12.8 Particulate Matter

Due to the variety of different sizes of particulate matter, some can cause damage to the biofilter while others may not. The very little and the very big are the two simplified enormous extremes. Smoke, viruses, and smog are examples of very small particles that can pass through the biofilter without ever coming into contact with a single water particle, medium particle, or biofilm. Due to their small size, these particles follow the course of the flue gas and avoid collisions. Sand, ash, sawdust, and pollen are examples of very large particles that can accumulate on the surface or fall out into the medium.

These substantial particles can eventually clog the air's passageways through the medium, causing an excessive pressure reduction. With the same outcome, these organisms fill the holes and cause an excessive pressure decrease. A secondary effect of these huge particles may be that they serve as food for other creatures like slime and fungi that are already present in the media. The grey area, or those particles that are unquestionably neither large nor small, lies in the 1 to 10 micron range, and its impact on the system varies depending on the particular biofilter.

12.12.9 Solubility

The breakdown of the chemicals is one of the essential elements for biological oxidation to be successful. The substance needs to be where the bacterium and its enzymes can influence it, which is in the biofilm, which is primarily made of water, for this breakdown to occur. A compound's molecules will choose to remain in the vapour phase rather than dissolve in the water that makes up the biofilm if it is not particularly water soluble.

12.12.10 Compound Concentration

The resultant population of bacteria is geared toward the degradation of the primary compound and not the secondary compound if gas streams contain high concentrations of a particular compound, the primary compound, and low concentrations of a different compound, unless, of course, the secondary compound is a step in the degradation process of the first compound. A chemical cannot penetrate the biofilm farther because it degrades more quickly the easier it is to break down.

12.12.11 Physical/mechanical Design

A biofilter's mechanical design affects everything from the inlet nozzle to the output nozzle. Air flow imbalances over the width of the biofilter caused by improper intake design may never be fixed without completely reconstructing the inlet. Inadequate packaging design can lead to greater pressure drop as well as lower efficiency removal in the bio-scrubber section. Unwanted microorganisms might grow on the surface of media due to improper demister design that permits water carryover. Inappropriate medium selection can lead to large pressure drops and promote the growth of undesirable microorganisms like fungus. A pressure drop that is too great can be caused by improper outlet design (i.e., in the 3" to 5" range).

12.13　OPERATION CONDITIONS OF BIOFILTERS

The following factors can be used to assess a biofilter's effectiveness:

(i) Empty bed residence time (EBRT)
The EBRT is the total volume that the filter bed takes up in relation to the gas flow rate. Taking into account the entire, unpacked, reactor volume, it shows how long the pollutant would typically stay in the filter bed (equation 12.1)

$$EBRT = V / Q \, [\text{s}] \tag{12.1}$$

where V is the volume of the reactor (m^3) that contains the filter bed and Q is the gas flow rate (m^3h^{-1}). However, the theoretical real residence time of the pollutant (τ) in a packed biofilter is determined by the void space of the filter media (θ) and can be computed as follows.

$$\tau = \frac{V \times \theta}{Q} \, [\text{g} / \text{m}^3\text{h}] \tag{12.2}$$

Here, θ represents the filter bed porosity.

Under many instances, it is challenging to exactly quantify the value of θ because of the biofilm formation that occurs and subsequently fluctuates with operational time. The EBRT makes it simple to compare the qualities and effectiveness of various biofilters (Rene et al., 2013).

(ii) Mass loading rate (MLR)
The mass loading rate is defined as the mass of the pollutant entering the biofilter per unit of time and volume of the filter bed (Rene et al., 2013). A biofilter can accommodate a range of mass loading rates, depending on the kind and characteristics of the waste gas. Depending on how well the pollutant can be used by the microorganisms, this value decreases along the biofilter's height as the pollutant is eventually destroyed. The following is an expression for the MLR:

$$MLR = \frac{Q \times C_i}{V} [\text{g} / \text{m}^3\text{h}] \tag{12.3}$$

Here, C_i represents the inlet pollutant concentration (gm^{-3}).

(iii) Volumetric loading rate (VLR)
The ratio of the gas flow rate to the volume of the filter bed is known as the volumetric loading rate, and it is calculated as follows (equation 12.4).

$$VLR = V / Q \left[\text{m}^3 / \text{m}^3\text{h} \right] \tag{12.4}$$

(iv) Elimination capacity (ELC)

The ability to eliminate pollutants is measured by the mass of pollutants degraded per unit volume of filter bed and unit time is known as elimination capacity. The efficiency of a biofilter is frequently represented by the ELC. The values of higher maximum ELC (ELCmax) are preferable since the design of smaller biofilters is enabled (Rene et al., 2013). ELC is expressed as follows (equation 12.5).

$$ELC = \frac{Q \times \left(C_i - C_0\right)}{V} \left[g / m^3 h\right] \tag{12.5}$$

Here, C_0 represents the outlet pollutant concentration (gm^{-3}).

(v) Rate of CO_2 production (P_{CO2})

Since carbon dioxide can also be produced by other processes, such as endogenous respiration in biofilters, performing intricate mass balance calculations is not always easy. Nevertheless, it is frequently used to assess the level of biodegradation and full mineralization of pollutants within the biofilter. Autotrophic bacteria, on the other hand, might use CO_2 as their primary carbon source (Rene et al., 2013).

The CO_2 production rate can be determined by monitoring the CO_2 concentrations at the filter bed's intake and outflow (equation 12.6)

$$P_{CO2} = \frac{Q \times \left(CO_{2,out} - CO_{2,in}\right)}{V} \left[g / m^3 h\right] \tag{12.6}$$

where $CO_{2,out}$ and $CO_{2,in}$ are the corresponding carbon dioxide concentrations (gm^{-3}) at the outlet and the inlet.

(vi) Removal effectiveness (RE)

The amount of pollution supplied into the biofilter divided by the amount of pollution removed is known as the removal efficiency. According to the pollutant concentration, gas flow rate, and microbial activity, RE values frequently change in a biofilter (Rene et al., 2013). The RE is generally stated as follows (equation 12.7).

$$RE = \frac{C_i - C_0}{C_i} \cdot 100 \left[\%\right] \tag{12.7}$$

12.14　BIOFILTRATION: ADVANTAGES AND DISADVANTAGES

Biofiltration has been emphasized as a practical method for eliminating odours, volatile organic compounds (VOCs), and air toxics from waste-gas streams due to its low initial and ongoing costs, low energy requirements, and lack of residual products that need extra processing or disposal. Biofiltration units, which are microbiological systems, use microorganisms grown on porous solid media, such as compost, peat, soil, or a mix of these materials. The filter media and the microbial culture are encircled by a thin water film called a biofilm. Waste gases containing biodegradable VOCs as well as inorganic air toxins are vented through this biologically active material. Solvable contaminants in this situation partition into the liquid film and are broken down by the local microorganisms in the biofilm. While having advantages like increased contaminant removal, when compared to traditional filtration systems, biofiltration systems are more efficient in

TABLE 12.2
Advantages and Disadvantages of Biofiltration

Advantages	Disadvantages
Low investment and operational costs	Bed clogging due to biomass growth
Easy operation and maintenance	Packing material compaction
Effective removal at low H_2S concentrations	Difficult to control pH drop
Results in a complete decomposition of the pcllutants and no production of hazardous byproducts	Need of filter bed replacement

eliminating inorganic, organic, and microbiolcgical pollutants. They also reduce DBP production (in wastewater); by efficiently removing organic carbon from wastewater, biofiltration systems reduce the likelihood that disinfection byproducts like haloacetic acids and trihalomethanes will occur.

Biofiltration is seen as more affordable, cleaner, and environmentally friendly than alternative air pollution management technologies:

1) Low running expenses. Biofilters use very little power because they work at ambient pressures and temperatures. Pressure decreases typically do not exceed a 10 cm water column.
2) There are no leftovers. Typically, compounds are broken down into safe byproducts like H_2O, CO_2, and inorganic ions such as NO_{3-} and SO_4^{2-}. Other alternative controls result in residuals like used activated carbon and chemical sludges that need to be further treated.
3) In contrast to thermal oxidation, biofil:ration emits very little CO_2 and NOx.

Biofiltration has drawbacks, such as moderate to high capital costs and relatively high area requirements. Additionally, biofiltration is only effective with relatively low quantities of soluble and biodegradable chemicals. However, it has certain disadvantages too. Table 12.2 presents a list of advantages and disadvantages.

12.15 ROLE OF MICROORGANISMS IN BIOFILTRATION

Recent research has focused on a variety cf topics related to the microbiological potential of biofilters, including isolation and characterization; the use of pure bacterial cultures; mixed microbial communities; the effect of culture enrichment, including the application of special strains; types of microorganisms and their metabolic activities; the effects of environmental factors on microbial activity; and the release of microorganisms from biofilters. The naturally occurring bacteria in the filter bed of soil, compost, and peat biofilters typically have the potential to break down the specific gas-phase contaminant (Sheoran et al., 2022).

Bacteria and mixed cultures have been the focus of much biofiltration research. Activated sludge has diverse microbial communities with a wide range of metabolic variability and can be used for inoculation. It works effectively for large-scale biofilters that treat new pollutants coming from composting or water treatment facilities. In biofilters that remove organic pollutants, yeasts, heterotrophic eubacteria, fungus, protozoa, and actinomycetes have all been discovered. Finding adaptable biocatalysts that can provide additional advantages like resistance to loads with high levels of pollution, maintaining the reactor's performance at its peak without causing blockage or other operational issues, and efficiency when transient-state biofilter processes are in progress has been a focus of research in recent years (Rene et al., 2013)

The production of biofilms in this system depends on the adhesion of microorganisms to the packing medium's surface. Biofilms are mixed cultures of fungi, bacteria, algae, and actinomycetes that are grown in bioreactors. They combine to create a sophisticated ecology. There are approximately bacteria, actinomycetes, and fungi ranging from 106 to 1010 CFU per gram of bed in a typical biofilter. Additionally, 1–15% of the overall microbial population in a biofilter is often made up of degrading bacteria. Communities are also impacted by changes in the environment brought on by biodegradation processes. In order to improve the design and operation of such biological systems, research into and analysis of their diversity is necessary (Cheng et al., 2016).

The existence of anthropogenic compounds as emerging contaminants in natural resources like air and water, and among other organic wastes, is inevitable due to the increasing quantity and use of these substances. A biological treatment like BAC (biological activated carbon) can aid in completely decomposing intermediates and residues to accomplish mineralization in combination with advanced chemical treatment for emerging contaminants. However, the high toxicity of emerging contaminants and their precursors may cause biofilms in BAC to separate and perish. Therefore, bioaugmentation with strains that are able to degrade such toxins may be the key for BAC to counter the toxicity of emerging contaminants.

12.16 SIGNIFICANCE OF THE TYPES OF MICROORGANISM

The most important elements of biofilters are microbes because they are used to transform or degrade pollutants. Most common biofiltering microorganisms are aerobic and need oxygen to digest organic substances (nutrients or pollutants). In general, fungi can absorb hydrophobic chemicals more quickly than bacterial biofilm, but bacteria are probably better at removing hydrophilic substances under optimal conditions (Devinny et al., 2017).

A study was conducted to investigate the comparison of biofilter performance for treating polluted air. There were two cultures, a pure bacterial culture and a pure fungal culture. For the bacterial culture, the strain *P. putida* PTCC was provided. It was developed in nutritional broth and incubated at 30 °C (125 rpm for 24 h) in a rotating shaker. *Pleurotus ostreatus* IRAN 1781C (Oyster mushroom) was used for the fungi culture. It was grown in a dextrose broth powder and allowed to incubate at an ambient temperature for two days (Ghasemi et al., 2020). Toluene, which can serve as an example of a VOC due to its hydrophobicity and slow rate of biodegradation, was selected as the model pollutant (Zhang et al., 2007). In accordance with the toluene removal results in bacterial and fungal BFs, sterilized beds (first phase) were only, at most, capable of removing 57.1% of toluene in the first stage and 54.9% in the second stage.

Following inoculation (the beginning of the second phase), the bacterial biofilter removed around 82.1% of the organic load after just five days, and the fungal BF removed about 82.5% after six days. Comparing the outcomes of the bacterial and fungal BFs, it is evident that the fungus was more effective at eliminating the contaminant toluene. Apart from that, Li et al. (2015) found that a *P. ostreatus*-inoculated fungal BF removed styrene, ethyl-mercaptan, sulphur compound combinations, and alpha pinene at a rate of 99%. Another example of a microorganism used for biofiltration is *pseudomonas putida* (Li et al., 2015).

12.16.1 *PSEUDOMONAS PUTIDA*

Pseudomonas putida typically dwells in water, like *E. coli*, and it is a Gram-negative bacterium. It can degrade aromatic compounds like toluene. It is widely used in oil spill decontamination or bioremediation. During the initial three days, there is rapid colonization. Contrary to active biomass and polymer growth, which has been seen to rise linearly, biofilm thickness has been observed to increase exponentially with time. In a study to look into toluene degradation, *Pseudomonas putida* was selected as a representative of the toluene-degrading population. From the multispecies biofilm, a number of toluene degraders were isolated. Using a specific rRNA oligonucleotide probe, the quantity and cellular rRNA content of the toluene-degrading *P. putida* in the multispecies biofilm in the filter were monitored. *P. putida* appeared to dissociate from the biofilm during the first three days of growth, but it was later shown to be present at a consistent level of 10% of the active biomass. Based on the rRNA content, it was expected that in situ activity would decline to 20% of cells cultivated at maximum conditions in batch culture. Estimates indicate that *P. putida* was responsible for just 11% of the overall toluene degradation (Pedersen et al., 1997).

## 12.17	EXISTING BIOLOGICAL TREATMENT TECHNOLOGIES AND THEIR INTEGRATION FOR THE ELIMINATION OF EMERGING CONTAMINANTS

The decomposition and elimination of EC residues have been explored using a variety of conventional and cutting-edge biological treatment techniques. It is one of the most used EC treatment methods because of its affordability, accessibility, and environmental friendliness. These procedures include membrane bioreactors (MBR), the activated sludge process, biofilters, and aerobic and anaerobic bioreactors. These biological treatment methods, however, fall short in their ability to effectively remove certain organic micropollutants that are not biodegradable.

These are some of the most popular biodegradation-based techniques for removing ECs. Microorganisms like bacteria, algae, and fungi biomineralize or break down big molecules into inorganic ones with lower molecular weights, like water and CO_2, during the process of biodegradation. Organic substances are often primarily absorbed by bacteria during biodegradation for the growth of their cells. The hazardous properties of ECs, however, stop biodegradation by inhibiting microbial development. In these conditions, the microbe employs a distinct method called co-metabolism, where they need additional growth substrates, including nitrogen and phosphate, as a source of electron acceptor. A number of factors, such as the relevant treatment method, operating settings, the ECs' physicochemical characteristics, and their biological persistence, can affect the removal rate or degradation of the ECs. There are two types of biological treatments: conventional and nonconventional. Most of these divisions are based on the traits of the organic waste that needs to be processed.

The main elements of traditional treatment methods are activated sludge, biological activated carbon, biofilters, and various aerobic/anaerobic/facultative microbiological treatments.

Microalgae/fungal-based therapies have reported successful removal of ECs through the processes of phytoremediation and degradation. These procedures demonstrated effective removal (95–100%) of numerous ECs, including EDCs (Endocrine disrupting compounds) and PPCPs (pharmaceuticals and personal care products), but they performed poorly when it came to removing pesticides. More thorough research should be done on these processes due to their promising potential, and they can be combined with other biological active processes (BAC) to enhance the elimination of pesticides. In BAC, where microorganisms, granular activated carbon (GAC), contaminants, and dissolved oxygen (DO) interact, both biodegradation and adsorption are occurring at the same time. Pesticides, beta blockers, and certain pharmaceuticals were shown to be more easily removed by this method than some EDCs, PCPs, and some pharmaceuticals. In order to remove a variety of ECs more efficiently, this procedure can be used in hybrid system in combination with the biofiltration technology.

The activated sludge process (ASP), which is the most extensively used method for the removal of numerous ECs worldwide, relies mostly on biodegradation carried out by microorganisms in an aeration tank. ASP treatment demonstrated excellent removal efficacy against EDCs (75–100%), surfactants (95–98%), and certain PCPs (78–90%), while pharmaceuticals displayed removal efficiency in the range of 65–100%. However, several beta blockers and insecticides performed poorly because their removal was mostly accomplished through adsorption onto suspended particles rather than biodegradation. Because of this, ASP and biofilters can be used to increase the removal effectiveness of such ECs. A good removal efficiency of ECs was observed in South Wales, UK, when trickling biofilters were studied in combination with ASP for the removal of 55 PPCPs. With or without standard biodegradation, non-conventional treatment methods employ removal mechanisms such as oxidation and sorption. The integration of these processes with biofiltration system can minimize the challenges faced by them.

Overall, both conventional and unconventional biological treatments have been found to be capable and effective for the elimination of numerous ECs, but the majority of the procedures only work for the chosen categories of ECs. These techniques can be integrated to create hybrid systems in order to remove ECs more effectively. The hybrid or integrated system works around the drawbacks of current therapies to remove ECs while overcoming their challenges and limitations. Advanced oxidation processes coupled with the biological filtration method is one of the effective hybrid systems being utilized currently (Sivaranjanee & Kumar, 2021).

The removal of the ECs was achieved through biofiltration using biological activated carbon (BAC) along with an ozonation process. The ozonation treatment causes the wastewater to produce hydroxyl radicals, which aid in the pollutants' breakdown. The majority of the examined ECs were successfully eliminated (90–98%) by the combined ozonation + BAC filtration method. Overall, the hybrid system, which uses ozonation and BAC filtration to remove ECs, has a promising future (Dhangar & Kumar, 2020).

12.18   LIMITATIONS OF THE EXISTING METHOD

The limitations associated with the biofiltration technique with respect to emerging pollutants are as follows.

The efficacy of biofilters is greatly influenced by the microbial community and its growth. When examining how microbial activity affects the breakdown of pollutants, we should consider the possibility of scaling up the process. After being completed at the pilot level in a laboratory setting, many studies are not translated into actual organic waste treatment applications.

The mechanisms of the metabolic pathways used by microorganisms to break down contaminants are not well understood. Monitoring the sights of bacterial metabolic pathways is crucial for the complete elimination of contaminants. Additionally, it depends on the genus, species, and community of the microbes that are present in the biofilter bed. Therefore, in order to effectively implement biofiltration technology, it is essential to gain a thorough understanding of the principles underlying microbial growth and degradation.

A thick biofilm layer starts to develop as a result of an increase in the concentration of pollutants. The bacteria that live on the surface of the biofilm carry out adsorption and thereafter it penetrates into the biofilm's depth. After a while, the contaminant's ability to diffuse and reach the deep of the biofilm is lost, resulting in the development of an inactive microbial biofilm with various pore diameters. Multi-microbial cultures must be used in this situation to improve the elimination of different contaminants with faster degradation.

The contaminants come from several sources, including oil refineries, municipal sewage systems, water treatment facilities, industrial wastewater treatment plants, and pharmaceutical waste disposal facilities. The features and quantity of the contaminants present in each source must be assessed in order to develop and optimize suitable biofiltration equipment and methods. The secure disposal, recycling, or manufacturing of valuable items should also take into account the treated wastes and acquired metals (Pachaiappan et al., 2022).

12.19   CHALLENGES AND FUTURE PROSPECTS

In the past two decades, biofiltration has grown in significance as a tool for reducing odorous emissions and other developing contaminants from composting operations, wastewater treatment facilities, confined animal facilities, the food industry, and additional sources for organic waste. The future of biofiltration depends on the industry and is subject to regulatory requirements. However, there are particular trends that may have an impact on the market for biofiltration technology and they are considered challenging with respect to the said technology.

Regulatory concern about nitrogen oxide emissions, which are produced during thermal treatment, has increased. Biofilters do not increase nitrogen oxide production. More people are complaining about the odours coming from publicly operated wastewater treatment facilities plants, manufacturing businesses, facilities for treating solid waste, etc. Utilizing pollution control measures has boosted the usage of biodegradable solvents, decreased air emission levels, and raised emphasis on achieving zero discharge operations. There has also been an increased focus on low-cost, environmentally friendly treatment options, worker exposure to organics, and air pollution emissions. Due to the ineffectiveness of present conventional technologies in treating new pollutants, a large number of them are entering our ecosystem. At present, there isn't a complete, integrated solution to handle these challenges.

It appears that there is no approach that is extremely effective for eliminating every EC under normal operating settings, leading to the conclusion that ideal treatment procedures should be used. It is now possible to measure the emerging contaminants due to the explosion of analytical tools. In our complex environment, new technologies should be used to identify new ECs and learn about their effective expulsion strategies. The development of hybrid frameworks for the degradation and removal of these foreign chemicals from organic wastes should be the focus of further study.

Future studies should focus heavily on treatment procedures, where these foreign substances end up in the sewage biomass, and their potential to transform during treatment into more dangerous or pharmacologically active metabolites. Using a combination of treatment methods, as opposed to a single procedure or conventional methods, is necessary for the expulsion of ECs. This should include integration of recently established knowledge systems, such as genetic engineering and nanoscale science, for the degradation of ECs, as well as in-depth analyses of the reaction kinetics, reactor designs, effects of operating parameters, and mechanisms underpinning the degradative processes of ECs. Future analysts should focus on creating risk-based screening models and algorithms to predict the sources, fates, and reactions of ECs in the aquatic environment.

By overcoming the aforementioned issues, it is possible to anticipate the development of technically superior biofilters with lower initial investment in the future. Machine intelligence paves a much broader path for the same. Artificial intelligence (AI) has demonstrated its viability in several fields today, including waste treatment, where it could forecast how different adsorbents (microbes) would behave in relation to various types and quantities of pollutants from diverse sources. Additionally, it would be anticipated that impurities will be removed simultaneously without causing any secondary pollutants or fouling effects and with value-added products. According to recent studies, hybridization techniques for the removal of contaminants from waste sources can be used to achieve the necessary biological-based filtration. Therefore, it will be possible to implement the best AI-controlled biobased water treatment method in the forthcoming days.

REFERENCES

Ashraf, M. A., Batool, S., Ahmad, M., Sarfraz, M., and Noor, W. S. A. W. M. 2016. Biopolymers as biofilters and biobarriers. *In Biopolymers and Biotech Admixtures for Eco-Efficient Construction Materials* 387–420. Woodhead Publishing.

Bunke, D., Brack, W., Moritz, S., López Herráez, D., and Posthuma, L. 2017. Developments in society and implications for emerging pollutants in the aquatic environment. *Oeko-Institut Working Paper* 1.

Cheng, Y., He, H., Yang, C., Zeng, G., Li, X., Chen, H., and Yu, G. 2016. Challenges and solutions for biofiltration of hydrophobic volatile organic compounds. *Biotechnology Advances* 34, 6: 1091–1102.

Christen, P., Domenech, F., Michelena, G., Auria, R., and Revah, S. 2002. Biofiltration of volatile ethanol using sugar cane bagasse inoculated with Candida utilis. *Journal of Hazardous Materials* 89: 253–265.

Cohen, Y. 2001. Biofiltration–the treatment of fluids by microorganisms immobilized into the filter bedding material: A review. *Bioresource Technology* 77:257–274.

Conserva, S., Tatti, F., Torretta, V., Ferronato, N., and Viotti, P. 2019. An integrated approach to the biological reactor–sedimentation tank system. *Resources* 8:94.

Detchanamurthy, S., and Gostomski, P. A. 2012. Biofiltration for treating VOCs: an overview. *Reviews in Environmental Science and BioTechnology* 113: 231–241.

Devinny, J. S., Deshusses, M. A., and Webster, T. S. 2017. *Biofiltration for Air Pollution Control*. CRC press.

Dhaka, S., Kumar, R., Deep, A., Kurade, M. B., Ji, S. W., and Jeon, B. H. 2019. Metal–organic frameworks (MOFs) for the removal of emerging contaminants from aquatic environments. *Coordination Chemistry Reviews* 380: 330–352.

Dhangar, K., and Kumar, M. 2020. Tricks and tracks in removal of emerging contaminants from the wastewater through hybrid treatment systems: a review. *Science of the Total Environment* 738: 140320.

Fast, S. A., Gude, V. G., Truax, D. D., Martin, J., and Magbanua, B. S. 2017. A critical evaluation of advanced oxidation processes for emerging contaminants removal. *Environmental Processes 4*, no.1: 283–302.

Forero, D., Pena, C., Acevedo, P., Hernandez, M., & Cabeza, I. 2018. Biofiltration of acetic acid vapours using filtering bed compost from poultry manure-pruning residues-rice husks. *Chemical Engineering Transactions* 64: 511–516

Fulazzaky, M. A., Talaiekhozani, A., Ponraj, M., Abd Majid, M. Z., Hadibarata, T., and Goli, A. 2014. Biofiltration process as an ideal approach to remove pollutants from polluted air. *Desalination and Water Treatment* 52: 3600–3615.

Gagnon, C., Lajeunesse, A., Cejka, P., Gagne, F., and Hausler, R. 2008. Degradation of selected acidic and neutral pharmaceutical products in a primary-treated wastewater by disinfection processes. *Ozone: Science and Engineering* 30: 387–392.

Ghasemi, R., Golbabaei, F., Rezaei, S., Pourmand, M. R., Nabizadeh, R., and Jafari, M. J. 2020. A comparison of biofiltration performance based on bacteria and fungi for treating toluene vapors from airflow. *AMB Express* 10: 1–9.

Hassard, F., Biddle, J., Cartmell, E., Jefferson, B., Tyrrel, S., and Stephenson, T. 2015. Rotating biological contactors for wastewater treatment–a review. *Process Safety and Environmental Protection* 94: 285–306.

Iorhemen, O. T., Hamza, R. A., and Tay, J. H. 2016. Membrane bioreactor (MBR) technology for wastewater treatment and reclamation: membrane fouling. *Membranes* 6: 33.

Ismail, W. N. W., and Mokhtar, S. U. 2021. Various Methods for Removal, Treatment, and Detection of Emerging Water Contaminants. *Emerging Contaminants* 27. IntechOpen; 2021. Available from: http://dx.doi.org/10.5772/intechopen.93375

Kumar, V., Agrawal, S., Bhat, S.A., Américo-Pinheiro, J.H.P., Shahi, S.K., Kumar, S. 2022. Environmental impact, health hazards, and plant-microbes synergism in remediation of emerging contaminants. *Cleaner Chemical Engineering* 2, 100030. https://doi.org/10.1016/j.clce.2022.100030

Kyzas, G. Z., and Kostoglou, M. 2014. Green adsorbents for wastewaters: a critical review. *Materials* 7: 333–364.

Lei, M., Zhang, L., Lei, J., Zong, L., Li, J., Wu, Z., and Wang, Z. 2015. Overview of emerging contaminants and associated human health effects. *BioMed Research International* 404796. doi: 10.1155/2015/404796. Epub 2015 Dec 2. PMID: 26713315; PMCID: PMC4680045.

Li, X., Shi, H., Li, K., and Zhang, L. 2015. Combined process of biofiltration and ozone oxidation as an advanced treatment process for wastewater reuse. *Frontiers of Environmental Science & Engineering* 9:1076–1083.

Lima, E. C. 2018. Removal of emerging contaminants from the environment by adsorption. *Ecotoxicology and Environmental Safety 150*:1–17.

Malakar, S., Saha, P. D., Baskaran, D., and Rajamanickam, R. 2017. Comparative study of biofiltration process for treatment of VOCs emission from petroleum refinery wastewater – A review. *Environmental Technology & Innovation* 8: 441–461.

Marcoux, M. A., Matias, M., Olivier, F., and Keck, G. 2013. Review and prospect of emerging contaminants in waste–Key issues and challenges linked to their presence in waste treatment schemes: General aspects and focus on nanoparticles. *Waste Management* 33: 2147–2156.

Mishra, V. K., and Kumar, A. (Eds.). 2021. *Sustainable Environmental Clean-up: Green Remediation*. Elsevier.

Naidu, R., Espana, V. A. A., Liu, Y., and Jit, J. 2016. Emerging contaminants in the environment: risk-based analysis for better management. *Chemosphere* 154:350–357.

Pachaiappan, R., Cornejo-Ponce, L., Rajendran, R., Manavalan, K., Femilaa Rajan, V., and Awad, F. 2022. A review on biofiltration techniques: recent advancements in the removal of volatile organic compounds and heavy metals in the treatment of polluted water. *Bioengineered* 13: 8432–8477.

Pedersen, A. R., Møller, S., Molin, S., and Arvin, E. 1997. Activity of toluene-degrading Pseudomonas putida in the early growth phase of a biofilm for waste gas treatment. *Biotechnology and Bioengineering* 54:131–141.

Raju, S. S. 2008. *Biofiltration-A Versatile Technique in Pollution Control*. Conference paper for "National Level Conference On Novel Products and Processes In Chemical Engineering" at St. Joseph's College of Engineering, Chennai, India.

Rene, E. R., Veiga, M. C., and Kennes, C. 2013. *Biofilters. Air Pollution prevention and control: bioreactors and bioenergy: 57–119.*

Rodriguez-Narvaez, O. M., Peralta-Hernandez, J. M., Goonetilleke, A., & Bandala, E. R. 2017. Treatment technologies for emerging contaminants in water: A review. *Chemical Engineering Journal 323*: 361–380.

Sarma, H., Dominguez, D. C., and Lee, W. Y. 2022. *Emerging Contaminants in the Environment: Challenges and Sustainable Practices*. Elsevier..

Setiadi, T., Harimawan, A., Sumampouw, G. A., and Indarto, A. 2022. The mixture of agricultural pesticides and their impact on populations: bioremediation strategies. *In Emerging Contaminants in the Environment*, Elsevier:511–546,

Sharma, S. K. (Ed.). 2014. *Heavy metals in water: Presence, removal and safety*. Royal Society of Chemistry.

Sheoran, K., Siwal, S. S., Kapoor, D., Singh, N., Saini, A. K., Alsanie, W. F., and Thakur, V. K. 2022. *Air Pollutants Removal Using Biofiltration Technique: A Challenge at the Frontiers of Sustainable Environment. ACS Engineering Au.*

Sivaranjanee, R., and Kumar, P. S. 2021. A review on remedial measures for effective separation of emerging contaminants from wastewater. *Environmental Technology & Innovation* 23:101741.

Swanson, W. J., and Loehr, R. C. 1997. Biofiltration: fundamentals, design and operations principles, and applications. *Journal of Environmental Engineering* 123:538–546

Thomaidis, N. S., Asimakopoulos, A. G., and Bletsou, A. A. 2012. Emerging contaminants: a tutorial mini-review. *Global NEST Journal* 14: 72–79.

Torretta, V., Raboni, M., Copelli, S., and Caruson, P. 2015. Effectiveness of a multi-stage biofilter approach at pilot scale to remove and VOCs. *International Journal of Sustainable Development and Planning* 10: 373–384.

Van Langenhove, H., Van Elst, T., and Verstraete, W. 1997. Enhancement of ethene removal from waste gas by stimulating nitrification. *Biodegradation* 8: 21–30.

Zhang, Y., Liss, S. N., and Allen, D. G. 2007. Enhancing and modeling the biofiltration of dimethyl sulfide under dynamic methanol addition. *Chemical Engineering Science* 62: 2474–2481

13 Microbe-assisted Enzymatic Degradation of Microplastic

Sougata Ghosh, Bishwarup Sarkar, and Sirikanjana Thongmee

13.1 INTRODUCTION

Microplastics are small plastic particles that have a size range of 1 μm t 5 mm and have various primary and secondary types, morphologies, and compositions (Miri et al., 2022). Microplastics are one of the most commonly used synthetic polymers because of their unique properties such as durability, cost-effectiveness, and light weight (Yuan et al., 2020). However, these microplastics are known as harmful pollutants that have widely contaminated atmospheric, terrestrial, and aquatic environments (Bhat et al., 2022). It has been proposed that global plastic pollutants will reach 13.2% by weight in the next 40 years (Liu et al., 2022). These microplastics are also very toxic to organisms as they inhibit the growth of plants and animals through the destruction of the intestinal flora and tissues. Various plastic degradation products result in the formation of certain toxic amines and phthalate esters that are carcinogenic and mutagenic (Zhang et al., 2021).

These heterogeneously mixed plastics are usually present in the form of fibers, granules, and fragments that contaminate every trophic level (Guo et al. 2020). Hence, this chapter emphasizes the importance of biodegradation of these microplastic pollutants that are found in soil and aquatic environments. Various bacterial and fungal species are reported to efficiently act as biodegradation tools that are summarized in Table 13.1. Bacterial species such as *Bacillus* sp., *Clostridium thermocellum, Enterobacter asburiae, Exiguobacterium* sp., *Lysinibacillus* sp., and *Rhodococcus ruber* exhibit considerable biodegradation of microplastics such as polyethylene, polypropylene, polystyrene, and polyethylene terephthalate that are isolated from various plastic contaminated sites (Dey et al., 2023). Likewise, certain fungal species such as *Fusarium solani, Penicillium chrysogenum, Purpureocillium lilacinum, Rhizopus oryzae, Scedosporium aurantiacum,* and *Trichoderma brevicompactum* were also studied for polyethylene biodegradation. Hence, further evaluation of the in situ biodegradation ability of these microbial isolates would be useful in highlighting the importance of the biodegradation of plastic contaminants in the environment.

13.2 MICROPLASTICS DEGRADED BY MICROBIAL ENZYMES

13.2.1 POLYETHYLENE

Yang et al. (2014) reported the bacteria-mediated biodegradation of polyethylene (PE) that was isolated from the guts of plastic-eating waxworms. Two bacterial strains, namely *Enterobacter asburiae* YT1 and *Bacillus* sp. YP1, were isolated from the larvae of *Plodia interpunctella* that were capable of eating PE films. A carbon-free basal agar medium (CFBAM) was used for the cultivation of bacterial isolates, which was supplemented with a PE film sheet. The tensile strength of these PE film sheets was reduced by 50% after culturing the two bacterial strains as compared to the control sheets. Moreover, the small pieces of PE film became suspended in liquid carbon-free basal medium

DOI: 10.1201/9781003408352-15

TABLE 13.1
Microplastic Degradation by Bacteria and Fungi

Microbe	Source of microbial strain	Microplastic degraded	Degradation efficiency	Reference
Enterobacter asburiae YT1 and *Bacillus* sp. YP1	larvae of *Plodia interpunctella*	polyethylene	6.1 ± 0.3 and 10.7 ± 0.2	Yang et al., 2014
Rhodococcus ruber	soil contaminated with PE waste	polyethylene	-	Gilan et al., 2004
Cunnighamella echinulata, Pestalotiopsis sp., *Hypoxylon anthochroum, Paecilomyces lilacinus, Aspergillus* sp., and *Lasiodiplodia theobromae*	leaf and stem of *Psychotria flavida* Talbot plant	polyethylene	4.0	Sheik et al., 2015
Fusarium solani, Fusarium oxysporum, Penicillium chrysogenum, Purpureocillium lilacinum, Rhizopus oryzae, Scedosporium aurantiacum, and *Trichoderma brevicompactum*	waste disposal site	polyethylene	-	Spina et al., 2021
Bacillus thuringiensis strain C15, *Pseudomonas* sp. strain B10, *Pseudomonas* sp. strain SWI36, *Bacillus albus* strain PFYN01, and *Pseudomonas* sp. strain SWI36	polluted soil sample	polyethylene terephthalate	-	Roberts et al., 2020
Clostridium thermocellum expressing LCC protein	-	polyethylene terephthalate	62.0	Yan et al., 2021
Penicillium funiculosum	waste dump site	polyethylene terephthalate	2.95	Nowak et al., 2011
Cephalosporium NCIM 1251	-	polystyrene	20.62 ± 1.47	Chaudhary et al., 2021
Exiguobacterium sp. DR11 and DR14	wetland soil samples	polystyrene	8.0 and 8.8	Chauhan et al., 2018
Pseudomonas and *Bacillus* sp.	soil samples near a plastic dump yard	polystyrene	-	Mohan et al., 2016
Rhodococcus ruber	soil samples contaminated with PE waste from agricultural use	polystyrene	0.5	Mor and Sivan, 2008
Bacillus cereus and *Sporosarcina globispora*	sediment samples from mangroves	polypropylene	12.0 and 11.0	Helen et al., 2017
Bacillus sp. and *Rhodococcus* sp.	soil samples from mangroves	polypropylene	6.4 and 4.0	Auta et al., 2018
Lysinibacillus sp. JJY0216	soil samples treated with kenaf-plastic complex	polypropylene	4.3	Jeon et al., 2021
Stenotrophomonas panacihumi PA3-2	soil sample from municipal waste	polypropylene	20.3 ± 1.39 and 16.6 ± 1.70	Jeon and Kim, 2016

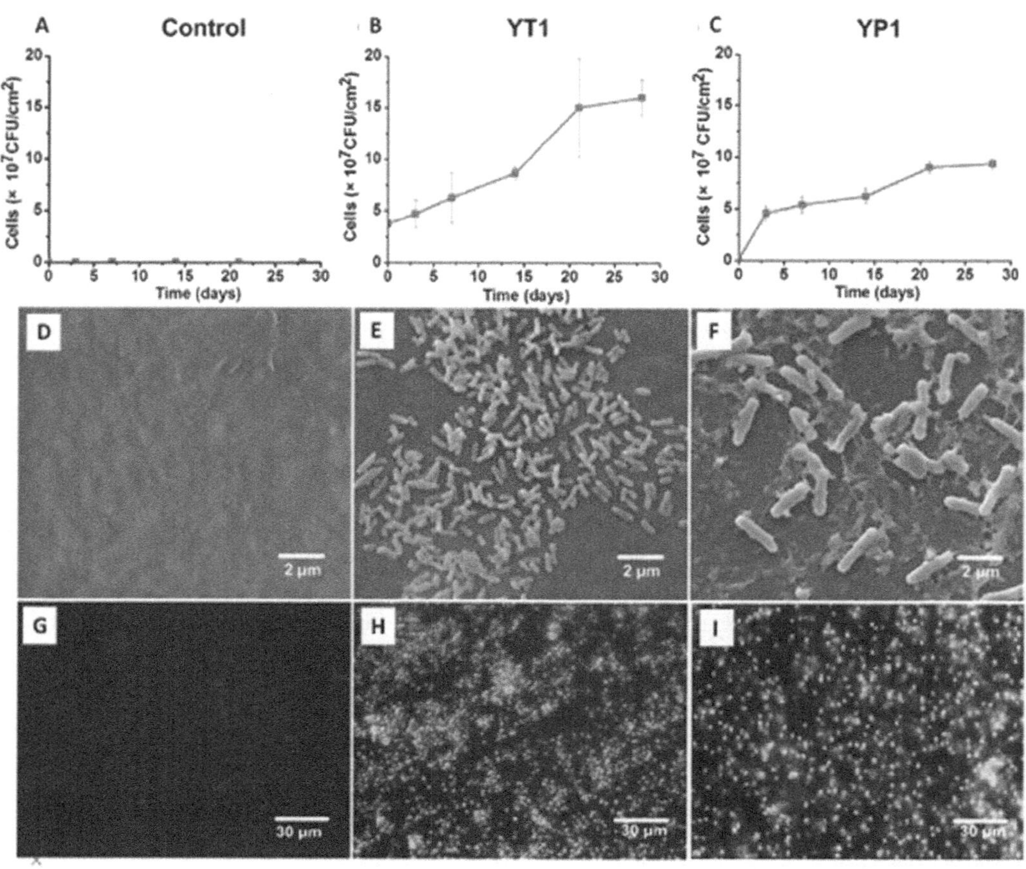

FIGURE 13.1 Biofilm formation and viability.

Source: Reprinted with permission from Yang, J., Yang, Y., Wu, W.M., Zhao, J., Jiang, L., 2014. Evidence of polyethylene biodegradation by bacterial strains from the guts of plastic-eating waxworms. Environ. Sci. Technol. 48(23), 13776–13784.Copyright © 2014 American Chemical Society.

Notes: (A-C) Changes in bacterial cell number on the PE film sheets; (D-F) Morphotypes of the cells in the mature biofilm on the PE sheet; (G-I) Fluorescent microscopic images of biofilms, which show cell viability after the 28-day incubation. Live cells are green and dead cells are red. The PE sheets were covered in CFBAM.

(LCFBM) after 28 days of incubation with the bacteria, which was attributed to the attachment of bacterial strains of the PE species that may have resulted in the formation of a suspension. The biofilm formation ability on the PE films was also evaluated wherein both the strains were able to form detectable biofilms within the first 3 h of incubation. The cell density of the YP1 and YT1 strains was 1.4×10^6 and 3.8×10^7 colony forming units (CFU)/cm^2, respectively. In addition, the biofilm densities were also increased after 28 days of incubation with cell densities of 16.0×10^7 and 9.4×10^7 CFU/cm^2 for YT1 and YP1 strains, respectively. LIVE/DEAD *Bac*Light bacterial viability staining further demonstrated the dominance of live cells in the biofilms of the two strains. In addition, scanning electron microscope (SEM) and atomic force microscope (AFM) images revealed the formation of pits and cavities on the surfaces of the PE films after 28 days of incubation with the bacteria as seen in Figure 13.1 and Figure 13.2. The degradation efficiencies after 60 days of incubation with YP1 and YT1 strains was $10.7 \pm 0.2\%$ and $6.1 \pm 0.3\%$, respectively.

In another study, Gilan et al. (2004) reported the biodegradation of PE by *Rhodococcus ruber*. The bacterial strain was isolated from soil samples contaminated with PE waste and then enriched

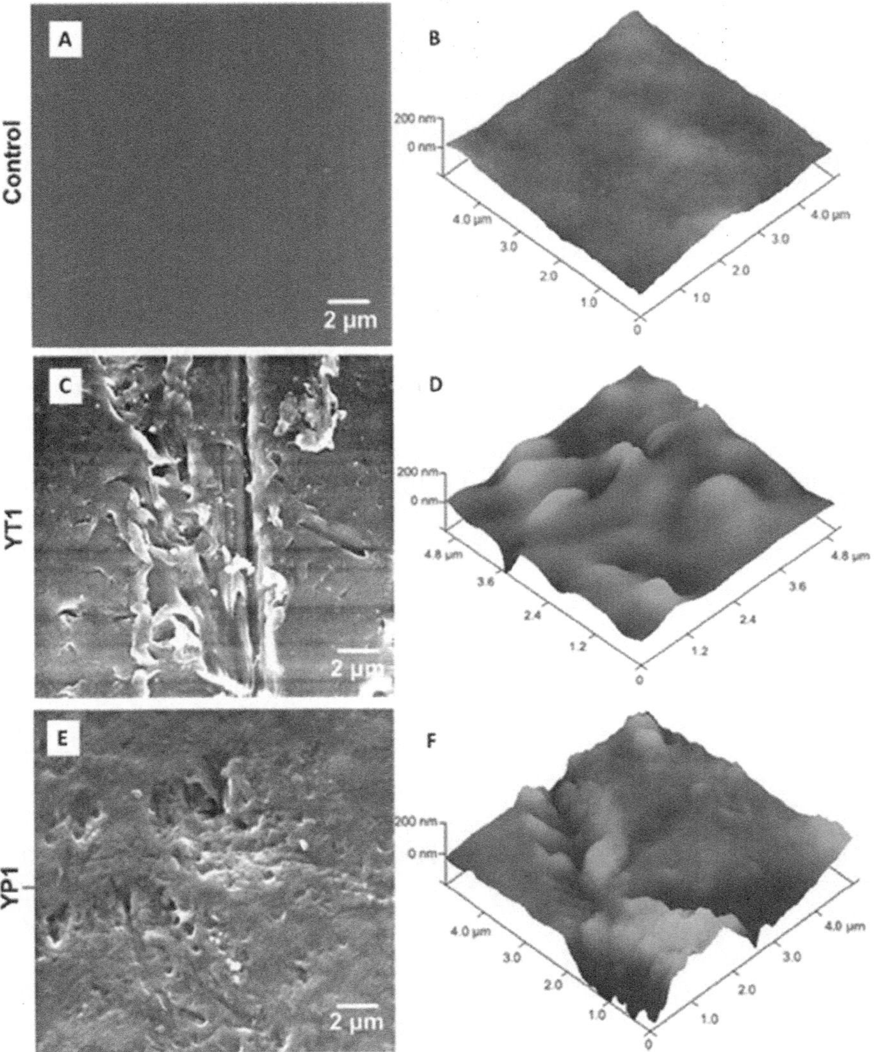

FIGURE 13.2   SEM and AFM observations of the physical surface topography of the sterile control versus the PE sheets incubated with strains YT1 and YP1 after 28 days.

Source: Reprinted with permission from Yang, J., Yang, Y., Wu, W.M., Zhao, J., Jiang, L., 2014. Evidence of polyethylene biodegradation by bacterial strains from the guts of plastic-eating waxworms. Environ. Sci. Technol. 48(23), 13776–13784. Copyright © 2014 American Chemical Society.

in a minimal synthetic medium (SM) supplemented with PE as the sole carbon source. Moreover, a two-step enrichment method was carried out wherein the initial cultivation was carried out in soil followed by further enrichment in liquid SM. The isolated *R. ruber* strain successfully reduced 8.0% of initial PE dry weight after 30 days of incubation. In addition, the bacterial adhesion to hydrocarbon (BATH) and salt aggregation test (SAT) was carried out to evaluate the bacterial strain's hydrophobicity. A higher hydrophobicity of the *R. ruber* strain was observed as compared to the other isolates obtained in this study. Thereafter, the addition of mineral oils demonstrated improvement of the colonization of *R. ruber* on PE whereas non-ionic surfactants such as Tween 60 and Tween 80 showed a negatory effect on biofilm formation. SEM images further confirmed an almost 50% increase in biofilm colonization of PE after the addition of 0.05% of mineral oil. Additionally, fluorescein diacetate (FDA) hydrolysis assay and analysis of the protein content of the bacterial strain showed rapid utilization of the mineral oil by the bacteria for PE adhesion.

In another study, low-density PE (LDPE) was biodegraded by Sheik et al. (2015) using endophytic fungi. The fungal strains were isolated from the leaf and stem samples of a *Psychotria flavida* Talbot plant. Six different fungal isolates, namely *Cunnighamella echinulata*, *Pestalotiopsis* sp, *Hypoxylon anthochroum*, *Paecilomyces lilacinus*, *Aspergillus* sp, and *Lasiodiplodia theobromae*, were isolated on the basis of laccase enzyme production and formation of biofilm on the plastic films after one month of incubation. *L. theobromae* produced 10.70 ± 1.53 U/mL, which was the highest among all of the six isolates. The laccase enzyme was proposed to mediate LDPE degradation through the coupled-oxidation process using its conserved copper binding sites. Moreover, *Aspergillus* sp, and *L. theobromae* showed a loss of weight in LDPE films after 90 days of incubation which further increased with increasing doses of gamma irradiation. The SEM images of the LDPE films exposed to fungi revealed the formation of fungal colonies and spores after 1000 kGy radiation wherein the surface of the films was brittle as seen in Figure 13.3. There was a significant reduction in the intrinsic viscosity and average molecular weight in the LDPE films after fungal inoculation and gamma irradiation with approximately 4% weight loss achieved by *Aspergillus* sp. after 1000 kGy irradiation.

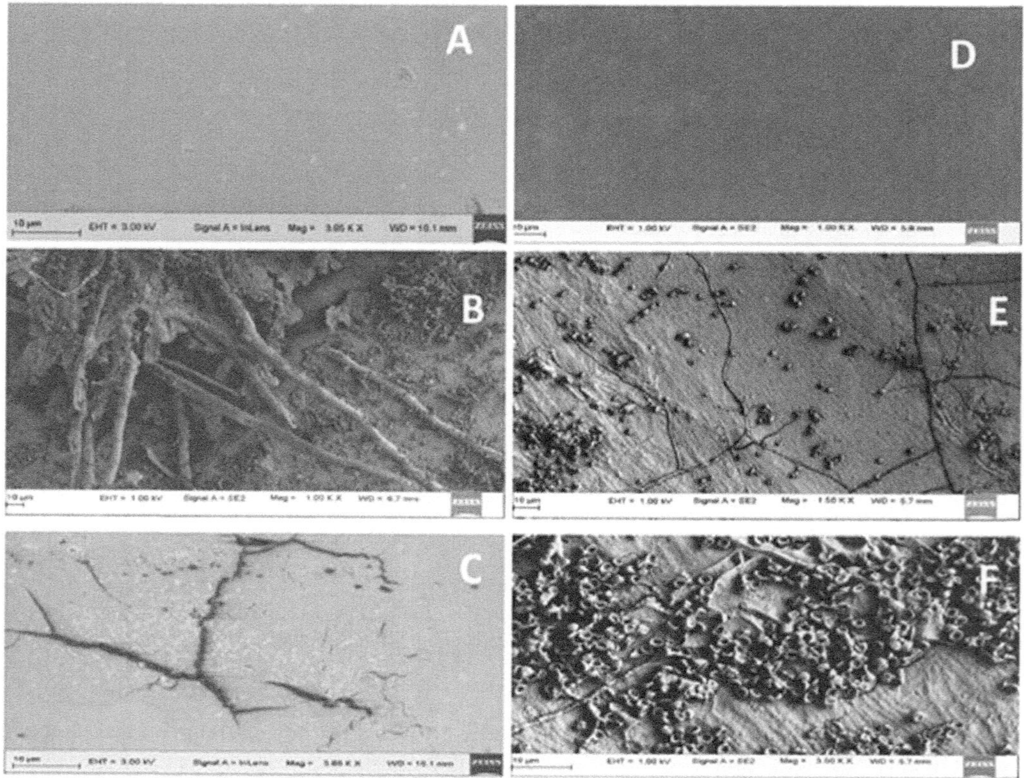

FIGURE 13.3   A. PP Gamma irradiated (100 kGy) control; B. Fungal growth forming biofilm on gamma irradiated PP (100 kGy); C. Fissures formed on gamma irradiated (100 kGy) PP after incubation with fungus; D. LDPE gamma irradiated (1000 kGy) control; E. *Mycelia* attached on gamma irradiated (1000 kGy) LDPE forming biofilm; F. Gamma irradiated (1000 kGy) LDPE with massive conidia of fungi.

Source: Reprinted with permission from Sheik, S., Chandrashekar, K.R., Swaroop, K., Somashekarappa, H.M., 2015. Biodegradation of gamma irradiated low density polyethylene and polypropylene by endophytic fungi. Int. Biodeterior. Biodegr. 105, 21–29. Copyright © 2015 Elsevier Ltd.

Various fungal strains are also reported as efficient tools for PE degradation. In a recent study, Spina et al. (2021) reported the fungi-mediated degradation of low-density PE. The fungal strain was isolated from a waste disposal site contaminated with several persistent plastic materials and cultivated in malt extract agar medium. Thereafter, a solid screening was performed of the fungal isolates wherein 97% of the isolates were able to utilize either 5 g/L or 10 g/L of PE as the primary source of carbon. Moreover, two of the fungal isolates, namely *Fusarium solani* and *Fusarium oxysporum*, showed better growth in the presence of 10 g/L of PE. Other fungal isolates such as *Penicillium chrysogenum*, *Purpureocillium lilacinum*, *Rhizopus oryzae*, *Scedosporium aurantiacum*, and *Trichoderma brevicompactum* also demonstrated comparable growth in the presence of 5 and 10 g/L of PE. Furthermore, the PE degradation activity of the fungal isolates was confirmed by the respirometry assay wherein a considerable amount of CO_2 was produced by the fungal strains after 10 days of incubation with PE, which subsequently resulted in the acidification of the medium and indicated the transformation of the microplastic. SEM images also showed the modification of PE surfaces cause by fungi as compared to the control samples as evident from

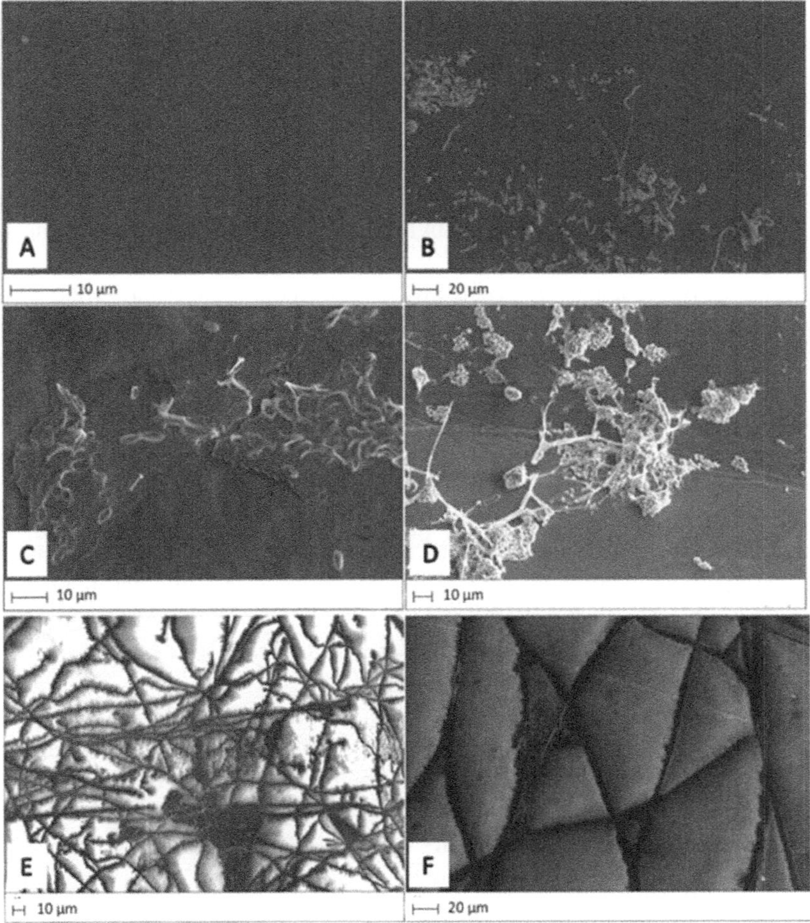

FIGURE 13.4 SEM images at high magnification for PE films subjected to fungal activity: A. control film; B. *P. chrysogenum*; C. *F. oxysporum*; D. *T. brevicompactum*; E. *P. lilacinum*; F. *F. falciforme*.

Source: Reprinted with permission from Spina, F., Tummino, M.L., Poli, A., Prigione, V., Ilieva, V., Cocconcelli, P., Puglisi, E., Bracco, P., Zanetti, M., Varese, G.C., 2021. Low density polyethylene degradation by filamentous fungi. Environ. Pollut. 274, 116548. Copyright © 2021 Elsevier Ltd.

Note: The differences in the magnification allow the specific superficial features to be observed.

Figure 13.4. Several cavities, grooves, and flaky structures were observed on the surface of the PE films. Moreover, *P. chrysogenum* displayed scanty and localized growth on the PE film whereas *F. oxysporum* formed bumps and partial exfoliations and *T. brevicompactum* demonstrated widespread colonization.

13.2.2 POLYETHYLENE TEREPHTHALATE

Roberts et al. (2020) recently reported the synergistic polyethylene terephthalate (PET) degradation by a consortium of *Pseudomonas* and *Bacillus* sp. that were isolated from polluted soil samples and cultivated in rhodamine B medium supplemented with olive oil as a substrate for confirming lipase activity. Later, general hydrolytic and polyesterase activities of the isolates were also evaluated using tributyrin and polycaprolactone diol (PCD) as the substrates, respectively. Thereafter, draft genomic sequences of the consortia strains were analyzed which revealed that the five strains were *Bacillus thuringiensis* strain C15, *Pseudomonas* sp. strain B10, *Pseudomonas* sp. strain SWI36, *Bacillus albus* strain PFYN01, and *Pseudomonas* sp. strain SWI36 that were grown synergistically. Moreover, the doubling time of the consortia was shorter along with a greater carrying capacity on PET samples treated with ultraviolet (UV) rays. The CFU per gram of PET for the consortia was higher than the CFUs of individual strains on both granular and postconsumer PET samples. SEM images of the PET surface treated with the consortia showed physical alteration of the material that was attributed to the utilization of PET as the carbon source by the bacterial strains. Furthermore, bis(2-hydroxyethyl) terephthalate (BHET) was observed as the by-product of PET utilization by the bacterial consortia which were further proposed to be converted into terephthalic acid (TPA) and ethylene glycol by the hydrolytic enzymes secreted by the consortia.

PET was also biodegraded by a genetically engineered thermophilic strain of *Clostridium thermocellum* by Yan et al. (2021). The bacterial strain was used for the heterologous expression of a thermophilic cutinase (LCC) enzyme, which is reported to exhibit efficient PET degradation activity at a high temperature of 70 °C as seen in Figure 13.5. Thereafter, both the parent and recombinant strains were cultivated in GS-2 medium, after which a Fast-Red dye was used for visualizing the LCC protein. The supernatant of the recombinant strain demonstrated the presence of a 28 kDa violet protein band, which suggested successful extracellular secretion of the heterologous protein. Later, 50 mg of PET films were supplemented in the culture medium wherein no change in the growth curve was observed after 32 h of incubation. Almost 62% of 50 mg of PET films were degraded after 14 days of incubation at 60 °C. High-performance liquid chromatography (HPLC) results then detected the presence of both terephthalate (TPA) and mono-(2-hydroxyethyl) terephthalate (MHET) in the culture supernatant. In addition, the esterolytic activity of the supernatant also increased to 1.01 U/mL after 1 day of incubation.

In another study, Nowak et al. (2011) reported fungus-mediated hydrolysis of PET films modified with Bionolle® polyester. A filamentous strain of *Penicillium funiculosum* was used in this study that was isolated from a waste dump. The fungal strain was cultivated in Czapek-Dox medium at 30 °C and 90% humidity for 2 weeks. Thereafter, the grown fungal cultures were centrifuged at 4000 rpm for the separation of the spores from the hyphae. The incubation of the fungal spores and the extracellular supernatant containing hydrolytic enzymes with PET films were then studied. The addition of PET films with the hydrolytic enzymes resulted in a 2.95% weight loss of the films after 84 days of incubation. The incubation of Bionolle® polyester with the fungal strain resulted in a 30-fold increase in PET weight loss as compared to the fungal enzymatic treatment. The fungal spores demonstrated scattered growth on the surface of the PET films, which subsequently resulted in the formation of small holes, whereas the addition of Bionolle® polyester resulted in the formation of multi-layered *P. funiculosum* hyphae.

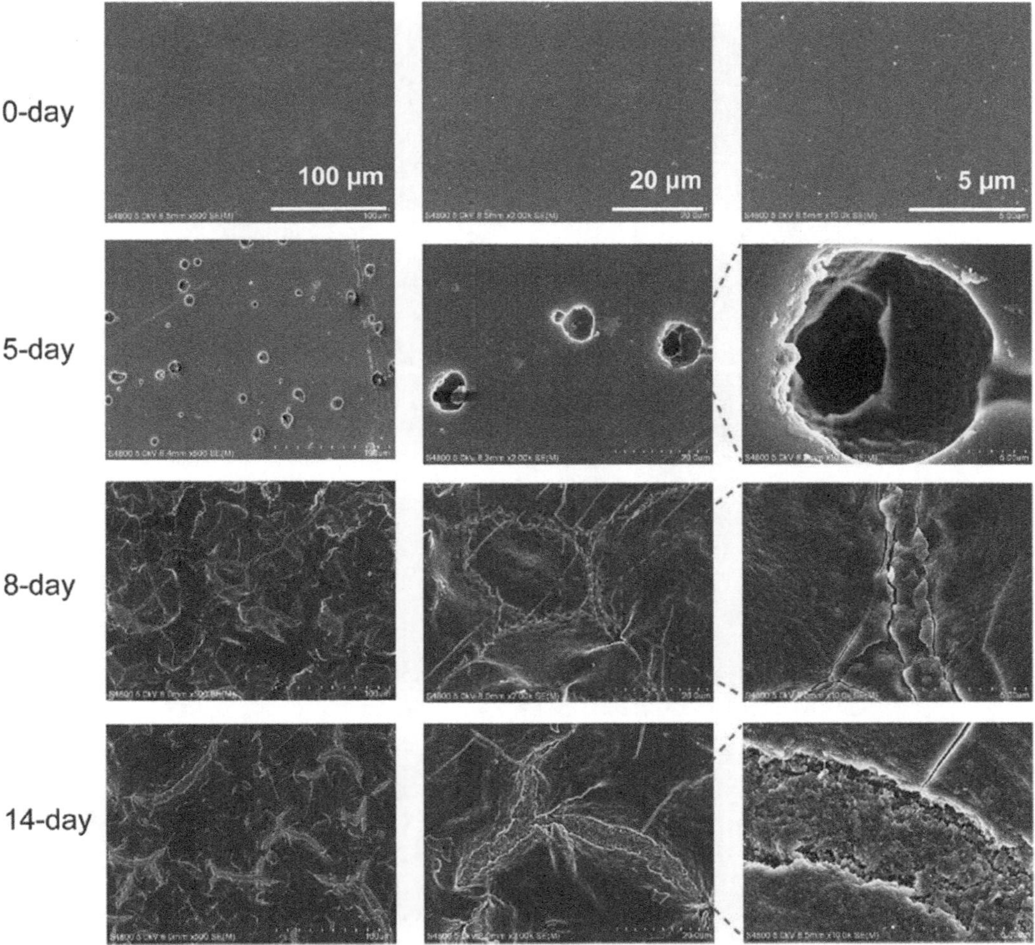

FIGURE 13.5 SEM analysis of PET films during biodegradation using *C. thermocellum* whole-cell biocatalyst at 60 °C.

Source: Reprinted with permission from Yan, F., Wei, R., Cui, Q., Bornscheuer, U.T., Liu, Y.J., 2021. Thermophilic whole-cell degradation of polyethylene terephthalate using engineered *Clostridium thermocellum*. Microb. Biotechnol. 14(2), 374–385. Copyright © 2020 The Authors. Microbial Biotechnology published by John Wiley & Sons Ltd and Society for Applied Microbiology.

Notes: The PET films were washed and dried by lyophilization before analysis. Cavities and holes appeared as an initial sign for the surface erosion. The number of holes increased along with the incubation time. From day 8, cracks and crevices appeared and became larger as a result of extensive enzymatic hydrolysis.

13.2.3 POLYSTYRENE

In another recent study, Chaudhary et al. (2021) showed the synergistic effect of UV rays on the biodegradation of polystyrene by *Cephalosporium* NCIM 1251 strain. The polystyrene films were irradiated with UV rays at 15 W and 50 Hz for a week followed by a 69% nitric acid treatment, after which the films were incubated with the bacterial isolate in a laboratory-made mineral salt medium. After 8 weeks of incubation, a 20.62 ± 1.47% weight loss in the pre-treated films was

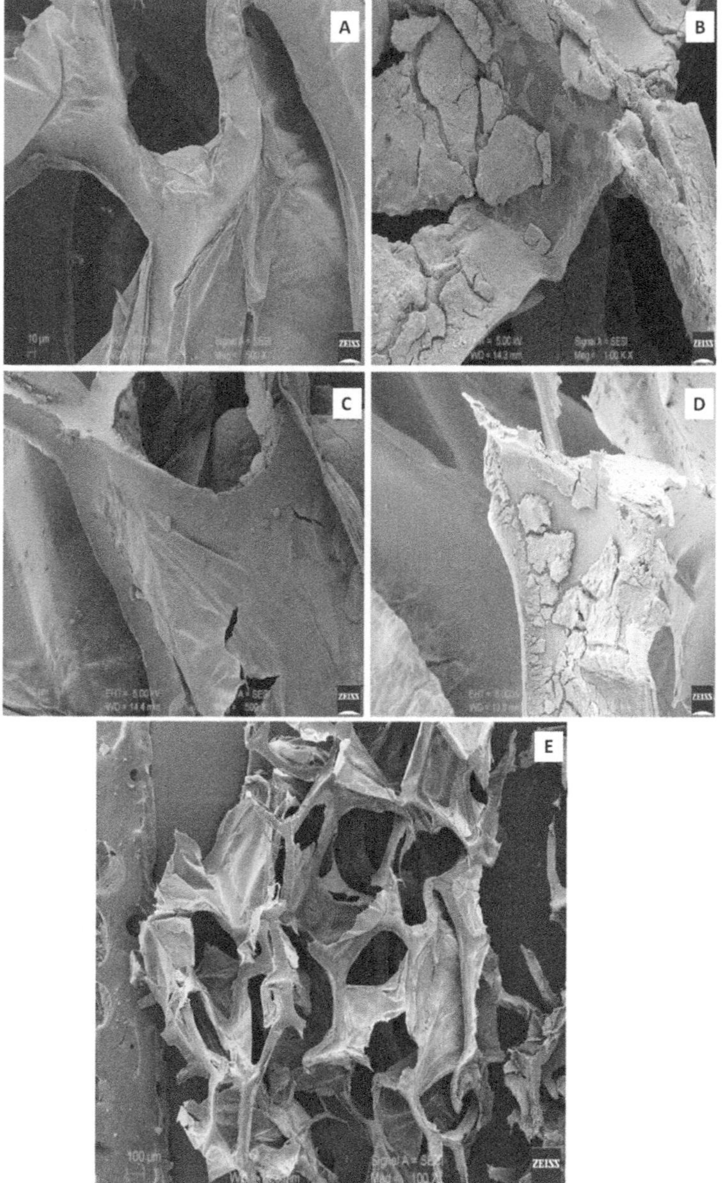

FIGURE 13.6 SEM images of: A. pure polystyrene; B. UV-treated polystyrene; C. UV- and chemical-treated polystyrene; D, E. pre-treated polystyrene after fungal treatment.

Source: Reprinted with permission from Chaudhary, A.K., Chaitanya, K., Vijayakumar, R.P., 2021. Notes: Synergistic effect of UV and chemical treatment on biological degradation of Polystyrene by *Cephalosporium* strain NCIM 1251. Arch. Microbiol. 203(5), 2183–2191.Copyright © The Author(s), under exclusive licence to Springer-Verlag GmbH, DE part of Springer Nature 2021.

observed. Furthermore, the pH value of the medium containing pre-treated polystyrene shifted from 7.01 ± 0.01 to 3.75 ± 0.09 after 8 weeks of bacterial incubation. Likewise, the total dissolved solids (TDS) increased from 0.575 ± 0.024 ppm to 26.53 ± 0.15 ppm after 9 weeks of exposure to *Cephalosporium* NCIM 1251 strain. Thermogravimetric analysis (TGA) further demonstrated a decrease in the thermal stability of polystyrene after incubation with *Cephalosporium* NCIM 1251 strain whereas SEM images in Figure 13.6 showed distinct alterations in the surface patterns of pre-treated polystyrene films. Thus, this study highlighted the possibility of bacteria-mediated degradation of polystyrene.

Chauhan et al. (2018) also reported polystyrene biodegradation by *Exiguobacterium* sp. DR11 and DR14 strains. The bacterial strains were isolated from the wetland soil and water samples which were then cultivated in different growth media such as Luria-Bertani, nutrient, and tryptic soy broths at 30 °C. These isolates were able to form robust biofilms on the polystyrene surface in the presence of nutrient-rich media. A highly robust biofilm formation was observed in tryptic soy broth compared to the other growth media. The morphology of the cells was altered in these biofilms as observed in AFM images wherein the cells were elongated and rod-shaped as compared to small rod shapes as seen in exponential phase cultures. The ability of the *Exiguobacterium* sp. to utilize polystyrene beads as a substrate was investigated in a liquid carbon-free basal medium. A longer lag phase was observed in the presence of polystyrene beads while the growth was also slower as compared to growth in the presence of glucose. However, steady growth was observed after 45 days, which highlighted sustained growth using polystyrene as the substrate. The weight reductions in 5 mg/mL of polystyrene beads were 8% and 8.8% for the DR11 and DR14 strains, respectively.

High-impact polystyrene (HIPS) biodegradation was studied by Mohan et al. (2016) using *Pseudomonas* and *Bacillus* sp. strains that were isolated from soil samples near a plastic dump yard. The isolates were initially cultivated in a nutrient medium until the optical density at 600 nm reached 0.6, after which the cultures were transferred to a mineral medium containing HIPS films. The protein content on the incubated HIPS films increased significantly on the third day, which suggested the cellular adhesion of the cells on the plastic film. SEM analysis further revealed the formation of incisions, pits, and holes on the surface of the HIPS film as compared to the control films, which had smooth and plane surfaces. A depolymerase assay was carried out using an emulsion of HIPS which showed 97.4% and 92.6% of turbidity clearance after incubating with the culture supernatant of *Pseudomonas* and *Bacillus* sp., respectively.

In another study *Rhodococcus ruber* was also reported to efficiently form biofilm and partially degraded polystyrene (Mor and Sivan, 2008). The bacterial culture was previously isolated from soil samples contaminated with PE waste from agricultural use and later cultivated in a nutrient medium at 35 °C. Thereafter, the biofilm formation of the *R. ruber* isolate was evaluated in a synthetic medium wherein maximum cell density was obtained after 4 days of incubation. The optimal conditions for the bacterial growth were evaluated, which revealed that a temperature of 35 °C, nitrogen source as KNO_3, and pH 7.0 resulted in similar growth of planktonic and sessile biofilm cultures. The addition of mannitol and/or glucose as additional carbon sources resulted in a 42.2 and 51.1% increase in biofilm formation, respectively. A biphasic pattern for biofilm formation was observed wherein the initial 4 days of incubation resulted in a significant increase in the protein content followed by a second phase where the protein concentration remained constant. The respiration level of the biofilm was also analyzed through the reduction of 5-cyano-2,3-ditolyltetrazolium chloride (CTC) fluorescent probe to CTC-formazan by the bacterial electron transport system. Bacterial biofilm respiration sharply increased during the initial 3 days of incubation after which it remained constant whereas planktonic cells showed a reduction in respiration activity with increasing incubation time. The effect of mineral oil on the biofilm formation ability was also evaluated. The growth of the bacterial strain remained the same in the presence of non-ionic detergents such as Tween 20, Tween 60, and Tween 80. However, the addition of mineral oil resulted in a 130% increase in the total biofilm bacterial count. Finally, 0.5% and 0.8% biodegradation efficiencies were obtained after 4 and 8 weeks of inoculation of *R. ruber* with polystyrene flakes.

13.2.4 POLYPROPYLENE

Helen et al. (2017) screened for bacteria from mangrove ecosystems that were able to degrade polypropylene. The mineral salt medium was used for the screening of microplastic degraders wherein *Bacillus cereus* and *Sporosarcina globispora* were the two spore-forming isolates that were identified. These two isolates successfully grew in a carbon-free medium supplemented with polypropylene granules. After 10–15 days of incubation, the microplastic granules showed bacterial biofilm growth on the surface. The growth kinetic studies of the bacterial isolates in the presence of polypropylene granules as the sole carbon source revealed a log-phase growth pattern for the initial 10 days, after which the growth was optimized on the 15th day of incubation. However, a slight difference in the growth pattern was observed between the two isolates as *B. cereus* showed optimal growth phase while *S. globispora* was in stationary phase after 34 days of incubation. The mass reduction efficiency of polypropylene granules was also evaluated wherein 12% and 11% reduction was observed by *B. cereus* and *S. globispora*, respectively.

Likewise, Auta et al. (2018) studied the biodegradation of polypropylene microplastics by bacterial isolates obtained from mangrove sediments. The bacterial isolates were cultured in a nutrient medium and identified as *Bacillus* sp. and *Rhodococcus* sp., which were able to degrade 6.4% and 4.0% of UV-treated polypropylene after 40 days of incubation. The growth kinetics evaluation of

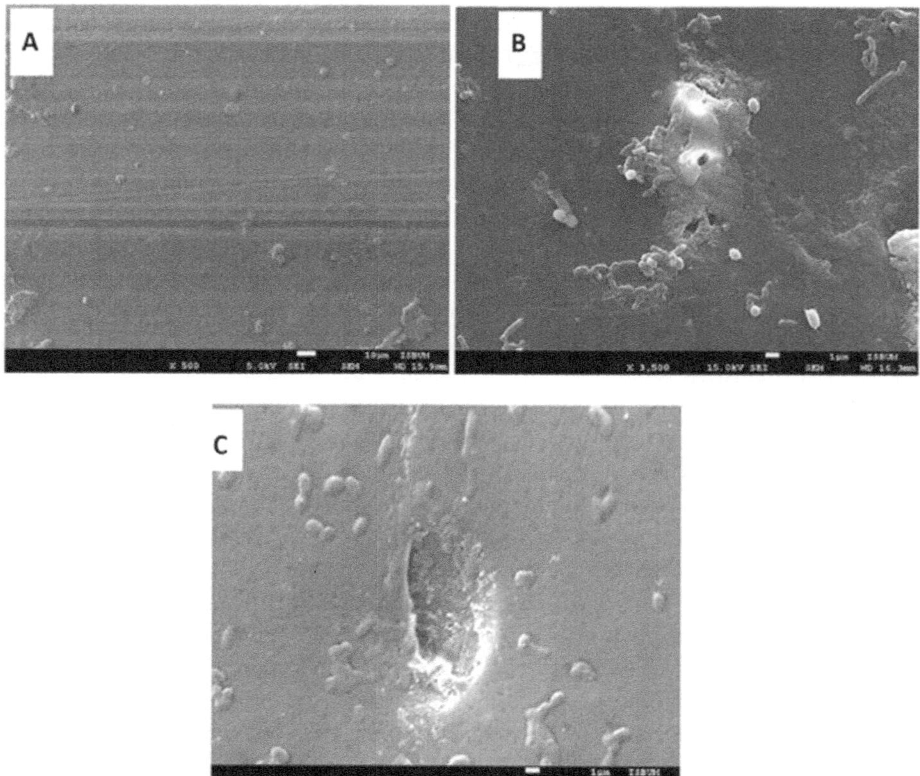

FIGURE 13.7 SEM micrographs of PP microplastics: A. un-inoculated (control) PP microplastic; B. PP microplastic inoculated with *Bacillus* sp. strain 27; C. PP microplastic inoculated with *Rhodococcus* sp. strain 36.

Source: Reprinted with permission from Auta, H.S., Emenike, C.U., Jayanthi, B., Fauziah, S.H., 2018. Growth kinetics and biodeterioration of polypropylene microplastics by *Bacillus* sp. and *Rhodococcus* sp. isolated from mangrove sediment. Mar. Pollut. Bull. 127, 15–21. Copyright © 2017 Elsevier Ltd.

Bacillus sp. and *Rhodococcus* sp. showed rapid metabolism for the initial 20 days, after which the number of cells was reduced. Moreover, the pH of the medium turned alkaline after the biodegradation of polypropylene microplastics. The optimal growth of *Bacillus* sp. and *Rhodococcus* sp. was observed at pH values of 8.51 and 8.96 after 10 and 20 days of incubation, respectively. SEM images then further confirmed the biodegradation of polypropylene as several pores and irregularities were observed on the surface of the microplastic incubated with the bacterial isolates as evident from Figure 13.7. However, the untreated polypropylene samples kept as control showed no such surface damage.

Lysinibacillus sp. JJY0216 isolated by Jeon et al. (2021) also exhibited microplastic biodegradation. The bacterial isolate was obtained from soil samples collected from surfaces treated with kenaf-plastic complex. The bacterial isolates were then cultured in an M9 minimal medium supplemented with 1% glucose and yeast extract along with pre-treated polypropylene films. After 21 days of incubation, a 4.3% reduction in the weight of oriented polypropylene films was observed. The weight loss ceased after 21 days of incubation. SEM images of the biodegraded polypropylene films then revealed the increase in the surface roughness as seen in Figure 13.8. The gas chromatography-mass spectrometry (GC-MS) results demonstrated the formation of various carboxylic acids after biodegradation. Hence, further studies on copper-binding oxidases were proposed in this study as they are known for the oxidation of fatty acids.

Jeon and Kim (2016) reported the isolation of a mesophilic strain of bacterium that demonstrated the potential degradation of polystyrene. The bacterial strain was isolated from a soil sample obtained

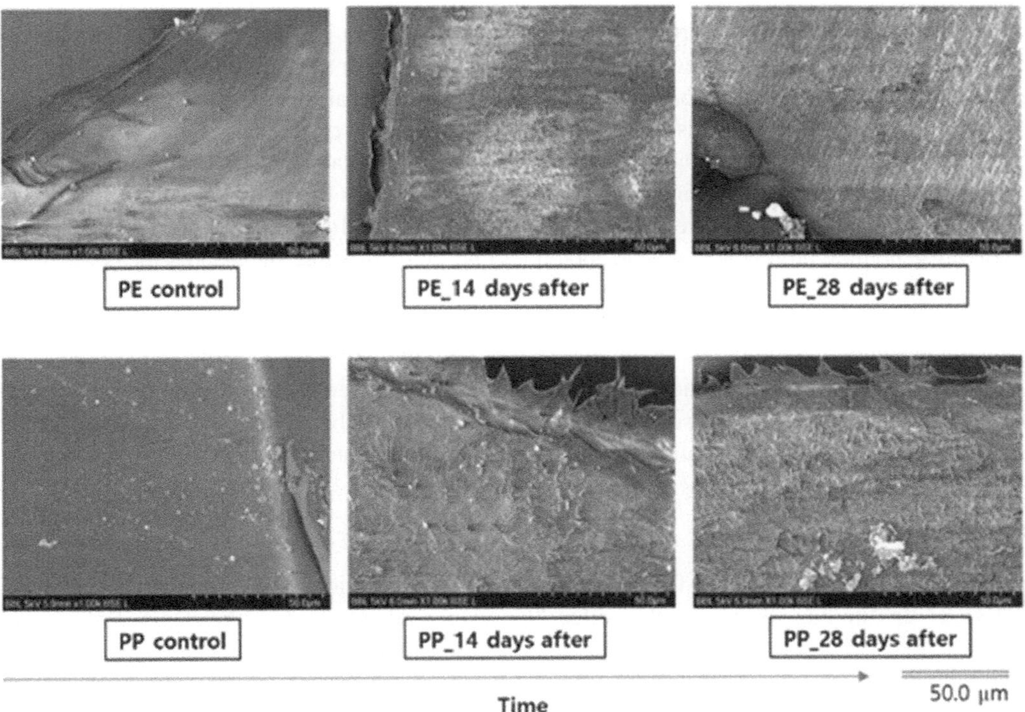

FIGURE 13.8   SEM images of polyethylene and polypropylene.

Source: Reprinted with permission from Jeon, J.M., Park, S.J., Choi, T.R., Park, J.H., Yang, Y.H., Yoon, J.J., 2021. Biodegradation of polyethylene and polypropylene by *Lysinibacillus* species JJY0216 isolated from soil grove. Polym. Degrad. Stab. 191, 109662. Copyright ©2021 The Author(s). Published by Elsevier Ltd.

from municipal solid waste which was cultivated in an enrichment medium supplemented with 0.1 g of low molecular weight polypropylene-1 (LMWPP-1) and incubated at 37 °C under shaking conditions (120 rpm) for 1 week. The isolated bacterium was identified as *Stenotrophomonas panacihumi* PA3-2 strain wherein the optimum pH and temperature range was 6–8 and 25–37 °C, respectively. Later, the polypropylene biodegradation ability of the PA3-2 strain was evaluated wherein polypropylene powder was mixed with sterilized compost and then inoculated with the bacterial strain at 37 °C. The biodegradation efficiency of LMWPP-1 was 20.3 ± 1.39% as compared to 16.6 ± 1.70% for LMWPP-2, which showed an inverse relationship between the molecular weight of polypropylene with that of biodegradability. A commercially manufactured polypropylene termed YUHWA was also biodegraded in this study. A wide distribution of molecular weight with a 5.7 polydispersity index (PDI) was observed for this sample wherein the biodegraded product was proposed to be a result of only low molecular weight fractions of YUHWA.

13.3   CONCLUSION AND FUTURE PERSPECTIVES

Microbe mediated biodegradation of polyethylene, polypropylene, polystyrene, and polyethylene terephthalate seems to be a promising strategy for controlling environmental pollution. Due to the recalcitrant nature, microplastics are almost nondegradable and take a long time to be removed from the surroundings. Several bacterial and fungal species have shown their promising potential in degrading microplastics, which is mainly attributed to their enzymatic activities. Continuous activity of microbial depolymerizing enzymes is considered as key player in microplastic degradation. Microbes from various environmental niches should be explored for microplastic degradation. The gut flora of the insects that live on plastic-based waste material may have untapped bacteria diversity that can extensively degrade microplastics. Further, the microbial enzymes can be functionalized on the nontoxic biogenic nanomaterials synthesized from bacteria, fungi, algae, or medicinal plants for enhancement of their activity (Ghosh et al. 2016a; Ghosh et al. 2016b; Shende et al. 2017; Ghosh, 2018; Shende et al. 2018; Bhagwat et al. 2018). This would ensure their reuse and recyclability. Further, the microbial enzymes responsible for microplastic degradation can be encapsulated or entrapped within polymeric beads made up of alginate, gelatin, chitosan, or other mesoporous biomaterials (Ghosh, 2019; Bloch et al. 2021). This would enhance their stability and can be used repeatedly. Integrated approaches involving genomics, proteomics, and metabolomics may help to explore and identify the key metabolic pathways and the metabolites involved in the microbial degradation of microplastics.

   In conclusion, microbe-based microplastic degradation is a most promising alternative to the available chemical and physical techniques. Identifying the genes responsible in the process will help to develop recombinant microbes with genetically engineered microplastic degrading traits. The transcriptomic analysis of such genetically modified microbes with a microplastic degrading ability would help to develop optimized bioprocess for environmental remediation.

ACKNOWLEDGMENT

The Program Management Unit for Human Resources & Institutional Development and Innovation (PMU-B) for funding the proposal entitled "Post-Doctoral Research Competency Enhancement Program for carbon recycling based on metal oxide graphene-liked nanosheet photocatalysts" under the Program of National Postdoctoral and Postgraduate System approved by PMU-B Board Committees (Contract No. B13F670067).

REFERENCES

Auta, H.S., Emenike, C.U., Jayanthi, B., Fauziah, S.H., 2018. Growth kinetics and biodeterioration of polypropylene microplastics by *Bacillus* sp. and *Rhodococcus* sp. isolated from mangrove sediment. *Mar. Pollut. Bull.* 127, 15–21.

Bhagwat, T.R., Joshi, K.A., Parihar, V.S., Asok, A., Bellare, J., Ghosh, S., 2018. Biogenic copper nanoparticles from medicinal plants as novel antidiabetic nanomedicine. *World J. Pharm. Res.* 7(4), 183–196.

Bhat, S.A., Cui, G., Yaseera, N., Lei, X., Ameen, F. and Li, F., 2022. Removal potential of microplastics in organic solid wastes via biological treatment approaches. *Microbial Biotechnology: Role in Ecological Sustainability and Research*, pp.255–263. Springer. https://onlinelibrary.wiley.com/doi/10.1002/978111 9834489.ch14

Bloch, K., Pardesi, K., Satriano, C., Ghosh, S. 2021. Bacteriogenic platinum nanoparticles for application in nanomedicine. *Front. Chem.* 9, 624344.

Chaudhary, A.K., Chaitanya, K., Vijayakumar, R.P., 2021. Synergistic effect of UV and chemical treatment on biological degradation of Polystyrene by *Cephalosporium* strain NCIM 1251. *Arch. Microbiol.* 203(5), 2183–2191.

Chauhan, D., Agrawal, G., Deshmukh, S., Roy, S.S., Priyadarshini, R., 2018. Biofilm formation by *Exiguobacterium* sp. DR11 and DR14 alter polystyrene surface properties and initiate biodegradation. *RSC Adv.* 8(66), 37590–37599.

Dey, S., Anand, U., Kumar, V., Kumar, S., Ghoraj, M., Ghosh, A., Kant, N., Suresh, S., Bhattacharya, S., Bontempi, E., Bhat, S.A., Dey, A., 2023. Microbial strategies for degradation of microplastic generated from COVID-19 healthcare waste. *Environmental Research* 216, 114438. https://doi.org/10.1016/j.env res.2022.114438

Ghosh, S., 2018. Copper and palladium nanostructures: a bacteriogenic approach. *Appl. Microbiol. Biotechnol.* 101(18), 7693–7701.

Ghosh, S., 2019. Mesoporous silica based nano drug delivery system synthesis, characterization and applications. In: Mohapatra, S.S., Ranjan, S., Dasgupta, N., Mishra, R.K., Thomas S. (Eds.), *Nanocarriers for Drug Delivery.* Elsevier Inc. Amsterdam, Netherlands. pp 285–317. ISBN: 978-0-12-814033-8.

Ghosh, S., Chacko, M.J., Harke, A.N., Gurav, S.P., Joshi, K.A., Dhepe, A., Kulkarni, A.S., Shinde, V.S., Parihar, V.S., Asok, A., Banerjee, K., Kamble, N., Bellare, J., Chopade, B.A., 2016a. *Barleria prionitis* leaf mediated synthesis of silver and gold nanocatalysts. *J. Nanomed. Nanotechnol.* 7, 4.

Ghosh, S., Gurav, S.P., Harke, A.N., Chacko, M.J., Joshi, K.A., Dhepe, A., Charolkar, C., Shinde, V.S., Kitture, R., Parihar, V.S., Banerjee, K., Kamble, N., Bellare, J., Chopade, B.A., 2016b. *Dioscorea oppositifolia* mediated synthesis of gold and silver nanoparticles with catalytic activity. *J. Nanomed. Nanotechnol.* 7, 5.

Gilan, I.H.Y.S., Hadar, Y., Sivan, A., 2004. Colonization, biofilm formation and biodegradation of polyethylene by a strain of *Rhodococcus ruber. Appl. Microbiol. Biotechnol.* 65(1), 97–104.

Guo, J.J., Huang, X.P., Xiang, L., Wang, Y.Z., Li, Y.W., Li, H., Cai, Q.Y., Mo, C.H., Wong, M.H., 2020. Source, migration and toxicology of microplastics in soil. *Environ. Int.* 137, 105263.

Helen, A.S., Uche, E.C., Hamid, F.S., 2017. Screening for polypropylene degradation potential of bacteria isolated from mangrove ecosystems in Peninsular Malaysia. *Int. J. Biosci. Biochem. Bioinform.* 7, 245–251.

Jeon, H.J., Kim, M.N., 2016. Isolation of mesophilic bacterium for biodegradation of polypropylene. *Int. Biodeterior. Biodegr.* 115, 244–249.

Jeon, J.M., Park, S.J., Choi, T.R., Park, J.H., Yang, Y.H., Yoon, J.J., 2021. Biodegradation of polyethylene and polypropylene by *Lysinibacillus* species JJY0216 isolated from soil grove. *Polym. Degrad. Stab.* 191, 109662.

Liu, L., Xu, M., Ye, Y., Zhang, B., 2022. On the degradation of (micro) plastics: Degradation methods, influencing factors, environmental impacts. *Sci. Total Environ.* 806, 151312.

Miri, S., Saini, R., Davoodi, S.M., Pulicharla, R., Brar, S.K., Magdouli, S., 2022. Biodegradation of microplastics: Better late than never. *Chemosphere* 286, 131670.

Mohan, A.J., Sekhar, V.C., Bhaskar, T., Nampoothiri, K.M., 2016. Microbial assisted high impact polystyrene (HIPS) degradation. *Bioresour. Technol.* 213, 204–207.

Mor, R., Sivan, A., 2008. Biofilm formation and partial biodegradation of polystyrene by the actinomycete *Rhodococcus ruber. Biodegradation* 19(6), 851–858.

Nowak, B., Pająk, J., Łabużek, S., Rymarz, G., Talik, E., 2011. Biodegradation of poly (ethylene terephthalate) modified with polyester "Bionolle®" by *Penicillium funiculosum. Polimery* 56(1), 35–44.

Roberts, C., Edwards, S., Vague, M., León-Zayas, R., Scheffer, H., Chan, G., Swartz, N.A., Mellies, J.L., 2020. Environmental consortium containing *Pseudomonas* and *Bacillus* species synergistically degrades polyethylene terephthalate plastic. *mSphere* 5(6), e01151–20.

Sheik, S., Chandrashekar, K.R., Swaroop, K., Somashekarappa, H.M., 2015. Biodegradation of gamma irradiated low density polyethylene and polypropylene by endophytic fungi. *Int. Biodeterior. Biodegr.* 105, 21–29.

Shende, S., Joshi, K.A., Kulkarni, A.S., Shinde, V.S., Parihar, V.S., Kitture, R., Banerjee, K., Kamble, N., Bellare, J., Ghosh, S., 2017. *Litchi chinensis* peel: A novel source for synthesis of gold and silver nanocatalysts. *Glob. J. Nanomedicine* 3(1), 555603.

Shende, S., Joshi, K.A., Kulkarni, A.S., Charolkar, C., Shinde, V.S., Parihar, V.S., Kitture, R., Banerjee, K., Kamble, N., Bellare, J., Ghosh, S., 2018. *Platanus orientalis* leaf mediated rapid synthesis of catalytic gold and silver nanoparticles. *J. Nanomed. Nanotechnol.* 9, 2.

Spina, F., Tummino, M.L., Poli, A., Prigione, V., Ilieva, V., Cocconcelli, P., Puglisi, E., Bracco, P., Zanetti, M., Varese, G.C., 2021. Low density polyethylene degradation by filamentous fungi. *Environ. Pollut.* 274, 116548.

Yan, F., Wei, R., Cui, Q., Bornscheuer, U.T., Liu, Y.J., 2021. Thermophilic whole-cell degradation of polyethylene terephthalate using engineered *Clostridium thermocellum*. *Microb. Biotechnol.* 14(2), 374–385.

Yang, J., Yang, Y., Wu, W.M., Zhao, J., Jiang, L., 2014. Evidence of polyethylene biodegradation by bacterial strains from the guts of plastic-eating waxworms. *Environ. Sci. Technol.* 48(23), 13776–13784.

Yuan, J., Ma, J., Sun, Y., Zhou, T., Zhao, Y., Yu, F., 2020. Microbial degradation and other environmental aspects of microplastics/plastics. *Sci. Total Environ.* 715, 136968.

Zhang, X., Li, Y., Ouyang, D., Lei, J., Tan, Q., Xie, L., Li, Z., Liu, T., Xiao, Y., Farooq, T.H., Wu, X., 2021. Systematical review of interactions between microplastics and microorganisms in the soil environment. *J. Hazard. Mater.* 418, 126288.

Section III

Environmental Nexus in Contaminated Soil Management

14 Different Technologies for Contaminated Soil Remediation
Hurdles and Perspectives

Stefanie Bernardette Costa-Gutierrez,
Enzo Emanuel Raimondo, Juan Daniel Aparicio,
Juliana María Saez, Analía Alvarez, Marta Alejandra Polti,
and Claudia Susana Benimeli

14.1 INTRODUCTION

14.1.1 Soil Generalities

Soil is a complex, structured, heterogeneous, and discontinuous system. Soil is an essential component of the environment that takes thousands of years to form and performs essential functions for the maintenance of life on earth. Soils are differentiated into horizons of mineral and organic components of varying depth, which differ in morphology, physical and chemical properties, and biological characteristics. Soil is composed by organic matter, minerals, and nutrients, which allows the development of organisms and microorganisms. Soil comprises three phases: solid (approximately 50% by volume mineral particles and 5% organic matter), liquid (25% water), and gaseous (20% air). Mineral particles vary in shape, size, and chemical composition and determine the soil texture, which is determined by the relative proportion of mineral particles; the largest are called sands, the medium ones are silts, and the smallest are clays (Raimondo et al., 2019).

In sandy soils, sand accounts for more than 70% of the solid fraction of the soil. Sand enhances permeability and aeration of soil and supports the root system of plants; but sandy soils have low water and nutrient holding capacity. Clay soils are characterized by a minimum clay content of 35%. The small size of clay particles hinders soil permeability and aeration. Clays, together with organic matter, increase water and nutrient retention in the soil, but also favor the adsorption and retention of pollutants. Loamy soils have high silt content, are characterized by retaining adequate amounts of water and nutrients, and have good aeration. Soil texture is an important physical characteristic, as it plays a role in water and nutrient retention, aeration, plant nutrition, and root system fixation, as well as in fixation and leaching of pollutants.

Soil is used by people for a wide variety of purposes, such as food production and extraction of materials, which modifies it and sometimes leads to its degradation and loss of quality. Soil should therefore be considered a natural resource to be maintained, protected, and restored.

14.1.2 ENVIRONMENTAL POLLUTION

Nowadays environmental pollution is a serious global concern. This major issue is not only worrying the scientific community, but society is beginning to demand action to address environmental pollution. Air, water, and soil are environments affected by the presence of pollutants. These environments are susceptible to variations, especially those caused by anthropogenic activities, such as: industry, mining, military activities, agriculture, and livestock. Although these activities have brought important benefits to human beings, improving living conditions in terms of health, communication, agriculture, among others, the inadequate management of the waste generated by these activities negatively affects the environment and ultimately human beings.

14.1.2.1 Soil Pollution

A concerning environmental problem is soil pollution, as soil can be a source and/or a sink of pollutants. Pollutants can reach the soil through different pathways, including: direct application of industrial waste, excess application of pesticides, herbicides, or fertilizers, atmospheric deposition, erosion of contaminated areas, irrigation with contaminated water, seepage from a landfill, percolation of contaminated water, and rupture of storage tanks (Bhat et al., 2022). Urban and industrial soils are strongly affected by human activities and are subject to a constant input of contaminants. Agricultural activity contributes to soil deterioration, as excessive or inappropriate use of fertilizers, organic amendments, and pesticides can lead to loss of soil quality. Currently, most pesticides used in agricultural fields are synthetic organic compounds, which can enter the soil through surface run-off or spillage during application; pesticide residues can subsequently contaminate groundwater aquifers by leaching.

Contamination can spread from one matrix to another. In this sense, atmospheric emissions from industry and traffic are also a source of soil contamination due, on the one hand, to the deposition of toxic particles or even radioactive substances and, on the other hand, to acid rain which lowers the soil pH, seriously altering the acid-base balance. Also, contaminated soils might be a source of contamination for groundwater through pollutants leaching, for freshwater and the marine environment through wind and water erosion, and for remote location through wind.

Soils act as an active filter where pollutants (depending on the soil type and the pollutant intrinsic properties) can degrade or exist in varying concentrations. The huge variability in soil composition and behavior makes it difficult to define a minimum contaminant concentration above which the soil is considered contaminated. The basic principles for soil management should therefore be based on the protection of human and ecosystem health, sustainable development, and the maintenance or restoration of the multi-functionality of soils in the long term.

14.1.2.2 Effect of Soil Pollution

Soil and marine contamination affects both terrestrial and aquatic biota. Pollutants can accumulate in animal tissues and organs, enter the terrestrial or aquatic food chain, and accumulate in several animals at the top of the food chain, including humans. Soil pollution can increase salinity and reduce soil fertility; this coupled with the toxic effect of pollutants has a detrimental effect on plant growth and agricultural yield. Plants can also accumulate toxic compounds in their tissues by absorbing them through the roots, or through aerial tissues when the pollutant volatilizes (Sabreena et al., 2022). Biodiversity is also affected by soil contamination, reducing the number of organisms or even eliminating entire populations due to toxicity, and changing community composition because species more tolerant to contamination can replace the more sensitive ones. Low pollutants concentrations in soil often lead to adaptive strategies through physiological changes to transform and remove the pollutant molecule (Naveed et al., 2014). Changes in the soil organisms' activity can ultimately lead to alteration of whole ecosystems. The presence of pollutants disrupts the biological, chemical, or physical characteristics of soil, and the presence of these components in hazardous doses can affect the environment and/or human health (Kaur et al., 2021).

14.1.3 Types of Pollutants: Organic Pollutants (OPs) and Potentially Toxic Elements (PTEs)

A soil contaminant is any compound able to affect the soil characteristics, such as its quality, texture, and/or mineral content, and/or to alter the biological balance of soil organisms. OPs are one of the main pollutants found in soil. They include fertilizers, pesticides, petroleum, and pharmaceutical and personal care products (PPCPs). PPCPs include: antibiotics, antimicrobials, analgesics, antipyretics, antidepressants, stimulants, steroids, disinfectants, fragrances, and cosmetics. The development and widespread use of PPCPs has increased considerably in the past two years, in the context of pandemic, and has raised concerns about potential environmental contamination from these products. Some emerging pollutants, such as bisphenol A, atenolol, ketoprofen, methyl parabens, and caffeine (one of the few that has natural origins), are found in high levels in agricultural soils, forest, and industrial areas (Gutiérrez et al., 2016).

In addition, a number of sources of plastics polluting the environment have been reported. These include domestic wastewater containing plastic microbeads from PPCPs, biosolids, fertilizers, sewage irrigation, landfills from industrial and urban centers, road fouling, vinyl mulch from agricultural activities, tire abrasion, and long-range airborne particles. These plastics enter the soil environment, are deposited on the surface, and penetrate the subsoil (Chae and An, 2018). Microplastic contamination is of increasing concern and was ranked as one of the most important scientific issues in the environmental and ecological field. Ingested microplastics can move through the food chain, inducing negative effects on organisms.

PTEs are ubiquitous elements, generally found at very low levels in the soil, and include heavy metals, such as cadmium (Cd), chromium (Cr), copper (Cu), cobalt (Co), nickel (Ni), mercury (Hg), lead (Pb), and zinc (Zn), and non-metals such as antimony (Sb), arsenic (As), and selenium (Se) (Gutiérrez et al., 2016). The wide variety of compounds makes their detection and control difficult.

The movement and removal of pollutants in soils involves processes in which the pollutant does not alter its chemical structure, such as volatilization, adsorption, desorption, and leaching; and those in which the chemical structure of the pollutant is altered to harmless compounds or mineralized, such as chemical, biological, or photochemical degradation.

14.1.4 Co-pollution

The effects of contamination are aggravated in sites in which the presence of different OPs is accompanied by PTEs in concentrations exceeding permitted levels, a phenomenon known as co-pollution or mixed contamination. This phenomenon, referring to the simultaneous presence of OPs and PTEs, released from several sources, is common in landfills and waste recycling areas, oil storage areas, industrial areas, among others (Aparicio et al., 2018). Increasing reuse of sewage sludge, irrigation with wastewater, and conventional pesticide use has led to the coexistence of PTEs and OPs in polluted soil.

Co-pollution is an even more complicated problem to address than single pollution, as it involves the degradation of two or more types of pollutants with different physicochemical characteristics. The negative impact of co-pollution is determined by the interactions (such as antagonistic, synergistic, and additive effects) between the different pollutants, their concentration, persistence, and bioavailability. The presence of organic and inorganic chemicals can modify some parameters as solubility, bioaccessibility, biotoxicity, and biological metabolic processes of pollutants. During the adsorption process, there would be competition over the binding sites between the pollutants.

One approach suggests the removal of OPs and the passivation of PTEs through several pathways (Aparicio et al., 2015).

14.2 REMEDIATION TECHNOLOGIES FOR POLLUTED SOILS: CHARACTERISTICS AND CLASSIFICATION

14.2.1 *In situ* and *ex situ* Technologies

The remediation technologies, intended to reduce the toxicity, volume, or mobility of pollutants by applying physicochemical and/or biological processes can be classified as *in situ* and *ex situ*, taking into account their form of application. *In situ* remediation involves treating pollutants at their location, removing them from the soil or sediment without moving the matrix itself. *Ex situ* remediation requires excavation of polluted soil, for its subsequent treatment or deposition elsewhere (off-site), or for surface treatment in the area where the pollution is present (on-site) (Naseri-Rad et al., 2020). Both technologies have advantages and disadvantages, and their selection requires analyzing the type of soil and characteristics of pollutants. *In situ* treatments are usually less costly, as the soil does not need to be extracted and moved to another location; however, treatment duration is longer, the success of the treatment is subject to soil uniformity, and the progress of the treatment is difficult to monitor. This is a good option when access to the contaminant is limited (e.g. for active industrial sites), access to the polluting center presents physical constraints for excavation facilities, or for large-scale sites. In *ex situ* technologies, the extraction and movement of the contaminated matrix can be costly, but some factors that enhance the process can be controlled. This is a good methodology to avoid the spread of contamination to surrounding areas and also for the biological degradation of soil pollutants.

14.2.2 Remediation Methods

A large number of physical-chemical and biological methods for soil remediation are being tested (Figure 14.1). Each method has its advantages and disadvantages, so the selection of the most suitable method for a specific case must take into account aspects such as the type and concentration of the contaminants, the soil characteristics, the volume and accessibility of the contaminated soil, the cost of the process, and the urgency of its resolution. This chapter summarizes and shows the state of the art of remediation methods as well as their combination, the evaluation of the remediation effectiveness, and the challenges of soil remediation research.

Urgent policy measures are needed to combat environmental pollution. While in some countries measures have been taken to prevent further contamination, in others, banned pollutants are still being used or obsolete pesticides stockpiles still exist, causing spills and leaks that pollute the soil. Hence, an enormous effort is being made to find effective, economical, and environmentally friendly ways to clean up contaminated sites and mitigate the hazardous effects of toxic chemicals on already contaminated soil.

14.3 PHYSICOCHEMICAL PROCESSES FOR SOIL REMEDIATION

Pollution can be contained and/or reversed by taking advantage of the chemical and physical properties of the contaminants and their environment through the application of physicochemical processes. These processes are usually targeted at certain groups of pollutants, as they can be complex and excessively expensive (Aparicio et al., 2022). Moreover, in some cases, their impact on the ecosystem is high. Some of the most frequently used techniques are presented below.

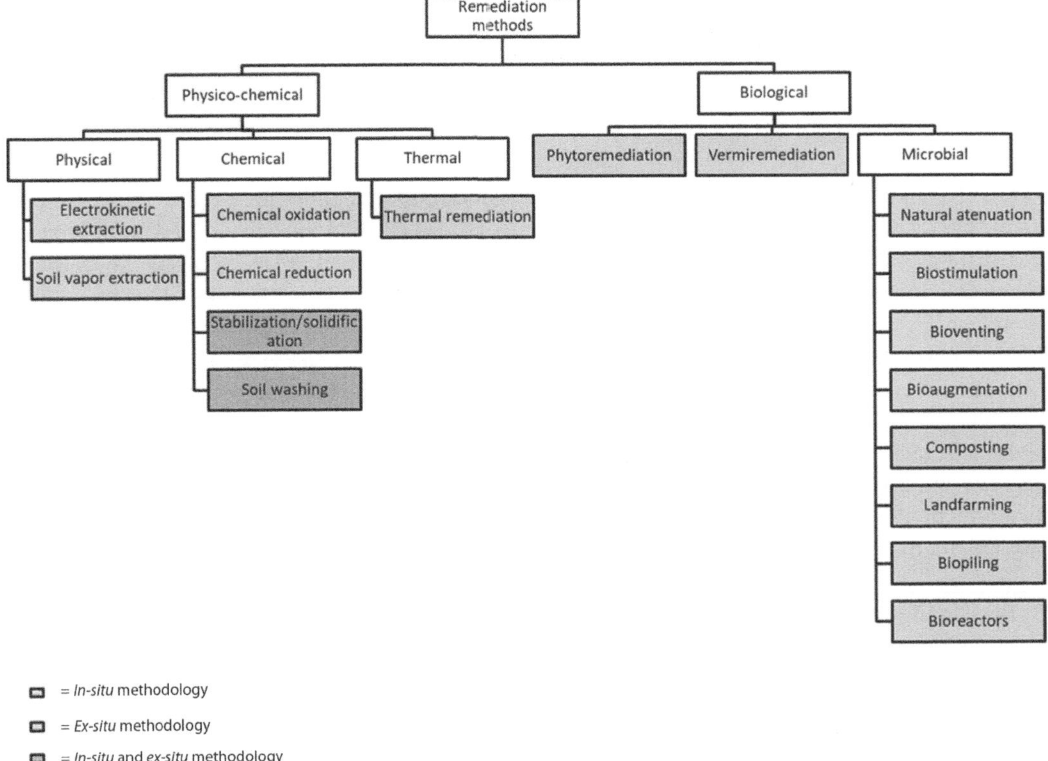

FIGURE 14.1 Classification of physicochemical and biological remediation methods.

14.3.1 Chemical Oxidation/reduction

Environmental redox processes are fundamental in the transformation of organic material and minerals, modifying their chemical speciation, bioavailability, mobility and, as a consequence, their toxicity (Tandon and Singh, 2016).

Chemical oxidation is widely used to decrease or avoid the migration of contaminants in subsurface sediments and groundwater. For *in situ* chemical oxidation (ISCO), different types of oxidants are used, such as permanganate (MnO_4), hydrogen peroxide (H_2O_2) and iron (Fe), persulfate ($S_2O_8^{2-}$), and ozone (O_3). These oxidizing agents are pumped into the contaminated matrix by direct injection into wells of different depths. In some cases, recirculation of the oxidants is applied, to increase treatment efficiency. Another alternative is the construction of a reactive permeable barrier, where the solid oxidizing agent is placed in a trench, delimiting the contaminated site. Water flows through this barrier, allowing the contaminants to react with the oxidant. *In situ* treatments can present challenges, mainly related to soil permeability and soil composition, which can generate a rebound effect at the end of the treatment; a variety of dissolved components in groundwater, such as oxygen, calcium, magnesium, carbonate, and sulfates, could react with the oxidant, modifying its reactivity (Wei et al., 2022).

ISCO is used to detoxify fuels, solvents, pesticides, and other pollutants from industrial and agricultural practices. Several researchers are developing ISCO methods. Although these technologies aim to counteract the aforementioned limitations, they have not yet been fully improved (Wei et al., 2022). For example, Song et al. (2019) tested the removal of polycyclic aromatic hydrocarbons (PAHs) with ammonium persulfate, activated with different types of zero-valent iron (ZVI), such as micro/nano-structured (nZVI), stearate-coated micro/nano-structured ZVI, and commercial micronsized ZVI, to show that nZVI is the most effective method.

Similar to ISCO, *in situ* chemical reduction (ISCR) includes redox reactions that chemically transform hazardous contaminants, but in this case, the reducing agents in contact with the medium rapidly oxidize and donate their electrons to the contaminants, reducing them (Dominguez et al., 2018). Some frequent intrinsic reducing agents are reduced species such as Fe^{2+}, Fe_3O_4, and FeO, HS^-, FeS, FeS_2, and organic compounds. More sophisticated ISCR technologies employ more powerful reductants, such as zero-valent metals (nano-remediation) or dithionite ($S_2O_4^{2-}$). The approaches involving nZVI are the only ones that have been sufficiently applied *in situ*. However, because the physicochemical transformations of nZVI are complex, their impact on the soil ecosystem could be significant.

The characteristics of the area to be treated are fundamental in choosing the type of treatment (Zhang et al., 2022). In moderately permeable zones, reducing agents are generally injected as an aqueous solution through subway wells. In high permeability media, the approach includes pumping with recirculation and/or injection-extraction: "push–pull". In low permeability materials, such as clays or mudflats, the low injection rate can be very inefficient, so hydraulic or pneumatic fracturing methods are being applied to improve the distribution of the reductants. Aquifer heterogeneity affects the release of the reducing reagents. The lifetime of the reductant is another important limitation.

Both ISCO and ISCR have several limitations; therefore, it is necessary to continue with research and development to overcome those problems for an extensive application of these technologies.

14.3.2 SOIL WASHING/FLUSHING

Soil washing consists of the removal of contaminants (organic or inorganic) by means of an extraction liquid. Soil washing involves excavating the soil and transporting it to a treatment site, where the oversized particles are subsequently screened out, and the washing solution, composed, generally, of water, surfactants, and acids or alcohols, is added. The washing solution has various effects, as it can affect acid-base reactions or generate redox reactions, causing desorption and/or solubilization of contaminants. The wash solution is drained at the end of each cycle and fresh wash solution is added. This procedure is repeated until the contaminant level is within the permitted limits or after an environmental risk assessment. Recovered effluent is treated to separate the pollutant. *Ex situ* soil washing is mainly employed to restore heavy metals polluted soil. Its effectiveness was demonstrated for the restoration of soils contaminated with arsenic, chromium, lead, mercury, nickel, among others (Kumar et al., 2022).

14.3.3 STABILIZATION/SOLIDIFICATION

Stabilization/solidification (S/S) is a remediation technology that uses binders or additives, which are mixed with the contaminated soil, generating a solid material with high structural integrity. As a result, contaminants are immobilized in the solidification matrix, becoming a stable product. Commonly used solidifying agents include different types of Portland cement, bitumen, lime, calcium alite, and asphalt (Yang et al., 2022).

This technology can be applied both *in situ* and *ex situ*. In the former case, binder or additives are injected into contaminated soils. In the second case, the soil is excavated, placed in containers, and machine-mixed with the solidifying agents; subsequently the treated soil is kept in the containers or returned to the original site.

Some of the problems associated with the use of cement-based S/S include massive CO_2 emission, low durability, decalcification, degradation, and potential leaching due to the low compatibility between clay and cement.

Vitrification is based on the use of electrical energy to significantly increase the temperature of the soil and melt it, and then form an inert gas upon cooling. To carry this out, graphite electrodes are inserted into the contaminated matrix, energy is supplied, and, due to high electrical resistance, temperatures above 1700 °C are achieved, resulting in the destruction or elimination of volatile organic compounds and the melting of the soil. Subsequently, due to cooling, a glassy solid is obtained with the metals immobilized inside. This product can be removed from the site. A disadvantage of this system is that it requires a high energy demand, making it very expensive.

S/S treatments have several limitations. During soil mixing, VOCs can diffuse and, therefore, they are not immobilized. Moreover, the depth of the contaminant limits some processes. However, this technology could be improved by employing more sustainable and efficient materials (Yang et al., 2022).

14.3.4 Soil Vapor Extraction

Soil vapor extraction (SVE) is applied in the remediation of soils with high vapor pressure and low molecular weight compounds. Fresh air is used to wash the contaminated soil, which is introduced through extraction wells using vacuum systems. Thus, volatile organic compounds are entrained into the vapor phase. SVE is used for the removal of light fraction hydrocarbons, which are often found in the subsurface of service stations, pipeline spills, and storage tanks (Cao et al., 2021).

SVE is an *in situ* technology, which can be applied to large volumes of soil, with minimal disturbance. However, it has several disadvantages. It is only applicable to heterogeneous unsaturated soils with high permeability and when the aquifer is not shallow. In addition, heavier fuels are not easily removed by soil venting. Finally, the energy requirements double the cost of the project, making it economically prohibitive (Cao et al., 2021).

14.3.5 Electrokinetic Remediation

Electrokinetic remediation removes pollutants using an electric field, to eliminate charged species (ions). Electrokinetics includes water transport (electroosmosis) as well as ion transport (electromigration) as a result of a low-intensity electric field. This technology is mainly used in poorly permeable soils. Moreover, contaminants must be in ionic form dissolved in the pore water to be transported to the anode or cathode, depending on the charge. During electrolysis different compounds are generated; while hydrogen and oxygen ions are produced at the anode, hydroxyl and hydrogen gas are obtained at the cathode. OH$^-$ and H$^+$ try to migrate through the ground to the anode and cathode, respectively. As a consequence, pH variations occur, which affect the main physicochemical processes – including oxidation-reduction, precipitation-dissolution, and adsorption-desorption – that drive the speciation and partitioning of pollutants in soils. Contaminants accumulated on the electrodes must be removed, for which different technologies are applied, such as adsorption on the electrode and pumping water near the electrodes or electroplating (Bessaim et al., 2019).

Electrokinetic remediation is highly efficient for metals and other inorganic anions removal, which can be solubilized by adjusting the soil pH or adding complexing agents. For instance, this technology was successfully applied on agricultural soils for the removal of Cd, Co, Cu and Zn, enhanced by using sulfuric or organic acids. Organic compounds with moderate to high water solubility could be removed by adjusting physicochemical conditions to increase desorption of the pollutant. The removal of Reactive black 5, an industrial organic dye, is challenging because of its stable adsorption to soil particles at neutral or acidic pH; however, at alkaline pH, it solubilizes, forming an anion, which can migrate to the anode. In general, organic compounds, especially hydrophobic organic compounds (HOCs), are xenobiotic and persistent contaminants. In these cases, it is necessary to increase the solubility of HOCs by using surfactants and co-solvents prior to the application of electrokinetics.

Electroremediation is a powerful tool but it has certain disadvantages: its efficiency can be low due to variations in soil electrical conductivity, the presence of mineral deposits that raise conductivity, extreme pH values, oxidoreduction processes, and the formation of more toxic products. In addition, it has a high cost. Therefore, it is important to continue with the study and development of this technology, with the objective of improving the control and evaluation of the process, optimizing the use of surfactants, and improving the control of the current per electrode, among other processes (Lysenko et al., 2022).

14.3.6 Thermal Remediation

Thermal desorption processes can be applied on soils with volatile and semi-volatile organic contaminants (VOCs and SVOCs, respectively) (Zhao et al., 2019). This technique consists of heating the soil, in the absence of oxygen, to vaporize the pollutants. The heat accelerates the release and transport of the contaminants through the soil, to subsequently direct them to a gas treatment system with the use of a carrier gas or vacuum system. As a result of thermal treatment, products are generated that can be used as fuels, such as gases (methane, ethane, and light hydrocarbons), aqueous and oily condensates, and solid carbonaceous residues. When the heating temperature is below 350 °C, the process is called low-temperature thermal desorption; it is useful for treating VOCs with low boiling points (benzene, gasoline). When the heating temperature is between 350 and 600 °C, the process is called high-temperature thermal desorption, and is applied to compounds with high boiling points or inorganic compounds. Unlike thermal desorption, pyrolysis employs a high-temperature ramp rate. The important rise of thermal energy in the system causes thermal cracking of macromolecules into more volatile monomers. Because its high efficiency, pyrolysis has been successfully applied to non-volatile compounds, including paints, rubbers, synthetic plastics, and PAHs. Finally, incineration processes use temperatures from 870 to 1200 °C to volatilize and burn organic compounds in the presence of oxygen. Generally, fuels are used to initiate the combustion process. It was applied for the removal of PCBs, chlorinated hydrocarbons, explosives, and dioxins (Zhao et al., 2019).

Thermal processes are fast, but are energy-intensive and require specialized equipment and a large amount of skilled labor. In addition, thermal separation technologies produce vapors, solid residues such as ash, and occasionally liquid residues that require treatment or disposal. On the other hand, after treatment of soils with PET and organic compounds, the gases generated must be cleaned, as metals can react with several compounds to form more toxic and volatile compounds (O'Brien et al., 2018).

14.4 BIOLOGICAL PROCESSES

14.4.1 Microbial Remediation

14.4.1.1 Natural Attenuation

Natural attenuation (NA) encompasses naturally occurring processes in soil, such as dispersion, dilution, sorption, volatilization, and biotic and abiotic degradation, for reducing the concentration of contaminants at polluted sites without human intervention. NA requires less equipment and labor than other remediation technologies, so its costs and generated remediation wastes are low. However, NA is a very slow process since only about 10% of the total soil microbial population is composed of degrading microorganisms. For this reason, NA is applied in some situations to complement an active system of remediation. Besides, environmental factors, such as the shortage of nutrients, pollutant type and concentration, the co-contamination occurrence, the size and depth of the contaminated area, and soil type, limit the effectiveness of NA.

This technique requires pollutants readily degradable so it may be used for treating soils contaminated with benzene, toluene, ethylbenzene, xylene, chlorinated hydrocarbons, and pesticides. However, factors such as subsurface geology, hydrology, and microbiology affect the success of NA (Raimondo et al., 2019).

14.4.1.2 Biostimulation

Biostimulation involves the modification of the physical and chemical properties of soils, adjusting certain environmental parameters to stimulate the development of the native microbiota and the degradation of the hazardous and toxic contaminants (Raimondo et al., 2020a). The addition of compounds such as rate limiting nutrients (carbon, nitrogen, or phosphorus), trace minerals, organic amendments, electron acceptors or donors, water, or oxygen improve the degradative activity of native microbiota or promote co-metabolism. Even surfactants can be used to increase the contaminant bioavailability. Biostimulation is successful for treating soils affected by petroleum hydrocarbons or pesticides. Although this technique is performed by native microorganisms suitable for the environment, which are spatially well distributed within the matrix, the main challenge lies in evenly distributing the additives throughout the affected area. Generally, biostimulation with nutrients must be done in a controlled way because they can alter the particle's surface, reducing the soil's interaction with the contaminant and favoring its presence in the soil-free phase and subsequent leaching when precipitation occurs.

14.4.1.3 Bioaugmentation

Bioaugmentation includes the inoculation of individual strains or microbial consortia with the necessary catalytic capacities to degrade the target pollutants. The addition of pure or mixed cultures complements the metabolic deficiencies of the indigenous microbial communities, favoring the elimination of the toxic compound or its conversion into less toxic or easily biodegradable products and improving the bioremediation process efficiency (Aparicio et al., 2018). Generally, microorganisms with bioremediation abilities are isolated from the polluted site and propagated in the laboratory, but non-indigenous or genetically modified microorganisms might be also used for soil remediation.

The bioaugmentation success is affected by the chemical structure, concentration, and availability of the pollutants, by the soil physicochemical properties, and by the interactions among indigenous and inoculated microorganisms (Raimondo et al., 2020b). Microorganisms used to bioaugment polluted soils must remove target pollutants, survive in an external and adverse habitat, and travel through the soil pores. For this reason, the introduced microorganisms might not compete with the indigenous population for nutrients, energy, and space, to maintain useful population levels or adapt to field conditions. Besides, the inoculum size will depend on the extension of the contaminated area, the dispersion of the toxic compound, and the growth rate of the degrading microorganisms. Bioaugmentation is a widely known approach to remediate soils contaminated with pesticides, among other pollutants (Raimondo et al., 2020a,b).

14.4.1.4 Bioventing

This technology is usually applied *in situ* and includes supplying oxygen to contaminated soil to encourage the biological activity of indigenous microbiota and increase the aerobic biodegradation of organic contaminants, such as petroleum hydrocarbons, POPs, and pesticides. Bioventing generally employs low airflow rates for proving the oxygen required to support the microbial activity, maximize contaminants biodegradation, and minimize pollutants volatilization and release of the generated volatile organic compounds (Niti et al., 2013).

The success of bioventing is affected by the air injection point numbers, which allow uniform air distribution in the soil matrix. Besides this, the efficiency depends on the capacity of the air to circulate through the soil, so bioventing may be inappropriate for low permeability or heterogeneous soils. For these reasons, it is effective for medium and coarse-textured soils (Niti et al., 2013). Despite this technology presenting minimal site disturbance, aeration is an energy-consuming and costly process, and can last between months and years, depending on the characteristics of the soil, the biological metabolic activity, and the contaminant type and concentration.

14.4.1.5 Composting

Composting is a controlled biological process appropriate for restoring soils contaminated with pesticides and total petroleum hydrocarbons, where organic pollutants are transformed to innocuous, stabilized by-products, at the same time that the soil properties are improved. It is performed at elevated temperatures, as a consequence of the heat generated during the microbial transformation of the organic matter.

In this process, soils are mixed with bulking agents and organic amendments, which decrease soil matrix bulk density, enhance porosity, oxygen diffusion, and soil water-holding capacity, and increase nutrient contents. Consequently, these organic materials stimulate the microbial population able to degrade the contaminants by co-metabolic pathways and provide microorganisms responsible for breaking up and transforming toxic compounds (Tomei and Daugulis, 2013).

This technology reuses the organic biodegradable fraction of wastes and generates a mature compost with potential applications in land restoration. However, mixing large amounts of organic amendments into polluted soil requires soil excavation and large amounts of space, resulting in high effort and costs. Since contaminant bioavailability is a significant parameter affecting biodegradation rates, the selected organic materials should overcome limitations that affect the process effectiveness.

14.4.1.6 Landfarming

Landfarming uses agricultural practices to promote biodegradation of organic contaminants. For this, contaminated soil is excavated, spread in a prepared bed, and microbial biodegradation activity is stimulated by adding nutrients and periodic tilling to increase aeration (Aparicio et al., 2022). Process performance can be enhanced by adding moisture, fertilizers, bulking agents, surfactants, and microbial inoculums. Even crushed limestone or agricultural lime can be employed to regulate the soil pH to 7 (Tomei and Daugulis, 2013).

Landfarming has been successful for cleaning up soils polluted with oily sludge, pesticides, and hydrocarbons at different work scales (laboratory, pilot, and field scale), this technology being appropriate when the contamination degree does not exert an inhibitory effect on the microbial population (Tomei and Daugulis, 2013).

Although less equipment is needed for the operation and, consequently, it has lower costs than other bioremediation alternatives, landfarming requires a large workspace and long incubation periods and presents low efficiency in removing inorganic pollutants.

14.4.1.7 Biopiles

Biopiling is an *ex situ* bioremediation strategy where the soil is extracted and heaped in piles in a suitable treatment area. The aerobic biological activity is favored by adding nutrients, providing oxygen, and adjusting pH and moisture levels, which results in the enhanced pollutant degradation. Additionally, biopiles can be amended with organic materials or bulking agents to improve the bioremediation process. Biopiles include different components such as an aeration system, an irrigation/nutrient system, a treatment bed, and a leachate collection system. Biopiles might even incorporate a heating system or be covered with plastic to increase soil temperature, which promotes pollutant biodegradation (Whelan et al., 2015). Biopiles include a perforated piping system, placed along the pile, through which air is delivered to the soil. Aeration can be passive or forced by using pumps that assist the air distribution in the pile.

The efficacy of biopiles for cleaning up different texture soils, including those with high contents of clay and sand, has been successfully demonstrated at laboratory and field scale (Whelan et al., 2015), so these bioremediation systems might be constructed to suit a variety of soil conditions. Although biopiles are employed for removing petrochemical pollutants in soils, they could not be efficient for the heavy components of petroleum.

14.4.1.8 Slurry Bioreactors

These systems are typically considered the most highly designed devices for soil bioremediation, in which the contaminated matrix is mixed with water and nutrients in suitable ratios, and the resulting suspension is mechanically or pneumatically agitated to maintain homogeneous conditions and advance the interaction among pollutants and microorganisms. These systems provide the possibility of adjusting the critical operating factors, e.g. temperature, aeration agitation rates, pH, and inoculum concentrations, to provide optimal bioremediation conditions and promote the microbial activity (Robles-González et al., 2012).

The reactor includes a reaction vessel provided with a mixing system, a nutrient and oxygen supply system, and influent and effluent pumps, and might be operated in batch, semi-continuous, or continuous feed modes. Batch and semi-continuous are the most common operating modes because they adapt easily to the handling of the polluted matrices and offer better operability.

Bioreactors have been employed for the biodegradation of polycyclic aromatic hydrocarbons, total petroleum hydrocarbons, pesticides, chlorophenols, and explosives, among others. The costs of this technology are usually higher than other conventional processes as a result of soil excavation and handling, groundwater pumping, and process control.

The addition of nutrients, exogenous degrading microorganisms, different electron acceptors, and solvents or surfactants represent alternatives for enhancing the effectiveness of slurry bioreactors. Besides, a successful performance in bioreactors implies soil sieving to separate the different particle types. Often, fine particles are fed into the reactor, since they mainly adsorb the pollutants, while coarse particles may be difficult to maintain in suspension (Robles-González et al., 2012).

14.4.2 Phytomanagement

Bioremediation of soils using phytoremediation techniques can restore fertility and biodiversity of the system. Plants promote the abundance of microbial species, particularly in the rhizosphere. These plant-microbial interactions can improve phytoremediation by promoting plant growth and by modifying the bioavailability of organic and inorganic pollutants. Plant growth-promoting microorganisms can improve plant nutrition by increasing the bioavailability of plant nutrients, such as nitrogen, phosphorus, or iron. They can also influence plant physiology through the production of phyto-hormones (Simón Solá et al., 2019).

Phytotechnologies include:

- Phytoextraction uses plants that take up pollutants and accumulate them in their biomass. This is mainly used for heavy metals. Two types of plants are useful for phytoextraction: (i) plants that produce little biomass but have a high capacity to concentrate metals in their tissues (hyperaccumulator species) and (ii) plants that produce abundant biomass but accumulate low amounts of elements.
- Phytostabilization uses plant species able to produce growth covering the area, reducing the bioavailability of pollutants. The technique becomes particularly important to immobilize organic and inorganic pollutants when concentrations are high. The processes that reduce the mobility of pollutants are: (i) uptaking by the root system, and (ii) production of root exudates.

- Phytovolatilization techniques use the ability of plants species to modify contaminants into volatile substances. Organic contaminants and metals with high volatility are suitable to be remediated using this technology.
- Phytodegradation uses plants and microorganisms to degrade organic contaminants (Giaccio et al., 2023). Root exudates promote the microbial growth and diversity of microorganisms in the rhizosphere. They can stimulate the expression of microbial genes involved in the bio-degradation process. The metabolic activities of rhizospheric microorganisms increase the concentrations of several compounds which are beneficial for the solubilization of plant nutrients (Alvarez et al., 2022).

Phytoremediation techniques may also provide biomass for the bio-economy. For instance, poplar and willow species are of particular interest since they can be harvested every 5 years to produce woody biomass.

14.4.3 VERMIREMEDIATION

Shi et al. (2019) recently defined vermiremediation as a bioremediation technology that uses earthworms and their interaction with the soil and the edaphic flora, fauna, and biota to remove contaminants. One of the most used earthworms in vermiremediation is *Eisenia fetida* (Bhat et al., 2017, 2018; Usmani et al. 2020).
The following processes and mechanisms could be involved in this technology.

- Vermiextraction and vermiaccumulation refer to collection of the contaminant through epidermal and dietary uptakes, digestion and storage in lipid-rich tissues of the earthworms (Shi et al. 2019). While there is uncertainty as to whether earthworms are capable of accumulating large amounts of pollutants, this process must be taken into account to ensure that there will be no transfer of the contaminant to nearby trophic levels.
- Vermitransformation and vermidegradation include processes by which earthworms transform organic contaminants into less toxic or harmless chemical species, using enzymes (e.g., cytochrome P450 and peroxidase) and/or their associated microorganisms (Shi et al. 2019). These terms are different from vermiconversion and vermicomposting, which refer to the conversion, through the slow degrading of solid waste by earthworms and their associated microorganisms, into vermicompost (Shi et al. 2019).

The life cycle of earthworms benefits the bioremediation of co-contaminated sites because it has a positive impact on the edaphic environment. The burrows resulting from movements of the earthworms assist soil aeration and the circulation of particles, water, and nutrients (Usmani et al. 2020). In addition, the digestion and excretion of earthworms break down the soil structures, increasing the availability of organic matter, which is favorable to microbial activity.

It is a fact that earthworms improve contact between organic pollutants and the edaphic microbiota, but two situations can occur. If the availability and degradability of the contaminant are high (e.g., phenanthrene), the effect of the earthworms on the removal is minimal, since the contaminant is rapidly degraded by the autochthonous microbiota. On the other hand, if the pollutant is poorly available, but has a high degradability (e.g., anthracene), the use of earthworms will release the compound, accelerating their removal.

Referring to the elimination of PTE, the worms absorb, excrete, and immobilize them through different mechanisms. Most of the absorbed metal is eliminated in the urine and mucopolysaccharides. This ability of earthworms to increase the bioavailability of PTEs is relevant in soil restoration, especially when conventional methods remove only part of the PTEs (i.e., soil leaching) or even remain immobilized in the soil (i.e., solidification/stabilization) (Usmani et al., 2020).

Although vermiremediation is an eco-friendly technology, it is only applicable in soils where the concentration of contaminants is low or moderate, and food availability is abundant. Also, this technology is limited to weather conditions (Usmani et al. 2020).

14.5 COMBINATION OF PHYSICOCHEMICAL AND BIOLOGICAL PROCESSES FOR SOIL RESTORATION

As described previously, biological and physicochemical technologies for soil remediation have several advantages but also present some restrictions. Physicochemical technologies are generally effective and fast; however, they require high costs and can negatively affect the soil microbiota and function. On the other hand, biological techniques are eco-friendly, have good public acceptance, and lead to efficient soil restoration; however, they generally require a longer time than physicochemical techniques. In this sense, to overcome the weaknesses of individual technologies for soil remediation and/or to achieve better results, chemical and biological treatments can be strategically combined. Some examples of soil remediation through this kind of combination are cited below.

Sometimes, biological remediation techniques such as phytoremediation could not be viable in areas with high heavy metal pollution. In such cases, to offset this defect, a double-stage remediation process has been proposed, taking advantage of both physicochemical and biological remediation techniques. The addition of chelating agents and organic acids has been reported to enhance metal uptake by plants. For instance, Guo et al. (2019) performed an assisted phytoremediation by *Brassica juncea* and found that chelators enhanced the bioavailability of Cd and Zn in soil, resulting in both metals being significantly amplified in shoot and root respect to the phytoremediation alone. However, the washing process generally reduces soil fertility, which represents a limitation. In this context, Xiao et al. (2019) tested multiple washing reagents for the assisted phytoremediation of heavy metals contaminated soils. They concluded that low molecular mass organic acids are better washing reagents than strong acids or chelators for enhanced phytoremediation since they lead to similar metal removal efficiencies and lower disturbance on soil fertility. Yan et al. (2017) also revealed that the combination of phytoremediation using *Pteris vittata* L. with soil flushing of phosphate is effective to remediate As-contaminated soils. This could be due to *P. vittata* producing root exudates containing organic carbon and acids that enhance As availability in the soil, as well as increasing soil permeability, through its extensive root system.

In the case of organic pollutants, high concentrations in the contaminant source can also cause a decrease in the microbial population and diversity. In this direction, Herrero et al. (2019) suggested that the combined lactic acid biostimulation and ISCR with ZVI resulted in the fastest and more sustained dehalogenation of PCE in comparison with individual techniques. Also, other physicochemical remediation techniques have been successfully combined with biological ones to counteract the limitations of each separate approach. For instance, Boutammine et al. (2020) have shown that the combined use of S/S and biostimulation can enhance the total petroleum hydrocarbons (TPH) degradation; furthermore, the incorporation of additives in the solidification formulation as partial replacements of cement also led to a decrease of TPH. Also, Ma et al. (2016) combined bioaugmentation with a microbial consortium and biostimulation with urea and K_2HPO_4 with SVE to enhance the removal efficiency of TPH. The low-volatile compounds were completely biodegraded and transformed into more volatile ones, which were then extracted by SVE.

Combining phytoremediation with physicochemical treatments has been also reported for the successful removal of pesticides. For instance, Sánchez et al. (2020) achieved an improvement of more than 20% in atrazine removal efficiency by electrokinetic-assisted phytoremediation, compared with the unplanted treatment. Along the same lines, Romeh and Ibrahim Saber (2020) integrated bio-nanotechnology, solubilizing agents, and phytoremediation for enhanced remediation of soil contaminated with chlorfenapyr.

In particular, combined physicochemical and biological processes could be highly efficient for the remediation of environments with mixed contamination, since the remediation technologies needed are different for each kind of pollutant. Besides, metal pollutants can negatively affect the microbial activity in the soil and thus hinder the biodegradation of the organic compounds. In this sense, Dong et al. (2013) evaluated the integration of electrokinetics and biostimulation as a remediation strategy to treat Pb-oil co-contaminated soil. The combination of EDTA and Tween 80 with the electrokinetic remediation resulted in the best option for the co-contaminated soil remediation. The chelating agent eliminated the toxicity from the soil, thus activating the microbial activity over the oil, whereas the surfactant enhanced microbial growth and biodegradation, probably serving as a second substrate. Gao and Zhou (2013) also reported that nZVI-phytoremediation was more effective for the remediation of soils co-contaminated with PCB and Pb than the single techniques. The nanoparticles improved the soil quality and contributed to the adjustment of soil pH, which both enhanced the growth of plants.

Furthermore, each contaminated site has unique edaphic characteristics, which end up affecting the effectiveness of different techniques. In this context, Aparicio et al. (2021) proposed different remediation techniques depending on the grade of pollution and or organic matter content of the soil. They combined nZVI with bioremediation techniques (bioaugmentation, phytoremediation, and vermiremediation) in soils with mixed contamination of Cr and lindane with and without organic amendments. They concluded that in soils with high concentrations of Cr(VI) and lindane the combination of organic amendments, biological treatments, and nZVI was the most effective strategy, whereas for moderate levels of chromium and organic matter, bioremediation was the best option, and nZVI was not necessary.

14.6 EVALUATION OF THE EFFECTIVENESS OF REMEDIATION PROCESSES

An important aspect in any process of remediating a contaminated soil involves the assessment of its efficiency by means of both biological and chemical analyses. This is because soils are complex matrices, which can result in low recoveries of investigated compounds or in the absence of precision and reproducibility of the results, due to interferences during stages of detection and quantification in analytical procedures (Kumpiene et al., 2014).

Currently, evaluations of the health and quality of soils, the damage caused by perturbations introduced into these systems, and efficacy of remediation processes mainly consider the physicochemical parameters of these matrices. However, for environmental purposes, these evaluations based exclusively on analytical determinations do not offer any signs of the serious effects that pollutants and remediation techniques can have on organisms and biodiversity. For this reason, ecologically relevant tools are important because they allow the impact on the environment of both contamination and restoration treatments to be evaluated. To achieve these goals, biological indicators offer an integrating character, due to their greater sensitivity and higher response speed to system fluctuations; in addition, they can react in advance to irreversible changes and disturbances, thus allowing a correct and integrated evaluation of soil quality and health (Kumpiene et al., 2014).

Biological toxicity tests, also called bioassays, are widely used ecotoxicological tools that study the effect of chemical and physical agents on living organisms, with particular attention to populations and communities of defined ecosystems. These toxicity tests are diagnostic parameters that allow experimental measurements to be made in biological systems, establishing dose-response relationships under controlled experimental conditions both in field or in laboratory. However, it is important to consider that there is no a single organism that can be used to assess all possible effects on the ecosystem, under the various biotic and abiotic conditions assayed. In practice, only a few species that fulfill important functions in ecosystems, commonly called model species or bioindicators, can be used for this purpose (Ronco et al., 2004).

The responses of living organisms against a disturbance are recorded on specific biomarkers, which can be expressed at different levels of biological complexity, from morphology of cellular structures or enzymatic activities, to complete organisms, populations, or even communities. Biomarkers are measurements that reveal whether there has been exposure to contaminants and/or the effects of that exposure, and they can be both inhibition and magnification, such as germination, death, growth, proliferation, multiplication, morphological, physiological, cellular or histological (depending on the species used). Since biomarkers are characterized by their ability to predict changes at complex biological levels (population, community, ecosystem), they have been proposed as early diagnostic tools for environmental health. Some of the advantages of bioassays are the sensitivity of the used species, the reduced exposure time of the test, the low associated costs, and the requirement of simple equipment, particularly in the application to environmental samples or in the monitoring of processes, detoxification, sanitation, and effluent control (Ronco et al., 2004).

14.6.1 Toxicological Bioassays with Plant Species

The growth and development of plants are essential processes that depend on the plant genotype and the environmental conditions in which they are cultured. In this sense, the sensitivity of plants to respond quantitatively to the toxic effect of pollutants present in the soil has made them excellent tools to assess the quality/health of terrestrial ecosystems. Therefore, *in vitro* ecotoxicological bioassays using plants have gained wide acceptance as monitoring strategies; in addition, they are economical, simple, fast, and provide reliable results.

Some of the vascular plants used in toxicological bioassays are *Brassica oleracea* (wild cabbage), *Zea mays* (corn), *Lactuca sativa* (lettuce), *Cucumis sativus* (cucumber), *Barbarea verna* (cress), *Lolium perenne* (English grass), *Raphanus sativus* (radish) and *Lycopersicon esculentum* (tomato). Phytotoxicity tests are useful tools for identifying the toxic effects of the contaminants present in soils, as well as testing the success of a remediation process (Kumpiene et al., 2014).

Based on that, Raimondo et al. (2020c) assessed the effectiveness of the bioremediation of lindane contaminated soils by means of ecotoxicity tests using three dicotyledonous seeds: *Lycopersicon esculentum*, *Raphanus sativus*, and *Lactuca sativa*. Among them, the most suitable specie was *L. esculentum*, due to its higher sensitivity to lindane compared to the other two. The vigor index (VI) is a plant parameter for quantitatively determining the effect of environmental and physiological conditions on their viability. VI simplifies the comparison between treatments because it combines the germination percentages and the length of hypocotyls and radicles in a single value. In all cases, the lowest VI were obtained in the tomato seedlings grown in samples from lindane contaminated soils, confirming the toxic effect of the pesticide on the bioindicators under the studied conditions. Moreover, the VI of tomato seedlings were higher in the bioremediated soils than in the contaminated ones. Furthermore, in most cases, the VI obtained for bioremediated and natural soils did not show significant differences between them, demonstrating that lindane toxic effects were reversed into the bioremediated soils.

14.6.2 Toxicological Bioassays with Earthworms

Earthworms are one of the most important terrestrial macroinvertebrates in the majority of soils in the world, both in terms of biomass (representing 60–80% of the total biomass) and in terms of activity (performing favorable actions for the structure and soil functions). These organisms are able to keep contributing to the productivity of the soil ecosystem by preserving its structure and regulating the renewal of organic matter. So, the contamination of soils can be detrimental for the earthworm populations. These organisms, due to their morphology and behavior, live in contact with both the aqueous and solid phases of the soil, generating quantifiable and repeatable responses in states of stress on a laboratory scale and in field conditions.

Earthworms have been used for some years for ecotoxicological tests in soils due to their easy maintenance in the laboratory and sensitivity to stress generated by anthropogenic contamination, providing early information on the level of expected affectation of the terrestrial ecosystem. Among the earthworms, *Eisenia fetida* is widely used as a model organism, since it presents sensitive bioindicators that permit evaluation of the alteration of soil ecosystems produced by polluting compounds (Aparicio et al., 2019).

Aparicio et al. (2019) showed that some biomarkers of *E. fetida*, such as mortality, weight loss, coelomocyte concentration, and cell viability, were the most sentient to evaluate the toxic impact of the tested concentrations of lindane and Cr(VI). Therefore, this bioindicator and their biomarkers were suitable for evaluating the effectiveness of the bioremediation process of soils affected by mixed contamination, under these conditions. Afterwards, Aparicio et al. (2020) demonstrated that ecotoxicity bioassays carried out with *Eisenia fetida* reflected high toxicity of the untreated soil impacted with moderate concentrations of Cr(VI) and lindane which was observed as a high mortality or the worms. However, in soils in which bioremediation by an actinobacteria consortium was carried out, significant alleviation of toxicity of the matrix to the earthworms was observed, with an increase of 97% in their survival rates. The actinobacteria consortium was able both to degrade lindane and to decrease the concentration of Cr(VI). This depletion of the pollution levels dropped the toxicity of the co-contaminated soils, as was revealed by higher *E. fetida* survival rates.

14.7 CONCLUSIONS AND PERSPECTIVES

Soil contamination has been increasing in recent decades due to the intensification in high-intensity industrial and agricultural activities, as well as deficient disposal practices in landfills, which negatively affect the soil, causing its degradation. Therefore, the high number of toxic and recalcitrant compounds reaching the soil represents a hazard to natural ecosystems and human health. In addition, the effects of pollution are intensified in sites where organic pollutants and inorganic compounds are found in concentrations higher than those allowed, a phenomenon known as mixed contamination or co-contamination. As a consequence of a greater global awareness of the scarce availability of high-quality soils and associated natural resources, there is an increase in research and development activities with the aim of cleaning up contaminated sites with an ecological and low-cost approach.

In this sense, there is a great diversity of strategies for the treatment of impacted soils, which present different efficiencies according to the characteristics of the soil and the contaminants, atmospheric conditions, and the extent of contamination, among other factors. The choice of one or another technology will also depend on the operating costs, the time required to complete the process, the by-products generated by the treatments, and the final destination of the restored soil. For these reasons, this chapter summarizes several restoration strategies for polluted soils based on the intrinsic characteristics of the remediation processes, taking into account the advantages and limitations of each technique. Currently, research needs in environmental remediation involve the improvement of the efficiency of available technologies, which can present certain shortcomings for the removal of some pollutants, as well as the enlargement of low costs engineered solutions suitable for their application. In other words, the goal of current trends is to find the most effective remediation methods with fewer weaknesses. For this purpose, the integration of biological treatments with physicochemical ones in a coordinated manner is a promising alternative since it allows the inherent restrictions of each type of treatment to be overcome by extending their remediation capabilities, as discussed in this chapter.

Since the application of coupled strategies for the restoration of contaminated soils allows for innovative remediation technologies, their practical application on a larger scale and in uncontrolled environmental conditions should be the next step for developing and implementating such coupled technologies, in order to demonstrate their technical efficacy compared to independent soil remediation technologies.

For the foregoing, it will be possible to have innovative soil remediation technologies that are more versatile in terms of efficiency, variety of contamination to be treated (even including cases of mixed pollution), and greater flexibility and applicability (actions in contamination sources or the specific areas affected), as well as providing an improvement in the costs, terms, and environmental impact of the treatments. Perhaps, the cost of combined processes is the limiting factor for applying them more widely in larger-scale fields. So, there should be further studies of the chance of doing these combined technologies more economically and profitably for their future development, for example through using agroindustrial residues in those techniques.

In summary, new research should be focused on improving operational parameters, evaluating the environmental incidence of remediation processes, estimating the residual toxicity of the treated soil and its leachates, and implementing simple or combined processes on a field scale. In addition, modern molecular and omic approaches, enzymatic engineering, and other emerging technologies could provide new and improved strategies for remediation of soils affected by harmful pollutants.

REFERENCES

Alvarez, A., Rodríguez-Garrido, B., Cerdeira-Pérez, A. Tomé-Pérez, A., Kidd, P. and A. Prieto-Fernández. 2022. Enhanced biodegradation of hexachlorocyclohexane (HCH) isomers by *Sphingobium* sp. strain D4 in the presence of root exudates or in co-culture with HCH-mobilizing strains. *Journal of Hazardous Materials* 433:128764.

Aparicio, J. D., Simón Solá, M. Z., Benimeli, C. S., Amoroso, M. J., and M. A. Polti. 2015. Versatility of *Streptomyces* sp. M7 to bioremediate soils co-contaminated with Cr(VI) and lindane. *Ecotoxicology and Environmental Safety* 116: 34–39.

Aparicio, J. D., Raimondo, E. E., Gil, R. A., Benimeli, C. S., and M. A. Polti. 2018. Actinobacteria consortium as an efficient biotechnological tool for mixed polluted soil reclamation: experimental factorial design for bioremediation process optimization. *Journal of Hazardous Materials* 342: 408–417.

Aparicio, J. D., Garcia-Velasco, N., Urionabarrenetxea, E., Soto, M., Álvarez, A., and M. A. Polti. 2019. Evaluation of the effectiveness of a bioremediation process in experimental soils polluted with chromium and lindane. *Ecotoxicology and Environmental Safety* 181: 255–63.

Aparicio, J. D., Lacalle, R. G., Artetxe, U., Soto, M., Álvarez, A., and M. A. Polti. 2020. Gentle remediation options for soil with mixed chromium (VI) and lindane pollution: biostimulation, bioaugmentation, phytoremediation and vermiremediation. *Heliyon* 6: e04550.

Aparicio, J. D., Lacalle, R. G., Artetxe, U., Urionabarrenetxea, E., Becerril, J. M., Polti, M. A., Garbisu, C., and M. Soto. 2021. Successful remediation of soils with mixed contamination of chromium and lindane: Integration of biological and physico-chemical strategies. *Environmental Research* 194:110666.

Aparicio, J. D., Raimondo, E. E., Saez, J. M., Costa-Gutierrez, S. B., Álvarez, A., Benimeli, C. S., and M. A. Polti. 2022. The current approach to soil remediation: A review of physicochemical and biological technologies, and the potential of their strategic combination. *Journal of Environmental Chemical Engineering* 107141.

Bessaim, M.M., Missoum, H., Bendani, K, Bekkouche M.S., and N. Laredj. 2019. Removal of hazardous cationic salt pollutants during electrochemical treatment from contaminated mixed heterogeneous saline soil. *Arabian Journal for Science and Engineering* 44: 4783–94.

Bhat, S.A., Singh, J. and A.P. Vig. 2017. Amelioration and degradation of pressmud and bagasse wastes using vermitechnology. *Bioresource Technology* 243: 1097–1104.

Bhat, S.A., Singh, S., Singh, J., Kumar, S. and A.P. Vig. 2018. Bioremediation and detoxification of industrial wastes by earthworms: vermicompost as powerful crop nutrient in sustainable agriculture. *Bioresource Technology* 252: 172–179.

Bhat, S.A., Cui, G., Yaseera, N., Lei, X., Ameen, F. and F. Li. 2022. Removal potential of microplastics in organic solid wastes via biological treatment approaches. *Microbial Biotechnology: Role in Ecological Sustainability and Research*: 255–263.

Boutammine, H., Salem, Z., and M. Khodja. 2020. Petroleum drill cuttings treatment using stabilization/ solidification and biological process combination. *Soil and Sediment Contamination: An International Journal* 29:369–383.

Cao, W., Zhang, L., Miao, Y., and Q. Liufan. 2021. Research progress in the enhancement technology of soil vapor extraction of volatile petroleum hydrocarbon pollutants. *Environmental Science: Processes and Impacts* 23: 1650–62.

Chae, Y., and Y. J. An. 2018. Current research trends on plastic pollution and ecological impacts on the soil ecosystem: A review. *Environmental Pollution* 240: 387–95.

Dominguez, C. M., Romero, A., Fernandez, J., and A. Santos. 2018. *In situ* chemical reduction of chlorinated organic compounds from lindane production wastes by zero valent iron microparticles. *Journal of Water Process Engineering* 26: 146–55.

Dong, Z. Y., Huang, W. H., Xing, D. F., and H. F. Zhang. 2013. Remediation of soil co-contaminated with petroleum and heavy metals by the integration of electrokinetics and biostimulation. *Journal of Hazardous Materials* 260:399–408.

Fuentes, M. S., Álvarez, A.., Cuozzo, S. A., and Benimeli, C. S. 2023. Combination of slurry-bioreactors and actinobacteria consortia as strategy to bioremediate chlordane-contaminated soils. *Chemosphere* 337: 139270.

Gao, Y. Y., and Q. X. Zhou. 2013. Application of nanoscale zero valent iron combined with *Impatiens balsamina* to remediation of e-waste contaminated soils. *Advanced Materials Research* 790:73–76.

Giaccio, G.C.M, Saez, J.M., Estévez, M.C., Salinas, B., Corral, R.A., De Gerónimo, E., Aparicio, V. and A. Álvarez. 2023. Developing a glyphosate-bioremediation strategy using plants and actinobacteria: potential improvement of a riparian environment. *Journal of Hazardous Materials*. https://doi.org/10.1016/j.jhazmat.2022.130675

Guo, D., Ali, A., Ren, C., Du, J., Li, R., Lahori, A. H., Xiao, R., Zhang, Z., and Z. Zhang. 2019. EDTA and organic acids assisted phytoextraction of Cd and Zn from a smelter contaminated soil by potherb mustard (*Brassica juncea*, Coss) and evaluation of its bioindicators. *Ecotoxicology and Environmental Safety* 167:396–403.

Gutiérrez, C., Fernández, C., Escuer, M., Campos-Herrera, R., Beltrán Rodríguez, M. E., Carbonell, G., and J. A. Rodríguez Martín. 2016. Effect of soil properties, heavy metals and emerging contaminants in the soil nematodes diversity. *Environmental Pollution* 213: 184–94.

Herrero, J., Puigserver, D., Nijenhuis, I., Kuntze, K., and J. M. Carmona. 2019. Combined use of ISCR and biostimulation techniques in incomplete processes of reductive dehalogenation of chlorinated solvents. *Science of the Total Environment* 648:819–829.

Kaur, J., Bhatti, S.S., Bhat, S.A., Nagpal, A.K., Kaur, V. and J.K. Katnoria. 2021. Evaluating potential ecological risks of heavy metals of textile effluents and soil samples in vicinit of textile industries. *Soil Systems* 5(4): 63.

Kumar, M. Bolan, N., Jasemizad, T., Padhye, L. P., Sridharan, S., Singh, L., Bolan, S., O'Connor, J., Zhao, H., Shaheen, S. M., Song, H., Siddique, K. H. M., Wang, H., Kirkham, M. B., and J. Rinklebe. 2022. Mobilization of contaminants: Potential for soil remediation and unintended consequences. *Science of The Total Environment* 839: 156373.

Kumpiene, J., Bert, V., Dimitriou, I., Eriksson, J., Friesl-Hanl, W., Galazka, R., Herzig, R., Janssen, J., Kidd, P., Mench, M., Müller, I., Neu, S., Oustriere, N., Puschenreiter, M., Renella, G., Roumier, P., Siebielec, G., Vangronsveld, J., and N. Manier. 2014. Selecting chemical and ecotoxicological test batteries for risk assessment of trace element-contaminated soils (phyto)managed by gentle remediation options (GRO). *Science of the Total Environment* 496: 510–522.

Lysenko, L., Mishchuk, N., and V. Kovalchuk. 2022. Basic principles and problems in decontamination of natural disperse systems. The electrokinetic treatment of soils. *Advances in Colloid and Interface Science* 310: 102798.

Ma, J., Yang, Y., Dai, X., Li, C., Wang, Q., Chen, C., Yan, G., and S. Guo. 2016. Bioremediation enhances the pollutant removal efficiency of soil vapor extraction (SVE) in treating petroleum drilling waste. *Water, Air, and Soil Pollution* 227:465.

Naseri-Rad, M., Berndtsson, R., Persson, K.M., and K. Nakagawa. 2020. INSIDE: An efficient guide for sustainable remediation practice in addressing contaminated soil and groundwater. *Science of The Total Environment* 740: 139879.

Naveed, M., Moldrup, P., Arthur, E., Holmstrup, M., Nicolaisen, M., Tuller, M., Herath, L., Hamamoto, S., Kawamoto, K., Komatsu, T., Vogel, H. and L. Wollesen de Jonge. 2014. Simultaneous loss of soil biodiversity and functions along a copper contamination gradient: When soil goes to sleep. *Soil Science Society of America Journal* 78: 1239–50.

Niti, C., Sunita, S., Kamlesh, K., and K. Rakesh. 2013. Bioremediation: An emerging technology for remediation of pesticides. *Research Journal of Chemistry and Environment* 17: 4.

O'Brien, P. L., DeSutter, T. M., Casey, F. X., Khan, E. and Wick, A. F. 2018. Thermal remediation alters soil properties–a review. *Journal of Environmental Management* 206: 826–835.

Raimondo, E. E., Aparicio, J. D., Briceño, G. E., Fuentes, M. S., and C. S. Benimeli. 2019. *Lindane bioremediation in soils of different textural classes by an actinobacteria consortium. Journal of Soil Science and Plant Nutrition* 19: 29–41.

Raimondo, E. E., Saez, J. M., Aparicio, J. D., Fentes, M. S., and C. S. Benimeli, 2020a. Coupling of bioaugmentation and biostimulation to improve lindane removal from different soil types. *Chemosphere* 238:124512.

Raimondo, E. E., Aparicio, J. D., Bigliardo, A. L., Fuentes, M. S., and C. S. Benimeli. 2020b. Enhanced bioremediation of lindane-contaminated soils through microbial bioaugmentation assisted by biostimulation with sugarcane filter cake. *Ecotoxicology and Environmental Safety* 190:110143.

Raimondo, E. E., Saez, J. M., Aparicio, J. D. Fuentes, M. S., and C.S. Benimeli. 2020c. Bioremediation of lindane-contaminated soils by combining of bioaugmentation and biostimulation: Effective scaling-up from microcosms to mesocosms. *Journal of Environmental Management* 276: 111309.

Robles-González, I. V., Ríos-Leal, E., Sastre-Conde, I., Fava, F., Rinderknecht-Seijas, N., and H. M. Poggi-Varaldo. 2012. Slurry bioreactors with simultaneous electron acceptors for bioremediation of an agricultural soil polluted with lindane. *Process Biochemistry* 47: 1640–1648.

Romeh, A. A., and R. A. Ibrahim Saber. 2020. Green nano-phytoremediation and solubility improving agents for the remediation of chlorfenapyr contaminated soil and water. *Journal of Environmental Management* 260:110104.

Ronco, A., Diaz Baez, M.C., and Y. Pica Granados. 2004. Conceptos generales. Ensayos de toxicidad aguda con semillas de lechuga (*Lactuca sativa* L.). En: *Ensayos toxicológicos y métodos de evaluación de calidad de aguas. Estandarización, intercalibración, resultados y aplicaciones*, ed. G. Castillo Morales, 17–30. Canada. IDRC, IMTA.

Sabreena., Hassan, S., Bhat, S.A., Kumar, V., Ganai, B.A. and F. Ameen. 2022. Phytoremediation of heavy metals: An indispensable contrivance in green remediation technology. *Plants* 11(9): 1255.

Sánchez, V., López-Bellido, F. J., Rodrigo, M. A., Fernández, F. J., and L. Rodríguez. 2020. A mesocosm study of electrokinetic-assisted phytoremediation of atrazine-polluted soils. *Separation and Purification Technology* 233:116044.

Shi, Z., Liu, J., Tang, Z., Zhao, Y., and C. Wang. 2019. Vermiremediation of organically contaminated soils: Concepts, current status, and future perspectives. *Applied Soil Ecology* 147: 103377.

Simón Solá, M.Z., Lovaisa, N., Davila Costa, J.S., Benimeli, C.S., Polti, M.A. and A. Alvarez. 2019. Multi-resistant plant growth-promoting actinobacteria and plant root exudates influence Cr(VI) and lindane dissipation. *Chemosphere* 222:679–687

Song, Y., Fang, G., Zhu, C., Zhua, F., Wu, S., Chen, N., Wu, T., Wang, Y., Gao., J and D. Zhou. 2019. Zero-valent iron activated persulfate remediation of polycyclic aromatic hydrocarbon-contaminated soils: An in situ pilot-scale study. *Chemical Engineering Journal* 355: 65–75.

Tandon, P. K., and S. B. Singh. 2016. Redox processes in water remediation. *Environmental Chemistry Letters* 14: 15–25. doi: 10.1007/s10311-015-0540-4

Tomei, M. C., and A. J. Daugulis. 2013. *Ex situ* bioremediation of contaminated soils: an overview of conventional and innovative technologies. *Critical Reviews in Environmental Science and Technology* 43: 2107–2139.

Usmani, Z., Rani, R., Gupta, P., and M. N. V. Prasad. 2020. Vermiremediation of agrochemicals. In *Agrochemicals Detection, Treatment and Remediation*, ed M. N. V. Prasad, 329–367. Butterworth-Heinemann.

Wei, K. H., Ma, J., Xi, B. D., Yu, M. D., Cui, J., Chen, B. L., Li, Y., Gu, Q. B. and X. S. He. 2022. Recent progress on *in-situ* chemical oxidation for the remediation of petroleum contaminated soil and groundwater. *Journal of Hazardous Materials* 432: 128738.

Whelan, M. J., Coulon, F., Hince, G., Rayner, J., McWatters, R., Spedding, T., and I. Snape. 2015. Fate and transport of petroleum hydrocarbons in engineered biopiles in polar regions. *Chemosphere* 131: 232–240.

Xiao, R., Ali, A., Wang, P., Li, R., Tian, X., and Z. Zhang. 2019. Comparison of the feasibility of different washing solutions for combined soil washing and phytoremediation for the detoxification of cadmium (Cd) and zinc (Zn) in contaminated soil. *Chemosphere* 230:510–518.

Yan, X., Liu, Q., Wang, J., and X. Liao. 2017. A combined process coupling phytoremediation and in situ flushing for removal of arsenic in contaminated soil. *Journal of Environmental Sciences* 57:104–109.

Yang, T., Xue, Y., Liu, X., and Z. Zhang. 2022. Solidification/stabilization and separation/extraction treatments of environmental hazardous components in electrolytic manganese residue: A review. *Process Safety and Environmental Protection* 157: 509–526.

Zhang, X., Wells, M., Niazi, N.K., Bolan, N., Shaheen, S., Hou, D., and Wang Z. 2022. Nanobiochar-rhizosphere interactions: implications for the remediation of heavy-metal contaminated soils. *Environmental Pollution* 299: 118810.

Zhao, C., Dong, Y., Feng, Y., Li, Y., and Y. Dong. 2019. Thermal desorption for remediation of contaminated soil: A review. *Chemosphere* 221: 841–55.

15 Polyaromatic Hydrocarbons

Sources, Detection, Risk Assessment, and Remediation

Bhumi M. Javia, Dushyant R. Dudhagara, and Anjana K. Vala

15.1 INTRODUCTION

Polycyclic aromatic hydrocarbons (PAHs) are a broad category of several hundred chemically related, persistent organic chemicals with varying toxicity and carcinogenicity. PAHs are major pollutants in environmental bodies due to their toxic effect and occurrence, as well as bioaccumulation potential. PAHs pollution is mostly caused by an increasing population, fast industrialization and extensive petroleum utilization. Under inefficient combustion conditions (such as an insufficient oxygen supply), PAHs are typically released from a variety of exhausts (Nam et al., 2003; Agrawal et al., 2021). The majority of them are formed through the heat decomposition (pyrolysis) and recombination (pyrosynthesis) of various organic compounds. PAHs reach the environment primarily through releases to soil, water and air from forest fires, volcanos, burning wood, automobile exhausts and factories (Lee et al., 2007). Additionally, PAHs are present in materials like coal, roofing tar, coal tar pitch, crude oil, creosote, and soot (Kumar et al., 2014). Currently, both individual and mixed reports of more than 100 different PAHs have been reported. Low molecular weight PAHs (LMW-PAH) contains three or fewer aromatic rings, whereas high molecular weight (HMW-PAH) contain four or more rings. The following are a few examples of PAHs: benzo[k] fluoranthene, benzo[a] pyrene, benzo[b] fluoranthene, benzo[j] fluoranthene, benzo[e] pyrene, naphthene, anthracene, benzo[a] anthracene, chrysene, fluoranthene and fluorine (Kuppusamy et al., 2017).

The carcinogenicity of PAH increases as its molecular weight increases, whereas acute toxicity decreases. The first chemical carcinogen to be identified was benzo[a] pyrene (Lee et al., 2007). PAHs are well-known for being teratogenic and carcinogenic (Ravindra et al., 2001). Terrestrial organisms are less affected by soil PAHs, while in aquatic biota, high toxicity of PAHs was observed. In aquatic organisms, toxicity of PAHs is influenced by photooxidation and metabolism. Typically, PAHs are more hazardous in the presence of UV. In mammals, PAH absorbs through dermal contact, inhalation and ingestion. Plants take up soil PAH via roots and translocate it to stems, leaves and other parts (Kumar et al., 2014). Many factors affect the PAHs absorption in plants include solubility in water, concentration, physicochemical state and type of soil. Some plants produce substances that protect them from PAHs. However, some plants can produce PAH as a growth-regulating hormone. In fish and shellfish, PAHs levels have been reported to be significantly high as compared to the natural environment and bioaccumulation of PAHs has been observed in terrestrial invertebrates (Kumar et al., 2014). Biomagnification of PAHs can be prevented by metabolism or degradation of PAHs by remediation. Microorganism-facilitated PAHs degradation involves the biological transformation of complex substances to simpler compounds for example H_2O, CO_2 (in aerobic condition), or CH_4 (in anaerobic condition) (Premrath et al., 2021).

DOI: 10.1201/9781003408352-18

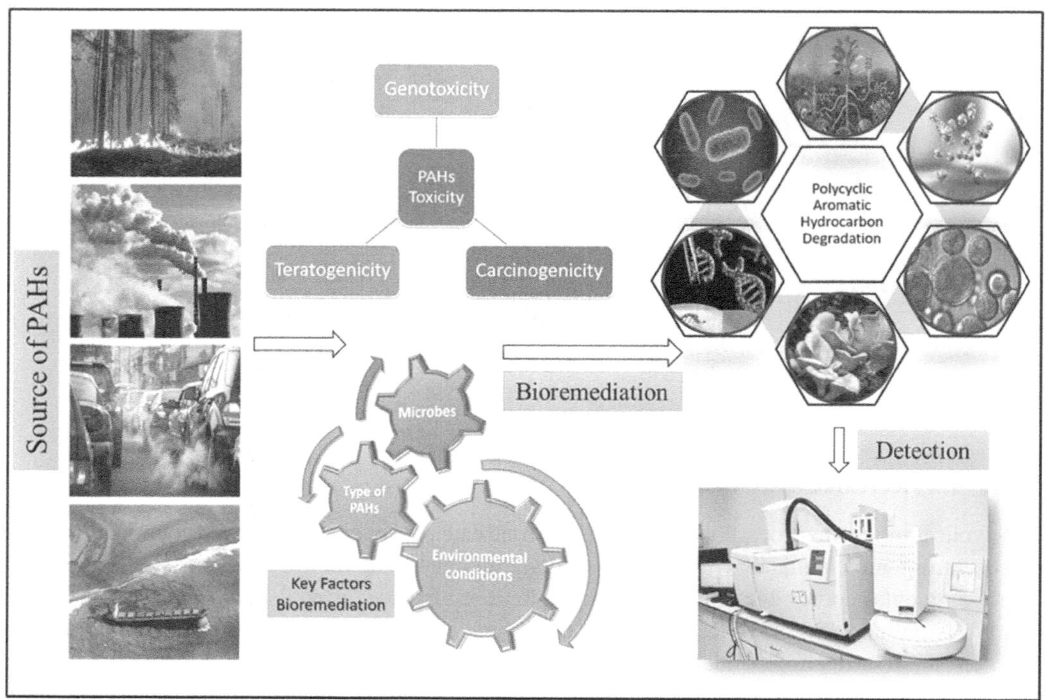

FIGURE 15.1 Polycyclic aromatic hydrocarbon sources, toxicity, remediation and detection.

A lot of research has been done on PAHs breakdown by plants, fungi, bacteria and algae due to their quick adaptability in an environment. Multiple factors, including microorganisms' population, chemical structures of pollutant, availability of nutrient, level of acclimatization, pH, temperature and oxygen affect the rate of PAHs breakdown (Figure 15.1) (Beyer et al. 2010; Hussain et al., 2018). The main aim of this chapter is to outline the source of PAHs, their identification, detection and risk assessment techniques, and several bioremediation technologies to remove or reduce level of PAHs in environment.

15.2 SOURCE OF PAHS

15.2.1 Natural and Anthropogenic Emission Sources

PAHs pollution sources are categorized into two types: natural emission source and anthropogenic emission source. Natural emission sources include natural petroleum, vegetative decay, erosion of sedimentary rocks containing PAHs, and bacterial, algal and plant synthesis. Anthropogenic emissions are the main cause of PAHs pollutions. Anthropogenic emission sources include incomplete combustion, such as incineration and industrial processes, automotive emissions, fungicide and insecticide production, petroleum product spills, cigarette and cigar smoke, jet aircraft exhausts, garbage burning, smoke from wood-burning stoves, and sewage sludge (Mojiri et al., 2019).

Anthropogenic emission (mobile, agricultural, domestic and industrial sources) is the main cause of PAHs pollution. Mobile emission sources include vehicle exhausts, such as off-road heavy-weight and lightweight vehicles, trains, aircrafts and ships. Agricultural emission sources involve the burning of biomass and waste material from agriculture. Domestic emission sources include cooking on oil/gas burners, burning of garbage, burning of wood, kerosene/wood stoves, coal coking and other residential heating. Industrial emission sources include electric arc furnaces, gasification of coals, oxygen furnaces, gasoline engines and diesel engines (Ravindra et al., 2008; Patel et al., 2020)

In a rural area, high PAHs pollution is observed as a consequence of domestic and agricultural sources, while mobile and industrial sources cause pollution in urban areas. PAHs levels are high in winter, followed by spring, autumn, summer. In winter, high PAHs concentration is observed by reason of atmospheric conditions like calm wind and temperatures below normal, which leads to lower photodegradation and poor diffusion (Miura et al., 2019).

15.2.2 PETROGENIC, PYROGENIC AND BIOGENIC SOURCES

Polycyclic aromatic hydrocarbon sources are classified into three groups based on their origin of production: pyrogenic, petrogenic and biogeric sources (Mojiri et al., 2019).

Pyrogenic PAH is formed during the process of pyrolysis. Pyrolysis occurs at high temperatures, about 350–1200 °C. Pyrogenic PAH is mostly found in urban areas that are near to major sources of PAHs. When an organic substance is exposed to high temperatures with a low oxygenic or no oxygenic condition, it will lead to formation of pyrogenic PAHs. Incomplete combustion of motor fuel or wood can also cause the formation of pyrogenic PAHs. Pyrogenic PAHs are also released during the conversion of coals into coke or tar or during the pyrolytic refinement of fuels into lower hydrocarbons (Guarino et al., 2019).

Petrogenic PAHs are formed through the maturation of crude oils and other similar processes. Major petrogenic PAH sources include ocean and river water oil spills, above/below ground oil storage tank leakages, and releases of motor oils, gasolines and other similar substances connected with transportation. PAHs can be synthesized biologically by bacteria and plants or through degradation of organic matter. Volcanic eruption, forest fire, and biological synthesis are considered natural source of PAH in the environment (Jiao et al., 2017; Guarino et al., 2019).

PAHs in the soil can accumulate in living biotic components through the food supply chain, causing indirect or direct human exposure (Bortey-Sam et al., 2014). Therefore, it is a critical step to identify the source of PAHs to reduce ecological and health issues.

15.3 SOURCE IDENTIFICATION METHODS

Identification of the relevant source of PAHs is important to know the cause of pollution, to develop new strategies for degradation of hydrocarbon and to reduces risk associated with health. Three approaches exist for PAHs identification: molecular diagnostic ratios (DRs), compound-specific isotopic analysis (CSIA) and receptor models. Receptor models include principal component analysis (PCA), positive matrix factorization (PMF), chemical mass balance (CMB) and Unmix. Use of multiple techniques is important to identify precise sources of PAHs (Figure 15.2) (Kumar et al., 2021).

Molecular DRs provide qualitative data while receptor model provides quantitative information. Pyrogenic PAHs consist of four or more aromatic rings while petrogenic PAHs have two to three aromatic rings. Therefore, a low molecular weight to high molecular weight (LMW/HMW) ratio of more than 1 indicates pyrogenic sources and less than 1 indicates petrogenic sources (Zheng et al., 2016).

In CSIA, a 13C/12C ratio produced through combustion processes from materials containing carbons is unique and diverse for specific emission sources. The ratio 13C/12C provides a stable signature of carbon isotope (δ13C) and an effective way for studying the carbon cycles at the molecular levels is to identify sources of PAHs. SCIA is more useful to determine the source of PAHs than DRs since the resultant 13C isotopes are not influenced by environmental deprivation (Okuda et al., 2002; Yang et al., 2021).

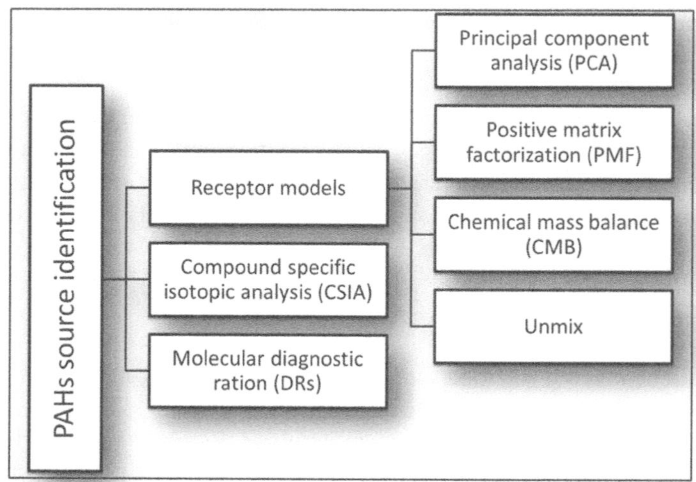

FIGURE 15.2 Method for identification of PAHs source.

The receptor models (CMB, PMF, Unmix, PCA) utilize statistical tools to recognize the PAHs source and it reduces the variable numbers in analytical data sets. The CMB model relies on database sources because it accepts relatively fewer observations but still needs previous information from reliable sources (Clarke et al., 2012). The CMB model has the potential of producing inaccurate results since the source databases' composition varies from the emission sources (Hopke, 2016). Limitations of the CMB model include: missing values, which has a significant impact on the outcome, and chemically related sources can create analytical problems; data collected over a short duration of time may not be a useful demonstration because the contamination data profile alters over time; and significant percentages of the measured pollutant load might not be considered since indeterminate and unknown sources cannot be directly recognized (Zheng et al., 2017; Famiyeh et al., 2021).

Principal component analysis (PCA) and the positive matrix factorization (PMF) model are based on multivariant analysis. The PMF is suggested for locating the source of PAHs since it gives a time-series data of the source's contributions. The PMF model measures point by point estimations of uncertain errors occurred in data and thus gives reliable results. In this model, statistical analysis is carried out by putting multiple variables in two matrices, i.e., factors contribution and characteristic or profile. The factor matrices interpretations are done by comparing matrices with already measured information of source emissions and discharge (Wang et al., 2015).

Principal component analysis with multiple linear regression (PCA-MLR) is useful in identifying the complex characteristic of sources and relative contribution of primary source. In PCA, a quantitative estimation of potential sources is carried out by combining multiple variables with latent factors (Dudhagara et al., 2016; Zheng et al., 2016).

In contrast to PMF and PCA, Unmix does not require any preliminary information about the sources (Zheng et al., 2017). The PCA is frequently used with the receptor model to identify the source of PAHs, after that PMF, which needs a higher sample size and 10–20 studied chemical types.

15.4 ECOLOGICAL RISK ASSESSMENT OF PAHS

15.4.1 Analysis of PAHs Toxicity

The USEPA (U.S. Environmental Protection Agency, 1993) established the toxicity equivalency factor (TEF) for use with PAHs to assess the health consequences of various contaminants (Nisbet and LaGoy, 1992). The TEF values recommended by Nisbet and LaGoy (Nisbet and LaGoy, 1992) were applied to determine the PAHs' carcinogenic effect and environmental hazards. TEFs values are frequently used to calculate the toxicities of mixtures of PAHs based on how poisonous each individual PAH constituent is in comparison to the reference chemical, benzo[a] pyrene (BaP). Among the 16 PAH chemicals, benzopyrenes have the maximum cancer risk. For this reason, toxic equivalent concentration of benzo[a] pyrene ($TEQ_{B[a]P}$) are used for assessing the ecological hazards of the compound PAHs in the soil and atmosphere (Hussain et al., 2018). The formula to calculate total toxic equivalent concentration of PAH is given below (Equation 15.1).

$$TEQ_{B(a)P} = \sum_{i=1}^{i=16} \left(W_i \times TEF_i \right) \tag{15.1}$$

Where, W_i is concentration of the 'i' PAHs compound, TEF_i is toxic equivalent factor of 'i' PAH. $TEQ_{B[a]P}$ is toxic equivalent concentration of total PAH compounds. Since $TEQ_{B[a]P}$ is the most carcinogenic of the 16 PAHs, it serves as a standard reference chemical. It has a value of 1 for its lethal equivalent factor. The TEFs values for the PAHs are given in Table 15.1 (Hussain et al., 2018; Liang et al., 2022).

15.4.2 Ecological Risk Assessment of PAHs in Sediments

The ecological risk of polycyclic aromatic hydrocarbons in sediment can be evaluated using sediment quality criteria (SQC). Commonly used indicators for assessing pollution are impact range low (ERL) and effect range median (ERM). According to ERL and ERM indicators, there are three stages of sediment pollution. When pollutant concentrations are below the ERL, the chance of ecological

TABLE 15.1
TEFs and NCs, MPCs Recommended Values of Individual PAHs Compound

Sr. No.	Compounds	TEFs value	Water (Recommended values)		Sediment (Recommended values)	
			NCs	MPCs	NCs	MPCs
1	Benzo(a)pyrene	1	0.5	50	27	2700
2	Dibenzo(a,h)anthracene					
3	Indeno(1,2,3-cd)pyrene	0.1	0.4	40	59	5900
4	Benzo(k)fluoranthene				24	2400
5	Benzo(b)fluoranthene		0.1	10	3.6	360
6	Benzo(a)anthracene					
7	Anthracene	0.01	0.7	70	1.2	120
8	Benzo(g,h,i)perylene		0.3	30	75	7500
9	Acenaphthylene	0.001	0.7	70	1.2	120
10	Acenaphthene					
11	Fluorene					
12	Pyrene					
13	Naphthalene		12	1200	1.4	140
14	Phenanthrene		3	300	5.1	510
15	Fluoranthene			300	26	2600
16	Chrysene		3.4	340	107	10700

Source: Hussain et al. (2018); Cao et al. (2010).

risk is decreased (10%); there is an environmental risk to living things and their ecosystems when a contaminant concentration is > ERL but < ERM (10% to 50%); when the pollutants amount is greater than ERM, it leads to greater incidence of biological and ecological risk (>50%). As well, based on the ERL-ERM value, exceedance coefficient (K) can be used to evaluate the environmental hazard of PAH in stream sediment (Long et al., 1998). The formula for calculating exceedance coefficient (K) is given below (Equation 15.2).

$$K = CPAHs / ERL \tag{15.2}$$

Where, K is exceedance coefficient and C_{PAHs} is the amount of PAH in sediments (ng/g). $K \le 0.1$, $K \le 3$, $K \le 7$, $K \le 10$ and $K > 10$ show none, low, medium, high, and higher ecological risk levels (Liang et al., 2022).

15.4.3 Ecological Risk Assessment of PAHs in Water

The presence of PAHs in an aquatic environment can pose a hazard to aquatic life and the aquatic bionetwork. As a result, an ecological risk assessment of water has been required to determine the risk that PAHs pose to nearby organisms and the ecosystem (Wu et al. 2011). In order to measure the risk that 10 PAHs posed to the ecosystem, Kalf et al. (1997) used the risk quotient (RQ). Later, this technique was refined by Cao et al. (2017) by using toxic equivalency factors (TEF) for such 16 PAHs listed by the USEPA (Cao et al. 2017). Researchers frequently use this improved technique to evaluate the overall hazard of PAHs in both water and sediment.

RQ values entitle the level of risk caused by certain PAH and calculated as formula given below (Equation 15.3).

$$RQ = C_{PAH_s} / C_{QV} \tag{15.3}$$

Where, C_{PAHs} is the amount of certain PAHs in the medium, C_{QV} is the corresponding quality value of certain PAH in the same medium. Instead of quality value (QV), the negligible concentrations (NCs) and the maximum permissible concentration (MPCs) in the medium as suggested by Kalf et al. (1997) have been used by researchers (Table 15.1). The value of $C_{QV(NCs)}$ and $C_{QV(MPCs)}$ represents the low-risk concentrations and high-risk concentrations of PAH respectively (Cao et al. 2017). The RQ_{NCs} and RQ_{MPCs} can be denoted as given below (Equations 15.4 and 15.5).

$$RQ_{NCs} = C_{PAH_s} / C_{QV(NCs)} \tag{15.4}$$

$$RQ_{MPCs} = C_{PAH_s} / C_{QV(MPCs)} \tag{15.5}$$

Where, $C_{QV(NCs)}$ is the quality values of the NC of PAH in medium and $C_{QV(MPCs)}$ is the quality values of the MPC of PAH in the same medium. $RQ_{NCs} < 1$ represents PAHs pollution may be ignored; $RQ_{NCs} > 1$ and $RQ_{MPC} < 1$ indicate that the study region is moderately polluted; $RQ_{NCs} > 1$ and $RQ_{MPC} < 1$ denotes that the study region is severely contaminated and appropriate steps are essential to reduce PAHs.

$$RQ_{\Sigma PAHs} = \Sigma RQ_i \left(RQ_i \ge 1 \right) \tag{15.6}$$

TABLE 15.2
Risk Classifications of Individual PAHs and ΣPAHs

Risk Classification	RQ_{NCs}	RQ_{MPCs}	ΣPAHs	$RQ_{\Sigma PAHs\,(NCs)}$	$RQ_{\Sigma PAHs\,(MPCs)}$
Risk-free	0		Risk-free	0	
			Low risk	≥1; <800	0
Moderate risk	≥1	<1	Moderate risk -1	≥800	0
			Moderate risk -2	<800	≥1
High risk		≥1	High risk	≥800	≥1

Source: Cao et al. (2010); Hussain et al. (2018).

Note: The RQ_{PAHs}, $RQ_{PAHs\,(NCs)}$, and $RQ_{PAHs\,(MPCs)}$ are well-defined as follows: (Equation 6, 7, 8).

$$RQ_{\Sigma PAHs(NCs)} = \Sigma RQi_{(NCs)} \left(RQi_{(NCs)} \geq 1 \right) \tag{15.7}$$

$$RQ_{\Sigma PAHs(MPCs)} = \Sigma RQi_{(MPCs)} \left(RQi_{(MPCs)} \geq 1 \right) \tag{15.8}$$

This technique provides a means to completely take into consideration each PAH's risk to the ecosystem. The risk categories for each individual are shown in Table 15.2 (Hussain et al., 2018).

15.5 DETECTION OF PAHS

PAHs with their derivatives are globally identified from different contaminated samples using analytical methods permitted by the International Organization for Standardization (ISO), the United States Environmental Protection Agency (USEPA) and the National Institute for Occupational Safety and Health (NIOSH). The methods include spectrometric, chromatographic and immunoassay methods. Gas and liquid chromatography are the most widely utilized techniques to analyse PAHs (Pavlova et al., 2003).

The liquid chromatography (LC) method uses mass spectrometer (MS) and high-performance liquid chromatography with fluorescence (FL) and UV detector, fluorescence detector, or fluorescence detector combined with PDA. GC methods used alongside MS include MS alone and MS with thermal extraction, flame ionization detector, or Fourier transform-IR detector (Adeniji et al., 2018).

Immunoassay detection mostly includes kit-based methods for detection of PAHs and is generally used for field screening in the study of PAHs from water and soil. Affinity and accuracy of immunoassay methods for detection of aromatic compounds are lower than other standard methods (Weisman et al., 1998).

Spectrometric, infrared (IR) and ultraviolet methods (UV) are most prominent for detection of hydrocarbons. The IR spectrometric approach is quick and inexpensive, but it requires cleaning the sample prior to analysis. UV spectrometric techniques are affected by the interference of lipids or other similar substances present in the sample (Pavlova et al., 2003; Adeniji et al., 2018).

15.5.1 SAMPLE EXTRACTION METHODS

15.5.1.1 Solvent-based Extraction

PAHs-contaminated samples are hydrophobic and polar. The polarity of hydrocarbons vary based on their type. Therefore, nonpolar solvents are used for extracting PAHs from the

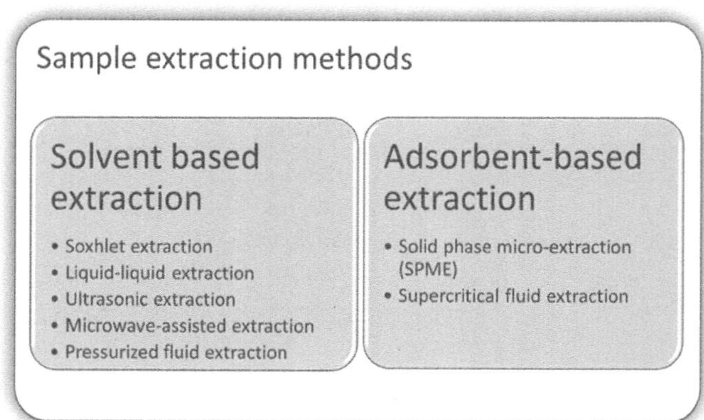

FIGURE 15.3 Sample extraction methods for analysis of PAHs.

polluted samples. Solvent-based extractions include Soxhlet extractions, microwave-assisted extractions, liquid–liquid extraction, pressurized fluid extractions and ultrasonic extractions as given in Figure 15.3.

Soxhlet extraction method is used to extract heat-stable and persistent organic compounds by using a volatile solvent inside a Soxhlet. This method uses dichloromethane and toluene, methanol and dichloromethane for extraction of PAH from oil-contaminated sites. Soxhlet extraction methods are appropriate for determining the degree of PAHs adsorption of the soil particles. Drawbacks are that volatile PAHs are lost within a short period of extraction, hazardous solvents are used, the process is time-consuming, and artefact peaks are produced either by GC or GC-MS techniques (Hawthorne et al., 2000).

In liquid–liquid extraction, immiscible solvents are employed for the extraction of PAHs from one phase to other. To achieve this, extraction using the appropriate nonpolar solvents is used directly on the soil. Solvents with moderate polarity, for example toluene, benzene, dichloromethane or their mixtures, are used to extract persistent polycyclic aromatic hydrocarbons. Methanol, Hexane, n-butanol, acetone/hexane, dichloromethane/ethanol and dichloromethane/acetone mixtures are also used (Heath et al., 1997).

Ultrasonic extraction uses shearing action of low energy sound waves. This method is simple, takes a very short time (3 min), does not harm heat sensitive compounds and trace levels can be extracted from a sample. The method's drawback is the cavitation effect caused by ultrasonic radiation pressure, which raises the molecular motion frequency of the particles present in solid samples.

In microwave-assisted extraction, the sample is heated by microwaves through ionic conduction and dipole rotations. Polar solvents like acetone, dichloromethane or hexane/acetone mixtures are used for extraction. This method uses less solvent volume and offers fast extraction of sample (Smith et al., 1993).

High temperature and pressure are used in pressurized fluid extraction to remove the heat-stable and persistent chemical from solid or semi-solid media and it works quickly with use of less solvents overall. It suggests using higher pressure (1500 to 2000 psi) and temperature (100 to 180 °C) to get recoveries that are comparable to those obtained using the Soxhlet extraction method. For the investigation of polycyclic aromatic hydrocarbon from soil samples, PFE, hyphenated with HPLC, demonstrated exceptional detection limits of 0.07 and 0.21 ppb with a fluorescence detector. However, the primary drawback associated with PFE is its initial equipment cost (Chen et al., 1999; Imam et al., 2019).

15.5.1.2 Adsorbent-based Extraction

A large quantity of solvent is used during the extraction process, which is harmful to the environment and human health and affects waste management. The direct extraction of soils using the appropriate solvent is the one benefit of solvent-based extractions for identifying hydrocarbons in bioremediation research. Several drawbacks of solvent-based extraction can be solved by adsorbent-based extraction techniques.

Typically, a membrane or tiny disposable syringe-type column packed with adsorbents is used in solid-phase extraction procedures. In the solid extracting phase, hydrophobic organic compounds are bonded to silica. The nature of adsorbents (like polar, moderately polar or nonpolar) determines interactions with the sample. In the commercially available cartridges of adsorbents, diverse affinity for analytes has been found. This approach is simple, automated, quick and environmentally friendly, and it has great productivity with little matrix interference, as shown by its superior ranges of quantitation and detection.

The solid-phase micro-extraction method (SPME) utilizes an extracting fibre at the end of the assemblage that can be dipped in an analyte for extraction and then immediately evaluated by using gas chromatography. The principles governing the mass transfer of samples in a multiphase system, as illustrated in the application of mass transfer kinetic and thermodynamics, served as the basis for the design of this approach. It employs a nonpolar polydimethylsiloxane coating on the fibre used for extraction. When compared to other extraction procedures, this extraction procedure is quick, easy, solvent-free, involves sampling in on-site application with no need of immediate analysis, provides excellent detection limits, and is best used in conjunction with the gas chromatography–mass spectrometry technique (Pawliszyn et al., 1997).

Supercritical fluid extraction uses supercritical fluid to extract samples from solid matrixes. CO_2 is typically used as a supercritical solvent that is held just above critical point, where no phase boundary exists. The advantage of this procedure stems from the characteristics of supercritical fluids, which combine the solubilization capacity of a liquid with diffusivity of gas. It takes less time and doesn't require organic solvents, but there are some restrictions, such as the procedure's selectivity and cost (Saari et al., 2007; Imam et al., 2019).

15.5.2 GAS CHROMATOGRAPHY AND GAS CHROMATOGRAPHY–MASS SPECTROMETRY (GC/MS)

Gas chromatography is used for the separation and detection of volatile, heat-stable, nonpolar organic compounds and certain semi-volatile compounds such as PAHs. This technique has high-resolution power with usage of capillary columns. Prerequisites for GC are the volatility of the sample and analyte preparation to extract them from the sample using optimum solvents. In GC, detectors used for determination of PAHs from samples include MS, Photoionization detector (PID), Fourier transform-infrared (FT-IR) and Flame ionization detector (FID). Identification of compounds is usually done by comparing it with standard hydrocarbons. PAHs containing >300 amu molecular weight are difficult to analyse through GC due to their low volatility, possibility of decomposition at higher temperature and adsorption to the GC inlet/column (Cai et al., 2009).

The GC/MS analysis of polycyclic aromatic hydrocarbons is a method of analysis that gives structural data in addition to analyte identification via retention time and mass spectral data. It is more expensive and requires high sensitivity for quantifying in selective Ion monitoring method than other non-selective techniques like GC with FID and GC with PID (SIM). However, the use of electron ionization (EI) to identify unknown compounds using MS is restricted because mass spectra alone are insufficient for accurate substance identification. For further elucidation, high-resolution mass spectrometry or chemical ionization is desirable. Two-dimensional gas chromatography, an application of multicolumn techniques for high-resolution mass spectrometry, uses a nonpolar column made entirely of polydimethylsiloxane in first dimensions to separate analytes according to boiling points before samples are eluted from it and sent to a second, shorter column for polarity-based separation. As a result, the second column in two-dimensional gas chromatography successfully separates every unresolved component (Sparkman et al., 2011; Adeniji et al., 2018).

15.5.3 Liquid Chromatography (LC) and Liquid Chromatography–mass Spectrometry (LC/MS)

HPLC systems are more widely employed than other detection techniques due to their specificity and reproducibility, and the simplicity of identifying thermal labile, less volatile, semi-volatile or non-volatile chemicals. The aqueous sample must be processed using liquid–liquid or solid phases extraction prior to the HPLC analysis per EPA 550, 610 and 8310. Solid materials are extracted by Soxhlet or ultrasonication, and samples should be dissolved in the extraction solvent prior to HPLC analysis. The MS is a detector used in LC to find and describe trace substances from a variety of chemically interesting substances. Target compounds are separated from the aqueous mobile phase using sophisticated interfaces such as electrospray, thermospray, moving belts and particle beams in LC/MS before being characterized by the mass spectrometer (Cai et al., 2009).

15.5.4 Infrared Spectroscopy

Infrared spectroscopic method (IR) has been utilized to characterize the precise structure of the target analytes. Hydrocarbon exhibits IR characteristic absorption between 2800 and 3300 cm^{-1} by reason of CeH stretching vibration, and hybridization of carbon slightly alters the exact position of the absorption. A benzene ring replacement results in distinctive absorbance peak at 680 to 900 cm^{-1}. Wu et al. identified the metabolites of phenanthrene such as organic acid and phenolic compounds using FTIR during study. This technique is simple, non-destructive and requires minimum sample preparation. Still, limitations include poor detection and lack of sensitivity. The study reported using infrared based oil content analyser to examine the rate for PAHs, TPHs and alkanes from petroleum-contaminated soil (Wu et al., 2010).

15.6 REMEDIATION OF PAHS-CONTAMINATED LAND

Remediation of a contaminated site encompasses three phases: (1) site investigation and risk assessment; (2) remedial option appraisal and (3) remediation and monitoring.

Phase I entails determining if a place has PAH contamination or not. The objective of phase I is to characterize the site and conduct a primary risk assessment. A site is declared contaminated and needs the implementation of such a soil management programme if the quantity of PAHs exceeds the allowable limit. The management programme begins with source control to prevent PAH releases. Following an awareness of the level of pollution, remediation aims are defined in many ways, including from broad national recommendations to risk assessments of specific sites that include human, ecological and environmental risks.

The site-specific risk assessment includes: (1) Formulation of the problem (PAH level and distribution in respect to receptors); (2) Exposures assessment (pathways through which PAHs are absorbed by the receptor); (3) Toxicity assessments (adverse effect caused by PAH to the receptor as well as the concentration at which adverse effects are caused) and (4) Risk characterization (comparison of data to determine adverse effect is probable to cause or not). Risk assessment methods may also be used to calculate the quantity of PAH in soil that is expected to have no detrimental effects.

Phase II encompasses identification, complete assessment of attainable and treatable remedial options and progression of remediation approaches. The results found in phase II are useful to describe the furthermost suitable remedial technology for PAHs polluted sites. Generally, laboratory testing of selected remedial techniques is applied before field application to predict the efficiency of selected remediation techniques. Determining the microbial enumeration of PAHs degraders, soil respirometry test, ecological effect and assessment of toxicity, quantity of PAH that desorbs from the soil matrix, bioluminescence-based biosensor test, rate of degradation, tracking of inoculated organisms, microbial survival test and dehydrogenase activity can be beneficial to assess the treatability and feasibility of soil remediation technology (Diplock et al., 2009; Kuppusamy et al., 2017).

In phase III, suitable remediation technology is implemented with long-term monitoring. Health and ecological risk assessments are also carried out to check if the risk-based method of remediation for the polluted site is accomplished or not. If the risk persists, additional remediation will need to be performed on the contaminated site (Kuppusamy et al., 2017).

15.6.1 BIODEGRADATION OF PAHs

Biodegradation is the use of living organisms to recycle waste material by breaking down organic and inorganic material into nutrients. The biodegradation process occurs in either the presence or absence of oxygen, called aerobic biodegradation and anaerobic biodegradation correspondingly. For PAHs degradation of a contaminated site, in situ (on site) and ex situ methods are used. On-site methods include land farming, composting and soil piles while ex situ approaches practice bioreactors to advance the degradation of PAH in soil.

15.6.1.1 Land Farming

Land farming includes the employment of conventional agricultural-based techniques for the degradation of pollutants, including tilling, bulking, irrigation and fertilizer application. In land farming, soil with PAHs pollution is ploughed periodically to increase oxygen rate and soil homogenization for biodegradation. To maximize the rate of PAHs degradation, soil properties are managed by measuring the pH, moisture levels and nutrients. PAHs-contaminated soil is mixed with nutrients, waste materials and bulking agent to enhance oxygen level and to improve degradation by aerobic microorganisms.

15.6.1.2 Composting/biopiling

Composting or bio piling is a procedure used to degrade solid waste. Composting has gained more attention as a result of its efficiency in degrading soil that has been contaminated with PAHs. Cajthaml et al. (2002) carried out removal of PAHs from soil by mixing mushroom compost with soil in a composting chamber; 20–60% degradation was observed after 54 days in the composting chamber under controlled conditions like optimal temperature and aeration. Further degradation of up to 37% to 80% was observed after hundreds more days of compost development in open ground.

15.6.1.3 Biostimulation

Biostimulation (addition of nutrients) is used to speed up the degradation rate by adding nutrients into soil to increase growth of microorganisms. This method is considered as a best option in the sites that lack nutrients. Taylor and Jones (2001) explained that adding biodiesel with inorganic nutrients into a contaminated sample led to an increase in degradation rate as compared to the nutrient-only treated sample. The rise in PAHs biodegradability was observed due to increasing PAHs bioavailability and uptake by indigenous degraders.

15.6.1.4 Bioaugmentation

Bioaugmentation is the process of adding microorganisms to remediate a contaminated place that contains a low number of indigenous PAH-degrading microorganisms. Microorganisms that are either aerobic or anaerobic can aid in the bioaugmentation process. Aerobic bioremediation is commonly used because aerobic bacteria and fungi account for the vast majority of laboratory cultivated PAH-degrading microorganisms. Using various microbial consortium of bacteria, fungus and bacterial-fungal complexes isolated from aged oil-polluted soil, Li et al. (2008) showed an enhanced level (about 45% to 56%) of PAHs decomposition in soil as well as slurry phase.

15.6.2 Remediation by Plants

Phytoremediation is a type of biological remediation method that uses plants to degrade environmental contaminants (Sabreena et al., 2022). Phytoremediation is eco-friendly, sustainable and can be used in larger scale to detoxify various types of hazardous contaminants from polluted sites (Shao et al., 2023). Plants have been extensively studied to understand several physiological, biochemical and genetic capabilities to mineralize toxic organic pollutants into non-toxic forms.

Plants are capable of degrading organic pollutants via various biophysical and biochemical mechanisms, including: manipulating plants' uptake of contaminant by providing an acidic environment in soil; adsorption of nutrients bound with pollutants; secreting enzymes that act as surfactants to enhance bioavailability of pollutants; sequestrating and detoxifying toxic chemicals into non-toxic form by phytodegradation; improving the nutrient level in soil; and interacting with the soil micro-environment to enhance degradation of xenobiotic compound by rhizospheric microbes. The benefits of the phytoremediation process include the maintenance of the soil's native properties, the use of sunlight as an energy source, and the presence of a greater extent of microorganism's biomass in the soil (Figure 15.4) (Mohan et al., 2006).

Sun et al. (2011) carried out a 7-month field experiment on PAHs dissipation by intercropping and found that, in intercropping, tall fescue and alfalfa were found higher than monoculture and unplanted soils. Lee et al. (2008) stated that *Panicum bisulcatum* (the native Korean grass spp.) and *Echinogalus crus galli* degraded 77–94% of pyrene and more than 99% of phenanthrene in

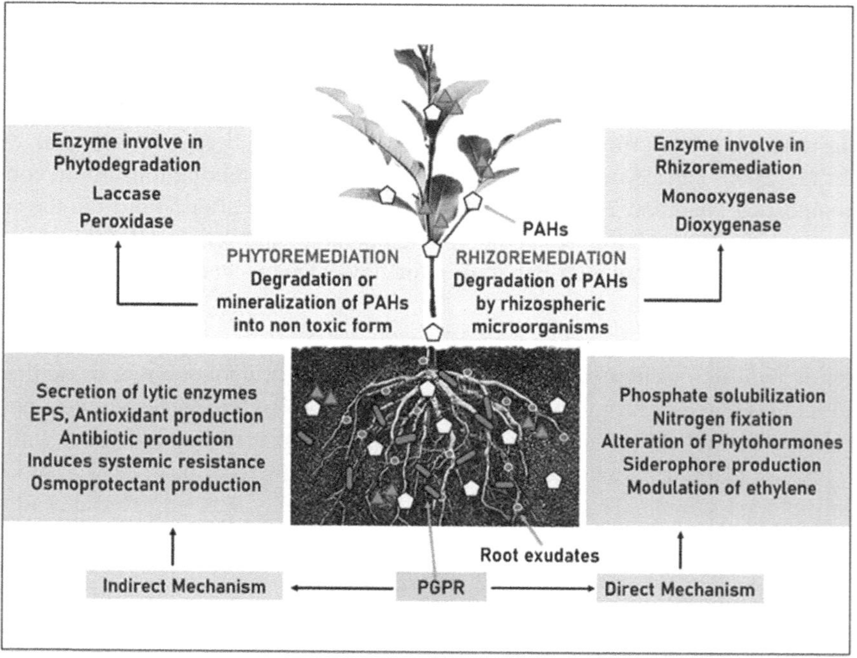

FIGURE 15.4 Remediation of PAHs pollution by plants and plant growth-promoting rhizobacteria (PGPR).

contaminated soil after 80 days of remediation process. Cofield et al. (2008), observed that *Panicum virgatum* (switchgrass) and *Festuca arundinacea* (tall fescue) removed all PAHs of 40% with the exception of indeno (1,2,3- c,d) pyrene, which only had 1.5% PAHs removal efficiency. Rezek et al. (2008), studied 15 PAHs degradation in soil vegetated with *Lolium perenne* (ryegrass). Benzo(g,h,i) perylene, indeno(1,2,3- c,d) pyrene and dibenzo(a,h) anthracene showed significant decrease up to 50%. Listeet al. (2000) observed that complete removal of phenanthrene from aged soil was most rapid as compared to pyrene and anthracene. Similarly, jack pines (*Pinus banksiana*), white pines (*Pinus strobus*) and red pines (*Pinus resinosa*) revealed 74% degradation of pyrene and 40% or less in unplanted soil within 8 weeks.

The drawbacks associated with phytoremediation are that it requires more time to remediate pollutants, due to the slow growth of plants, it is limited by the effect of environmental changes and soil characteristics etc. However, using plant growth-promoting rhizobacteria (PGPR), which can reduce environmental stress and increase plant life, is another technique to enhance phytoremediation (de Boer and Wagelmans, 2016; Bisht et al., 2015; Gan et al., 2009).

15.6.3 REMEDIATION BY RHIZOSPHERIC MICROORGANISMS

Rhizoremediation comprises of phyto-stimulation and rhizo-degradation, involving the positive interaction between plant and the rhizospheric bacteria. PGPR enhances plant growth along with soil fertility by stimulating plant root development, siderophore and phytohormone production, nitrogen fixation and mineral solubilization. Plant and microorganism's interactions in the rhizosphere is vital for remediating recalcitrant PAHs compounds from polluted environments (Figure 15.4) (Chaudhry et al., 2005). Plants secrete various organic compounds derived from photosynthesis called root exudates, which include amino acids, sugar, alcohol, proteins, nucleotides, organic acids, phenolic and flavanone compounds and certain enzymes. The nature, time and quantity of exudates from roots are vital for the process of rhizoremediation. Enzymes produced by PGPR can work as biosurfactants to raise the bioavailability of pollutants for the degradation process. The rhizoremediation process is dependent on the branching root system of plants to harbour diverse microbial flora and their metabolism, survival and environmental interaction with other microbes. Mucigel is released from the root, from the root cap or by the decaying of the complete root, which provides nutrients to rhizospheric microbes (Kuiper et al., 2004).

Rhizobacteria decrease the phytotoxic effect of the root-bound PAHs and helps in the sustainable growth of plants in PAHs polluted sites (Alagi´c et al., 2015). Alves and Grimalt (2018) found degradation of phenanthrene by alfalfa cultivar Crioulo up to 95% in a few days and pyrene in 40 days. Generally, mixed populations of bacteria can remove contaminants more efficiently than a single bacterium as a result of use of various intermediates of the degradation pathway produced by another species. Verane et al. (2020) studied *Rhizophora mangle* plants for remediation and observed degradation of 16 PAHs in sediment as compared to normal conditions. Teng et al. (2011) studied how a synergic relationship between *Rhizobium* bacteria and alfalfa plant could stimulate PAHs degradation of about 51% in 90 days. Combined cultivation of maize, white clover and ryegrass exhibited degradation of pyrene 95% and phenanthrene 98%. During the pilot scale experiment, *H. verticillate* and *V. spiralis* improved degradation of pyrene and phenanthrene hydrocarbons by 9% to 23% in 108 days (He and Chi, 2019). *Sphingomonas yanoikuyae* bacteria on root exudates was found to remove benzo[a] pyrene by constitutively produced degradative enzyme (Paquin et al., 2002).

Huang et al. (2004) devised the multi-process phytoremediation process to improve the phytoremediation process and to reduce the remediation time. Steps in the multi-process phytoremediation were land farming and mixing PAHs-degrading microbial culture into the soil. Treatment resulted in 78% degradation of the contaminant, which was 23% greater than phytoremediation alone. Over 120 days, this multi-process remediation degraded more than 5 rings containing high molecular weight PAHs as well (Bisht et al., 2015).

15.6.4 Remediation by Bacteria

Bioremediation is an effective, environmentally friendly method to degrade organic compounds by using bacteria, plants, fungi, algae and microbial enzymatic reaction to detoxify and degrade pollutants from the environment (Ghosal et al., 2016). Generally, indigenous microbes found in water and soil have the ability to degrade different hydrocarbons from oil waste (Das and Chandran, 2011). Among other microbes, bacteria are the most active potent degraders of PAHs due to diversity, adaptability and formation of less toxic metabolites (Meckenstock et al., 2016).

15.6.4.1 Aerobic and Anaerobic Degradation of PAHs by Bacteria

Biodegradation of PAHs by bacteria can be characterized into two types on the basis of either presence or absence of oxygen i.e., aerobic and anaerobic degradation. In aerobic degradation, microorganisms use oxygen as an e⁻ acceptor for PAHs ring cleavage and hydroxylation, breakdown it into CO_2 and H_2O as shown in Figure 15.5 (Gan et al., 2009). While, in an anaerobic process, reductive reaction occurs for PAHs degradation. Anaerobic microorganisms utilize nitrate, sulphate, iron, manganese and carbon dioxide in place of O_2 as an e⁻ acceptor while in some cases, anaerobic microorganisms utilize organic pollutants as e⁻ acceptor during fermentation process and release CO_2 and CH_4 as a final product during degradation of PAHs (Figure 15.5).

The genes that are responsible for catabolism of PAHs get activated in the presence of PAHs and produce enzymes that degrade hydrocarbons. In aerobic degradation, *Pseudomonas*, *Klebsiella*, *Sphingomonas*, *Mycobacteria*, *Flavobacterium*, *Burkholderia*, *Acinetobacter* and *Rhodococcus*

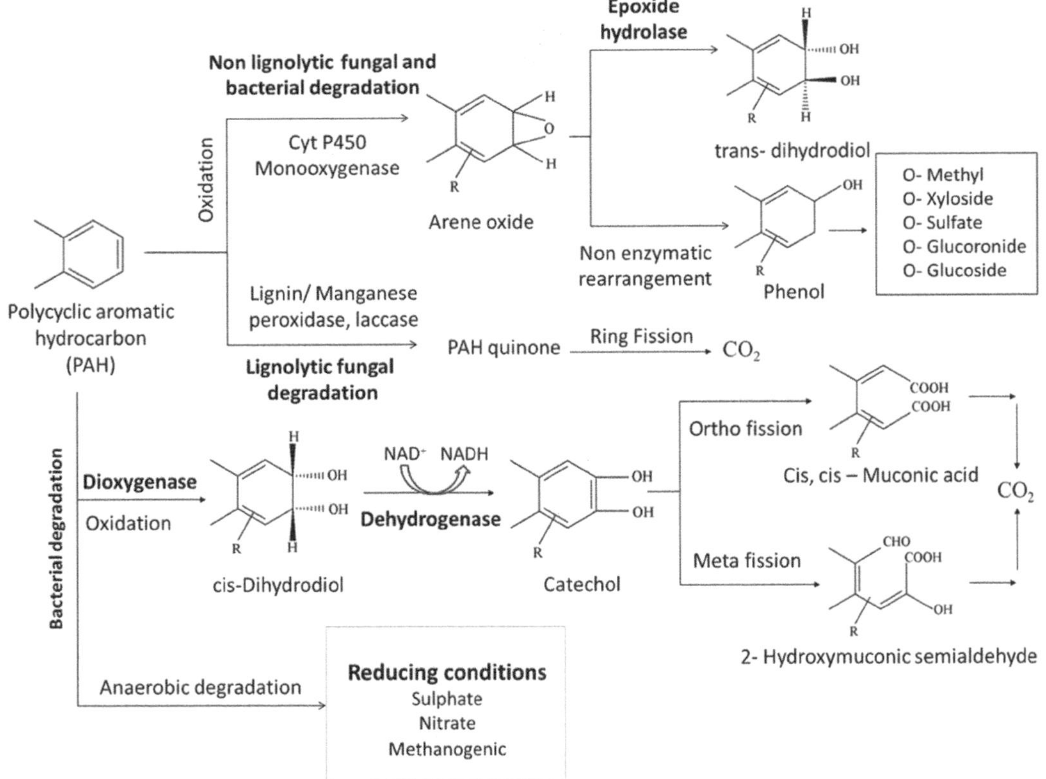

FIGURE 15.5 PAHs degradation pathway in bacteria and fungi.

Source: Modified from Haritash et al. (2009).

were reported to degrade PAHs aerobically by using PAHs as a carbon source (Dhar et al., 2022). It involves dioxygenase and dehydrogenase enzyme for degradation of PAHs. The dioxygenase enzyme promotes aromatic ring hydroxylation, diol production via dehydrogenase enzyme, ortho/ meta cleavage of diols, and catechol formation via the tricarboxylic acid cycle (TCA) (Mallick et al., 2011).

Clostridium spp. and *Paracoccus denitrificans* were reported for 50–93% degradation of pyrene under anoxic conditions by (Yuan and Chang, 2007). Anaerobic bacterial consortium was used by Chang et al. (2006) to degrade spiked PAHs in soil. The spiked soil, nutrients and consortium were mixed and then incubated at 37 °C for 90 days. Significant degradation was observed under sulphate-reducing conditions, with further improvement in degradation rates noted upon the addition of nutrients such as lactate, pyruvate, and acetate. Under iron-reducing circumstances, the *Hydrogenophaga* spp. was observed to degrade pyrene and benzo[a] pyrene at a rate of 22–25% (Yan et al., 2017). Ambrosoli et al. (2005) used a mixed population of microbes from paddy soil and obtained 30–60% degradation in denitrifying conditions. The highest degradation rate of PAHs was detected under sulphate reduction conditions and subsequently methanogenic and nitrate reduction environments (Chang et al., 2002; Gou et al., 2020).

15.6.4.2 Degradation of High Molecular Weight PAHs

The different degradative pattern and biological mechanisms are used by bacteria for the degradation of low and high molecular weight containing PAHs. Bacteria utilizes PAHs as a sole source of carbon and energy. HMW PAHs are hydrophobic, thermodynamically stable, recalcitrant and less bioavailable due to complex fused aromatic rings in their structure. Bacteria utilizes the substrates through emulsification or an enzyme like oxygenase that aids addition of O_2 atom into the hydrocarbons to facilitate degradation of complex hydrocarbons (Kumari et al., 2018; Xia et al., 2019).

Benzo[a] pyrene (BaP) is one of the most carcinogenic, high molecular weight containing and toxic form of PAHs. Maoet al. (2017) conducted a liquid culture experiment to degrade BaP by immobilization of *Sphingomonas paucimobilis* on N-agar with glucose as a carbon source. A 5% reduction in BaP content was obtained after 168 hrs of incubation. Yuan et al. (2014) isolated *Pseudomons fluoresens* and *Haemophilus* spp. and another four rod-shaped bacteria having capacity for degrading 70–100% of pyrene, acenaphthene, fluorene, phenanthrene and anthracene in 40 days. The researchers examined the impact of the medium's nutritional composition on removal efficiency and found that using fructose as the carbon source resulted in removal efficiencies of 42.5% BaP and 75% Phe (Kannan and Krishnamoorthy, 2006). According to Pozdnyakova et al. 2010a, *Beijerinckia* spp. play a significant role in the conversion of benzo[a] anthracene and benzo[a] pyrene to dihydrodiols. Dean-Ross et al. (2002) isolated *Mycobacterium flavescens* and *Rhodococcus* and observed pyrene 0.044 mg/L by *M. flavescens* and 0.470 mg/L anthracene mineralization by *Rhodococcus*. Lang et al. (2016) studied *Rhodococcus erythropolis* derived from sediment of mangrove, which was able to remove 99% of Phenanthrene, fluoranthene and anthracene in 5 weeks with treatment of biostimulation and bioaugmentation, using nitrogen and phosphorous sources (de Almeida et al., 2021). *Bacillus flexus* and *B. pumilus* and other halotolerant bacterial taxa have been isolated from different contaminated sites, including PAHs, primarily with pyrene and naphthalene; they are able to break down the pollutants, enabling the remediation of polluted places (Khanna et al., 2011). Govarthanan et al. (2020) observed that *Halomonas* halophilic bacteria in the presence of the 4% NaCl concentration degraded 74.5% of Pyrene (Imam et al., 2019).

15.6.4.3 Degradation of Low Molecular Weight PAHs

Mohapatra and Phale (2021) reported *Pseudomonas*, *Bacillus*, *Rhodoccous* and *Mycobacterium* as naphthalene degraders and observed degradation of 95% of 7 mM naphthalene by *Bordetella avium* and 77% of 800 mg/L naphthalen by *Pseudomonas* in 48 hrs of incubation time.

Mangrove sediments are closely tied to anthropological activity and are exposed to PAHs pollution. Bacteria found in mangrove sediment can remove 42% to 78% phenanthrene depending on the type of sediments (Tam et al., 2002). The genus *Alcanivorax*, which dominated at high levels of salt, and bacteria from the genus *Acinetobacter*, as per Jiang et al. (2018), are both capable of decomposing PAHs despite the media containing a variety of different substances. *Cycloclasticus* strains were capable of growing on media containing aromatic hydrocarbons such as naphthalene, Ph and anthracene (Crisafi et al., 2016). Daane et al. (2001) isolated bacteria from the rhizosphere of salt marshes plant and categorized them into Gram-positive, spore-forming *Paenibacillus*, non-spore-forming *Nocardioforms* and Gram-negative *Pseudomonas* groups. Study revealed that phenanthrene-enriched isolates were able to degrade more PAHs than naphthalene-enriched isolate. With the initial oxygenation intermediate dihydriol, bacterial species such as *Nocardia*, *Pseudomonas, Mycobacterium, Beijerinckia, Rhodococcus* and *Sphingomonas* can completely degrade anthracene hydrocarbon (Moody et al., 2001). Hedlund and Staley (2006) found that a *Pseudoalteromonas* strain with the same gene sequence (16S rRNA) differed in the range of PAHs that they could degrade and had acquired a gene coding for an enzyme through horizontal gene transfer that degrades naphthalene. Bisht et al. (2014) reported that *P. deltoidesa* was able to remove 80–90% naphthalene and anthracene.

15.6.4.4 Microbial Consortium for PAHs Degradation

Microbial consortia produced better results than single microbial species because they were more able to adapt to the heterogeneous matrices. *Microbacterium, Rhodococcus* and *Sphingomonas*, three genera of bacteria, were used in a consortium by Li et al. (2011) that breaks down PAHs in mangrove sediments. The study found that bacteria including *Bacillus, Arthrobacter, Rhodococcus* and *Paenibacillus* effectively used HMW hydrocarbons. Other bacterial spp. can only consume a limited quantity of high molecular weight hydrocarbon under identical conditions, while reversal of breakdown were found under co-metabolic settings (Zeng et al., 2019). Within 28 days, 88.5% of the hydrocarbons were degraded by a co-culturing of several bacteria, like *Bacillus subtilis*, *Pseudomonas aeruginosa* and *Acinetobacter lwoffii* (Al-Wasify and Hamed, 2014).

15.6.5 REMEDIATION BY ALGAE

Remediation of environmental pollutants by algae is called phycoremediation. BaP is known to be converted by marine algae into diols and quinones over the course of 5 to 6 days. *Selenastrum capricornutum*, a freshwater green alga, uses a dioxygenase enzyme analogous to that seen in heterotrophic prokaryotes to convert BaP into cis-dihydrodiols, according to Warshawsky et al. (1995).

The elimination of pyrene, fluoranthene and a fluoranthene-pyrene by algae was reported. *S. capricornutum* and *C. vulgaris* successfully removed 78% and 48% of the PAH in seven days of incubation time. The degradation rate of fluoranthene-pyrene mixture was higher than that of the individual PAH, showing the removal of one PAH was enhanced by the presence of the other. Microalgal species *Prototheca zopfii* immobilized on polyurethane foam has shown to degrade a combination of PAHs in the matrix (Ueno et al., 2008).

A study reported that synthesis of BaP quinone increased in *S. capricornutum* when light energy in the PAHs-absorbing area increased as gold to white to UV-A. According to the study's findings, only green algae successfully converted BaP into dihydrodiols, while yellow as well as blue green algae were unable to do so (Warshawsky et al., 1988).

In a study by Hong et al. (2008), bioaccumulation and degradation of fluoranthene as well as phenanthrene by algal species derived from mangrove ecosystems was investigated. *S. costatum* and *Nitzschia* spp., two isolated microalgal species, could accumulate and degrade PAHs. *Nitzschia* spp. has better capacities for accumulation and decomposition than *S. costatum*. The two algal species degraded fluoranthene more slowly than expected, demonstrating its resistance. It is evident that the presence of one PAH speeds up the breakdown of the other because the elimination of the mixture of PAHs by the microalgal species was equally effective as the removal of phenanthrene or fluoranthene alone. Algae is suited for PAHs remediation and functions co-metabolically with bacterium, according to research and results.

The first occurrence of photosynthesis-enhanced biological degradation of hazardous aromatic contaminants by algal-bacterial microcosms in a single phase of treatment was reported by Borde et al. (2017). Salicylate, phenol, and phenanthrene biodegradation capacity in algal-bacterial microcosms was investigated, and it reported that systems that were cultured under continuous lighting and inoculated with both algae and bacteria significantly eliminated >85% PAHs (Haritash et al., 2009).

15.6.6 Remediation by Fungi

Mycoremediation is the term used to describe degradation of persistent, recalcitrant contaminants by fungi. The type and abundance of fungi, as well as the microorganisms present, all influence the rate at which pollutants degrade (Cutright et al., 1995). Fungi are classified based on lignin metabolism into two types: ligninolytic fungi and non-lignolytic fungi. Lignolytic fungi chiefly belong to the *Basidiomycetes* group, which have the capacity to degrade lignin, while non-ligninolytic fungi include fungi that cannot degrade lignin.

15.6.6.1 Catabolism of PAHs by Ligninolytic Fungi

Lignolytic fungi have been extensively researched to degrade different pollutants. Because of the uneven structure of lignin, enzymes from lignolytic fungus have relatively poor substrate specificity that leads to more suitability for the degradation of a varied type of contaminants. The three primary enzyme groups that make up the lignolytic system are peroxidase, phenoloxidases (lacases, tyrosinases), and enzyme that produce H_2O_2. The capability of enzymes to break down PAHs was demonstrated in an experiment using isolated enzymes from lignolytic fungus (Hofrichter et al., 1998). PAH degraders include *Basidiomycetes* species of white-rot fungi (*Phanerochaete chysosporium*, *Bjerkandera adjusta*, *Trametes versicolor*, *Irpex lacteus* and *Pleurotus*) (Ma et al., 2021).

White-rot fungus produces extracellular lignin-degrading enzymes that are probably involved in the oxidation of PAH. Examples of these enzymes include manganese peroxidase (MnP), lignin peroxidases (LiP) and laccases. These enzymes typically break down the components of wood, but rarely they may also oxidize PAHs to produce transitory PAH diphenol, which then spontaneously oxidize to quinones. In the presence of H_2O_2, LiP oxidizes PAH. MnP oxidizes PAHs via peroxiding unsaturated lipids in a Mn-dependent manner. The laccase oxidizes PAH in the presence of mediators such as methionine, cysteine, phenol, aniline. 4-hydroxybenzoic acid, 4-hydroxybenzyl alcohol or reduced glutathione. Some white-rot fungi have the ability to split the ring of PAH quinones, which eventually produces CO_2 as shown in Figure 15.5. It's also possible that intracellular enzymes like cytochrome P450 monooxygenase and epoxide hydrolase contribute to the breakdown of PAH by ligninolytic fungus (Golan-Rozen et al. 2011).

According to Pozdnyakova et al. (2013), white-rot fungi *P. chrysosporium* degraded PAHs but still only partially transformed radiolabelled benzo[a] pyrene to carbon dioxide in addition to some other polar metabolite. These enzymes could break down lignin in wood using lignin peroxidases and manganese peroxidases.

When *P. chrysosporium* is grown in nitrogen-rich, non-ligninolytic environments, it metabolizes phenanthrene to a variety of metabolites, including phenanthrene trans-dihydrodiols. However, *P. chrysosporium* metabolizes phenanthrene 9,10-quinones as a substitute of phenanthrene when it is grown in low-nitrogen and ligninolytic environments, resulting in the production of 2,20-diphenic acid and ultimately carbon dioxide. In summary, *P. chrysosporium* has at least two metabolic routes for the breakdown of PAH, one of which results in trans-dihydrodiol and the other in quinone and other oxidation products (Syed et al. 2010).

Shahi et al. (2016) explored the microbial degradation of hydrocarbons by *T. versicolor* as well as *Bjerkandera adusta* in petroleum-contaminated soil as a post-treatment approach following the biostimulation procedure. The impact of fungus on the intrinsic bacterial population during the remediation method were also investigated, and no major variations in the populations of Gram-negative and Gram-positive bacteria were observed. Although neither *T. versicolor* nor *B. adusta* were able to completely biodegrade petroleum residues, it has been demonstrated that *T. versicolor* was more active as compared to *B. adusta* at increasing the degradation of oil-polluted soil when utilized during the post-treatment stage.

Cajthaml et al. (2008) investigated the degradation product of PAH by the lignin-degrading fungi *Irpex lacteus* and observed that phenanthrene and anthracene were primarily degraded into phenanthrene-9,10-dihydrodiols and anthraquinones, respectively. White-rot fungi have lignin-degrading systems that enable them to break down a variety of organic contaminants. White-rot fungus exposed to PAHs degraded benzo[a] pyrene in liquid media more quickly, according to Boyle et al. (1998). As a result, the degradative enzyme system appears to be inducible. *Trichoderma longibrachiatum* was used by Cobas et al. (2013) to demonstrate the production of permeable reactive bio barriers on nylon sponge for the purpose of removing PAH from polluted groundwater. Phenanthrene content was decreased by 90% in an aqueous media after 14 days, while 70% in soil after 28 days.

Clemente et al. (2001) demonstrated 13 deuteromycete ligninolytic fungus, different lignolytic enzymes have different effects on the breakdown of PAHs; 65% of anthracene degradation was found by strain 710. By using the enzymes Mn-peroxidase and lignin peroxidase-laccase, strain 870 was shown to degrade naphthalene by 69% and 17%, respectively. In the same strain with active laccase and Mn-peroxidase, 12% phenanthrene degradation was seen.

P. ostreatus, a lignin-degrading fungus, has the ability to metabolize as well as mineralize phenanthrene, fluorene, pyrene and anthracene, when cultured either in nitrogen-low or nitrogen-rich environments. The *P. ostreatus* laccase may additionally oxidize perylene and fluoranthene (Pozdnyakova et al. 2010). It indicates that *P. ostreatus* uses the suggested pathways to degrade PAH in a high stereo- and regio-selective way (Purnomo et al. 2010).

15.6.6.2 Catabolism of PAHs by Non-ligninolytic Fungi

Casillas et al. (1996) investigated *Cunninghamella elegans* non-ligninolytic Zygomycetes fungi for capability for the production of enzyme cytochrome P_{450} to degrade PAH. During study, it was found that *Rhodotorula glutinis*, yeast isolated from a polluted stream, degraded phenanthrene at the same rate as found by *Pseudomonas aeruginosa* (Romero et al., 1998). Chigu et al. (2010) reported the first transition of PAH in *Aspergillus*, *Penicillium* and *Cunninghamella* species involves an oxidation process by enzyme cytochrome P_{450} monooxygenases to generate arene oxides and H_2O. Eggen and Majcherczyk (1998) stated that *P. ostreatus* fungus removed benzo[a] pyrene efficiently from creosote-contaminated soil in one month. The chloroperoxidases produced by mould *Caldariomyces fumago* can greatly increase the mutagenic effect of PAH in the presence of H_2O_2 by adding 1 to 3 chlorine atoms (Ning et al. 2010). Sharma et al. (2016) evaluated the increased bio-degradation of PAH in soil supplemented by compost, ammonium sulphate and paddy straws under in situ condition utilizing a microbial consortium composed of *Streptomyces rochei* PAH-1, *Serratia marcescens* L-11, 3, and *P. chrysosporium* VV-18 and found that 56% to 98% PAHs were degraded in just 7 days, with compost-amended soil.

15.6.6.3 Aerobic and Anaerobic PAHs Degradation Pathway in Fungi

The oxidation of PAH in an aerobic environment during PAH degradation leads to the synthesis of intermediates such as protocatechuate and catechol, which is initiated by the formation of dihydrodiols by multi-component dioxygenase enzymatic activity (Figure 15.5) (Sipilä et al. 2010). In anaerobic environments, aromatic substances donate electrons, and bacteria proliferate by oxidizing these substances in the presence of terminal electron acceptors. Anaerobic processes transform organic materials into methane, which can be used to generate energy. Anaerobic PAH degradation occurs frequently in oxygen-depleted environments polluted with tar wastes and petroleum that are produced unintentionally by industrial application (Qiu et al. 2004; Lu et al. 2012). Boonchan et al. 2000 observed that by adding fungal and bacterial cocultures to soil that has been contaminated with PAHs, the breakdown of PAH containing high molecular weight, such as chrysene, dibenzo[a, h] anthracene and benzo[a] anthracene significantly increased (Aydin et al. 2017; Haritash et al., 2009)

15.6.7 NANOREMEDIATION

Recently, nanotechnology has increased immense scientific relevance in the 21st century as it can generate nano-sized materials with unique characteristics and enormous applications in biosensors and pharmaceuticals, along with the remediation of soil, water and air pollutants (Adeola and Forbes, 2019). The intrinsic features of nanoparticles, which range in size 1–100 nm, have a large surface area, and have a better surface coating, allowing them to be more broadly spread than larger-sized particles, makes them ideal for in situ application.

Nanoremediation has emerged as a significant area of study and development, with enormous promise for cleaning up polluted places. Hence, nanoremediation technology should be integrated with conventional remediation methods such as first and foremost adsorption methods by new, efficient, eco-friendly nano adsorbents and the chemical oxidation through nano oxidizers for oxidation of PAHs (Guerra et al., 2018).

Yang et al. (2019) found 93.9% degradation of benzo[a] pyrene by using carbon dots/fatty acid/ Fe3O4. Moussawi and Patra (2016) observed 100% removal of chrysene using Curcumin conjugated ZnO nanoparticles in 2.2 h by photodegradation mechanism. Hassan et al. (2018) described an eco-friendly technique to synthesize iron oxide nanoparticles (IONPs) to eliminate pyrene and benzo[a] pyrene from PAHs polluted water. Factors like nanoparticle amount, temperature, pH and initial amount of PAH affecting degradation were assessed. The adsorption rate of iron nanoparticles toward benzo[a] pyrene and pyrene were 0.029 mg/g with 99% removal efficiency and 2.8 mg/ g with 98.5% removal efficiency respectively. It was observed that IONPs can be reusable up to five times and have antimicrobial activity. The sorption mechanism was exothermic, followed the pseudo-second-order kinetic reaction route, and characterized through Langmuir adsorption isotherm. The quantity of nanoparticles needed to treat a significant volume of PAHs polluted water is excessive, and it may induce toxicity if not thoroughly removed from treated water because of their small size, which aids movement into living tissues. The removal of fine nanoparticles from treated water using membrane-filter techniques has been established. However, this leads to increases in operating costs (Adeolaet al., 2021).

Shanker et al. (2017a,b) used FeHCF and KZnHCF to degrade different PAHs and found degradation of anthracene (90%) phenanthrene (80%), fluorene, chrysene and benzo[a] pyrene ~70–80%. Mukwevho et al. (2019) observed 99.6% phenanthrene degradation under ultraviolet exposure within 3 h using $Gd_2O_2CO_3.ZnO.CuO$ nanocomposite. Zhang et al. (2010) found that an adsorbent made of iron (III) oxide magnetic nanoparticles ($Fe_3O_4@C_{18}$) and hydrophilic barium alginates (Ba^{2+} -ALG) polymer efficiently adsorbs PAHs contaminants. Zhao et al. (2016) found 98.6% removal of phenanthrene in 12 h by using Co-doped titanate nanotubes as a nanomaterial. Liao et al. (2015) used Fe_3O_4 coated with fatty acids to remove BaP by adsorption mechanisms. Complete adsorption was observed within 5 min at room temperature from water. Huang et al. (2016) reported that acenaphthene from contaminated water was removed by using tetraethyl orthosilicate magnetized with maghemite nanoparticles and recorded 85% of PAHs removal efficiency.

15.7 CONCLUSION

Polycyclic aromatic hydrocarbon (PAHs) pollution is a worldwide concern because of its toxicity to the environment. Bioremediation is emerging as a potential and effective technology for removing PAHs from the polluted environment by utilizing microorganisms such as bacteria, fungi and algae. Because of its high reclaiming efficiency, this process is environmentally beneficial and provides a novel way to lower the hazard of PAHs to health and the environment. PAHs can be eliminated from a polluted area by indigenous microorganisms. Environmental factors, like a deficit in nutrients, might, however, limit microbial activity and, as a result, the degradation of the pollutant. These limitations can be overcome and the elimination of PAHs can proceed more quickly through bioaugmentation and biostimulation. Furthermore, phytoremediation speeds up the degradation of PAHs from soil as the rhizosphere provides a favourable environment for microbial growth. As a result of its widespread use, microbial degradation is associated with several disadvantages. These disadvantages include high costs, a long processing time, difficulty handling and a limited selection of substrates. The use of nanomaterials is also an eco-friendly way to turn toxic PAHs into less hazardous form in an ecologically sustainable manner with minimal time and expense, and they may be reused several times.

15.8 FUTURE DIRECTIONS

The pollution in soil by PAHs is frequently larger than the microorganisms' ability to remove them. Recent approaches are too expensive and take a long time to fully function, often up to a month. As a result, they cannot be used on a large scale. The constraints encountered with using native microorganisms could then be overcome by new technologies that enhance a microorganism's capacity to break down PAHs. Furthermore, it is important to take precautions to prevent the spread of harmful traits among the population of soil microbes or to prevent the genetically modified organisms (GMOs) outcompeting the native ones. Therefore, when genetically engineered microbes are released into the surroundings to clean up PAH-contaminated soil, the ecological impact is latent. Moreover, the limited research increases the need for investigating novel strategies, particularly the usage of nanomaterials, which have the unique virtue of having a high surface-to-volume ratio, making them effective adsorbents. Additionally, environmentally friendly nanoparticles that were created have been an alternative way for PAHs degradation, and an integrated system of physical, chemical and biological degradation should also be used to remove contaminant from the environment.

REFERENCES

Adeniji, A. O., Okoh, O. O., & Okoh, A. I. (2018). Analytical methods for polycyclic aromatic hydrocarbons and their global trend of distribution in water and sediment: A review. *Recent Insights in Petroleum Science and Engineering*, 10.

Adeola, A. O., & Forbes, P. B. (2019). Optimization of the sorption of selected polycyclic aromatic hydrocarbons by regenerable graphene wool. *Water Science and Technology*, 80(10), 1931–1943.

Adeola, A. O., & Forbes, P. B. (2021). Advances in water treatment technologies for removal of polycyclic aromatic hydrocarbons: Existing concepts, emerging trends, and future prospects. *Water Environment Research*, 93(3), 343–359.

Agrawal, N., Kumar, V., Shahi, S.K. (2021). *Biodegradation and detoxification of phenanthrene in in-vitro and in-vivo conditions by a newly isolated ligninolytic fungus Coriolopsis byrsina strain APC5 and characterization of their metabolites for environmental safety.* Environmental Science and Pollution Research. https://doi.org/10.1007/s11356-021-15271-w

Alagić, S. Č., Maluckov, B. S., & Radojičić, V. B. (2015). How can plants manage polycyclic aromatic hydrocarbons? May these effects represent a useful tool for an effective soil remediation? A review. *Clean Technologies and Environmental Policy*, 17(3), 597–614.

Alves, R., & Grimalt, R. (2018). A review of platelet-rich plasma: history, biology, mechanism of action, and classification. *Skin Appendage Disorders*, 4(1), 18–24.

Al-Wasify, R. S., & Hamed, S. R. (2014). Bacterial biodegradation of crude oil using local isolates. *International Journal of Bacteriology*, 2014.

Ambrosoli, R., Petruzzelli, L., Minati, J. L., & Marsan, F. A. (2005). Anaerobic PAH degradation in soil by a mixed bacterial consortium under denitrifying conditions. *Chemosphere*, 60(9), 1231–1236.

Aydin, S., Karaçay, H. A., Shahi, A., Gökçe, S., Ince, B., & Ince, O. (2017). Aerobic and anaerobic fungal metabolism and Omics insights for increasing polycyclic aromatic hydrocarbons biodegradation. *Fungal Biology Reviews*, 31(2), 61–72.

Beyer, J., Jonsson, G., Porte, C., Krahn, M. M., & Ariese, F. (2010). Analytical methods for determining metabolites of polycyclic aromatic hydrocarbon (PAH) pollutants in fish bile: a review. *Environmental Toxicology and Pharmacology*, 30(3), 224–244.

Bisht, S., Kumar, V., Kumar, M., & Sharma, S. (2014). Innoculant technology in Populus deltoides rhizosphere for effective bioremediation of Polyaromatic hydrocarbons (PAHs) in contaminated soil, Northern India. *Emirates Journal of Food and Agriculture*, 25, 786–799.

Bisht, S., Pandey, P., Bhargava, B., Sharma, S., Kumar, V., & Sharma, K. D. (2015). Bioremediation of polyaromatic hydrocarbons (PAHs) using rhizosphere technology. *Brazilian Journal of Microbiology*, 46, 7–21.

Boonchan, S., Britz, M. L., & Stanley, G. A. (2000). Degradation and mineralization of high-molecular-weight polycyclic aromatic hydrocarbons by defined fungal-bacterial cocultures. *Applied and Environmental Microbiology*, 66(3), 1007–1019.

Bordes, É., Costa, A. J., Szala-Bilnik, J., et al. (2017). Polycyclic aromatic hydrocarbons as model solutes for carbon nanomaterials in ionic liquids. *Physical Chemistry Chemical Physics*, 19(40), 27694–27703.

Bortey-Sam, N., Ikenaka, Y., Nakayama, S. M., Akoto, O., Yohannes, Y. B., Baidoo, E., et al. (2014). Occurrence, distribution, sources and toxic potential of polycyclic aromatic hydrocarbons (PAHs) in surface soils from the Kumasi Metropolis, Ghana. *Science of the Total Environment*, 496, 471–478.

Boyle, D., Wiesner, C., & Richardson, A. (1998). Factors affecting the degradation of polyaromatic hydrocarbons in soil by white-rot fungi. *Soil Biology and Biochemistry*, 30(7), 873–882.

Cai, S. S., Syage, J. A., Hanold, K. A., & Balogh, M. P. (2009). Ultra-performance liquid chromatography–atmospheric pressure photoionization-tandem mass spectrometry for high-sensitivity and high-throughput analysis of US environmental protection agency 16 priority pollutants polynuclear aromatic hydrocarbons. *Analytical Chemistry*, 81(6), 2123–2128.

Cajthaml, T., Bhatt, M., Šašek, V., & Matějů, V. (2002). Bioremediation of PAH-contaminated soil by composting: A case study. *Folia Microbiologica*, 47(6), 696–700.

Cajthaml, T., Erbanová, P., Kollmann, A., Novotný, Č., Šašek, V., & Mougin, C. (2008). Degradation of PAHs by ligninolytic enzymes of Irpex lacteus. *Folia Microbiologica*, 53(4), 289–294.

Cao, X., Hao, X., Shen, X., Jiang, X., Wu, B., & Yao, Z. (2017). Emission characteristics of polycyclic aromatic hydrocarbons and nitro-polycyclic aromatic hydrocarbons from diesel trucks based on on-road measurements. *Atmospheric Environment*, 148, 190–196.

Cao, Z., Liu, J., Luan, Y., Li, Y., Ma, M., Xu, J., & Han, S. (2010). Distribution and ecosystem risk assessment of polycyclic aromatic hydrocarbons in the Luan River, China. *Ecotoxicology*, 19(5), 827–837.

Casillas, R. P., Crow Jr, S. A., Heinze, T. M., Deck, J., & Cerniglia, C. E. (1996). Initial oxidative and subsequent conjugative metabolites produced during the metabolism of phenanthrene by fungi. *Journal of Industrial Microbiology and Biotechnology*, 16(4), 205–215.

Chang, B. V., Shiung, L. C., & Yuan, S. Y. (2002). Anaerobic biodegradation of polycyclic aromatic hydrocarbon in soil. *Chemosphere*, 48(7), 717–724.

Chang, K. F., Fang, G. C., Chen, J. C., & Wu, Y. S. (2006). Atmospheric polycyclic aromatic hydrocarbons (PAHs) in Asia: a review from 1999 to 2004. *Environmental Pollution*, 142(3), 388–396.

Chaudhry, Q., Blom-Zandstra, M., Gupta, S. K., & Joner, E. (2005). Utilising the synergy between plants and rhizosphere microorganisms to enhance breakdown of organic pollutants in the environment (15 pp). *Environmental Science and Pollution Research*, 12(1), 34–48.

Chen, C. W., Wang, Y. Y., & Wu, G. J. (1999). Interfacing ASE and HPLC for the determination of PAH in soil. *Journal of the Chinese Chemical Society*, 46(2), 245–251.

Chigu, N. L., Hirosue, S., Nakamura, C., Teramoto, H., Ichinose, H., & Wariishi, H. (2010). Cytochrome P450 monooxygenases involved in anthracene metabolism by the white-rot basidiomycete Phanerochaete chrysosporium. *Applied Microbiology and Biotechnology*, 87(5), 1907–1916.

Clarke, K., Romain, A. C., Locoge, N., & Redon, N. (2012). Application of chemical mass balance methodology to identify the different sources responsible for the olfactory annoyance at a receptor-site. *In 3rd International Conference on Environmental Odour Monitoring and Control NOSE2012*. AIDIC, Milan, Italy.

Clemente, A. R., Anazawa, T. A., & Durrant, L. R. (2001). Biodegradation of polycyclic aromatic hydrocarbons by soil fungi. *Brazilian Journal of Microbiology*, 32, 255–261.

Cobas, M., Ferreira, L., Tavares, T., Sanromán, M. A., & Pazos, M. (2013). Development of permeable reactive biobarrier for the removal of PAHs by Trichoderma longibrachiatum. *Chemosphere*, 91(5), 711–716.

Cofield, N., Banks, M. K., & Schwab, A. P. (2008). Lability of polycyclic aromatic hydrocarbons in the rhizosphere. *Chemosphere*, 70(9), 1644–1652.

Crisafi, F., Giuliano, L., Yakimov, M. M., Azzaro, M., & Denaro, R. (2016). Isolation and degradation potential of a cold-adapted oil/PAH-degrading marine bacterial consortium from Kongsfjorden (Arctic region). *Rendiconti Lincei*, 27(1), 261–270.

Cutright, T. J. (1995). Polycyclic aromatic hydrocarbon biodegradation and kinetics using Cunninghamella echinulata var. elegans. *International Biodeterioration & Biodegradation*, 35(4), 397–408.

Daane, L. L., Harjono, I., Zylstra, G. J., & Haggblom, M. M. (2001). Isolation and characterization of polycyclic aromatic hydrocarbon-degrading bacteria associated with the rhizosphere of salt marsh plants. *Applied and Environmental Microbiology*, 67(6), 2683–2691.

Das, N., & Chandran, P. (2011). *Microbial degradation of petroleum hydrocarbon contaminants: An overview*. Biotechnology Research International, 2011, 1–13.

de Almeida, F. F., Freitas, D., Motteran, F., Fernandes, B. S., & Gavazza, S. (2021). Bioremediation of polycyclic aromatic hydrocarbons in contaminated mangroves: Understanding the historical and key parameter profiles. *Marine Pollution Bulletin*, 169, 112553.

De Boer, J., & Wagelmans, M. (2016). Polycyclic aromatic hydrocarbons in soil–practical options for remediation. *CLEAN–Soil, Air, Water*, 44(6), 648–653.

Dean-Ross, D., Moody, J., & Cerniglia, C. E. (2002). Utilization of mixtures of polycyclic aromatic hydrocarbons by bacteria isolated from contaminated sediment. *FEMS Microbiology Ecology*, 41(1), 1–7.

Dhar, K., Panneerselvan, L., Venkateswarlu, K., & Megharaj, M. (2022). Efficient bioremediation of PAHs-contaminated soils by a methylotrophic enrichment culture. *Biodegradation*, 33(6), 575–591.

Diplock, E. E., Mardlin, D. P., Killham, K. S., & Paton, G. I. (2009). Predicting bioremediation of hydrocarbons: laboratory to field scale. *Environmental Pollution*, 157(6), 1831–1840.

Dudhagara, D. R., Rajpara, R. K., Bhatt, J. K., Gosai, H. B., Sachaniya, B. K., & Dave, B. P. (2016). Distribution, sources and ecological risk assessment of PAHs in historically contaminated surface sediments at Bhavnagar coast, Gujarat, India. *Environmental Pollution*, 213, 338–346.

Eggen, T., & Majcherczyk, A. (1998). Removal of polycyclic aromatic hydrocarbons (PAH) in contaminated soil by white rot fungus Pleurotus ostreatus. *International Biodeterioration & Biodegradation*, 41(2), 111–117.

Famiyeh, L., Chen, K., Xu, J., et al. (2021). A review on analysis methods, source identification, and cancer risk evaluation of atmospheric polycyclic aromatic hydrocarbons. *Science of the Total Environment*, 789, 147741.

Gan, S., Lau, E. V., & Ng, H. K. (2009). Remediation of soils contaminated with polycyclic aromatic hydrocarbons (PAHs). *Journal of Hazardous Materials*, 172(2–3), 532–549.

Ghosal, D., Ghosh, S., Dutta, T. K., & Ahn, Y. (2016). Current state of knowledge in microbial degradation of polycyclic aromatic hydrocarbons (PAHs): a review. *Frontiers in Microbiology*, 1369.

Golan-Rozen, N., Chefetz, B., Ben-Ari, J., Geva, J., & Hadar, Y. (2011). Transformation of the recalcitrant pharmaceutical compound carbamazepine by Pleurotus ostreatus: Role of cytochrome P450 monooxygenase and manganese peroxidase. *Environmental Science & Technology*, 45(16), 6800–6805.

Gou, Y., Zhao, Q., Yang, S., Wang, H., Qiao, P., Song, Y., ... & Li, P. (2020). Removal of polycyclic aromatic hydrocarbons (PAHs) and the response of indigenous bacteria in highly contaminated aged soil after persulfate oxidation. *Ecotoxicology and Environmental Safety*, 190, 110092.

Govarthanan, M., Khalifa, A. Y., Kamala-Kannan, S., Srinivasan, P., Selvankumar, T., Selvam, K., & Kim, W. (2020). Significance of allochthonous brackish water Halomonas sp. on biodegradation of low and high molecular weight polycyclic aromatic hydrocarbons. *Chemosphere*, 243, 125389.

Guarino, C., Zuzolo, D., Marziano, M., Conte, B., Baiamonte, G., Morra, L…., & Sciarrillo, R. (2019). Investigation and assessment for an effective approach to the reclamation of polycyclic aromatic hydrocarbon (PAHs) contaminated site: SIN Bagnoli, Italy. *Scientific Reports*, 9(1), 1–12.

Guerra, F. D., Attia, M. F., Whitehead, D. C., & Alexis, F. (2018). Nanotechnology for environmental remediation: materials and applications. *Molecules*, 23(7), 1760.

Haritash, A. K., & Kaushik, C. P. (2009). Biodegradation aspects of polycyclic aromatic hydrocarbons (PAHs): A review. *Journal of Hazardous Materials*, 169(1–3), 1–15.

Hassan, H. M., Castillo, A. B., Yigiterhan, O., Elobaid, E. A., Al-Obaidly, A., Al-Ansari, E., & Obbard, J. P. (2018). Baseline concentrations and distributions of Polycyclic Aromatic Hydrocarbons in surface sediments from the Qatar marine environment. *Marine Pollution Bulletin*, 126, 58–62.

Hawthorne, S. B., Grabanski, C. B., Martin, E., & Miller, D. J. (2000). Comparisons of Soxhlet extraction, pressurized liquid extraction, supercritical fluid extraction and subcritical water extraction for environmental solids: recovery, selectivity and effects on sample matrix. *Journal of Chromatography A*, 892(1–2), 421–433.

He, Y., & Chi, J. (2019). Pilot-scale demonstration of phytoremediation of PAH-contaminated sediments by Hydrilla verticillata and Vallisneria spiralis. *Environmental Technology*, 40(5), 605–613.

Heath, D. J., Lewis, C. A., & Rowland, S. J. (1997). The use of high temperature gas chromatography to study the biodegradation of high molecular weight hydrocarbons. *Organic Geochemistry*, 26(11–12), 769–785.

Hedlund, B. P., & Staley, J. T. (2006). Isolation and characterization of Pseudoalteromonas strains with divergent polycyclic aromatic hydrocarbon catabolic properties. *Environmental Microbiology*, 8(1), 178–182.

Hofrichter, M., Scheibner, K., Schneegaß, I., & Fritsche, W. (1998). Enzymatic combustion of aromatic and aliphatic compounds by manganese peroxidase from Nematoloma frowardii. *Applied and Environmental Microbiology*, 64(2), 399–404.

Hong, Y. W., Yuan, D. X., Lin, Q. M., & Yang, T. L. (2008). Accumulation and biodegradation of phenanthrene and fluoranthene by the algae enriched from a mangrove aquatic ecosystem. *Marine Pollution Bulletin*, 56(8), 1400–1405.

Hopke, P. K. (2016). Review of receptor modeling methods for source apportionment. *Journal of the Air & Waste Management Association*, 66(3), 237–259.

Huang, X. D., El-Alawi, Y., Penrose, D. M., Glick, B. R., & Greenberg, B. M. (2004). A multi-process phytoremediation system for removal of polycyclic aromatic hydrocarbons from contaminated soils. *Environmental pollution*, 130(3), 465–476.

Huang, Y., Fulton, A. N., & Keller, A. A. (2016). Simultaneous removal of PAHs and metal contaminants from water using magnetic nanoparticle adsorbents. *Science of the Total Environment*, 571, 1029–1036.

Hussain, K., Hoque, R. R., Balachandran, S., Medhi, S., Idris, M. G., Rahman, M., & Hussain, F. L. (2018). Monitoring and risk analysis of PAHs in the environment. *Handbook of Environmental Materials Management*, 2, 1–35.

Imam, A., Suman, S. K., Ghosh, D., & Kanaujia, P. K. (2019). Analytical approaches used in monitoring the bioremediation of hydrocarbons in petroleum-contaminated soil and sludge. *TrAC Trends in Analytical Chemistry*, 118, 50–64.

Jiang, Y., Zhang, Z., & Zhang, X. (2018). Co-biodegradation of pyrene and other PAHs by the bacterium Acinetobacter johnsonii. *Ecotoxicology and Environmental Safety*, 163, 465–470.

Jiao, H., Wang, Q., Zhao, N., Jin, B., Zhuang, X., & Bai, Z. (2017). Distributions and sources of polycyclic aromatic hydrocarbons (PAHs) in soils around a chemical plant in Shanxi, China. *International Journal of Environmental Research and Public Health*, 14(10), 1198.

Kalf, D. F., Crommentuijn, T., & van de Plassche, E. J. (1997). Environmental quality objectives for 10 polycyclic aromatic hydrocarbons (PAHs). *Ecotoxicology and Environmental Safety*, 36(1), 89–97.

Kannan, S. K., & Krishnamoorthy, R. (2006). Isolation of mercury resistant bacteria and influence of abiotic factors on bioavailability of mercury—a case study in Pulicat Lake north of Chennai, south east India. *Science of the Total Environment*, 367(1), 341–353.

Khanna, P., Goyal, D., & Khanna, S. (2011). Pyrene degradation by Bacillus pumilus isolated from crude oil contaminated soil. *Polycyclic Aromatic Compounds*, 31(1), 1–15.

Kuiper, I., Lagendijk, E. L., Bloemberg, G. V., & Lugtenberg, B. J. (2004). Rhizoremediation: A beneficial plant-microbe interaction. *Molecular Plant-Microbe Interactions*, 17(1), 6–15.

Kumar, A. V., Kothiyal, N. C., Kumari, S., et al. (2014). Determination of some carcinogenic PAHs with toxic equivalency factor along roadside soil within a fast-developing northern city of India. *Journal of Earth System Science*, 123(3), 479–489.

Kumar, M., Bolan, N. S., Hoang, S. A., et al. (2021). Remediation of soils and sediments polluted with polycyclic aromatic hydrocarbons: To immobilize, mobilize, or degrade?. *Journal of Hazardous Materials*, 420, 126534.

Kumari, S., Regar, R. K., & Manickam, N. (2018). Improved polycyclic aromatic hydrocarbon degradation in a crude oil by individual and a consortium of bacteria. *Bioresource Technology*, 254, 174–179.

Kuppusamy, S., Thavamani, P., Venkateswarlu, K., Lee, Y. B., Naidu, R., & Megharaj, M. (2017). Remediation approaches for polycyclic aromatic hydrocarbons (PAHs) contaminated soils: Technological constraints, emerging trends and future directions. *Chemosphere*, 168, 944–968.

Lang, F. S., Destain, J., Delvigne, F., Druart, P., Ongena, M., & Thonart, P. (2016). Biodegradation of polycyclic aromatic hydrocarbons in mangrove sediments under different strategies: Natural attenuation, biostimulation, and bioaugmentation with Rhodococcus erythropolis T902. 1. *Water, Air, & Soil Pollution*, 227(9), 1–15.

Lee, B. M., & Shim, G. A. (2007). Dietary exposure estimation of benzo [a] pyrene and cancer risk assessment. *Journal of Toxicology and Environmental Health, Part A*, 70(15–16), 1391–1394.

Lee, S. H., Lee, W. S., Lee, C. H., & Kim, J. G. (2008). Degradation of phenanthrene and pyrene in rhizosphere of grasses and legumes. *Journal of Hazardous Materials*, 153(1–2), 892–898.

Li, C. H., Ye, C., Wong, Y. S., & Tam, N. F. Y. (2011). Effect of Mn (IV) on the biodegradation of polycyclic aromatic hydrocarbons under low-oxygen condition in mangrove sediment slurry. *Journal of Hazardous Materials*, 190(1–3), 786–793.

Li, X., Li, P., Lin, X., Zhang, C., Li, Q., & Gong, Z. (2008). Biodegradation of aged polycyclic aromatic hydrocarbons (PAHs) by microbial consortia in soil and slurry phases. *Journal of Hazardous Materials*, 150(1), 21–26.

Liang, S., Yan, X., An, B., Yang, Y., & Liu, M. (2022). Distribution and ecological risk assessment of polycyclic aromatic hydrocarbons in different compartments in Kenya: A review. *Emerging Contaminants*, 8, 351–359.

Liao, W., Ma, Y., Chen, A., & Yang, Y. (2015). Preparation of fatty acids coated Fe3O4 nanoparticles for adsorption and determination of benzo (a) pyrene in environmental water samples. *Chemical Engineering Journal*, 271, 232–239.

Liste, H. H., & Alexander, M. (2000). Plant-promoted pyrene degradation in soil. *Chemosphere*, 40(1), 7–10.

Long, E. R., & MacDonald, D. D. (1998). Recommended uses of empirically derived, sediment quality guidelines for marine and estuarine ecosystems. *Human and Ecological Risk Assessment*, 4(5), 1019–1039.

Lu, X. Y., Li, B., Zhang, T., & Fang, H. H. (2012). Enhanced anoxic bioremediation of PAHs-contaminated sediment. *Bioresource Technology*, 104, 51–58.

Ma, X., Wan, H., Zhao, Z., et al. (2021). Source analysis and influencing factors of historical changes in PAHs in the sediment core of Fuxian Lake, *China. Environmental Pollution*, 288, 117935.

Mallick, S., Chakraborty, J., & Dutta, T. K. (2011). Role of oxygenases in guiding diverse metabolic pathways in the bacterial degradation of low-molecular-weight polycyclic aromatic hydrocarbons: A review. *Critical Reviews in Microbiology*, 37(1), 64–90.

Mao, YE, Mingming, SUN, Shanni, XIE, Kuan, LIU, Yanfang, FENG, Yu, ZHAO, ... & Xin, JIANG (2017). Tea saponin for improving elution and repair of coking in soybean oil-water system Feasibility study on polycyclic aromatic hydrocarbons, Cd, Ni in plant soil. *Pedosphere*, 27 (3), 452–464.

Meckenstock, R. U., Boll, M., Mouttaki, H., et al. (2016). Anaerobic degradation of benzene and polycyclic aromatic hydrocarbons. *Microbial Physiology*, 26 (1–3), 92–118.

Miura, K., Shimada, K., Sugiyama, T., et al. (2019). Seasonal and annual changes in PAH concentrations in a remote site in the Pacific Ocean. *Scientific Reports*, 9(1), 1–10.

Mohan, S. V., Kisa, T., Ohkuma, T., Kanaly, R. A., & Shimizu, Y. (2006). Bioremediation technologies for treatment of PAH-contaminated soil and strategies to enhance process efficiency. *Reviews in Environmental Science and Bio/Technology*, 5(4), 347–374.

Mohapatra, B., & Phale, P. S. (2021). Microbial degradation of naphthalene and substituted naphthalenes: Metabolic diversity and genomic insight for bioremediation. *Frontiers in Bioengineering and Biotechnology*, 9, 602445.

Mojiri, A., Zhou, J. L., Ohashi, A., Ozaki, N., & Kindaichi, T. (2019). Comprehensive review of polycyclic aromatic hydrocarbons in water sources, their effects and treatments. *Science of the Total Environment*, 696, 133971.

Moody, J. D., Freeman, J. P., Doerge, D. R., & Cerniglia, C. E. (2001). Degradation of phenanthrene and anthracene by cell suspensions of Mycobacterium sp. strain PYR-1. *Applied and Environmental Microbiology*, 67(4), 1476–1483.

Moussawi, R. N., & Patra, D. (2016). Nanoparticle self-assembled grain like curcumin conjugated ZnO: curcumin conjugation enhances removal of perylene, fluoranthene and chrysene by ZnO. *Scientific Reports*, 6(1), 1–13.

Mukwevho, N., Fosso-Kankeu, E., Waanders, F., Kumar, N., Ray, S. S., & Yangkou Mbianda, X. (2019). Photocatalytic activity of Gd2O2CO3· ZnO· CuO nanocomposite used for the degradation of phenanthrene. *SN Applied Sciences*, 1(1), 1–11.

Nam, J. J., Song, B. H., Eom, K. C., Lee, S. H., & Smith, A. (2003). Distribution of polycyclic aromatic hydrocarbons in agricultural soils in South Korea. *Chemosphere*, 50(10), 1281–1289.

Ning, D., Wang, H., Ding, C., & Lu, H. (2010). Novel evidence of cytochrome P450-catalyzed oxidation of phenanthrene in Phanerochaete chrysosporium under ligninolytic conditions. *Biodegradation*, 21(6), 889–901.

Nisbet, I. C., & Lagoy, P. K. (1992). Toxic equivalency factors (TEFs) for polycyclic aromatic hydrocarbons (PAHs). *Regulatory Toxicology and Pharmacology*, 16(3), 290–300.

Okuda, T., Kumata, H., Naraoka, H., & Takada, H. (2002). Origin of atmospheric polycyclic aromatic hydrocarbons (PAHs) in Chinese cities solved by compound-specific stable carbon isotopic analyses. *Organic Geochemistry*, 33(12), 1737–1745.

Paquin, D., Ogoshi, R., Campbell, S., & Li, Q. X. (2002). Bench-scale phytoremediation of polycyclic aromatic hydrocarbon-contaminated marine sediment with tropical plants. *International Journal of Phytoremediation*, 4(4), 297–313.

Patel, A. B., Shaikh, S., Jain, K. R., Desai, C., & Madamwar, D. (2020). Polycyclic aromatic hydrocarbons: sources, toxicity, and remediation approaches. *Frontiers in Microbiology*, 11, 562813.

Pavlova, A., & Ivanova, R. (2003). Determination of petroleum hydrocarbons and polycyclic aromatic hydrocarbons in sludge from wastewater treatment basins. *Journal of Environmental Monitoring*, 5(2), 319–323.

Pawliszyn, J. (1997). *Solid phase microextraction: Theory and practice*. John Wiley & Sons.

Pozdnyakova, N. N., Nikiforova, S. V., & Turkovskaya, O. V. (2010). Influence of PAHs on ligninolytic enzymes of the fungus Pleurotus ostreatus D1. *Central European Journal of Biology*, 5(1), 83–94.

Pozdnyakova, N., Makarov, O., Chernyshova, M., Turkovskaya, O., & Jarosz-Wilkolazka, A. (2013). Versatile peroxidase of Bjerkandera fumosa: substrate and inhibitor specificity. *Enzyme and Microbial Technology*, 52(1), 44–53.

Premnath, N., Mohanrasu, K., et al. (2021). A crucial review on polycyclic aromatic Hydrocarbons-Environmental occurrence and strategies for microbial degradation. *Chemosphere*, 280, 130608.

Purnomo, A. S., Mori, T., Kamei, I., Nishii, T., & Kondo, R. (2010). Application of mushroom waste medium from Pleurotus ostreatus for bioremediation of DDT-contaminated soil. *International Biodeterioration & Biodegradation*, 64(5), 397–402.

Qiu, Y. L., Sekiguchi, Y., Imachi, H., et al. (2004). Identification and isolation of anaerobic, syntrophic phthalate isomer-degrading microbes from methanogenic sludges treating wastewater from terephthalate manufacturing. *Applied and Environmental Microbiology*, 70(3), 1617–1626.

Ravindra, K., Mittal, A. K., & Van Grieken, R. (2001). Health risk assessment of urban suspended particulate matter with special reference to polycyclic aromatic hydrocarbons: A review. *Reviews on Environmental Health, 16(3), 169–189.*

Ravindra, K., Sokhi, R., & Van Grieken, R. (2008). Atmospheric polycyclic aromatic hydrocarbons: source attribution, emission factors and regulation. *Atmospheric Environment*, 42(13), 2895–2921.

Rezek, J., in der Wiesche, C., Mackova, M., Zadrazil, F., & Macek, T. (2008). The effect of ryegrass (Lolium perenne) on decrease of PAH content in long term contaminated soil. *Chemosphere*, 70(9), 1603–1608.

Romero, M. C., Cazau, M. C., Giorgieri, S., & Arambarri, A. M. (1998). Phenanthrene degradation by microorganisms isolated from a contaminated stream. *Environmental Pollution*, 101(3), 355–359.

Saari, E., Perämäki, P., & Jalonen, J. (2007). A comparative study of solvent extraction of total petroleum hydrocarbons in soil. *Microchimica Acta*, 158(3), 261–268.

Sabreena., Hassan, S., Bhat, S.A., Kumar, V., Ganai, B.A. and Ameen, F., 2022. Phytoremediation of Heavy Metals: An Indispensable Contrivance in Green Remediation Technology. *Plants*, 11(9), p.1255.

Shahi, A., Aydin, S., Ince, B., & Ince, O. (2016). The effects of white-rot fungi Trametes versicolor and Bjerkandera adusta on microbial community structure and functional genes during the bioaugmentation process following biostimulation practice of petroleum contaminated soil. *International Biodeterioration & Biodegradation*, 114, 67–74.

Shanker, U., Jassal, V., & Rani, M. (2017a). Green synthesis of iron hexacyanoferrate nanoparticles: potential candidate for the degradation of toxic PAHs. *Journal of Environmental Chemical Engineering*, 5(4), 4108–4120.

Shanker, U., Jassal, V., & Rani, M. (2017b). Degradation of toxic PAHs in water and soil using potassium zinc hexacyanoferrate nanocubes. *Journal of Environmental Management*, 204, 337–348.

Shao, H., Cui, G. and Bhat, S.A., 2023. Phytoremediation a green technology for treating heavy metal contaminated soil. In *Bioremediation of toxic metal(loid)s* (pp. 81–94). CRC Press.

Sharma, A., Singh, S. B., Sharma, R., et al. (2016). Enhanced biodegradation of PAHs by microbial consortium with different amendment and their fate in in-situ condition. *Journal of Environmental Management*, 181, 728–736.

Sipilä, T. P., Väisänen, P., Paulin, L., & Yrjälä, K. (2010). Sphingobium sp. HV3 degrades both herbicides and polyaromatic hydrocarbons using ortho-and meta-pathways with differential expression shown by RT-PCR. *Biodegradation*, 21(5), 771–784.

Smith, B. L., & Carpentier, M. H. (1993). The microwave engineering handbook, *Microwave Technology Series*. Vol. 1–3, pp. 1–500. Springer.

Sparkman, O. D., Penton, Z., & Kitson, F. G. (2011). *Gas chromatography and mass spectrometry: A practical guide*. Academic press.

Sun, M., Fu, D., Teng, Y., Shen, Y., Luo, Y., Li, Z., & Christie, P. (2011). In situ phytoremediation of PAH-contaminated soil by intercropping alfalfa (Medicago sativa L.) with tall fescue (Festuca arundinacea Schreb.) and associated soil microbial activity. *Journal of Soils and Sediments*, 11(6), 980–989.

Syed, K., Doddapaneni, H., Subramanian, V., Lam, Y. W., & Yadav, J. S. (2010). Genome-to-function characterization of novel fungal P450 monooxygenases oxidizing polycyclic aromatic hydrocarbons (PAHs). *Biochemical and Biophysical Research Communications*, 399(4), 492–497.

Tam, N. F. Y., Guo, C. L., Yau, W. Y., & Wong, Y. S. (2002). Preliminary study on biodegradation of phenanthrene by bacteria isolated from mangrove sediments in Hong Kong. *Marine Pollution Bulletin*, 45(1–12), 316–324.

Taylor, L. T., & Jones, D. M. (2001). Bioremediation of coal tar PAH in soils using biodiesel. *Chemosphere*, 44(5), 1131–1136.

Teng, Y., Shen, Y., Luo, Y., et al. (2011). Influence of Rhizobium meliloti on phytoremediation of polycyclic aromatic hydrocarbons by alfalfa in an aged contaminated soil. *Journal of Hazardous Materials*, 186(2–3), 1271–1276.

Ueno, R., Wada, S., & Urano, N. (2008). Repeated batch cultivation of the hydrocarbon-degrading, micro-algal strain Prototheca zopfii RND16 immobilized in polyurethane foam. *Canadian Journal of Microbiology*, 54(1), 66–70.

United States. Environmental Protection Agency. Environmental Criteria, & Assessment Office (Cincinnati). (1993). Provisional guidance for quantitative risk assessment of polycyclic aromatic hydrocarbons (Vol. 600). Environmental Criteria and Assessment Office, Office of Health and Environmental Assessment, US Environmental Protection Agency.

Verâne, J., Dos Santos, N. C., da Silva, at al. (2020). Phytoremediation of polycyclic aromatic hydrocarbons (PAHs) in mangrove sediments using Rhizophora mangle. *Marine Pollution Bulletin*, 160, 111687.

Wang, F., Lin, T., Feng, J., Fu, H., & Guo, Z. (2015). Source apportionment of polycyclic aromatic hydrocarbons in PM2. 5 using positive matrix factorization modeling in Shanghai, China. *Environmental Science: Processes & Impacts*, 17(1), 197–205.

Warshawsky, D., Radike, M., Jayasimhulu, K., & Cody, T. (1988). Metabolism of benzo (A) pyrene by a dioxygenase enzyme system of the freshwater green alga selenastrumcapricornutum. *Biochemical and Biophysical Research Communications*, 152(2), 540–544.

Warshawsky, D., Cody, T., Radike, M., et al. (1995). Biotransformation of benzo [a] pyrene and other polycyclic aromatic hydrocarbons and heterocyclic analogs by several green algae and other algal species under gold and white light. *Chemico-Biological Interactions*, 97(2), 131–148.

Weisman, W. (Ed.). (1998). *Analysis of petroleum hydrocarbons in environmental media* (Vol. 1). Amherst Scientific Publishers.

Wu, B., Zhang, R., Cheng, S. P., Ford, T., Li, A. M., & Zhang, X. X. (2011). Risk assessment of polycyclic aromatic hydrocarbons in aquatic ecosystems. *Ecotoxicology*, 20(5), 1124–1130.

Wu, M. L., Nie, M. Q., Wang, X. C., Su, J. M., & Cao, W. (2010). Analysis of phenanthrene biodegradation by using FTIR, UV and GC–MS. *Spectrochimica Acta Part A: Molecular and Biomolecular Spectroscopy*, 75(3), 1047–1050.

Xia, M., Fu, D., Chakraborty, R., Singh, R. P., & Terry, N. (2019). Enhanced crude oil depletion by constructed bacterial consortium comprising bioemulsifier producer and petroleum hydrocarbon degraders. *Bioresource Technology*, 282, 456–463.

Yan, S., & Wu, G. (2017). Reorganization of gene network for degradation of polycyclic aromatic hydrocarbons (PAHs) in Pseudomonas aeruginosa PAO1 under several conditions. *Journal of Applied Genetics*, 58(4), 545–563.

Yang, D., Tammina, S. K., Li, X., & Yang, Y. (2019). Enhanced removal and detection of benzo [a] pyrene in environmental water samples using carbon dots-modified magnetic nanocomposites. *Ecotoxicology and Environmental Safety*, 170, 383–390.

Yang, L., Zhang, H., Zhang, X., Xing, W., Wang, Y., Bai, P., ... & Tang, N. (2021). Exposure to atmospheric particulate matter-bound polycyclic aromatic hydrocarbons and their health effects: a review. *International Journal of Environmental Research and Public Health*, 18(4), 2177.

Yuan, H., Li, T., Ding, X., Zhao, G., & Ye, S. (2014). Distribution, sources and potential toxicological significance of polycyclic aromatic hydrocarbons (PAHs) in surface soils of the Yellow River Delta, China. *Marine Pollution Bulletin*, 83(1), 258–264.

Yuan, S. Y., & Chang, B. V. (2007). Anaerobic degradation of five polycyclic aromatic hydrocarbons from river sediment in Taiwan. *Journal of Environmental Science and Health Part B*, 42(1), 63–69.

Zeng, S., Ma, J., Ren, Y., Liu, G. J., Zhang, Q., & Chen, F. (2019). Assessing the spatial distribution of soil PAHs and their relationship with anthropogenic activities at a national scale. *International Journal of Environmental Research and Public Health*, 16(24), 4928.

Zhang, S., Niu, H., Cai, Y., & Shi, Y. (2010). Barium alginate caged Fe3O4@ C18 magnetic nanoparticles for the pre-concentration of polycyclic aromatic hydrocarbons and phthalate esters from environmental water samples. *Analytica Chimica Acta*, 665(2), 167–175.

Zhao, X., Cai, Z., Wang, T., O'Reilly, S. E., Liu, W., & Zhao, D. (2016). A new type of cobalt-deposited titanate nanotubes for enhanced photocatalytic degradation of phenanthrene. *Applied Catalysis B: Environmental*, 187, 134–143.

Zheng, B., Wang, L., Lei, K., & Nan, B. (2016). Distribution and ecological risk assessment of polycyclic aromatic hydrocarbons in water, suspended particulate matter and sediment from Daliao River estuary and the adjacent area, *China. Chemosphere*, 149, 91–100.

Zheng, H., Yang, D., Hu, T., Li, Y., Zhu, G., Xing, X., & Qi, S. (2017). Source apportionment of polycyclic aromatic carbons (PAHs) in sediment core from Honghu Lake, central China: comparison study of three receptor models. *Environmental Science and Pollution Research*, 24(33), 25899–25911.

16 Organic Amendments in Soil Systems for a Sustainable Ecosystem

Madhu Rani and Sonia Kapoor

16.1 INTRODUCTION

Soil is a natural, living, dynamic and also non-renewable part of the terrestrial ecosystem (Doran J W et al. 1996). In previous decades farmers recognized the health of soil from its taste, colour, smell and touch. Since the dawn of agriculture about 10,000 years ago, these methods are only tools to recognize the condition of soil and also the potential of soil for crop production (Magdoff F et al. 2000). Soil is the food source of humans as well as many insects, birds, nematodes, fungi etc. Thus, a healthy soil can help to stabilize a diverse ecosystem. Along with this, if an ecosystem is stable then the food chain and food web also becomes stable. Hence, the importance of a healthy soil is a main concern nowadays. The soil food web in Figure 16.1 describes the importance of a healthy soil. At the start of 20th century, soil quality was described by its organic matter, nutrition value, crop yield and biological properties (Harris R. F. et al. 1994). A soil quality concept first evolved and was discussed during 1990 because of sustainable land use and a serious concern about the limitation of yield in production (Doran J.W. et al. 1996, Warkentin BP et al. 1977, NRC (National Research Council 1993, USA). As this concept evolved, different tools and methods for assessment and testing of soil were developed. These different tools and methods help in increasing the production of soil along with sustainable land use (Karlen D.L. et al. 1997, Doran J.W. et al. 1996, Andrews S.S. et al. 2004, Karlen D. L. et al. 1994, Smith J.L. et al. 1993). But the main purpose was to design the method in such a way that it can help to increase multiple soil functions (Andrews SS et al. 2004). Along with these broader goals, there are other objectives which have to be fulfilled in an assessment of healthy soil (Karlen D.L. et al. 1997, Parr J. et al. 1992, Doran et al. 1994). These include plant health, erosion control, enhanced air and water quality, and support for habitats of various living creatures, as well as human health and reduction of greenhouse gases.

Several parameters define the soil health profile, which can be improved by different methods (Figure 16.2). One such method is organic amendment. There are many types of organic amendments for improving soil organic matter, like the use of biofertilizers, vermicompost, green manure, biocompost, nano biofertilizer, and biostimulants (Raza et al. 2022). These methods can help to meet the above described goal of a healthy soil and also fulfil a sustainable development goal. In addition to this, different types of organic manures, for example compost, vermicompost and farmyard manure, are used as organic fertilizers to enhance crop yield and improve soil fertility (Yatoo A.M. et al. 2022). Many natural products comprising of different metabolites and bioactive compounds from decomposed bacteria and earthworms are present in vermicompost and its derivatives, which may help to improve stress tolerance mainly caused by soil pathogens in plants and also improve soil fertility (Gudeta K. et al. 2022). Therefore, vermicompost provides multidimensional benefits for improving soil fertility as compared to the use of chemical fertilizers (Bhat

DOI: 10.1201/9781003408352-19

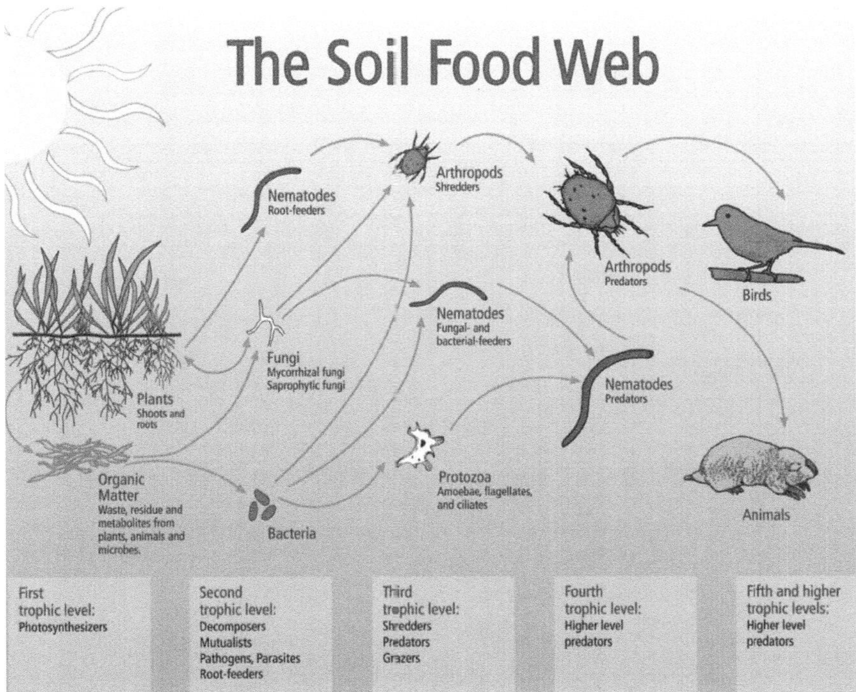

FIGURE 16.1 Soil food web.

Notes: The web contains habitats of different organisms like bacteria, fungi, plants, nematodes, arthropods, birds and animals in the form of soil organic matter. Different trophic levels define the stages of soil synthesis and functions like photosynthesis, decomposers, grazers, predators.

Source: Free from google images.

et al. 2018). Biofertilizers are also used as an alternative to chemical fertilizers for a healthy soil. A biofertilizer contains living microorganisms that when applied to the seeds increase the nutrient level of the soil as well as that of the crop. The main examples are bacteria, fungi, cyanobacteria, *Azospirillum*, *Azotobacter* and *Rhizobium*, which can fix atmospheric nitrogen in soil. Fertilizer use in soil can increase the nutrient level of different plants and also increase the function and diversity of the number of microorganisms (Xiayan liu et al. 2022, Marscher P. et al. 2003). The use of chemical fertilizers in excess is associated with a number of environmental issues, such as degradation of soil layers and deterioration of soil quality, and it also decrease the utilization efficiency of nutrients (Savci, et al. 2012). Further, research studies have proved that excessive use of chemical fertilizers generally disturb the ecological balance of soil and the related nutrient and microbial content of soil (Ramirez K. S. et al. 2010, Sun R.B. et al. 2015). So in these cases organic amendment in soil has been found to be more effective than many chemical or inorganic fertilizers for increasing the nutrient level of soil and also to increase the number of microorganisms for plant growth (Rehman et al. 2019). Hence, organic amendment of soil in place of inorganic fertilizer is an effective way to increase the fertility of soil (Cesarano G. et al. 2017, Tao R. Hu et al. 2020, Ayilara M.S. et al. 2020). Therefore, compost along with fertilizer can be an effective method that helps to improve crop growth and soil properties like its nutrient level, number of microorganisms, texture, moisture and humidity (Coelho et al. 2020, Corato U.D et al. 2020).

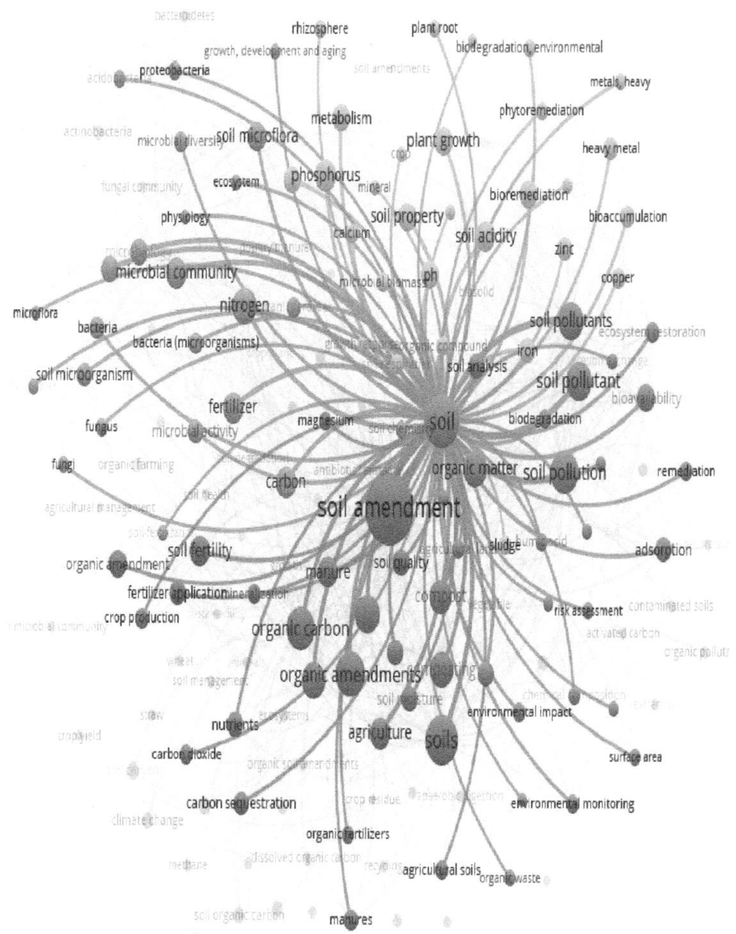

FIGURE 16.2   Bubble map of published research.

Notes: The words represent keywords of soil organic amendment types in various publications from 2019 to 2022. The size of each bubble represents the frequency of each word that appears in publications. Interaction between any two bubbles describes their linkage with each other, and the colour of each bubble represents the citation of the publication.

16.2 BIOCOMPOST

According to research studies, the addition of compost can help to improve soil microbial diversity and activity (Tao R. et al. 2020, Chen L. et al. 2016). Biocompost is a type of compost that uses a specific type of microorganism using different organic wastes, like straw, animal manure (such as cow dung) and the peel of different vegetables (Corato U.D. et al. 2020) (see Figure 16.3). Mainly, the diversity of microorganisms increases with the application of manure (Li P.F. et al. 2019, Hamm A.C. et al. 2016). Soil microorganisms can also increase the N cycle as well as organic matter with the application of biocompost (Alami M. M. et al. 2020, He H.B. et al. 2020). The application of biocompost increased the amount of many types of bacteria like proteobacteria and its genera (Longa et al. 2017). Also, biocompost applications can increase the number of many functional bacteria *Skermanella* & *Sphingomonas* (Pan H. et al. 2020). In a particular experiment, biocompost increased the abundance of many soil bacteria like *Pseudoarthrobacter, Streptomyces* and *Nocardiode*s within Actinobacteria (Pan H. et al. 2020, Yang Y.R. et al. 2019).These are all

FIGURE 16.3 The conversion of organic waste into bio-compost.

Source: Free from Google images.

Note: Microorganisms present in soil help in the conversion of waste into compost, which is then used as a fertilizer in agriculture.

beneficial bacteria and also secrete actinomycin, which helps to reduce the growth of soil pathogens (Pan H. et al. 2020, Yang Y. R. et al. 2019). Besides this, biocompost applications can increase the number of beneficial fungi like *Trichoderma*, *Chrysosporium*, *Acremonium* and *Chaetomium*. These fungi help in biocontrol by inhibiting plant pathogenic microorganisms (Corato U.D et al. 2020, Blaya J. et al. 2013). Soil organic matter is also a main indicator of fertility, yield and crop growth (Chang et al. 2007). Gabriel et al. 2010 also reported that based on crop, weather and soil, the yield of organic crops on a per ha basis was equal to conventional agriculture. Different studies have found that organic farming can increase soil microbial community (Forge et al. 2008), increase diversity in soil (Gabriel et al. 2010), increase the fertility of organic matter and the structure and function of the soil food web and also decrease many plant diseases (Clark et al. 1998).

16.3 BIOSTIMULANTS

Many types of substances other than fertilizers have the capability to increase the nutrient level in soil, even when they are applied in a very low amount; these are mainly known as biostimulants (Du Jardin et al. 2015). Generally biostimulants are used to increase crop production in organic methods of farming. According to a research study, the total yield and overall production in organic farming is low by 5–32% in comparison to the conventional farming approach (Ponisio et al. 2015). Also, some research studies suggest that the nutrient supply, mainly of nitrogen (N) and phosphorus (P), is low in the case of organic methods (Berry et al. 2002). Most organic fertilizers are low in nutrients so, due to their composition and characteristics, they are not capable to supply all the nutrients for soil and plant growth in organic farming (Tuomisto et al. 2012, Zhao et al. 2009). One more important factor that is under consideration is the supply of other nutrients like zinc (Zn), copper (Cu), iron (Fe), manganese (Mn), and P. These elements can all absorb in soil at a neutral pH but they form insoluble compounds in acidic and alkaline soil (Barbieri et al. 2015). Du Jardin et al. (2015) define biostimulant mainly as a substance that has to be supplied to plants with the purpose to increase the nutrient uptake efficiency but they can also enhance crop quality and soil's organic matter and also increase efficiency in abiotic stress tolerance. The most common plant biostimulants (PB) are seaweed extract, humic and fulvic acid, protein hydrolysates, chitosan, beneficial bacteria and fungi (Ruzzi and Aroca et al. 2015). Biostimulants can enhance the activity of soil both enzymatically and microbiologically, plus they can affect the structure of roots in plants and change the transportability and solubility of macro as well as micronutrients (Rouphael and Colla; 2015, Ertani et al. 2009,

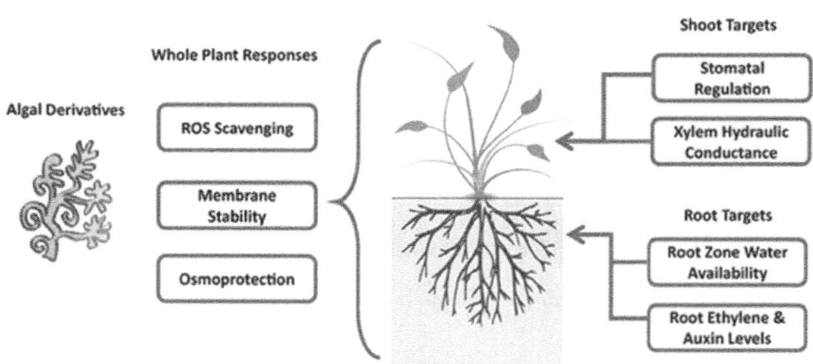

FIGURE 16.4   Algal-based biostimulants.

Source: Free from Google images.

Notes: Biostimulants can play a role in plant protection from different biotic and abiotic stresses. They may help in membrane stability, osmoprotection, root growth and various other targets like stomata regulation and conductive tissues of plants.

Lucini et al. 2015) (see Figures 16.4 and 16.5). Biostimulant can be mixed with silica to increase the ability of soil to deal with biotic stress (Etesami et al. 2018).

Different types of biostimulants have been used for soil amendment nowadays and seaweed is the most commonly used biostimulant (Colla et al. 2015a). It can be used in combination with vermicompost to increase the production of crops and the fertility of the soil. Seaweed is a complex mixture that mainly consists of phytohormones, polysaccharides, vitamins, mineral nutrients, and some fatty acids. Many research studies suggest that the addition of seaweed extract (SWE) in the form of foliar application thrice (2 ml/litre) can help to increase fruit yield, root and plant growth (Rana et al. 2023). Along with this, many biostimulants like humic substances can increase root growth and increase soil nutrient efficiency by an increase in the area of contact between soil and root (Canellas et al. 2015). Along with this, humic substances can help to neutralize the pH of soil by increasing soil cation exchange capacity (Garcia Mina et al. 2004). Protein hydrolysate (PH) is also another type of biostimulant that consists of amino acids, oligopeptides, and polypeptides which is formed by the process of protein hydrolysis or proteins (Schaafsma et al. 2009). They can be used in dose form or in the form of foliar application (Colla et al. 2015b). They also help in improving different soil properties like respiration and help as growth stimulants for different microorganisms present in the soil. Along with this, they help in easy access of soil macro and micronutrients to absorb easily in the soil (Farrell et al. 2014, du Jardin et al. 2015).

16.4   VERMICOMPOST

Vermicompost is a natural process in which earthworms, which belong to subclass Oligochaeta of phylum Annelida, convert waste material into compost (Figure 16.6). Epigenic earthworms mainly employed for vermicomposting are *Eisenia fetida, Eudrilus eugeniae, Perionyx excavates* and *Lumbricus rubellus* (Bhat et al 2013, 2014, 2015; Zainudin et al. 2014). Vermicompost mainly acts as a natural fertilizer for increasing the growth of plants (Rupani et al. 2022, Kalra et al. 2013, Wang et al. 2021). The prefix "vermi" comes from the Latin word that means "worm". The main process in vermicompost is to convert biodegradable matter into a high quality manure with the help of earthworms. Microorganisms present in the gut of earthworms produce some types of exoenzymes

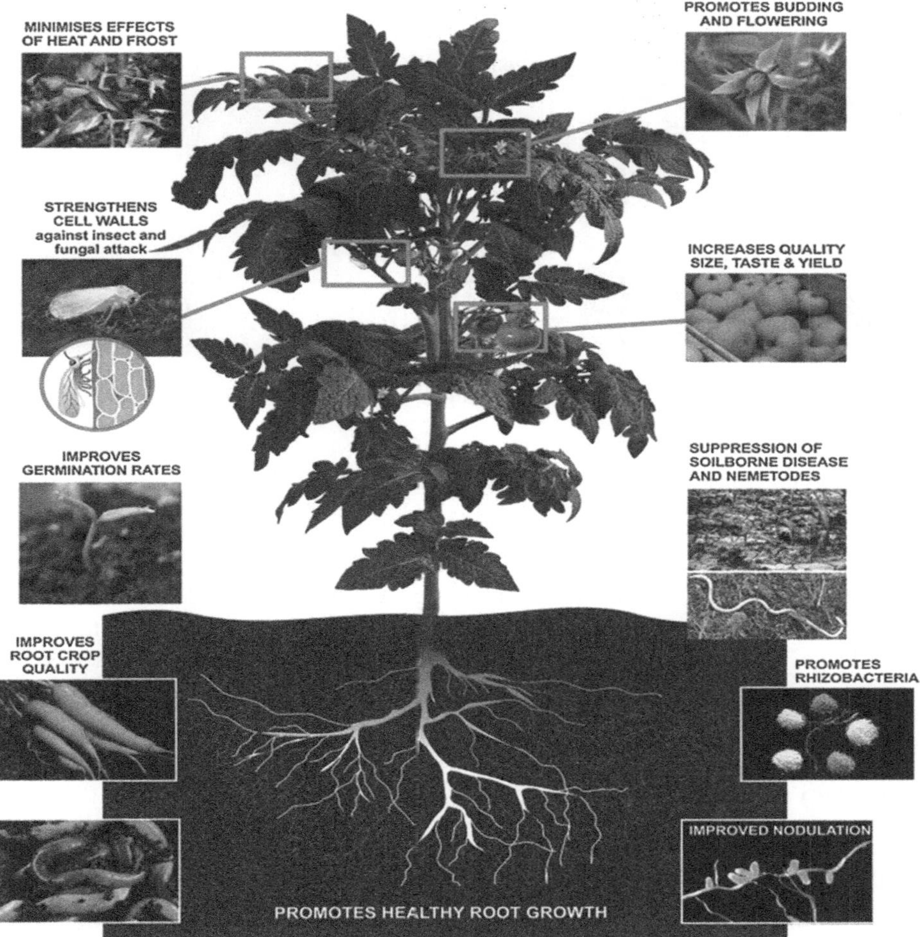

FIGURE 16.5 Seaweed fertilizer.

Source: Free from Google images.

Notes: Seaweed fertilizer is used to promote root growth, improve crop quality, taste and yield, help in germination and also fix nitrogen in soil by fixing nodulation in root nodules.

that mostly help in the degradation of organic matter in the form of specific nutrients. These nutrients in the soil help to improve plant growth and root growth also (Drake and Horn et al. 2007). However, the ingestion system of earthworm depends on water content, initial feed substance, topography and temperature (Chandna et al. 2013). The distribution and growth of microbial species present in the gut of earthworms are mainly directed by changes in moisture, pH, temperature and nutrients. So the community of microbes present in the gut and used in vermicompost also varies with time (Gomez-Brandon et al. 2012). Eco-friendly methods for the extraction of earthworms that are efficient in the production of vermicompost comprise of Allyl-isothiocynate solution (AITC), *Allium cepa* solution and the mustard method (Singh J. et al. 2018). Many researchers also used rice straw (Lim et al. 2014) as a carbon source with vermicompost to increase the fertility of soil for the production of palm oil. Different types of waste are also treated with vermicompost to convert them into organic matter (Singh et al. 2022). It has been shown that vermicompost made by earthworms is more stable than the soil and also contains some enzymes, organic materials, microorganisms and plant growth hormones (Singh J. et al. 2018). According to some research studies there is evidence of

FIGURE 16.6   Vermicompost: earthworms converting organic waste into organic manure.

Source: Free from Google images.

Notes: Microbes present in earthworm guts degrade organic waste into refined manure, which can be employed to enhance soil fertility and crop production.

successful utilization of vermicompost for stabilizing various organic wastes such as paper industry sludge, herbal pharmaceutical industry, sugarcane trash, tea waste, seaweed, vegetable wastes and distillery sludge (Bhat S.A. et al. 2017, 2022). The use of fresh EAS (excess activated sludge) with vermicompost was found to be beneficial fertilizer for improving soil fertility (Wenjaio Li. et al. 2021).

16.5   NANO BIOFERTILIZER

Nano fertilizers are primarily molecules that contain a size of <100 nm, which can enter plant systems through soil application or in the form of a foliar spray (Seleiman et al. 2021) (Figure 16.7). They have a very small size and also carry a large surface area as well as a very large surface area to volume ratio (Hussain et al. 2022), which helps these materials to increase the retention and absorption capacity of soil. Nano fertilizers can be synthesized through biological and chemical methods. Nano fertilizers can be synthesized from biological sources, like different types of plant waste and peat, as well as by chemical methods with the help of urea, ammonia, zinc, iron, lead etc. Since they have more surface area than any type of conventional fertilizer, they contain more nutrients and release them to plants in a gradual manner without any negative effect on the environment (Siddiqi and Husen et al. 2017). As there is very low efficiency in the uptake of chemical fertilizers, lots of nutrients go to waste due to emission of gases, which also results in the destruction of the environment (Dimkpa et al. 2015 a, b). Strong ionizing power, increased absorbability, high surface tension, improved chemical stability and mobility are some of the important characteristics of nano fertilizers (Seleiman et al. 2021). NF have the capability to reduce application frequency, improve nutrient use efficiency (NUE), and reduce soil chemical load (El Ghamry et al. 2018). In the context of organic amendment in soil, nano-biofertilizers essentially contain bacteria, algae and fungal formulation, for example, many fungal mycorrhizae, many nitrogen-fixing bacteria like *Rhizobium* and *Azotobacter*, phosphate-stabilizing bacteria (PSB), *Azospirillum*, *Bradyrhizobium* and blue green algae (BGA) (Galbashy et al. 2017). Major bioorganic compositions in biofertilizer help in phosphate solubilization, soil nutrient supplement, and fixation of nitrogen in root nodules. Along with this, a coating of nanoparticles with the biofertilizer provides a gradual and slow release of nutrients with more efficiency and in a controlled manner (Morales-Diaz et al. 2017). Different nanomaterials such as zeolite, chitosan and polymer in the form of nano encapsulation also help in the slow and gradual release of nutrients into the soil (El Ghamry et al. 2018). A nano-structured

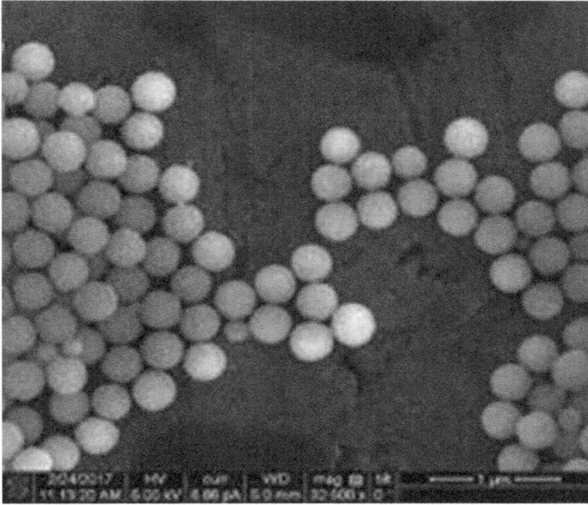

FIGURE 16.7   Nanoparticles from different plants being used for soil amendment applications.
Source: Free from Google images.

fertilizer fundamentally contains PGPR (Plant growth promoting regulator) and neem cake was found to increase the production of some members of the Fabaceae family by fixing the nutrients and increasing the germination of seeds (Rahman and Zhang et al. 2018). Some titanium nanoparticles coated with nano biofertilizer help in the growth of beneficial bacteria in mustard roots (Mishra and Kumar et al. 2009). Nano biofertilizer based on microbes like *Bacillus subtilis*, *pseudomonas putida*, *Pseudomonas fluorescence* and *Paenibacillus elgii* help in reducing the bacterial and fungal infection in the rhizosphere of Fabaceae family plants (Dikshit et al. 2013, Gould et al. 2018, Mukhopadhyay et al. 2014). Also, a foliar spray of nano-biofertilizer on tomatoes was shown to increase pest resistance capacity (Gatahi et al. 2015).

16.6   CONCLUSION AND FUTURE PROSPECTS

Soil is a non-renewable natural resource for all biotic as well as abiotic ecosystems. It is the habitat of different classes of insects, nematodes, algae and fungi and also the main food source for such a large galloping global population. There is a burden on the soil for feeding this ever-increasing population. Hence, there is indiscriminate use of chemical fertilizers for increasing production, which results in the degradation of soil health as well as destruction of the environment. Since all the sources depend on the soil for different purposes, there is an urgent need to take some realistic steps for its protection and conservation. Different tools and techniques need to be developed for restoring soil health in terms of nutrients, minerals, soil organic matter and micro-organisms. Soil organic amendment is one such method to improve soil health and reduce soil erosion as well as preventing soil degradation. The use of biofertilizers, bio-compost, vermicompost, biostimulants, green manure and nano-biofertilizers have been demonstrated to be effective and important tools for the restoration of soil health. Among these, biostimulants and vermicompost have been found to be more effective in improving the soil fertility as well as playing a role in plant protection from different stresses. Lots of research is being carried out in the area of application of biostimulants. Seaweed, in particular, has been most commonly used and found to be highly effective in enhancing the nitrogen content of soil, improving the germination rate, and minimizing heat and frost stress. Similarly, the application of vermicompost has also been found to enhance soil quality in terms of organic matter. It is pertinent to mention that agriculture using organic amendments have exhibited comparable crop yield to that of conventional farming using chemical fertilizers. Hence, employing low cost and environmental friendly organic amendments can play a central role in the reclamation of soils and production of crops without the use of chemical fertilizers. Future research studies may be carried out in the area of nano biofertilizers, biostimulants and biopesticides for reducing the use of chemical fertilizers and pesticides.

REFERENCES

Alami, M.M.; Xue, J.Q.; Ma, Y.T.; Zhu, D.Y.; Abbas, A.; Gong, Z.D.; Wang, X. (2020) Structure, Function, Diversity, and Composition of Fungal Communities in Rhizospheric Soil of Coptis Chinensis Franch Under a Successive Cropping System. *Plants, 9*, 244.

Andrews, S.S.; Karlen, D.L.; Cambardella, C.A. (2004) The Soil Management Assessment Framework: A Quantitative Soil Quality Evaluation Method. *Soil Science Society of America Journal, 68*, 1945–1962.

Ayilara, M.S.; Olanrewaju, O.S.; Babalola, O.O.; Odeyemi, O. Waste. (2020) Management through Composting: Challenges and potentials. *Sustainability 12*, 4456.

Barbieri, G.; Colonna, E.; Rouphael, Y.; De Pascale, S. (2015) Effect of the Farming System and Postharvest Frozen Storage on Quality Attributes of Two Strawberry Cultivars. *Fruits, 70* ,351–360. https://doi.org/10.1051/fruits/2015036.

Berry, P.M.; Sylvester-Bradley, R.; Philipps, L.; Hatch, D.J.; Cuttle, S.P.; Rayns, F.W.; Gosling, P. (2002) Is The Productivity of Organic Farms Restricted by the Supply of Available Nitrogen? *Soil Use and Management, 18*, 248–255.

Bhat, S.A., Singh, J. and Vig, A.P. (2013). Vermiremediation of Dyeing Sludge from Textile Mill with the Help of Exotic Earthworm Eisenia Fetida Savigny. *Environmental Science and Pollution Research*, 20(9), pp.5975–5982.

Bhat, S.A.; Singh, J.; and Vig, A.P. (2014) Genotoxic Assessment and Optimization of Pressmud with the help of Exotic Earthworm Eisenia Fetida. *Environmental Science and Pollution Research*, 21(13), pp.8112–8123.

Bhat, S.A.; Singh, J.; and Vig, A.P. (2015) Vermistabilization of Sugar Beet (Beta Vulgaris L) Waste Produced from Sugar Factory using Earthworm Eisenia Fetida: Genotoxic Assessment by Allium Cepa Test. *Environmental Science and Pollution Research*, 22(15), pp.11236–11254.

Bhat, S.A.; Singh, J.; and Vig, A.P. (2017) Amelioration and Degradation of Pressmud and Bagasse Wastes using Vermitechnology. *Bioresource Technology*, 243, pp.1097–1104.

Bhat, S. A., Singh, J., & Vig, A. P. (2018). Earthworms as Organic Waste Managers and Biofertilizer Producers. *Waste and Biomass Valorization*, 9(7), 1073–1086.

Bhat, S. A., Bashir, O., Haq, S. A. U., Amin, T., Rafiq, A., Ali, M., ... & Sher, F. (2022). Phytoremediation of heavy metals in soil and water: An eco-friendly, sustainable and multidisciplinary approach. *Chemosphere*, 303, 134788

Blaya, J.; López-Mondéjar, R.; Lloret, E.; Pascual, J.A.; Ros, M. (2013) Changes induced by Trichoderma Harzianum in Suppressive Compost Controlling Fusarium Wilt. *Pesticide Biochemistry and Physiology, 107*, 112–119.

Canellas, L. P., Olivares, F. L., Aguiar, N. O., Jones, D. L., Nebbioso, A., Mazzei, P., & Piccolo, A. (2015). Humic and fulvic acids as biostimulants in horticulture. *Scientia Horticulturae, 196*, 15–27.

Cesarano, G.; Filippis, F.D.; Storia, A.L.; Scala, F.; Bonanomi, G. (2017) Organic Amendment Type and Application Frequency Affect Crop Yields, Soil Fertility and Microbiome Composition. *Applied Soil Ecology*, 120, 254–264.

Chandna, P., Nain, L., Singh, S., & Kuhad, R. C. (2013). Assessment of bacterial diversity during composting of agricultural byproducts. *BMC Microbiology*, 13, 1–14.

Chang, E.H.; Chung, R.S.; & Tsai, Y.H. (2007) Effect of Different Application Rates of Organic Fertilizer on Soil Enzyme Activity and Microbial Population. *Soil Science and Plant Nutrition*, 53(2), 132–140.

Chen, L.; Redmile-Gordon, M.; Li, J.W.; Zhang, J.B.; Xin, X.L.; Zhang, C.Z.; Ma, D.H.; Zhou, Y.F. (2016) Linking Cropland Ecosystem Services To Microbiome Taxonomic Composition and Functional Composition in a Sandy Loam Soil with 28-Year Organic and Inorganic Fertilizer Regimes. *Applied Soil Ecology, 139*, 1–9.

Clark, M. S., Horwath, W. R., Shennan, C., Scow, K. M. (1998). Changes in soil chemical properties resulting from organic and low-input farming practices. *Agronomy Journal, 90*(5), 662–671.

Coelho, L.; Reis, M.; Guerrero, C.; Dionísio, L. (2020) Use of Organic Composts to Suppress Bentgrass Diseases in Agrostis Stolonifera. *Biological Control, 141*, 104154.

Colla, G.; Rouphael, Y.; Di Mattia, E.; El-Nakhel, C.; Cardarelli, M. (2015a) Co-Inoculation of Glomus Intraradices and Trichodermaatroviride Acts as a Biostimulant to Promote Growth, Yield and Nutrient Uptake of Vegetable Crops. *Journal of the Science of the Food and Agriculture, 95*, 1706–1715. https://doi.org/10.1002/jsfa.6875.

Colla, G.; Nardi, S.; Cardarelli, M.; Ertani, A.; Lucini, L.; Canaguier, R.; Rouphael, Y. (2015b) Protein Hydrolysates as Biostimulants in Horticulture. *Scientia Horticulturae, 196*, 28–38.

Corato, U.D. (2020).Disease-suppressive Compost Enhances Natural Soil Suppressiveness against Soil-Borne Plant Pathogens: *A critical review. Rhizosphere, 13*, 100192.

Dikshit, A.; Shukla, S.K.; Mishra, R.K. (2013) *Exploring Nanomaterials with PGPR in Current Agricultural Scenario.* Lap Lambert Academic Publishing, Saarbrücken, 978–3–659–36774–8.

Dimkpa, C.O.; Hansen, T.; Stewart, J.; McLean, J.E.; Britt, D.W.; Anderson, A.J. (2015a) Zno Nanoparticles and Root Colonization by a Beneficial Pseudomonad Influence Essential Metal Responses in Bean (Phaseolus Vulgaris). *Nanotoxicology, 9*, 271–278.

Dimkpa, C.O.; McLean, J.E.; Britt, D.W.; Anderson, A.J. (2015b) Nano-CuO and Interaction with Nano-Zno or Soil Bacterium provide Evidence for the Interference of Nanoparticles in Metal Nutrition of Plants. *Ecotoxicology 24*, 119–129.

Doran, J.W.; Jones, A.J. (1996) *Methods for Assessing Soil Quality; Soil Science Society of America: Madison,* WI, USA.

Doran, J.W.; Parkin, T.B. (1994) Defining and assessing soil quality. In *Defining Soil Quality for Sustainable Environment*; Doran, J.W., Coleman, D.C., Bezdicek, D.F., Stewart, B.A., Eds.; Soil Science Society of America: Madison, WI, USA; pp. 3–21.

Doran, J.W., Coleman, D.C., Bezdicek, D.F., Stewart, B.A. (1994), Eds.; Soil Science Society of America: Madison, WI, USA; pp. 53–72.

Doran, J.W.; Sarrantonio, M.; Liebig, M.A. (1996) Soil Health and Sustainability. *Advances in Agronomy,* 56, 1–54.

Drake, H. L.; & Horn, M.A. (2007) As the Worm Turns: The Earthworm Gut as a Transient Habitat for Soil Microbial Biomes. *Annual Review of Microbiology, 61*, 169–189.

Du Jardin, P. (2015). Plant biostimulants: Definition, concept, main categories and regulation. *Scientia Horticulturae*, 196, 3–14.

El-Ghamry, A.; Mosa, A.A.; Alshaal, T.; El-Ramady, H. (2018) Nanofertilizers vs. Biofertilizers: New Insights. *Environment, Biodiversity and Soil Security, 2*, 51–72.

Ertani, A.; Cavani, L.; Pizzeghello, D.; Brandellero, E.; Altissimo, A.; Ciavatta, C.; Nardi, S. (2009) Biostimulant Activity of Two Protein Hydrolyzates in the Growth and Nitrogen Metabolism of Maize Seedlings. *Journal of Plant Nutrition and Soil Science, 172*, 237 244. https://doi.org/10.1002/jpln.200800174.

Etesami, H., Jeong, B. R., & Glick, B. R. (2023). Potential use of Bacillus spp. as an effective biostimulant against abiotic stresses in crops – A review. *Current Research in Biotechnology*, 100128.

Farrell, M.; Prendergast-Miller, M.; Jones, D.L.; Hill, P.W.; Condron, L.M. (2014) Soil Microbial Organic Nitrogen Uptake is Regulated by Carbon Availability. *Soil Biol Biochem, 77*, 261–267. https://doi.org/10.1016/j.soilbio.2014.07.003.

Forge, T.A., Hogue, E.J.; Neilsen, G.; & Neilsen, D. (2008) Organic Mulches Alter Nematode Communities, Root Growth and Fluxes of Phosphorus in the Root Zone of Apple. *Applied Soil Ecology, 39* (1), 15–22. doi:10.1016/j.apsoil.2007.11.004.

Gabriel, D.; Sait, S.M.; Hodgson, J.A.; Schmutz, U.; Kunin, W.E;, & Benton, T. G. (2010) Scale Matters: The Impact of Organic Farming on Biodiversity at Different Spatial Scales. *Ecology Letters, 13*(7), 858–869.

García-Mina, J.M.; Antolín, M.C.; Sanchez-Diaz, M. (2004) Metalhumic Complexes and Plant Micronutrient Uptake: A Study Based on Different Plant Species Cultivated in Diverse Soil Types. *Plant Soil, 258*, 57–68. https://doi.org/10.1023/B:PLSO.0000016509.56780.40.

Gatahi, D.M., Wanyika, H., Kihurani, A.W., Ateka, E., Kavoo, A. (2015) Use of Bio-nanocomposites in Enhancing Bacterial Wilt Plant Resistance, and Water Conservation in Greenhouse Farming. In: *The JKUAT Scientific Conference on Agricultural Sciences, Technologies and Global Networking*, pp. 41–52.

Golbashy, M.; Sabahi, H.; Allahdadi, I.; Nazokdast, H.; Hossein, M. (2017) Synthesis of Highly Intercalated Urea-Clay Nanocomposite via Domestic Montmorillonite as Eco-Friendly Slow-Release Fertilizer. *Archives of Agronomy and Soil Science, 63*, 84–95.

Gómez-Brandón, M.; Lores, M.; & Domínguez, J. (2012) Species-Specific Effects of Epigeic Earthworms on Microbial Community Structure During First Stages of Decomposition of Organic Matter. *PloS one, 7*(2), e31895.

Gould, S.; Kerry, R.G.; Das, G.; Paramithiotis, S.; Shin, H.-S.; Patra, J.K. (2018) Revitalization of Plant Growth Promoting Rhizobacteria for Sustainable Development in Agriculture. *Microbiology Research, 206*, 131–140.

Gudeta, K.; Bhagat, A.; Julka, J.M.; Sinha, R.; Verma, R.; Kumar, A.; … & Sharma, M. (2022) Vermicompost and Its Derivatives against Phytopathogenic Fungi in the Soil: A Review. *Horticulturae, 8*(4), 311.

Hamm, A.C.; Tenuta, M.; Krause, D.O.; Ominski, K.H.; Tkachuk, V.L.; Flaten, D.N. (2016) Bacterial Communities of an Agricultural Soil Amended with Solid Pig and Dairy Manures, and Urea Fertilizer. *Applied Soil Ecology, 103*, 61–71.

Harris, R.F.; Bezdicek, D.F. (1994) Descriptive Aspects of Soil Quality/Health. In *Defining Soil Quality for a Sustainable Environment*; Doran, J.W., Coleman, D.C., Bezdicek, D.F., Stewart, B.A., Eds.; *Soil Science Society of America: Madison*, WI, USA, pp. 23–35.

He, H.B.; Li, W.X.; Zhang, Y.W.; Cheng, J.K.; Jia, X.Y.; Li, S.; Yang, H.R.; Chen, B.M.; Xin, G.R. (2020) Effects of Italian Ryegrass Residues As Green Manure on Soil Properties and Bacterial Communities Under an Italian Ryegrass (Lolium Multiflorum l.)-Rice (oryza sativa l.) Rotation. *Soil and Tillage Research, 196*, 104487.

Hussain, N.; Bilal, M.; Iqbal, H.M. (2022) Carbon-based Nanomaterials With Multipurpose Attributes for Water Treatment: Greening the 21st-Century Nanostructure Materials Deployment. *Biomaterial Polymer Horizons, 1* (1), 1–11.

Kalra, A.; Shukla, S.; Singh, R.; Verma, R.K.; Chandra, M.; Singh, S.; … & Patra, D.D. (2013) Contribution and Assessment of Recycled Menthol Mint Vermicompost on Productivity and Soil Quality in Mint and Mint–Rice–Wheat Rotation: A Case Study. *Renewable Agriculture and Food Systems, 28*(3), 212–219.

Karlen, D.L.; Mausbach, M.J.; Doran, J.W.; Cline, R.G.; Harris, R.F.; Schuman, G.E. (1997). Soil Quality: A Concept, Definition, and Framework for Evaluation. *Soil Science Society of America Journal, 61*, 4–10.

Li, P.F.; Liu, M.; Ma, X.Y.; Wu, M.; Jiang, C.Y.; Liu, K.; Liu, J.; Li, Z.P. (2019) Responses of Microbial Communities to a Gradient of Pig Manure Amendment in Red Paddy Soils. *Science of the Total Environment, 705*, 135884.

Li, W., Li, J., Bhat, S.A., Wei, Y., Deng, Z., & Li, F. (2021) Elimination of Antibiotic Resistance Genes from Excess Activated Sludge Added for Effective Treatment of Fruit and Vegetable Waste in a Novel Vermireactor. *Bioresource Technology, 325*, 124695.

Lim, S.L.; Wu, T.Y.; & Clarke, C. (2014) Treatment and Biotransformation of Highly Polluted Agro-Industrial Wastewater from a Palm Oil Mill into Vermicompost using Earthworms. *Journal of Agricultural and Food Chemistry, 62*(3), 691–698.

Liu, X.; Shi, Y.; Kong, L.; Tong, L.; Cao, H.; Zhou, H.; Lv, Y. (2022) Long-Term Application of Bio-Compost Increased Soil Microbial Community Diversity and Altered Its Composition and Network Microorganisms,10, 462.

Longa, C.M.O.; Nicola, L.; Antonielli, L.; Mescalchin, E.; Zanzotti, R.; Turco, E.; Pertot, I. (2017) Soil Microbiota Respond To Green Manure in Organic Vineyards. *International Journal of Applied Microbiology, 123*, 1547–1560.

Lucini, L., Rouphael, Y., Cardarelli, M., Canaguier, R., Kumar, P., & Colla, G. (2015). The effect of a plant-derived biostimulant on metabolic profiling and crop performance of lettuce grown under saline conditions. *Scientia Horticulturae*, 182, 124–133.

Magdoff, F.; Van Es, H. (2000) *Building Soils for Better Crops; Sustainable Agriculture Network Publications: Burlington*, VT, USA; p. 241.

Marschner, P.; Kandeler, E.; Marschner, B. (2003) Structure and Function of the Soil Microbial Community in a Long-term Fertilizer Experiment. *Soil Biology & Biochemistry, 35*, 453–461.

Mishra, V.K.; Kumar, A. (2009) Impact of Metal Nanoparticles on the Plant Growth Promoting Rhizobacteria. *Digest Journal of Nanomaterials and Biostructures, 4*, 587–592.

Morales-Díaz, A.B.; Ortega-Ortíz, H.; Ju´arez-Maldonado, A.; Cadenas-Pliego, G.; Gonz´alez-Morales, S.; Benavides-Mendoza, A. (2017) Application of Nanoelements in Plant Nutrition and its Impact in

Ecosystems. *Advance Natural Sciences – Nanoscience and Nanotechnology, 8,* 013001 https://doi.org/10.1088/2043-6254/8/1/013001.

Mukhopadhyay, R.; De, N. (2014) Nano Clay Polymer Composite: Synthesis, Characterization, Properties and Application in Rainfed Agriculture. *Global Journal of Bioscience and Biotechnology, 3,* 133–138.

National Research Council (NRC). *Soil and Water Quality; An Agenda for Agriculture;* National Academy Press: Washington, DC, USA, 1993.

Pan, H.; Chen, M.M.; Feng, H.J.; Wei, M.; Song, F.P.; Lou, Y.H.; Cui, X.M.; Wang, H.; Zhuge, Y.P. (2020). Organic and Inorganic Fertilizers Respectively Drive Bacterial and Fungal Community Compositions in a Fluvo-Aquic Soil in Northern China. *Soil and Tillage Research, 198,* 104540.

Parr, J.; Papendick, R.; Hornick, S.; Meyer, R. (1992) Soil Quality: Attributes and Relationship to Alternative and Sustainable Agriculture. *American Journal of Alternative Agriculture, 7,* 5–11.

Ponisio, L.C.; M'gonigle, L.K.; Mace, K.C.; Palomino, J.; De Valpine, P.; Kremen, C. (2015) Diversification Practices Reduce Organic to Conventional Yield Gap. *Proceedings of the Royal Society B.* https://doi.org/10.1098/rspb.2014.1396.

Rahman, K.M.A.; Zhang, D. (2018) Effects of Fertilizer Broadcasting on the Excessive Use of Inorganic Fertilizers and Environmental Sustainability. *Sustainability, 10,* 759.

Ramirez, K.S.; Craine, J.M.; Fierer, N. (2010) Nitrogen Fertilization Inhibits Soil Microbial Respiration Regardless of the Form of Nitrogen applied. *Soil Biology & Biochemistry,42,* 2336–2338.

Raza, S.H.; Shafiq, F.; Parveen, A.; & Iqbal, M. (2022) Potential of Organic Fertilisers. *Introduction and Application of Organic Fertilizers as Protectors of Our Environment, 436.*

Rehman, M., Liu, L., Bashir, S., Saleem, M. H., Chen, C., Peng, D., Siddique, K. H. (2019). Influence of rice straw biochar on growth, antioxidant capacity and copper uptake in ramie (Boehmeria nivea L.) grown as forage in aged copper-contaminated soil. *Plant Physiology and Biochemistry, 138,* 121–129.

Rouphael, Y., & Colla, G. (2020). Biostimulants in agriculture. *Frontiers in Plant Science, 11,* 511937.

Rupani, P. F., Zabed, H. M., & Domínguez, J. (2022). Vermicomposting and bioconversion approaches towards the sustainable utilization of palm oil mill waste. In *Advanced Organic Waste Management* (pp. 193–205). Elsevier.

Ruzzi, M.; Aroca, R. (2015) Plant Growth-Promoting Rhizobacteriaact as Biostimulants in Horticulture. *Scientia Horticulturae, 196,* 124–134.

Savci, S. (2012) An Agricultural Pollutant: Chemical Fertilizer. *International Journal of Environmental Science & Development, 3,* 77–80.

Schaafsma, G. (2009) Safety of Protein Hydrolysates, Fractions Thereof and Bioactive Peptides in Human Nutrition. *European Journal of Clinical Nutrition, 63,* 1161–1168. https://doi.org/10.1038/ejcn.2009.56

Seleiman, M.F.; Almutairi, K.F.; Alotaibi, M.; Shami, A.; Alhammad, B.A.; Battaglia, M.L. (2021) Nano-fertilization as an Emerging Fertilization Technique: Why Can Modern Agriculture Benefit from its Use? *Plants 10,* 2.

Siddiqi, K.S.; Husen, A. (2017) Plant Response to Engineered Metal Oxide Nanoparticles. *Nanoscale Research Letter, 12,* 1–18.

Singh, J.; Singh, S.; Bhat, S.A.; Vig, A.P.; & Schädler, M. (2018) Eco-Friendly Method for the Extraction of Earthworms: Comparative Account Of Formalin, Aitc and Allium Cepa as Extractant. *Applied Soil Ecology, 124,* 141–145.

Singh, J.; Singh, S.; Vig, A.P.; Bhat, S.A.; Hundal, S.S.; Yin, R.; & Schädler, M. (2018) Conventional Farming Reduces the Activity of Earthworms: Assessment of Genotoxicity Test of Soil and Vermicast. *Agriculture and Natural Resources, 52*(4), 366–370.

Singh, S.I.; Singh, W.R.; Bhat, S.A.; Sohal, B.; Khanna, N.; Vig, A.P.; … & Jones, S. (2022) Vermiremediation of Allopathic Pharmaceutical Industry Sludge Amended with Cattle Dung Employing Eisenia Fetida. *Environmental Research, 214,* 113766.

Smith, J.L.; Halvorson, J.J.; Papendick, R.I. (1993) Using Multiple-Variable Indicator Kriging for Evaluating Soil Quality. *Soil Science Society of America Journal, 57,* 743–749.

Sun, R.B.; Zhang, X.X.; Guo, X.S.; Wang, D.Z.; Chu, H.Y. (2015) Bacterial Diversity in Soils Subjected to Long-term Chemical Fertilization can be more Stably Maintained with the Addition of Livestock Manure than Wheat Straw. *Soil Biology & Biochemistry, 88,* 9–18.

Tao, R.; Hu, B.; Chu, G.X. (2020) Impacts of Organic Fertilization with a Drip Irrigation System on Bacterial and Fungal Communities in Cotton Field. *Agricultural Systems, 182,* 102820.

Tuomisto, H.L.; Hodge, I.D.; Riordan, P.; Macdonald, D.W. (2012) Does Organic Farming Reduce Environmental Impacts? *Meta-Analysis of European Research. Journal of Environmental Management,* 112, 309–320.

Wang, N.; Wang, W.; Jiang, Y.; Dai, W.; Li, P.; Yao, D.; ... & Wang, H. (2021) Variations in Bacterial Taxonomic Profiles and Potential Functions in Response to the Gut Transit of Earthworms (Eisenia Fetida) Feeding on Cow Manure. *Science of the Total Environment, 787,* 147392.

Warkentin, B.P.; Fletcher, H.F. (1997) Soil quality for intensive agriculture. In *Intensive Agriculture Society of Science, Soil and Manure. Proceedings of the International Seminar on Soil Environment and Fertilizer Management;* National Institute of Agricultural Science: Tokyo, Japan; pp. 594–598.

Yang, Y.R.; Li, X.G.; Liu, J.G.; Zhou, Z.G.; Zhang, T.L.; Wang, X.X. (2019) Fungal Community Structure in Relation to Manure Rate in Red Soil in Southern China. *Applied Soil Ecology, 147,* 103442.

Yatoo, A.M.; Bhat, S.A.; Ali, M N.; Baba, Z.A.; & Zaheen, Z. (2022) Production of Nutrient-Enriched Vermicompost from Aquatic Macrophytes Supplemented with Kitchen Waste: Assessment of Nutrient Changes, Phytotoxicity, and Earthworm Biodynamics. *Agronomy, 12*(6), 1303.

Zainudin, M.H.M.; Zulkarnain, A.; Azmi, A.S.; Muniandy, S.; Sakai, K.; Shirai, Y.; & Hassan, M.A. (2022) Enhancement of Agro-Industrial Waste Composting Process via the Microbial Inoculation: A Brief Review. *Agronomy, 12*(1), 198. (2014, 2017).

Zhao, X.; Nechols, J.R.; Williams, K.A.; Wang, W.; Carey, E.E. (2009) Comparison of Phenolic Acids in Organically and Conventionally Pac Choi (Brassica rapa L. chinensis). *Journal of the Science of the Food and Agriculture, 89,* 940–946. https://doi.org/10.1002/jsfa.3534.

17 Environmental Risk and Management of Microplastics in Soil

Babita Thakur, Pardeep Kaur, Rakesh Kumar, Joginder Singh, Adarsh Pal Vig, Deachen Angmo, and Jaswinder Singh

17.1 INTRODUCTION

Microplastics are considered as a ubiquitous constituent of almost all the compartments of different ecosystems. Microplastic particles are small in size, being less than 5mm in diameter. Due to its small size they are easily transferred to different ecosystems. Every year millions of tons of plastics are produced, facilitating all aspects of people's lives. The European Association of Plastics Recycling and Recovery Organizations (EPRO) and the Association of Plastic Manufacturers in Europe (2016) reported incomplete data on the production of plastics and estimated 335 million tons with an annual growth rate of 8.6% since the 1950s, when it was 1.7 million tons (UNEP, 2015). Plastic waste is increasing due to its low recovery rate, present level of production and demographic data (Dahlbo et al., 2018; Hahladakis et al., 2018). Plastics are both durable and recyclable; only about 5% of total plastics are recycled (Sutherland et al., 2014) and approximately 4.8–12.7 million tons of plastics enter the oceans (Jambeck et al., 2015). According to the current global estimate, 93–236 thousand tons of microplastics particles or fragments float on the ocean surface (Sebille et al., 2015). In the Arctic region, 38–234 particles m^{-3} of frozen ice has been reported (Obbard et al., 2014). The primary causes of microplastic contamination comes from the soil with sewage sludge, weathering and degradation of plastic items, industrial products, silt from irrigation waterways, atmospheric deposition, laundry effluents and road traffic dust (Kumar et al., 2020; Bhat et al., 2022). Particularly agro-ecosystems are coming into focus as a major entry point for microplastics into the continental systems (Nizzetto et al., 2016b) where contamination might occur via different sources as sludge amendment or plastic mulching (Steinmetz et al., 2016). Many studies on field data on measured microplastic presence in agricultural soils are still not widely available but nevertheless this material is certain to arrive at the soil surface. Hereby, the fate of material deposited at the soil surface is not clear as particles may be removed by wind or water erosion and become airborne or may be lost by surface runoff (Nizzetto et al., 2016a). Many terrestrial sources are directly linked to agro-ecosystems. Recent available reports have demonstrated that microplastics inhibit plant growth, decrease shoot and root biomass, reduce germination rates, promote oxidative damage and impair absorption and translocation of micro and macro elements (Wang et al., 2021; Lee et al., 2022). Researchers have observed that different factors play an important role in the distribution of microplastics, such as human activity and climatic conditions, and have a direct effect on the abundance of microplastics (Wu et al., 2020; Dey et al., 2023). These findings add to our understanding of how important agricultural crops react to microplastic toxicity.

DOI: 10.1201/9781003408352-20

17.2   SOURCES OF MICROPLASTICS IN SOIL

Microplastics are small plastic particles ranging from 1 nm to a size less than 5 mm (Domenech and Marcos, 2021). Generally, there are two main sources of microplastics in the natural environment: (i) primary sources comprise plastic pellets, microbeads in personal care and cosmetic products, paint, sewage sludge, domestic washing wastewater, artificial turf, plastic running tracks in schools and automobile tire wear etc. (Figure 17.1); (ii) secondary sources of microplastics are larger fragments of plastics that have broken down into smaller ones, such as plastic bottles and bags, farming film, fishing wastes and other bulky plastic wastes (An et al., 2020).

Microplastics are extensively dispersed across the globe, even in the snow of the arctic region and remote mountainous areas. The atmospheric deposition is considered as a source for its dispersal near urban sites. A number of research studies have pointed towards waste water as the key medium for the spread of microplastic fibers and particles (Woodward et al., 2021; Wu et al., 2022). Previous studies reflected microplastics in marine ecosystems while recent research has provided the majority of evidence claiming contamination of surface water and groundwater (Selvam et al., 2021; Fan et al., 2022). The untreated wastewater containing small hydrophobic contaminants leads to the spread of microplastics in soil and terrestrial ecosystems (Zurier et al., 2021). Plastic goods are widely employed in modern agricultural systems for a variety of purposes such as cultivation, fertilization, plastic mulching and wastewater irrigation. The fragmentation of larger plastic pieces increases the concentration of micro particles in soil, which are even accumulated by crops (Tian et al., 2022). Agricultural soils have been shown to contain a variety of polymers, including polyethylene, polypropylene, polystyrene and polyvinyl chloride, at varying detectable amounts (Tian et al., 2022; Shi et al., 2022). Hence, sources of microplastics are ubiquitous and their occurrence, transport and fate in the different environmental matrices are affected by various natural aspects as well as their own physicochemical characteristics.

17.3   ABUNDANCE OF MICROPLASTICS IN AGRICULTURE SOIL

Agriculture soils are more abundant with different types of microplastics and distort the soil activities. Previous study investigated that the abundance of MPs in the upper layer of soil (0–3cm) is more as compared to the lower layer (3–6cm) (Liu et al., 2018). Another study also investigated the

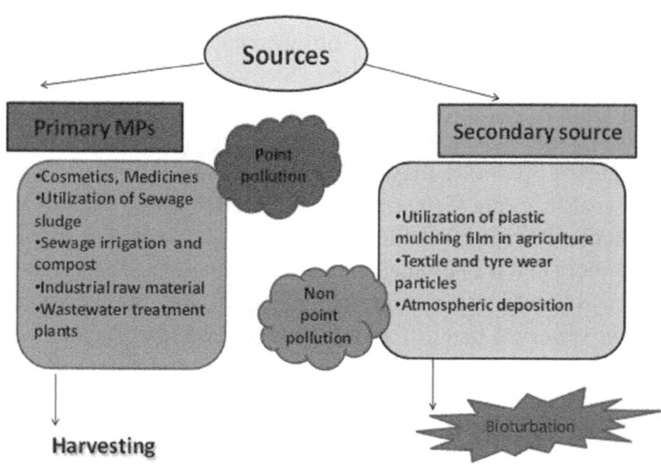

FIGURE 17.1   Schematic diagram of sources of microplastics in soil. It is the schematic diagram showing sources of microplastics including point source in red color and non-point pollution source in yellow color.

TABLE 17.1
Abundance of Different Types of Microplastics in Agriculture Soil

Country	Soil	Abundance of MP (items/kg)	Types of MPs	Reference
Germany	Agriculture soil	0.34	PE, PS	Piehl et al. 2018
China		78.00	PP, PE	Liu et al. 2018
Wuhan (China)		43000–620000	PP	Yanfei et al. 2019
Korea		160	PP, PE	Kim 2019
European union		26156–15137	PE, PP, PS	Mohajerain and Karabtak 2020
Canada		1581–8770	PS, PE	Mohajerain and Karabtak 2020

abundance of MPs in the upper and lower layer of soil 0–5cm and 5–10 cm respectively but there was no significant difference between two layers. Huerta et al. (2017) also studied the upper and lower layer of soil at 0–10 cm and 10–20 cm respectively and observed the texture of soil that can affect the migration and MPs accumulation in soil. Table 17.1 shows different types of MPs found in agriculture soil with its abundance.

17.4 ECOLOGICAL RISK OF MICROPLASTICS ON SOIL PROPERTIES AND SOIL BIOTA

17.4.1 Effect on Soil Chemical Properties

Microplastics cause several deleterious effects on soil nutrient cycling by affecting the activities of different soil enzymes, including urease, diacetic acid and luciferin hydrolase (FDAse) (Fei et al., 2020). The enzymes play a major role in both soil biochemical activity and soil nutrient cycling (Hu et al., 2020). Previous studies reported that a high concentration of polypropylene microplastics affects the different nutrients such as carbon, nitrogen and phosphorus, while low density polyethylene (LDPE) affects the pH, conductivity and carbon to nitrogen ratio value (Liu et al., 2017; Qi et al., 2020). Plastic mulch improves specific soil quality indicators in agricultural production (Ma et al., 2018). Plastic residues have the potential to harm soil properties, resulting in unsustainable farmland use (Steinmetz et al. 2016). The role of MPs in soil quality assessment systems is still unknown. However, few studies have quantified critical thresholds (or critical points) for MP contamination at which the negative effects can be observed. Determining soil MP thresholds is useful for assessing the spatial scale of pollution and the load rate of MPs, as well as predicting the loading capacity of agricultural ecosystems.

17.4.2 Effect on Soil Physical Properties

The effect of microplastics on soil's physical properties depends upon the shape, size, type and exposure time of microplastics. Microplastics cause changes in the soil's structure, humidity, minerals and pore size distribution. MPs also cause changes in several physiological indicators, such as photosynthetic efficiency, and they diminish the plant growth as well as root growth (De Souza Machado et al., 2019; Gao et al., 2019). Large amounts of microplastics cause alterations in the soil water percolating capacity (WPC) as well as altering the evaporation rate (Boots et al., 2019).

17.4.3 Direct Effect on Soil Biota

Like other pollutants, microplastics are also an organic pollutant and cause severe toxic effects on animals that live in soils. Toxic effects mainly depend upon the concentration, shape and size. One study shows low density polyethylene (LDPE) could affect the survival rate of earthworms due to small-size MPs being easily ingested by them (*Lumbricus terrestris*). A high concentration of LDPE alters the growth and shows mortality of earthworm species. Other studies also show that low concentrations of MPs have no significant toxic effects on growth and survival rate (Hodson et al., 2017; Rodriguez-Seijo et al., 2017; Wang et al. 2019). Zhu et al. (2018) studied the toxicity of polyvinyl chloride (PVC) microplastics in *Folsomia candida* (soil collembolan), which distorts feeding activity, growth and reproduction. Lei et al. (2018) checked the toxic effect of polystyrene (PS) in nematodes and it reduced the growth, reproduction and survival rate. Small microplastics particles of polystyrene easily accumulate in the intestines of nematodes. MP particles in the intestines of earthworms cause a strong association and show many toxic effects. Rodriquez-Seijo et al. (2017) studied the direct effects of polyethylene PE pellets on earthworm species (*Eisenia anderi*) tissue and intestine damage. Wang et al. (2019) studied the toxicity of polystyrene and polyethylene particles on both types of microplastics and produced oxidative stress on earthworm species (*Eisenia fetida*). Zhu et al. (2018) studied the effect of PVC particles on earthworms, which enhances the potential of collembolan tissues.

17.4.4 INDIRECT EFFECT ON SOIL BIOTA

Microplastic is an organic or toxic pollutant and gains its bioavailability in the intestinal tract of soil biota. Microplastic acts as the vectors for the environment to gain pollutants' exposure to soil animals. Various studies have been undertaken with different findings. Hodson et al. (2017) checked the adsorption and desorption of Zn on high-density polyethylene bags with a particular size less than 400μm (HDPE). After a limited time period they observed that Zn cannot accumulate in earthworms. The same study was conducted by Rodriquez-Seijo et al. (2019) on low density polyethylene (LDPE) but it could not clarify LDPE microplastics act as a source of pesticides to soil animals. Wang et al. (2019) reported microplastic particles minimize the gathering of As, PAHs and PCB in the intestine of earthworm species.

17.5 RECENT CASE STUDIES ON PHYTOTOXIC EFFECTS ON CROPS

The recent case studies related to the abundance of microplastics in different land use practices, and agricultural farmlands are summarized in Table 17.2. Choi et al. (2020) analyzed the distribution and abundance of microplastics in agrarian soil and found it to be 664 pieces/kg. The abundance varied in the agriculture farmlands, with orchards having the highest abundance followed by the upland, greenhouse and rice field locations reflecting utilization of the mulching and vinyl films. Van den Berg et al. (2020) detected 2130±950 particles/kg in agrarian fields employed to cultivate cereals in Eastern Spain. Similarly, according to Beriot et al. (2021) there were 21,161,024 particles/kg of light density microplastic in the vegetable fields. A case study conducted by Isari and coworkers (2021) revealed the presence of black-colored microplastics in the agricultural fields used for cultivating watermelons and canning tomatoes. It was found that microplastics originated from black agricultural mulch film and watermelon fields had more than four times the quantity of MPs (301,140 items/kg) than canning tomato farms (6938 items/kg). Another study analyzed the microplastic contamination in four different agricultural farmlands viz, rural, sub-urban, urban and near metropolitan in central Bangladesh. Analysis revealed that fragments were dominant as compared to films, fibers and spherules and about 22 microplastic particles of size ranging 1–5 mm were observed in different farmlands (Himu et al., 2022). Zhang et al. (2022) investigated the abundance of microplastics in the top and deep soil of farmland (wheat, maize, potato) and intercropping (potato +apple). The researchers noticed that the highest abundance was detected in the topsoil i.e. 3077.3 items/kg in the potato field, followed by 3044.7 items/kg in the maize field and 2718.4 items/

TABLE 17.2

Recent Case Studies Indicating Abundance of Microplastics in Different Types of Agricultural Farmlands and Land Use Practices

Number of soil samples/land use types	Microplastic (MPs) abundance (items/kg)	Type of crop/ farmland	Origin of microplastic / polymer type	Country	References
100	664 ± 83	Orchard, upland, greenhouse, paddy	Polyethylene, polypropylene,	Korea	Choi et al., 2020
30	301±140 69±38	Watermelons and canning tomatoes	Plastic film	Greece	Isari et al., 2021
6	2116 ± 1024	Melons, lettuce, broccoli and fennel	Plastic mulch	Southeast Spain	Beriot et al., 2021
32	0.12 ± 0.23 0.499 ± 0.48 0.355 ± 0.553 0.315 ± 0.476	Rural, Sub-urban, Urban Metropolitan	Acrylonitrile butadiene styrene and polycarbonate Polyethylene, Polypropylene and polyethylene terephthalate Polycarbonate, ethylene vinyl acetate, polypropylene, polymethyl methacrylate	Central Bangladesh	Himu et al., 2022
4	Potato: 3077.3 items/kg Maize: 3044.7 items/kg Intercropping: 2718.4 items/kg	Wheat, maize potato, intercropping (potato + apple)	Polyethylene, polypropylene	China	Zhang et al., 2022

kg in the intercropping. It is summarized from their study that microplastic abundance was substantially greater in potato and maize (plastic mulched) farmland in comparison to wheat (no mulched) farmland, while the intercropping was intermediate.

17.6 PHYTOTOXIC EFFECT OF MICROPLASTICS ON AGRICULTURAL CROPS

Table 17.3 provides an overview of the toxic effects of microplastics originating from various sources. Wang et al. (2021) assessed the adverse impacts of polyethylene on *Glycine max* (soybean) and *Vigna radiata* (mung bean) germination rates. Both these crops were given treatment with two different sizes (6.5 and 13 µm) of microplastics and six different concentrations up to 500 mg/L. It was found that the phytotoxicity was greater in soybean than that of mung bean. The dry weight, root length, germination index, germination energy and vigor index of soybean were reduced, but root length of mung bean was increased. The accumulation and dispersion of microplastics in rice fields were investigated by Zhang et al. (2021). The research team examined how varied polystyrene (200 nm) concentrations affected rice seed germination, root development, activity of several antioxidant enzymes and transcriptome. The findings exhibited that microplastics had no significant

TABLE 17.3
Phytotoxic Effect of Microplastics on Different Agricultural Crops Reported in Recent Research Studies

Crops	Microplastic	Size/concentration	Phytotoxic effect on crops	References
Soybean (*Glycine max*) and mung bean (*Vigna radiata*)	Polyethylene	6.5 and 13 µm; 0, 10, 50, 100, 200, and 500 mg/L	Decrease germination energy, Inhibition of shoot and root length	Wang et al., 2021
Rice (*Oryza sativa*)	Polystyrene	0.1, 10 and 1000 mg/L	Accretion of reactive oxygen species in roots	Zhang et al., 2021
Wheat (*Triticum aestivum*)	High-density polyethylene	2%	Reduction in root and shoot lengths of seedlings Enhanced lipid peroxidation production	Pflugmacher et al., 2021
Tomato (*Lycopersicon esculentum*)	Polystyrene, polyethylene and polypropylene	0, 10, 100, 500 and 1000 mg/L	Decreased content of soluble sugar and proteins	Shi et al., 2022
Mung bean (*Vignaradiata*)	Benzothiazole, ethyl-vinyl-acetate	204 ± 131 µm (trekking shoes) 57 ± 46 µm(slippers) 156 ± 144 µm (sneakers) and 229 ± 108 µm (running shoes)	Impaired photosynthetic activities Oxidation stress and hindered plant growth	Lee et al., 2022
Cucurbitapepo L.	Polypropylene, polyethylene, polyvinylchloride and polyethyleneterephthalate	0.02, 0.1 and 0.2%	Reduction of photosynthetic efficiency and chlorophyll content Alterations in the composition of microelements and macroelements	Colzi et al., 2022

influence on the seed germination while root length was endorsed and antioxidative activities were significantly reduced. Additionally, genes associated with the production of flavonoids and flavonols were elevated, whereas those involved in the metabolism of linolenic acid and nitrogen were down regulated. Pflugmacher et al. (2021) studied the impact of microplastics on *Triticum aestivum* (L.) in two climatically distinct locations: Singapore (tropical rainforest environment) and Lahti, Finland (continental climate). The study concluded that seedling root and shoot lengths were decreased while lipid peroxidation and catalase activity were increased when exposed to microplastics derived from the high-density polyethylene bottle caps. Shi et al. (2022) investigated the effects of polystyrene, polyethylene, and polypropylene microplastics at concentrations ranging 0–1000 mg/L on tomato plants (*Lycopersicon esculentum* L.). The findings showed that all three microplastics had an adverse effect on seed germination and growth rate. Moreover, polyethylene was observed to be more toxic than polystyrene and polypropylene. Further, these microplastics cause oxidative stress leading to a decrease in antioxidant enzyme activities. The study conducted by Lee et al. (2022)

evaluated the toxicity of shoe sole fragments on mung bean. Shoe sole fragments of size ranging 57–229 µm were obtained from various shoe types and the fragments were leached for 30 days. The microplastic fragments increased the bulk density of soil, reduced its water holding capacity and directly affected plant growth, flavonoid content and impaired photosynthetic activities. *Cucurbitapepo* L. has been employed as a model plant in a recent study to investigate the harmful effects of the four most prevalent microplastics, i.e., polypropylene, polyethylene, polyvinylchloride and polyethylene terephthalate. These contaminants hampered root and shoot growth, decreased leaf size and chlorophyll content, lowered photosynthetic efficiency and altered the micro- and macro-elemental profile. Polyvinylchloride was detected to be the most toxic microplastic while polyethylene exhibited the least toxicity (Colzi et al., 2022).

17.7 MANAGEMENT OF MICROPLASTICS IN SOIL

The introduction of microplastics in soil has significant implications for policy and the management of this issue. As a result, it is suggested that the plastic pollution problem be viewed as a concept. Plastic pollution became a serious problem to the environment and biogeochemical cycle. Based on atmospheric sciences and biogeochemistry, trophic transfer, and human health and exposure, we can identify the continuous and complex movement of plastic materials (Bank and Hansson, 2019) (a diverse contaminant suite of large, medium, micro-, and nanoplastic) between different abiotic and biotic ecosystem compartments (including humans). Governments should introduce several policies and campaigns to manage microplastic pollution in soil. Strict rules and regulations must be implemented by pollution control legislation, and rules need to be developed. Government authorities can help to overcome the microplastics in soil by taking certain steps (Guo et al., 2020): a) make people aware by organizing seminars on microplastics; b) clarify responsibilities and issue penalties to people and businesses involved in the production of microplastics; c) establish polluter-pays and beneficiary-compensation as basic principles when designing environment taxes; d) consult the public, ranging from individuals to non-profit environmental groups, improving public participation, and use this to develop new policies.

17.8 CONCLUSION

The extensive usage of plastic mulching, sewage sludge and other contemporary agricultural methods has led to the detection of microplastics in agricultural soils worldwide ranging from 10 to 12,560 items/k. Microplastics are widely distributed in the terrestrial ecosystem and produce several deleterious effects in soil properties and soil biota. Microplastic particles and fragments may contribute to the alteration of soil's structure and properties, nutritional contents and microbial community. Microplastics may accumulate in the plant roots and are transported to leaves, flowers and fruits, contributing to the toxic effects. The impact varies depending on the plant species and kind of contamination. The toxic effects of different types of microplastic on the most widely grown, eaten and economically significant crops (wheat, rice) raise concerns about their potential entry into the food chain and long-term effects on nutritional quality and human wellbeing. To overcome the microplastics pollution in agriculture, the excessive use of plastic mulching and sewage must be avoided. The current literature review suggests that, considering the impacts of microplastics on soil health and crop growth, it is imperative to formulate policies pertaining to agricultural sustainability under global change. Future research is needed to establish the standardized protocol for extraction and detection of micron-sized microplastics particles, and the respective mechanisms to understand the transformation and transport of microplastics. It is critical to study how microplastics and its additives have harmful effects on agricultural soil and plants and to introduce mechanisms for microplastic degradation.

REFERENCES

An, Lihui, Qing Liu, Yixiang Deng, Wennan Wu, Yiyao Gao, and Wei Ling. "Sources of microplastic in the environment." *Microplastics in Terrestrial Environments* 95 (2020): 143–159.

Bank, Michael S., and Sophia V. Hansson. "The plastic cycle: A novel and holistic paradigm for the Anthropocene." *Environmental Science and Technology* 53, no.13 (2019): 7177–7179.

Beriot, Nicolas, Joost Peek, Raul Zornoza, Violette Geissen, and Esperanza Huerta Lwanga. "Low density-microplastics detected in sheep faeces and soil: A case study from the intensive vegetable farming in Southeast Spain." *Science of the Total Environment* 755 (2021): 142653.

Bhat, S.A., Cui, G., Yaseera, N., Lei, X., Ameen, F. and Li, F. Removal potential of microplastics in organic solid wastes via biological treatment approaches. *Microbial Biotechnology: Role in Ecological Sustainability and Research*, 1(2022): 255–263.

Boots, Bas, Connor William Russell, and Dannielle Senga Green. "Effects of microplastics in soil ecosystems: Above and below ground." *Environmental Science & Technology* 53, no. 19 (2019): 11496–11506.

Choi Yu Ri, Young-Nam Kim, Jung-Hwan Yoon, Nicholas Dickinson and Kye-Hoon Kim. "Plastic contamination of forest, urban, and agricultural soil: a case study of Yeoju City in the Republic of korea." *Journal of Soils and Sediments*, no.21 (2020): 1962–1973.

Colzi, Ilaria, Luciana Renna, Elisabetta Bianchi, Maria Beatrice Castellani, Andrea Coppi, Sara Pignattelli, Stefano Loppi, and Cristina Gonnelli. "Impact of microplastics on growth, photosynthesis and essential elements in Cucurbita pepo L." *Journal of Hazardous Materials* 423 (2022): 127238.

Dahlbo, Helena, Valeria Poliakova, Ville Mylläri, Olli Sahimaa, and Reetta Anderson. "Recycling potential of post-consumer plastic packaging waste in Finland." *Waste Management* 71 (2018): 52–61.

de Souza Machado, Anderson Abel, Chung W. Lau, Werner Kloas, Joana Bergmann, Julien B. Bachelier, Erik Faltin, Roland Becker, Anna S. Görlich, and Matthias C. Rillig. "Microplastics can change soil properties and affect plant performance." *Environmental Science & Technology* 53, no. 10 (2019): 6044–6052.

Dey, S., Anand, U., Kumar, V., Kumar, S., Ghorai, M., Ghosh, A., Kant, N., Suresh, S., Bhattacharya, S., Bontempi, E. and Bhat, S.A., 2023. Microbial strategies for degradation of microplastics generated from COVID-19 healthcare waste. *Environmental Research* 216, p.114438.

Domenech, J., & Marcos, R. (2021). Pathways of human exposure to microplastics, and estimation of the total burden. *Current Opinion in Food Science 39*, 144–151.

Fan, Jianxin, Lan Zou, Ting Duan, Liang Qin, Zenglin Qi, and Jiaoxia Sun. "Occurrence and distribution of microplastics in surface water and sediments in China's inland water systems: A critical review." *Journal of Cleaner Production* 331 (2022): 129968.

Fei, Yufan, Shunyin Huang, Haibo Zhang, Yazhi Tong, Dishi Wen, Xiaoyu Xia, Han Wang, Yongming Luo, and Damià Barceló. "Response of soil enzyme activities and bacterial communities to the accumulation of microplastics in an acid cropped soil." *Science of The Total Environment* 707 (2020): 135634.

Gao, Minling, Yu Liu, and Zhengguo Song. "Effects of polyethylene microplastic on the phytotoxicity of di-n-butyl phthalate in lettuce (Lactuca sativa L. var. ramosa Hort)." *Chemosphere* 237 (2019): 124482.

Guo, Jing-Jie, Xian-Pei Huang, Lei Xiang, Yi-Ze Wang, Yan-Wen Li, Hui Li, Quan-Ying Cai, Ce-Hui Mo, and Ming-Hung Wong. "Source, migration and toxicology of microplastics in soil." *Environment International* 137 (2020): 105263.

Hahladakis, John N., and Eleni Iacovidou. "Closing the loop on plastic packaging materials: What is quality and how does it affect their circularity?." *Science of the Total Environment* 630 (2018): 1394–1400.

Himu, M. M., Afrin, S., Akbor, M. A., Siddique, M. A. B., Uddin, M. K., & Rahman, M. M. "Assessment of microplastics contamination on agricultural farmlands in central Bangladesh." *Case Studies in Chemical and Environmental Engineering*, 5 (2022): 100195.

Hodson, Mark E., Calum A. Duffus-Hodson, Andy Clark, Miranda T. Prendergast-Miller, and Karen L. Thorpe. "Plastic bag derived-microplastics as a vector for metal exposure in terrestrial invertebrates." *Environmental Science & Technology* 51, no. 8 (2017): 4714–472

Hu, Rui, Xin-ping Wang, Jun-shan Xu, Ya-feng Zhang, Yan-xia Pan, and Xue Su. "The mechanism of soil nitrogen transformation under different biocrusts to warming and reduced precipitation: From microbial functional genes to enzyme activity." *Science of the Total Environment* 722 (2020): 137849.

Huerta Lwanga, Esperanza, Jorge Mendoza Vega, Victor Ku Quej, Jesus de los Angeles Chi, Lucero Sanchez del Cid, Cesar Chi, Griselda Escalona Segura et al. "Field evidence for transfer of plastic debris along a terrestrial food chain." *Scientific Reports 7*, no. 1 (2017): 1–7.

Isari, Ekavi A., Dimitrios Papaioannou, Ioannis K. Kalavrouziotis, and Hrissi K. Karapanagioti. "Microplastics in agricultural soils: a case study in cultivation of watermelons and canning tomatoes." *Water* 13, no. 16 (2021): 2168.

Jambeck, Jeena R, Roland Geyer, Chris Wilcox, Theodore R Siegler, Miriam Parryman, Anthony Andrady, Ramani Narayan, and Kara Lavender Law. "Plastic waste inputs from land into the ocea." *Science*. 347, no. 6223 (2015): 768–771, 2015.

Kim, Shin Woong, and Youn-Joo An. "Soil microplastics inhibit the movement of springtail species." *Environment international* 126 (2019): 699–706.

Kumar, Manish, Xinni Xiong, Mingjing He, Daniel CW Tsang, Juhi Gupta, Eakalak Khan, Stuart Harrad, Deyi Hou, Yong Sik Ok, and Nanthi S. Bolan. "Microplastics as pollutants in agricultural soils." *Environmental Pollution* 265 (2020): 114980.

Lee, Tae-Yang, Lia Kim, Dokyung Kim, Sanghee An, and Youn-Joo An. "Microplastics from shoe sole fragments cause oxidative stress in a plant (Vigna radiata) and impair soil environment." *Journal of Hazardous Materials* 429 (2022): 128306.

Lei, Lili, Mengting Liu, Yang Song, Shibo Lu, Jiani Hu, Chengjin Cao, Bing Xie, Huahong Shi, and Defu He. "Polystyrene (nano) microplastics cause size-dependent neurotoxicity, oxidative damage and other adverse effects in Caenorhabditis elegans." *Environmental Science: Nano* 5, no. 8 (2018): 2009–2020.

Liu, Hongfei, Xiaomei Yang, Guobin Liu, Chutao Liang, Sha Xue, Hao Chen, Coen J. Ritsema, and Violette Geissen. "Response of soil dissolved organic matter to microplastic addition in Chinese loess soil." *Chemosphere* 185 (2017): 907–917.

Liu, Mengting, Shibo Lu, Yang Song, Lili Lei, Jiani Hu, Weiwei Lv, Wenzong Zhou et al. "Microplastic and mesoplastic pollution in farmland soils in suburbs of Shanghai, China." *Environmental Pollution* 242 (2018): 855–862.

Ma, Dedi, Lei Chen, Hongchao Qu, Yilin Wang, Tom Misselbrook, and Rui Jiang. "Impacts of plastic film mulching on crop yields, soil water, nitrate, and organic carbon in Northwestern China: A meta-analysis." *Agricultural Water Management* 202 (2018): 166–173.

Mohajerani Abbas, and Karabatak Bojana. "Microplastics and pollutants in biosolids have contaminated agricultural soils: An analytical study and a proposal to cease the use of biosolids in farmlands and utilise them in sustainable bricks." *Waste Management* 107 (2020): 252–265.

Nizzetto, Luca, Gianbatistta Bussi, Martiyan N. Futter, Dan Butterfield, and Paul G. Whitehead, "A theoretical assessment of microplastic transport in river catchments and their retention by soils and river sediments." *Environmental Science: Processes & Impacts*, *18*(8) (2016a): 1050–1059.

Nizzetto, Luca, Martyn Futter, and Sindre Langaas. "Are agricultural soils dumps for microplastics of urban origin?." (2016b): 10777–10779.

Obbard Rachel, Saeed Sadri, Ying Qi Wong, Alexandar A. Khitun, Ian Baker, and Richard C.Thompson. "Global warming releases microplastic legacy frozen in Arctic Sea ice." *Earth's Future* 2, no. 6 (2014): 315–320.

Pflugmacher, Stephan, Saila Tallinen, Simon M. Mitrovic, Olli-Pekka Penttinen, Young-Jun Kim, Sanghun Kim, and Maranda Esterhuizen. "Case study comparing effects of microplastic derived from bottle caps collected in two cities on Triticum aestivum (Wheat)." *Environments* 8, no. 7 (2021): 64.

Piehl, Sarah, Anna Leibner, Martin G. Löder, Rachid Dris, Bogner, C., and Christian Laforsch. "Identification and quantification of macro-and microplastics on an agricultural farmland." *Scientific Reports* 8, no. 1 (2018): 1–9.

Qi, Yueling, Adam Ossowicki, Xiaomei Yang, Esperanza Huerta Lwanga, Francisco Dini-Andreote, Violette Geissen, and Paolina Garbeva. "Effects of plastic mulch film residues on wheat rhizosphere and soil properties." *Journal of Hazardous Materials* 387 (2020): 121711.

Rodríguez-Seijo Andres, Santos Bruna, da Silva Eduardo Ferreira, Cachada Anabela, and Pereira Ruth. "Low-density polyethylene microplastics as a source and carriers of agrochemicals to soil and earthworms." *Environmental Chemistry* 16 no.1 (2019): 8–17.

Rodriguez-Seijo, A., J. Lourenço, T. A. P. Rocha-Santos, J. Da Costa, A. C. Duarte, Helena Vala, and R. Pereira. "Histopathological and molecular effects of microplastics in Eisenia andrei Bouché." *Environmental Pollution* 220 (2017): 495–503.

Sebille, Erik van, Chris Wilcox, Laurent Lebreton, Nikolai Maximenko, Britta Denise Hardesty, Jan A. van Franeker, Marcus Eriksen, David Siegel, Francois Galgani, and Kara Lavender Law. "A global inventory of small floating plastic debris." *Environmental Research Letters* 10, no.12 (2015):124006.

Selvam, S., K. Jesuraja, S. Venkatramanan, Priyadarsi D. Roy, and V. Jeyanthi Kumari. "Hazardous microplastic characteristics and its role as a vector of heavy metal in groundwater and surface water of coastal south India." *Journal of Hazardous Materials* 402 (2021): 123786.

Shi, Ruiying, Weitao Liu, Yuhang Lian, Qi Wang, Aurang Zeb, and Jingchun Tang. "Phytotoxicity of polystyrene, polyethylene and polypropylene microplastics on tomato (Lycopersicon esculentum L.)." *Journal of Environmental Management* 317 (2022): 115441.

Steinmetz, Zacharias, Claudia Wollmann, Miriam Schaefer, Christian Buchmann, Jan David, Josephine Tröger, Katherine Muñoz, Oliver Frör, and Gabriele Ellen Schaumann. "Plastic mulching in agriculture. Trading short-term agronomic benefits for long-term soil degradation?." *Science of the total environment* 550 (2016): 690–705.

Sutherland, William J., Rosalind Aveling, Thomas M. Brooks, Mick Clout, Lynn V. Dicks, Liz Fellman, Erica Fleishman et al. "A horizon scan of global conservation issues for 2014." *Trends in Ecology & Evolution* 29, no. 1 (2014): 15–22.

Tian, Lili, Cheng Jinjin, Rong Ji, Yini Ma, and Xiangyang Yu. "Microplastics in agricultural soils: sources, effects, and their fate." *Current Opinion in Environmental Science & Health* 25 (2022): 100311.

UNEP. *Massive Online Open Course (MOOC) on Marine Litter, 2015.* (2015) UNEP University of the Netherlands. www.unep.org/explore-topics/oceans-seas/what-we-do/addressing-land-based-pollution/global-partnership-marine-2.

van den Berg, Pim, Esperanza Huerta-Lwanga, Fabio Corradini, and Violette Geissen. "Sewage sludge application as a vehicle for microplastics in eastern Spanish agricultural soils." *Environmental Pollution* 261 (2020): 114198.

Wang, Jie, Scott Coffin, Cheng Liang Sun, Daniel Schlenk, and Jay Gan. "Negligible effects of microplastics on animal fitness and HOC bioaccumulation in earthworm Eisenia fetida in soil." *Environmental Pollution* 249 (2019): 776–784.

Wang, Lin, Yi Liu, Mandeep Kaur, Zhisheng Yao, Taizheng Chen, and Ming Xu. "Phytotoxic effects of polyethylene microplastics on the growth of food crops soybean (glycine max) and mung bean (vigna radiata)." *International Journal of Environmental Research and Public Health* 18, no. 20 (2021): 10629.

Woodward, Jamie, Jiawei Li, James Rothwell, and Rachel Hurley. (2021). "Acute riverine microplastic contamination due to avoidable releases of untreated wastewater." *Nature Sustainability*, 4(9), 793–802.

Wu, Mengjie, Chunping Yang, Cheng Du, and Hongyu Liu. "Microplastics in waters and soils: Occurrence, analytical methods and ecotoxicological effects." *Ecotoxicology and Environmental Safety* 202 (2020): 110910.

Wu, Xiaowei, Xiaoli Zhao, Rouzheng Chen, Peng Liu, Weigang Liang, Junyu Wang, Miaomiao Teng, Xia Wang, and Shixiang Ga. "Wastewater treatment plants act as essential sources of microplastic formation in aquatic environments: A critical review." *Water Research* (2022): 118825.

Yanfei, Zhou, Liu Xiaoning, and Wang Jun. "Characterization of microplastics and the association of heavy metals with microplastics in suburban soil of central China [J]." *The Science of the Total Environment* 694 (2019): 133789.

Zhang, Haixin, Yimei Huang, Shaoshan An, Junfeng Zhao, Li Xiao, Haohao Li, and Qian Huang. "Microplastics trapped in soil aggregates of different land-use types: A case study of Loess Plateau terraces, China." *Environmental Pollution* 310 (2022): 119880.

Zhang, Qiuge, Mengsai Zhao, Fansong Meng, Yongli Xiao, Wei Dai, and Yaning Luan. "Effect of polystyrene microplastics on rice seed germination and antioxidant enzyme activity." *Toxics* 9, no. 8 (2021): 179.

Zhu, Dong, Qing-Lin Chen, Xin-Li An, Xiao-Ru Yang, Peter Christie, Xin Ke, Long-Hua Wu, and Yong-Guan Zhu. "Exposure of soil collembolans to microplastics perturbs their gut microbiota and alters their isotopic composition." *Soil Biology and Biochemistry* 116 (2018): 302–310.

Zurier, Hannah S., and Julie M. Goddard. "Biodegradation of microplastics in food and agriculture." *Current Opinion in Food Science* 37 (2021): 37–44.

18 Bioindicators

Natural Biotic Sensors of Environmental Pollution and Ecological Disturbance

Seema Sangwan, Mukesh Kumar, Renu Lamba, Sushila Singh, Anita Kumari, and Leela Wati

18.1 INTRODUCTION

Over the last few decades, the world has witnessed an exponential development phase of industries, transport, agriculture and urbanization involving uncontrolled use of resources, machinery and other multiple human activities. These development practices and the increasing advancements of technologies have also led to the subsequent amplification of various environmental pollutions involving water, air and soils. The increased level of pollution has strongly affected human as well as environmental health. The severe consequences of a polluted environment towards living organisms have presented the unavoidable circumstances for humankind to monitor and control the situation. Monitoring various pollutants and their originating processes is the first step towards controlling pollution.

At present, different methods employed for monitoring environmental contaminants involve techniques such as gas chromatography, high-performance liquid chromatography and mass spectroscopy. However, as a drawback, these methodologies require high cost of operation, time consumption, and sophisticated instrumentation with trained operators (Sharma et al. 2021). On the contrary, the concept of biological indication is solely based on the stress response model of living organisms to assess the health of environment. The strategy mainly includes the presence or abundance of some indicator organisms, such as plants, animals, microbes and planktons in a particular geographical location.

Indicators, by definition, are the agents or events that appear in response to any alteration in the targeted process sample or arena. Indicators of environmental contamination or biogeographical changes include detectable levels of a certain compound, product, or by-product in solid, liquid or gaseous forms, along with any stimulatory or inhibitory variations in the level of biological agents. These could be classified as physical, chemical and biological indicators depending upon the nature of the target entity. The term 'bioindicator' specifically refers to the sum of all the biological agents, including plants, animals and microorganisms, along with the metabolites showing a response with varying concentration of any particular contaminant or nutrient. Their presence or absence, abundance and natural state infer environmental quality directly or through laboratory tests in a more practical and economical way. Bioindicators should be sensitive enough to inform precisely about the cumulative impacts and persistence of various pollutants in the ecosystem which sometimes may not be assessed via physical and chemical testing. Bioindicator species generally have specific inhabitation requirements that differ significantly from the rest of the species found in that ecosystem. The information can be obtained in the form of population size and structure, behaviour,

TABLE 18.1
Desired Characteristics of a Good Bioindicator

Characteristic features	Remarks
Sensitivity, specificity and good indication ability	The response of the bioindicator towards the contaminant reflects the reaction of the entire population/community/ecosystem
	It should respond proportionally to the level of contamination or degradation
	It should provide a detectable and demonstrable response. It should be sensitive to disturbance or stress but should avoid the direct accumulation of pollutants from the environment
Abundance	It should be a common species, having distribution within the field of study. A sufficient local population density should be present as rare species are not optimal
	It should be relatively stable despite moderate climatic and environmental variability
	It should have a high reproductive rate
Well-studied	The bioindicator should be cheap and easy for surveying
	It should be well-documented and have a stable taxonomy
	It should have a well-understood ecology and life history
	It should be harmless to the environment

elemental composition and biochemical or metabolic processes. Bioindication, however, depends upon several other factors such as the level of moisture, temperature, transmission of light and total suspended solids.

Human activities augment multiple types of pollutants or nutrients in environment, which potentially affects the availability, physiology and metabolic activities of microorganisms; hence, these are regarded as the best indicators of soil health. Being the smallest organism in size, microbes have a larger surface area of interaction with amended components in the environment, thereby responding quickly by increasing or decreasing its population. Further, the environment also has the ability to revitalize the situation by neutralizing the waste if it is below the threshold carrying capacity of environment.

Bioindicators differ from biomonitors in that there is a qualitative assessment of biotic responses to environmental stress in the case of the former and a quantification of any response in the latter case. For instance, lichen, like *Lecanora conizaeoides*, acts as a bioindicator of poor air quality while a reduction in its chlorophyll content or diversity is the component of biomonitoring indicating the presence and severity of air pollution. Desired characteristics of a good bioindicator are listed in Table 18.1. These are considered as beneficial over chemical and physical methods of pollutant determination, in several ways. Such as, the bioindicators determine the synergistic and antagonistic effects of any contaminant on a biological system. Bioindicators such as microbes are easily enumerated or assessed due to their prevalence. Harmfulness of toxic components on humans could be assessed indirectly via recording their biomagnification in animals or plant bioindicator species. Furthermore, in comparison to other specialized measuring systems, bioindicators offer a more cost-effective alternative.

18.2 PRINCIPLE AND BASIS OF BIOINDICATION AND BIOMONITORING

Organisms (individual or in community) comprising information related to the quality of the environment or ecosystem are termed as bioindicators. The collection of data of such qualitative information is termed as bioindication or biological indication. Biomonitoring, on the other hand, includes the organisms or their physiological or metabolic components that contain information on the quantitative aspects of the changes occurring in the environment. It tells us about the severity of the impact of an environmental contaminant on the health of living beings. Shifts in biological communities due to pollution, growth stimulation or inhibition of indicator organisms and the emergence of resistant species are the components of bioindication while the increment in the level of primary or secondary metabolites is the module of biomonitoring. These components governing principles of bioindication are elaborated in this section.

18.2.1 SHIFT IN BIOLOGICAL DIVERSITY

Populations of living beings and the interaction among different groups of organisms are naturally influenced by biotic and abiotic stresses or any sort of climatic fluctuations. Some organisms have physiological or genetic potentialities to withstand environmental variation and contaminations, but not all the communities or biological processes are blessed with these qualities. Physical (pH, light intensity, temperature etc.), chemical (substrate or inhibitors) and biological (positive or negative interaction) factors differ among environments. Populations evolve tactics to maintain growth and reproduction with varying ranges of environmental factors. Beyond its optima or tolerance range, however, physiology, metabolism or behaviour of an individual may be adversely affected, ultimately reducing the overall biological communities and population dynamics, in an effected area. The species designated as bioindicators might represent robust organisms with a moderate to high level of tolerance to environmental variations or in some cases appear strikingly sensitive to a specific contaminant. Bioindicators are not just restricted to a single species with a limited environmental tolerance; the entire community or even group of communities of different types of organisms, having a broad range of environmental sensitivities, could serve as bioindicators, which could offer multiple pieces of information to assess environmental health. For example, persistent organic contaminants and heavy metals such as mercury are prevalent throughout the environment, affecting human health, aquatic life and even wildlife inhabiting distant locations. Once released, they keep circulating in various environments including soil, air, water, sediments and macro as well as micro-biota. In particular, exposure to levels of heavy metals and polychlorinated biphenyls is a matter of concern to Arctic biota, such as polar bears, whales and seals, and sea, shore or prey species of birds. These chemicals are harmful and cause carcinogenic, immune-system suppressing, and infertility-inductive effects. Nevertheless, the community shift or biodiversity changes are inevitable in response to events of pollution, precipitation or even natural disturbances.

18.2.2 GROWTH STIMULATION OR INHIBITION OF INDICATOR ORGANISM

The most desirable trait of a biological indicator is its sensitivity towards the slightest change in natural surroundings. Which means the indicator organism must possess a high level of sensitivity that could be reflected in the form of its growth stimulation or inhibition. Response of pollutant in environments showing richness of species is an unrealistic task. However, a clear bioindication can be concealed by responses extended by an excessive population of a divergent species. In order to compile the direct and indirect effects of a contaminant, a subset of the biota or even single species may be targeted. This type of conical approach makes bioindication more relevant and cost-effective.

18.2.3 EMERGENCE OF RESISTANT SPECIES

Natural selection is the basis of survival under harsher conditions. Genetically strengthened species evolve spontaneously in nature by inculcating the ability to tolerate, metabolize or detoxify the potential contaminants. These serve as both bioindicator and pollution scavengers. Higher plants like *Anthroxanthum odoratum* are tolerant to lead and zinc (Qureshi et al. 1981). Some plants (e.g. *Viola lutea* subsp. *Calaminaria*) are designated as metallophytes due to their tolerance towards metalliferous substrates (Baker et al. 2010). Therefore, the majority occurrence of tolerant species signifies the conditions of heavy metal presence. Similarly, the increased pH could be recognized by assessing the occurrence of acidophytes like common ling (*Calluna vulgaris*), hair grass (*Deschampsia flexuose*) and sunflower (*Drosera rotundifolia*) while an increase in salt concentration can be monitored by growing halophytes like lead grass (*Salicornia europeae*) and wild march beet (*Statice limonium*).

18.3 TYPE OF BIOINDICATORS

Bioindicators could be classified in different ways. On the basis of their application, these could be classified as ecological, environmental, pollution and biodiversity indicators. Ecological indicators detect changes in their natural surroundings and their impact on the ecosystem e.g. increased water acidity causes changes in community of diatoms. Various environmental indicators such as coral reefs symptomize any kind of physical or chemical alteration in the environment and other coastal animals and macro invertebrates indicate climate change and rises in sea level. Pollution indicators detect the presence or absence of pollutants e.g., lichens are sensitive to SO_2 levels in the air, and biodiversity indicators signify the phenotypic and genetic changes in the biotic community of a specified zone. Bacterium *Vogesella indigofera* produces blue pigmentation as a morphological indication of the absence of heavy metals like chromium.

Another good example is in earthworms where shallow-burrowing earthworms benefit from organic matter redistributed to the surface soil by surface-dwelling and deep-burrowing earthworms. Modification in the soil biodiversity of surface-dwelling and deep-burrowing earthworms due to any physical or chemical factor leads to changes in population index of shallow-burrowing earthworms, thereby indicating soil quality and health. On the basis of various types of biological agent used for the determination of contaminants, bioindicators can be categorized as follows.

18.3.1 Microorganisms as an Indicator of Environmental Pollution

Microbial populations, due to their exponential growth patterns and rapid response towards their changing surroundings, can act as a valuable, sensitive and fast acting bioindicator for different pollutants (Parmar et al. 2016). Microorganisms grow quickly and respond to a minor surge of the levels of contaminating compounds as well as other physico-chemical and biological changes. In order to utilize microbes as an indicator, genus-level ecotype identification and an understanding of microorganism ecophysiology are required.

Some bacteria, such as *Escherichia coli* and other *Coliforms*, *Streptococcus* sp., *Clostridia* sp., *Pseudomonas* sp., *Thiobacillus* sp., *Vibrio* sp., *Arcobacter* sp. and *Bifidobacterium pseudolongum*, have been utilized as potent indicators of environmental quality monitoring processes (Table 18.2). Microbial indicators can be used to perceive the presence of environmental pollutants in water in a variety of ways such as an indication including bioluminescent bacteria. Toxins in water can be easily detected by observing changes in the metabolic system of microbes, which is hampered or disrupted by the presence of toxins and may result in variations in the amount of light emitted by bacteria e.g. *Photobacterium fisceri*, *P. phosphoreum* (Butterworth et al. 2012). Bacterium *Vogesella indigofera* reacts quantitatively to heavy metals. In the absence of metal pollution, this bacterium produces blue pigmentation, which could be visibly recorded as a morphological indication. Alternatively, in the presence of hexavalent chromium, pigment production is inhibited, therefore pigment production could be attributed to chromium concentration (Jain et al. 2010; Aslam et al. 2012; Malik and Bharti 2012).

Bacteria like *Clostridium perfringens*, *Arcobacter butzleri*, *A. cryaerophilus* and *A. skirrowii* show abundant presence in contaminated water. Wastewater makes an important reservoir for these pathogenic microbes, which will enter into marine water through polluted fresh water bodies and co-exist with other native marine species like *A. marinus* and *A. halophilus*. *Escherichia coli*, *Bifidobacterium*

TABLE 18.2

Microorganisms as Biological Indicator of Various Contaminants

Type of organism	Organism	Indication	References
Bacteria	*Vogesella indigofera*	Sensitive to heavy metals contamination (mainly Cr)	Malik & Bharti (2012)
	Chlamydomonas reinhardtii	Heavy metal contamination (cadmium, lead and methyl mercury)	Hanikenne, M. (2003)
	E.coli, Yersinia enterocolitica, Streptococcus spp, Salmonella spp., Listeria monocytogenes, Shigella spp., Staphylococccus aureus, Vibrio spp	Indicators of food contamination	Bintsis, T. (2017)
	Coliform, Escherichia coli, Streptococcus sp., Clostridia sp., Pseudomonas sp., Thiobacillus sp., Vibrio sp., Arcobacter sp	Indicators of environmental quality	Sumampouw and Risjani (2014)
	Arcobacter butzleri, A. cryaerophilus and *A. skirrowii*	Water contamination	Ghaju et al. (2022)
	Vibrio harveyi	Mutagenic marine pollution	Podgórska et al. (2007)
	Coliform, Escherichia coli, Clostridium perfringens	Indicators of faecal matter contamination	McKee and Cruz (2021)
	Chromatium spp.	Crude oil contamination of environment	Essien et al. (2009)
	Bacillus subtilis	Toxins and pentachlorophenols (organochlorine compounds used as pesticides and disinfectants); Hypersalinity	Kour et al. (2022)
	Methanococcus spp.	Chlorinated aliphatic hydrocarbons	Ogunseitan (2000)
	Halobacterium spp.	Hydrothermal shift	Knoll and Bauld (1989)
	Phormidium laminosum	UV light irradiation	Nicholson et al.(1987)
	Cyanobacteria, Pseudomonas sp.	Nitrogen cycling	Mateo et al. (2015) Sanyal et al. (2019)
	Mytilinidales, Sebacinales, Chitinophagales, Cytophagales, Saccharimonadales	Phosphorous richness	Mason et al. (2021)
	Phanerocheate chryosporium, phytoplanktons	Organic matter	Yildirim et al. (2019)
	Salmonella typhimurium and *Clostridium spp*	Aquatic logging conditions	Zaghloul et al. (2020)
	Bioluminiscent Bacteria; Photobacterium fisceri, P. phosphoreum	Sensitive to environmental pollutants in water; Toxic chemical pollution	Hassan and Oh. (2010)
	Thiobacillus spp., Pseudomonas spp.	Mercury pollution indicator (resistant to mercury as they oxidize mercury)	Kargar et al. (2012)
	Serratia marcescens	Cd and Pb pollution indicator	Sumampouw and Risjani (2014)

(*continued*)

TABLE 18.2 (Continued)
Microorganisms as Biological Indicator of Various Contaminants

Type of organism	Organism	Indication	References
Protozoan (Parasites)	*Cyclospora cayetanensis, Toxoplasma gondii* and *Trichinella spiralis*	Indicators of food contamination	Bintsis (2017)
Phytoplanktons	*Phacus tortus, Trachelon anas, Euglena clastica*	Pollution indicators of marine system	Walsh 1978
Zooplanktons	*Alona guttata, Mesocyclops edax, Cyclops, Aheyella*	Zone-based indicators of pollution	Hosmani (2014)
Algae	*Chlamydomonas sp., Scenedesmus sp.,* and *Chlorella sp*	Aquatic ecosystems pollution	Zancan et al. (2006) Parmar et al. (2016)
	Euglena clastica, Phacus tortus, and *Trachelon anas*	Marine ecosystem degradation	
	Euglena gracilis, Chlorella vulgaris (Fresh water algae)	Pollutant indicator, sensitive to pollutants, and responds quickly to stresses and organic waste	
Lichen	Lichens (*Parmelia spp., Lecanora conizaeoides, Loabaris plumonaria*)	Forest structure, climate pollution (SO_2, H_2S, and other air pollutants)	Nash et al. 2002
Mycorrhizal Fungi	*Septoglomus constrictum*	Indicator species for grassland and reduced-tillage arable land	Oehl et al. (2016)
	Funneliformis caledonius	Indicator species for regularly ploughed, acidic arable land	

pseudolongum, Streptococcus spp., *Salmonella* spp., *Listeria monocytogenes* are the main indicators of food-borne illness and contamination. *Escherichia coli* is a major bacterium among the faecal coliform group comprising several other bacteria, such as *Klebsiella, Citrobacter* and *Enterobacter*, which are referred to as indicator organisms of sewage contamination in drinking water. These are rod-shaped Gram-negative, non-spore forming, motile or nonmotile bacteria commonly found in the faecal matter of warm-blooded animals. Heat and disinfection-resistant hardy spores of *Clostridium perfringens* have also been intended as faecal pollution indicators as these are entirely derived from human faeces. However, their ubiquity and extraordinary persistence in nature, along with a lack of evidence about its multiplication in water, restricts the acceptance of *C. perfringens* as an indicator in countries like the USA, especially in the absence of other indicators such as *E. coli*. In a given ecosystem *Salmonella typhimurium* and *Clostridium* spp. serve as biological indicators for aquatic logging (anaerobic) conditions. *Chromatium* spp. behave as bioindicators of crude oil contamination. *Bacillus subtilis* is a biomarker for toxicity and the presence of pentachlorophenols, which are organo-chlorine compounds generally used as pesticides and disinfectants.

Another bacterium, *Thiobacillus* sp., is frequently used as a bioindicator of heavy metal pollution such as mercury in marine environments (Sumampouw and Risjani 2014). Water-soluble and highly poisonous mercury ions (Hg^{2+}) are generated by oxidation of mercury sulphide, which enters the food chain, thereby endangering humans and other animals. The advanced mechanisms of bacteria like *Thiobacillus* convert Hg^{2+} into methyl mercury (CH_3-Hg) and dimethyl mercury (CH_3-Hg-CH_3), which are more toxic than its ionic form (i.e. Hg^+ or Hg^{2+}). Similarly, *Serratia marcescens* acts as an indicator of cadmium and lead pollution (Cristani et al. 2012), while *Vibrio* spp. acts as a heavy metal pollution indicator in sea water. *Vibrio harveyi* is a free-living non-pathogenic bacterium found in diverse marine environments. Its neomycin-resistant mutants are easily isolated and act as bioindicators of mutagenic marine pollutants. Their frequency increases when a mutagen is present in marine water.

Plankton (phytoplankton and zooplanktons) are significant biological agents found in many fresh as well as marine water bodies, including sea, lakes, streams and swamps. These are the communities that float along the currents and tides causing the blending and cycling of large amounts of energy, which is conceded to the next trophic level. Plankton, being extremely sensitive to change occurring in the natural environment, are the best indicators of water quality, especially lakes and other stagnant water bodies. The changes occurring within planktonic communities help in determining the general health of the water body under review. High phosphorus and nitrogen contents, for instance, increase reproduction in certain plankton thus indicating poor water quality. The accumulation of coloured components, industrial effluents and suspended solids on the water surface hampers the light absorption by plankton reducing the growth rate. Obstructing light reduces the uptake of nitrate and ammonia by aquatic phytoplankton (Walsh 1978) hence leading to a reduction in their levels in water bodies. Besides serving as an indicator of water quality, plankton form a primary food source for fish and, therefore, are essential to oceanic organisms (Thakur et al. 2013). Their reduced levels therefore affect many other trophic levels of food chains. Light-stimulated oxygen evolution in the case of a fresh water species of *Chlamydomonas reinhardtii* was found to be sensitive to cadmium, lead and methyl mercury. Organo-chlorine compounds reduce bicarbonate utilization by estuarine phytoplankton while the presence of zooplankton like *Alona guttata*, *Mesocyclops edax*, *Cyclops* and *Aheyella* is also reduced by these pollutants.

An algal response towards contamination is reflected in terms of their population size as well as photosynthetic rate. The occurrence of algal blooms indicate sewage, organic matter and chemical fertilizer pollution in water bodies. Algae belonging to *Chlamydomonas, Euglena, Chlorella* and *Scenedesmus* are indicators of pollutants in aquatic ecosystems (Hosmani 2013). *Chlorella vulgaris*, a unicellular algae, is widespread in fresh water. Increases in algal diversity, particularly of *Euglena, Phacus* and *Trachelomonas*, indicate marine ecosystem degradation (Jain et al. 2010). For instance, *Euglena gracilis*, a motile, freshwater, photosynthetic, flagellated, acid-tolerant microalga is sensitive to contaminants and responds quickly to stresses, related to potentially toxic elements (PTEs) and persistent organic pollutants (POPs).

Mycorrhizal fungi act as a bioindicators of soil radioactivity and heavy metal pollution. Fungi readily reflect changes in their environments, particularly in the soil. These accumulate radionuclides in their fruiting bodies in a manner distinct from that of heavy metals. In addition to known species (like *Septoglomus constrictum* and *Funneliformis caledonius*, the indicator species for reduced-tillage, arable grassland and frequently ploughed, acidic arable land, respectively), species like *Acaulospora alpina, Glomus macrocarpum, Gigaspora margarita* and *Cetraspora helvetica* are identified as indicator species for grassland (Oehl et al. 2016). The symbiotic organisms, lichens comprised of mycobiont and phycobiont, have been observed to be sensitive to environmental changes such as air pollution. These have the ability to accumulate airborne contaminants such as particulate matter and gases from the environment, including pollutants. Lichens are sensitive to SO_2, H_2S and other air pollutants and even trace amount of toxic substances can impair growth and cause death (Nash et al. 2002). The number, distribution and variability of lichen species around the contaminated area is assessed in order to estimate the pollution level of a zone.

18.3.2 Plants as Bioindicator of Environmental Pollution

Bryophytes, the second largest group of terrestrial plants, have been identified as a potential bioindicator of air pollution. Bryophytes consist of characteristic features in the form of their diverse habitats, structural simplicity and metal-accumulation capability, rendering them an ideal organism for pollution studies (Table 18.3). Mosses, a type of Bryophytes, are useful indicators of atmospheric deposition as they lack a root system, similar to lichens, and rely primarily on atmosphere for nutrient

TABLE 18.3
Plants as Biological Indicator of Air Pollution

Type of pollutant	Indicator plant	Symptoms	References
Ozone	Petunia, Tobacco (*Nicotiana tabacum* cv. Bel-W3), White pine, Watermelon (cv. Sugar Baby), Salvia, Dahlia	Sharp, dot-like lesions on the adaxial side of the leaf caused by death of groups of palisade cells. Spots or streaks of red or brown on the upper surface of the leaves. Curled leaf margins in the presence of high pollution levels. Wilting of the apical part of the pine needles do occur.	Cheng and Sun 2013
Peroxyacetyl Nitrate (PAN)	Chrysanthemum, Primrose, Petunia, Salvia	Destroyed apical tissues and leaf margins. Chlorophyll destruction and tissue death on the lower leaf surface.	Jain et al. 2022 Sase 2017
SO_2	Alfalfa, White birch, White pine (certain strains only), Bean (cv. OSU 1604), Ficus, Xenia	Light spots on the leaf margins or near the veins, resulting in tissue death.	Cotrozzi 2020 Uka et al. 2017
Fluoride	Gladiolus, Pinus, Iris, Ponderosa pine	Destroyed apical tissues and leaf margins.	Kalugina et al. 2017

supply. This character enables the mosses to be a precise bioindicating organism for the detection of heavy metals in the surrounding environment (Grodzinska and Szarek-Lukaszewska 2001). In order to evaluate the effects of air quality during the COVID 19 pandemic, moss *Pleurosiums hreberi* was collected in June 2020 from 19 polluted sites in the Moscow region (Yushin et al. 2020). Atomic absorption spectrometry was used to determine the levels of Cu, Cd, Cr, Ni, Pb and Fe in moss. The comparative analysis of the obtained values with a study carried out in 2019 represented drastic changes in metal concentrations. The cadmium content was found to have decreased by 2–46% as compared to the data obtained in the year 2019, while the iron content increased by a range of 3–127%. Except for the eastern part of Moscow, where a considerable number of engineering and metal processing plants operate, the content of Cu, Pb and Ni in mosses decreased at most sampling sites, indicating the effect of heavy metals on the environment. Similarly, *Atrichum undulatum* was also found to be extremely sensitive to air pollution and served as an excellent bioindicator. Similarly, the tundra ecosystem of Alaska is home to a moss *Hylocomium splendens* acting as a fungal-based indicator for the detection of potentially toxic elements (Hasselbach et al. 2005).

Liverworts are important contributors to the global carbon budget, plant succession and nutrient cycling. Heavy metals such as Cd, Cr, Co, Cu, Hg, Ni, Pb, Sr, V and Zn along with nitrogen, phosphorus, potassium, calcium and magnesium were found in *Conocephalum conicum*, *Pellia epiphylla* and *Marchantia polymorphia*, the liverworts sampled across Poland. The cationic equilibrium analysis of these species of liverworts revealed variation in ionic balance, caused most likely by increased concentrations of micro-elements, mainly iron, cobalt, lead and copper (Samecka et al. 1997).

Other plants like alfalfa, sesame, spinach, sweet potato, carrot, cucumber, tobacco, oat, soybean, cotton, pepper, pine, cedar, rose, chrysanthemum, apple tree, plum, poplar and birch show different levels of sensitivity towards SO_2 concentration in the air. Sulphur dioxide induces plasmolysis of spongy and palisade cells followed by shrinkage or collapsing along with chlorophyll decomposition (Cen 2015). Some other common air borne pollutants also cause similar effects. For example, fluoride induces plasmolysis of mesophyll and cell while ozone oxidizes glucose and breaks down cell walls of palisade tissue and epidermal cells. Peroxyacyl nitrates (PAN) is a chemical compound that induces leaf shrinkage and loss of water, while NO_2 causes cell breakage and chlorine and chloride damage chlorophyll, leading to a severe reduction in the photosynthetic rate.

Gladiolus is a most commonly used plant for monitoring fluoride levels and, similarly, *Equistem* serves as a bioindicator of the presence of gold in soil (Zhou et al. 2015). *Bougainvillea glabra* represents a popular ornamental plant of tropical and subtropical areas as a shrub, which is used as a bioindicator for accumulation of phenolics, flavonoids and metals in their leaves (Azzazy 2020). *Wolffia globosa* is another important plant demonstrating cadmium contamination. *Monotropa, Neottia* and mushrooms are signs of humus in the soil (Table 18.4). *Strobilanthes* and *Impatiens*

TABLE 18.4

Plants and Animals as Biological Indicators of Common Environmental Pollutants

Type of organism	Organism	Indication	References
Bryophytes	Mosses; *Pleurosium shreberi, Sphaganum* & Liverworts; *Conocephalum conicum, Marchantia polymorphia*, and *Pellia epiphylla*)	Heavy metal contamination; mosses also indicate water acidity	Yushin et al. (2020), Samecka et al. (1997)
Plants	*Bougainvillea glabra*	Bioindicator and assessing the accumulation of phenolic, flavonoids, and metals in their leaves	Azzazy (2020) Uttah et al. (2008)
	Wolffia globosa	Cadmium toxicity	
	Spermacoce sticta	Iron-rich soil	
	Lippia nodiflora and *Rumex species*	Nitrate-rich soil	
	Monotropa, Neottia, Strobilanthes and *Impatien*	Indicator of humus-rich soil	
	Caesalpinia pulcherrima and grass *Cyndon dactylon*	Heavy metals like Cu, Pb, Cd, Mn	
	Equistem (Horsetail)	Indicator of mineral content of the soil (gold indicator)	
	Rumex acetosa, Rhododendron, Polytrichum and *Sphagnum*	Indicator of acidic soil	
	Shorea robusta and *Pinus roxburghii*,	Ca-rich soil indicator	
	Suaeda fruticosa, Tamarix auriculata, Salicornea, Chenopodium and *Salsola foetida*	Salty soil indicator	
Earthworms	Earthworms (*Eisenia fetida, Lumbricus rubella*)	Heavy metal, ciprofloxacin accumulation (mainly cadmium, copper)	Sturzenbaum et al. (2004) Huang et al. (2009) Burgos et al. (2005)
Frogs & toads	Red-crowned toadlets (*Pseudophryne australis*) Bibron's toadlets (*Pseudophryne bibroni*)	Terrestrial/freshwater contamination (reduced population	White et al. (2006)
Fish	Salmon fish (*Oncorhynchus tshawytscha*) and red knot (*Calidris canutus rufa*)	Aquatic bioindicator (sensitive to human action such as deforestation, dam construction, and pollution)	Burger et al. (2015)

(continued)

TABLE 18.4 (Continued)
Plants and Animals as Biological Indicators of Common Environmental Pollutants

Type of organism	Organism	Indication	References
Bird	*Corvus brachyrhynchos*	Susceptible to the West Nile Virus	Hopf et al. (2022)
	Northern Bobwhite	Declining population due to change in land usage	
	Ospreys	Determining health of waterways	
	Brown pelicans	Pesticide indicator (declining population; egg shell thinning)	
Invertebrates	Caddisfly	Heavy metal contamination in marine water	Agouridis et al. (2015) de Almeida Rodrigues et al. (2021)
	Stone fly nymph, mayfly nymph	Fresh water indicator	
	Crayfish, shrimp and pond skaters	Moderately polluted water	
	Crustaceans	Heavy metal accumulation	
	Sludge worm, rat-tailed maggot, pouch snails	Indicators of heavily polluted water	
Insects	Honey bee (*Apis mellifera L.*)	Heavy metal and environmental quality	Zhelyazkova. 2012

species indicate a high humus or litter content in the soil, which prevents regeneration of tree species, indicating that the soil humus level is incompatible for forest growth. Many plants indicate the soil pH condition, for example *Rumex acetosa*, *Rhododendron*, *Polytrichum* and *Sphagnum* plants are indicators of acidic soils. Many forest trees, such as *Shorea robusta* and *Pinus roxburghii*, grow in calcium-rich soils. Halophytes such as *Suaeda fruticosa*, *Tamarix auriculata*, *Salicornea*, *Chenopodium* and *Salsola foetida* thrive as the bioindicators of salty soil.

18.3.3 ANIMALS AS BIOINDICATOR OF ENVIRONMENTAL POLLUTION

Some of the coral indicators were found to have significant correlations with a water quality index. Corals are invertebrates mainly found in tropical oceans all over the world. Coral polyps are tiny animals inhabiting calcium carbonate shelters that have accumulated over a period of thousands and millions of years. These skeletons grow in colonies and keep accumulating with other colonies to form a reef. Their vibrant colours are provided by symbiotic algae, *zooxanthellae*, which thrive alongside corals, providing oxygen and essential nutrients. Because corals build and grow in clean, clear water with stable environmental conditions, biologists conclude that the water quality of that specific water body is pure and full of life. As a result, corals serve as indicators of water purity. Under their influence, almost all of the water parameters and other chemical constituents remain stable. The absence of coral reefs in the depths of the Bay of Bengal has led to the conclusion that the water is too familiar to human interference. For example, the Great Barrier Reef in Australia has clear water around the reefs having visibility of around 30 metres, implying that corals have favourable conditions there. Corals are also potential indicators of climate change as it is difficult for them to survive as temperatures rise (Table 18.4). Coral bleaching is one of the most visible indicators of thermal stress due to climate change (www.globalcoralbleaching.org). This is also consistent with the fact that aquatic ecosystems are extremely sensitive to temperature fluctuations.

Macroinvertebrates are valuable and practical indicators of the ecological fitness of aquatic and terrestrial ecosystems. These organisms include aquatic insects, crustaceans, worms and molluscs that resides in river vegetation and stream beds and are simple to sample and identify. Benthic macroinvertebrates are found in a stream or river's benthic zone. Clean water invertebrates include stone fly nymph and mayfly nymph while shrimp and pond skaters inhabit moderately polluted water, and sludge worm and rat-tailed *maggot* invertebrates are found in heavily polluted water. Being easily distinguishable in the laboratory and having restricted mobility and longer survival for more than a year, these tiny animals are considered a powerful indicator of watershed health. Many invertebrate insects, such as mayflies and stoneflies, can detect the amount of oxygen in rivers. Similarly, pollinator insects, such as butterflies and bees, are able to determine the health of plants. The honeybees are an efficient bioindicator that are sensitive towards changing surroundings and could act as a quality-assessing factor. For instance, bees provide a measure of radioactive compounds such as Strontium in the environment when used in atmospheric nuclear testing. In a similar way, earthworms are particularly valuable biological indicators because of their sensitivity towards heavy metals and contaminants like fertilizer and pesticide. Further, the monitoring of earthworms is also an easy process. Earthworms are also sensitive to physical processes such as tillage, land use change and crop rotations. Tillage practices have a negative impact on surface dwelling earthworms (epigeic) and deep-burrowing earthworms (anecic). Shallow-burrowing earthworms (endogeic) are benefitted by organic matter redistribution caused by surface-dwelling and deep-burrowing earthworms. Any factor that negatively impacts the surface-dwelling and deep-burrowing earthworm population also reduces shallow-burrowing earthworms. Since earthworm biodiversity is positively related to soil biodiversity, the presence of species from all three groups indicates a high quality of soil.

In pollution studies, the populations of some amphibians such as frogs and toads are subjected for monitoring processes providing a measure of contaminants of their vicinity. They have permeable skin that led to the aggregation of toxic contaminants and subsequent evaluation of their effects in the form of a decline in the amphibian population. Additionally, they also represent the state of pesticide contamination of foods due to a similar accumulation of these chemicals in their bodies. Hence, these organisms are potential candidates for various ecotoxicological studies and monitoring programs as a bioindicator of altered habitats caused by human interventions. Salmon fish is a common indicator species of aquatic ecosystems worldwide, particularly in North America. The Environmental Protection Agency has designated salmon fish as an indicator species for determining the health of the greater Pacific Rim, which has been impacted by human action such as deforestation, dam construction and pollution, among other factors. Some of the birds such as American crows (*Corvus brachyrhynchos*) show susceptibility towards viral contaminants and this factor is utilized as a bioindicator mechanism for detection of possible incidences of virus-caused diseases in a particular area.

18.3.4 Soil Health Parameters for Bioindication

Soil health is prone to changing soil physico-chemical properties, micro/macronutrient availability and microbial diversity along with various interactions of plants with soils, which are also critical factors of agricultural productivity. Soil biota represents the status of soil health and quality based on factors such as their diversity and abundance along with its functions, such as plant and root development (competition), organic matter decomposition, heavy metal sequestration, detoxification (Nakatsu et al. 2005) of pesticides and other contaminants and disease-suppressiveness. Further, the soil texture and fertility, nutrient cycling and prevalence of earthworms and other eukaryotic habitants act as indicators of soil quality. The metabolic activities such as respiration (a measure of microbial activity in relation to organic matter decomposition), metabolic quotient (qCO_2), microbial biomass ratio and mineralization of organic substrate also represent soil health-based factors as biological indicators (Bastida et al. 2008). Springtails and mites are large, diverse inhabitants adapted to specific soil environments and are found to be sensitive towards alterations in soil quality.

Similarly, nematodes and potworms are valuable indicators of ecosystem disturbances and chemical stress. These groups are extremely diverse, abundant in soils and found at all trophic levels of the soil ecosystem. These organisms indicate the status of microbial communities and energy flows, such as nutrient cycling and carbon sequestration, in addition to soil biodiversity.

18.3.5 Microbial Metabolites as Indicator of Environmental Amendments

Microbial metabolites are biological products of metabolic pathways. These compounds play diverse roles in metabolism. For instance, enzymes such as cellulases, phosphatases, dehydrogenase, urease and arylsulfatase are involved in specific processes of substrate degradation or mineralization of organic matter. These metabolites along with many other metabolic products, such as sterols, antibiotics and protein, have the potential to be used as soil-quality indicators (Table 18.5). For instance, Ergosterol is the principal sterol found in fungal membranes that defines the characteristics of plasma membrane in the form of their fluidic consistency, biogenesis and its function (Yang et al. 2015). Its availability in soil, therefore, makes it a good indicator of fungal growth and mineralization in both soil and water. Similarly, Glomalin is a glycol-proteinaceous substance produced only by arbuscular mycorrhizal fungi (obligate biotroph belonging to phylum *Glomeromycota*) that lives symbiotically with the roots of approximately two-thirds of the higher plants. Arbuscular mycorrhizal fungi hyphae expand beyond nutrient-depleted zones found around root hairs. Hyphae create

TABLE 18.5
Primary and Secondary Metabolites as Soil-health Indicators

Metabolites		Origin	Function	Indication	References
Enzymes	Cellulase	Bacteria, fungi and actinomycetes	Cellulose hydrolysis	C-cycling	Utobo et al. (2015)
	Dehydrogenase	Bacteria	Electron transport chain	C-cycling	Utobo et al. 2015
	Phosphatases	Bacteria, fungi, plants	Release of PO_4^-	P-cycling	Pandey et al. 2014
	B-glucosidase	plants, animals, fungi, bacteria and yeasts	Cellobiose hydrolysis	C-cycling	Merino et al. 2016
	Urease	Bacteria, yeasts, fungi, algae, animal waste and plants	Urea hydrolysis	N-cycling	Das & Varma 2010
	Arylsulphatase	Bacteria, fungi, plants	Release of SO	S-cycling	Utobo et al. 2015
Proteins	Ergosterol	Fungi	Regulates membrane fluidity, plasma membrane biogenesis	Indicator of fungal growth and mineralization activity in both soil and water	Yang et al. 2015
	Glomalin	Arbuscular mycorrhizal fungi	Stabilizes soil and glues together the root aggregates, increased nutrient uptake	Good indicator for AFM and soil quality	Lovelock et al. 2004
Pigments	Phycocyanin	Blue green algae	Concentration of pigment has direct correlation to amount of algal biomass	Biological indication of eutrophication	Horvath et al. 2012

a framework for soil particles to collect Glomalin-coated aggregates. It stabilizes soil, glues root aggregates together and helps in the prevention of nutrient losses from hyphae and soil erosion, therefore acting as a good indicator of soil quality. Phycocyanin is another water-soluble accessory pigment produced by cyanobacteria that detects relative abundance and presence of blue green algae (Horvath et al. 2012). The concentration of phycocyanin can be directly corelated to amount of algal biomass, hence it could be utilized as a biological indication of eutrophication.

Microbial enzymes are responsible for transformation and mineralization processes in the environment. Additionally, many enzymes are involved in the control of soil toxicity and the bio-transformation of pollutants (Kaur et al. 2021). These enzymes are associated intracellularly or extracellularly in microbial cells. The soil enzymes-based microbial activities maintain the biogeo-chemical cycles of nutrients, providing direct support for plant fertility, growth and development. For instance, Dehydrogenases are the most crucial and significant microbial-activity representa-tive enzymes, catalysing reduction reactions by transferring hydrogen ions (protons) from the sub-strate to a co-enzyme or acceptor. The activity of the dehydrogenase enzyme, therefore, indicates the number of viable microorganisms in soil along with any disturbance in the biogeochemical cycles, climate or ecosystem due to any anthropogenic or natural activities. Deviation in the level of dehydrogenase enzymes thereby could act as an indicator of deterioration of soil health. Dehydrogenase enzyme activity, for instance, is studied in obligatory aerobic microbes belonging to genus *Pseudomonas* (Utobo et al. 2015). Similarly, Phosphatases are the enzymes catalysing hydrolysis of phosphoric acid esters and anhydrides, playing an important role in phosphorus cycles in agricultural soils. Its activity is sensitive to management practices like tillage and, therefore, could be an indicator of soil quality (Makoi and Ndakidemi 2008). Different studies revealed that legumes like chickpea, cyclopia and cowpea produce higher levels of phosphatase enzymes in comparison to non-leguminous plants (Makoi et al. 2010; Maseko and Dakora 2013) due to the increased require-ment of phosphate at the time of symbiotic nitrogen-fixation by legumes over cereals (Makoi and Ndakidemi 2008). Further, the mycorrhiza are also reported to enhance the phosphatase activity-enhancing release of soil-bound phosphorus hence representing itself as a reliable soil-quality indi-cator (Van Aarle and Plassard 2010; Pandey et al. 2014).

Likewise, the urease enzyme involved in hydrolysis of urea results in an increase in soil pH and N losses in the atmosphere due to NH_3 volatilization (Das and Varma 2010). The enzyme is found throughout nature and is derived from bacteria, yeasts, fungi, algae, animal waste and plants (Follmer 2008). Urease activities successfully distinguishing a wide range of soil management practices is indicative of environmental factors like temperature, pH, moisture and N cycling. It increases with amendments of organic fertilizer and decreases with management practices like tillage, thereby providing information on soil fertility and biodiversity. The enzyme β-glucosidase is abundantly present in various organisms and specifically in soil biota and plants. It plays its part in various processes such as the degradation of cellulose, which releases glucose monomers to be utilized as a source of energy for maintaining the metabolic activities (Gil-sotres et al. 2005; Merino et al. 2016). Further, its participation in the carbon cycle has been used as a key soil-quality indicator. Increased soil salinity affects the enzyme and leads to reduction of β-glucosidase activity (Rietz and Haynes 2003). Plant residues in heavy metal-polluted soil undergo lesser decomposition and reduced β-glucosidase activity (Geiger et al. 1993), resulting in lesser availability of sugar for multiplication of microbial communities. The interference caused by soil salinity and the presence of heavy metal defines the enzymatic activity of β-glucosidase along with an indication of soil quality. Similarly, enzyme cellulase is responsible for the hydrolysis of cellulose present abundantly in plant cell walls. The hydrolysis yields linear chains of glucose linked with glycosidic bonds. It accounts for nearly 50% of the total biomass synthesized by CO_2 fixation via photosynthetic means. Fungal cellulolytic enzyme systems are classified into three types: the soft-rot (*Aspergillus oryzae, A. niger, Fusarium solani, Trichoderma reesei, T. atroviride, T. harzianum, Mucor circinelloides*), brown-rot (*Poria pla-centa, Tyromyces palustris, Coniophora puteana, Fomitopsis spp., Lanzitestrabeum*) and white-rot

(*Phanerochaete chrysosporium, Agaricus arvensis, Sporotrichum thermophile, Pleurotus ostreatus*) fungi (Kleman-Leyer et al. 1996). Bacteria belonging to aerobic (*Bacillus subtilis, Pseudomonas cellulose, Acinetobacter junii etc.*) and anaerobic (*Butyrivibrio fibrisolvens, Acetivibrio cellulolyticus, Clostridium thermocellum etc.*) clusters also produce cellulase enzymes (Sadhu and Maiti 2013). The cellulolytic enzymes are also produced by actinomycetes, particularly *Cellulomonas fimi, Streptomyces drozdowiczii* and *Thermomonospora fusca.* The enzymes play a vital role in the recycling of the most profuse polymer in nature and hence is used as a predictor of soil fertility.

18.4 BIOLOGICAL MONITORING OF ENVIRONMENTAL POLLUTION

The great expansion of science and technology in recent times has opened the doors for the development of novel methodologies for environmental monitoring. The development of biosensor technology for the detection of environmental contaminants has modified the concept of traditional biological indicators. Further, the molecular tools and technologies are in line for the prospective future of environmental monitoring systems.

18.4.1 TRADITIONAL APPROACHES OF ENVIRONMENTAL MONITORING

The increasing burden of pollutants on the environment heavily affected the biological systems. This ever-growing evil of manmade origin requires stringent methodologies to monitor and control the adversities caused towards human wellbeing and on environmental aspects. Biosensors, as the name suggests, are devices with a sensing system made up of a biological component (such as nucleic acids, proteins, enzymes, or whole cell) and a transducer that converts the sensed biological signal into a readable event (such as optical, electrochemical or piezoelectrical activity) (Hameed et al. 2018). The biorecognition systems of biosensors show extraordinary specificity towards the biological molecule present in the sample. Furthermore, biosensor-mediated monitoring processes show more rapid responses along with high sensitivity and reliability. The construction of a potential biosensor for the aspect of environmental monitoring requires more emphasis towards the reusability, sensitivity and specificity with less susceptibility towards environmental fluctuations such as temperature changes (Mehrotra 2016).

Therefore, biosensors as technological advances of signal-detection systems, including the physico-chemical perception of a biological component along with fast and cost-effective analytical abilities, are a potential candidate for extensive monitoring programs. These biophysical sensing systems can be classified according to their recognition element such as immunosensors, aptasensors, genosensors and enzymatic biosensors. These sensors satisfy the requirements of basic techniques for environmental monitoring processes. The increasing level of environment pollution endangering the human health has elevated the immediate need of monitoring systems, therefore boosting the development of novel, fast and efficient biosensor technologies for real-time monitoring of various pollutants, toxic compounds and pathogens. Some of the major applications of biosensor-mediated monitoring of various parameters are discussed further.

18.4.1.1 Biosensor-mediated Monitoring of Biochemical Oxygen Demand (BOD)

BOD is a critical water-quality indicator. The oxidation of organic substrates present in water bodies by aerobic microbes leads to the depletion of dissolved oxygen of water. The reduced oxygen content corresponds to increased microbial load in the water bodies that also affects the aquatic organisms in negative ways. Reduced BOD can severely affect the ecosystem as it can lead to mass killing of fish. The five-day BOD test involves the quantification of oxygen consumption by the microbial metabolism after 5 days of cultivation in the dark and a temperature range of 20 ± 1 °C. This method is conventionally used for BOD concentration detection of water bodies because of the minimized requirements of any expensive equipment. However, the conventional methods being cumbersome and time-consuming are disadvantageous for water BOD monitoring. Therefore, a rapid and suitable method of real-time BOD detection is required. Biosensors, a self-contained integrated device with their ability of real-time data generation, are a suitable alternative for conventional procedures for BOD monitoring of water bodies. Microbial cells-based biosensors as a BOD detector have huge potential due to the simplicity, easy handling and real-time monitoring. For instance, a microbial fuel cell-based biosensor was used for the real-time monitoring of wastewater BOD under different operating temperatures (Ma et al. 2022). The results indicated that the biosensor was successful in detection of BOD concentrations in actual wastewaters with a reasonably small response time of 0.25 h.

18.4.1.2   Biosensor-mediated Detection of Heavy Metals

Heavy metals have severely contaminated water, air and food along with their presence in the upper fertile layer of earth in the form of oxides, sulphate, phosphates, silicates and organic compounds. These metallic contaminants are being added into the environment by natural events as well as human activities. Among these, the anthropogenic activities are the major contributor of heavy metals as contaminants of the ecosystem. The sources of major proportions include the industrial and automobile emissions due to urbanization and rapid industrialization. The existence of toxic and persistent heavy metals in higher quantities in the environment pose serious threats for human and animal health. There are different methods for the detection of heavy metals. Among these, biosensors are the novel approach for the real-time detection and quantification of heavy metals in a fast, in situ, cost-effective and efficient way. Mahbub et al. (2017) constructed a biosensor by integrating a fluorescent protein-producing reporter gene into the bacterial chromosome. The fluorescent gene inserted into the chromosome, based on the events of homologous recombination, converted the bacterial cell into a whole-cell Hg biosensor, as it was able to produce green fluorescence in the presence of nanomolar concentrations of mercury. The results indicated a linear positive correlation among the Hg concentrations and fluorescence intensity. However, the linearity of the detection was inconsistent at greater concentrations of mercury; therefore, this sensor was only suitable for immediate qualitative detection. Another study resulted in development of a whole-cell microbial biosensor for the concurrent detection of mercury in soluble and insoluble inorganic form. The bacterium *Escherichia coli* was converted into a biodetector of mercury by gene MerR as a detector along with the inducible red fluorescent protein working as the signal of sensing (Guo et al. 2020). The fluorescence intensity of this biosensor showed a positive linear relationship with varying concentrations of Hg (II) from 50 nM to 10 µM. Further, this biosensor also detected the insoluble Hg_2Cl_2. Similarly, the bioavailability of arsenic in water samples was detected by an *Escherichia coli* whole-cell-based biosensor with the help of a green fluorescent protein (GFP) reporter gene coupled with the regulatory sequence ArsR1 from *Geobacter sulfurreducens* (Li et al. 2021). The bacterial cell was able to display green fluorescence in the presence of arsenite with a detection limit of 0.01 µM. Similar genetic expression was also used for detection of cadmium Cd(II) in milk samples based on *Escherichia coli* DH5α (pNV12) with GFP gene expression under the influence of cad promoter and the *cad*C gene of *Staphylococcus aureus* plasmid pI258 (Kumar et al. 2017). This biosensor was able to detect cadmium ion concentrations of 10–50 µg/l. In the same way, a novel biosensor led to successful detection of mercury ions in aqueous solutions based on a homogeneous photochemical chemiluminescence assay (Mei et al. 2021). In the presence of $Hg2^+$, a decreased fluorescence signal was obtained due to the elimination of singlet oxygen channel led by the cleavage of a probe. This biosensor showed high selectivity towards $Hg2^+$ ions and limit of detection of 1.4 nM. However, the colorimetric sensors for the detection of heavy metals outweighed the fluorescence-based biosensors due to their easy detection with the naked eye. For instance, in a study, a whole-cell biosensor was engineered to sense the Pb (II) ion colorimetrically. Pb (II)-dependent mettaloregulator *Pbr* was found regulating the biosynthesis of violacein, therefore, providing the colour signal for the specific detection of lead with high sensitivity and selectivity (Hui et al. 2020). Further, the colour change by the intracellular accumulation of violacein was visible to unaided eyes directly resulting from the sensing of a high amount of lead and was further quantified by the spectrophotometric method. A good linear range for Pb(II) concentrations of 0.1875–1.5 µM was obtained. In another study, a similar biosensor was engineered metabolically to produce violacein pigment under the response to Cd(II) ions (Hui et al. 2022). The presence of cadmium ion as low as 0.049 µM was visualized by the colour change along with an increase in pigment signal.

18.4.1.3 Role of Biosensors in Pesticides Detection

Pesticides are chemically produced compounds employed in agricultural systems for pest management processes. This diverse group of chemicals eliminates agricultural as well as non-agricultural pests, controlling the quality and quantity of field crops and preventing the spread of disease in humans caused by insects and rodent vectors. However, pesticides are associated with some major disadvantages such as high toxicity, mobility and recalcitrance. Therefore, use of these products poses a serious threat for human health problems, and has damaging impacts on the ecosystems. The distinctive characteristics of pesticides, such as environmental persistence, bioaccumulation and toxicity, are the dangers that correlate with increased occurrence of diseases and poisoning in human beings. Extensive use of chemicals such as organophosphorus insecticides in agricultural fields have resulted in their significant presence in the environment, which, therefore, is heavily affecting the environmental balance due to their high toxicity. These pesticides contribute a major part among various environmental pollutants. Therefore, the increasing need of regulation and monitoring to develop evasion strategies for this environmental evil seems justifiable. Therefore, fast, simple and sensitive detection and monitoring strategies have been presented by utilizing downsized technologies such as biosensors, circumventing the need of extensive sample pre-treatment. For instance, a biosensor based on electrochemiluminescence (ECL) covalent organic framework (COF) coupled with a CRISPR/Cas12 a-assisted signal accumulation approach was utilized for the detection of acetamiprid residues (Li et al. 2022). Similarly, an electrochemical biosensor was constructed for evaluation of the residual level of organophosphorus (OPs) residues by a point-of-care test, which included the coupling of a laser-induced graphene (LIG) electrode on polyimide (PI) foil and MnO_2 nanosheets. The working principle was based on disintegration of MnO_2 nanosheets catalysed by acetylcholinesterase enzyme, releasing assistant DNA and initiation of recycling amplification. However, the activity of enzyme acetylcholinesterase was seized by the intervention of organophosphorus residues. Subsequently, the residues were detected by the persistent electrochemical signal by the failed cleavage of the probe labelled with electroactive molecules (Liu et al. 2022). A liquid chromatography-based capillary-sensing system developed for the quantification of acetylcholinesterase and pesticides utilizes the optical response either in the form of two lines of high contrast or a four-petal shape. Pesticides, such as fenobucarb and malathion, are inhibitors of the enzymatic activity of acetylcholinesterase. The inhibition was detected by the microcapillary sensor at a limit of detection (LOD) of 2.5 µg/mL for acetylcholinesterase, as the LC droplets sustained the four-petal appearance deviating from the expected two bright-lined texture (Nguyen et al. 2021). The presence of methyl parathion-like pesticides on environmentally exposed parts of crops and farm products such as leaves of spinach and surfaces of fruits were subjected for an in situ detection by a smart plant-wearable biosensor. The detection mechanism utilized LIG technology on the commercial polyimide film with a three-electrode system. The characteristic flexibility and bending ability of the LIG-based electrode produced the adequate electrochemical response, despite the process of stretching and bending. (Zhao et al. 2020).

18.4.1.4 Biosensor-mediated Detection of Pathogens

Pathogens in the environment, mainly in aquatic systems, pose a serious danger for human health as sole causative agents of various infectious diseases, which cause the loss of a large number of lives globally. The important aspect of pathogens lies with their epidemiology. The accurate and rapid detection methodologies are the solution to avoid the spread of infectious diseases. These methodologies need to be fast, reliable, specific and cheaper to make it accessible for resource-limited circumstances.

The detection and monitoring of pathogens mainly consisting of bacteria from food products, water and the environment can be led to fruition by using the conventional methods; however, a novel approach is to utilize the biosensor-based techniques. The conventional or culture-based techniques are readily used for the detection of bacterial pathogens as these techniques carry the advantage in terms of isolation and further characterization of the detected viable bacterial pathogens (Rajapaksha et al. 2019). However, the technique is disadvantageous due to it being labour-intensive and time-consuming, and lacking specificity and sensitivity, which render it ineffective for the monitoring purpose. Conversely, biosensors have the upper hand compared to traditional techniques in the detection of bacterial pathogens from various environments because of their rapidness, sensitivity, specificity, selectivity and less laborious application. Furthermore, culture-based means of identification requires characteristic identification features of the pathogenic organisms and their ability to be grown in an artificial environment (Ray et al. 2017). Therefore, these techniques are only suitable for culturable microbes, restricting our reach to explore just about one percent of the diversity and physiology of vast microbial communities. The microscale level of pathogen monitoring processes requires the specific sensitivity of biosensors for detection. The required sensitivity of biosensors depends on two vital parameters: the limit of quantification (LOQ) and limit of detection (LOD). The LOD specifies the lowest concentration of pathogenic microbes detected by the biosensor whereas LOQ denotes the lowest concentration to provide an accurate and consistent detection and quantification of the pathogens. However, LOD has been a more emphasized parameter as a standard measure of analytical sensitivity (Forootan et al. 2017). For instance, a microfluidic-based disposable impedance biosensor rapidly detected *Salmonella*, *Legionella* and *E.coli* O157:H7 simultaneously in tap water and wastewater samples. This biosensor was able to differentiate between viable and non-viable bacteria at a minimum concentration of 3 cells/ml for live *E. coli* O157: H7. Furthermore, the sensitivity of detection was boosted by 5–6 times by the accompanying of two sets of focusing electrodes (Mushin et al. 2022). Similarly, rapid and sensitive detection of four human-associated pathogens (HPV, HBV, EBV and U. *urealyticum*) in clinical samples at a level of 18 fM was done by fluorescent biosensor by using t-PER amplification of pathogen nucleic acids (Zhang et al. 2022). Qi et al. (2022) developed a microfluidic biosensor for detection of Salmonella with an on-chip micropump and micromixer in combination with gold-platinum nanocatalyst amplified detection mechanism. The biosensor was also coupled with smartphone imaging to successfully achieve rapid, sensitive and in-field detection of pathogen. A simple, rapid, safe, sensitive and accurate whole-cell-based biosensor for the quantification of autoinducer 2 (AI-2) produced from *Clostridium jejuni* improved the quality and safety of food production (Ramić et al. 2022). In a similar way, a novel MCL biosensor developed for fast-paced detection of *E. coli* O157:H7 was successful in detection of the pathogen up to a minimum concentration of 130 CFU/mL using a lowest sample volume of 10 μL. Further, the biosensor displayed decent selectivity for the target pathogen with a response time of less than two hours for simple operations. With these characteristics, it was suggested for its potential role in the detection of various food-borne pathogens (Sun et al. 2022). Similarly, a biosensor developed by Bacchu et al. (2022) exhibited high selectivity for the detection of *Salmonella typhi* with a wider detection array. The biosensor utilized the increased peak current response of a signal indicator AQMS.

18.4.1.5 Biosensors in Detection of Potentially Toxic Elements (PTEs)

Heavy metals and various chemical residues such as phenols increase the threat of adverse impacts on human health by polluting the major natural water bodies. This burning issue mandated the need of fast, economical and reusable methodologies for their timely detection. Phenols and their derivatives belong to a group of highly toxic elements that pollute several waterbodies such as the underground water table as well as the surface water reservoirs. The absence of any active functional groups hampers the monitoring processes of such aromatic pollutants. Conventional techniques for detection of phenols involves GC-MS (Gas Chromatography-Mass Spectrometry) and LC-MS (Liquid Chromatography-Mass Spectrometry). However, with the requirements of pre-treatment processing, these processes tend to be tedious, costly, time-consuming and inefficient for monitoring purposes. Therefore, the last couple of years have witnessed a rise in the development of biosensors for this purpose. In a study by Ray et al. (2018), a protein sensor strip was constructed for the detection of phenols based on the binding-mediated ATP hydrolysis activity. The biosensor employed the recombinant MopR protein and a ATPase domain. The sensing system used the optical sensing method with malachite green. The quantity of hydrolysed ATP directly linked with amount of pollutant present in the sample. Subsequently, the enzyme system was converted into a viable strip form using a physical adsorption technique with silica nanoparticles as solid support based on the conformational flexibility. The biosensor strip was able to detect real samples from two different water bodies with a minimum level of 12 ppb/0.12μM of phenol content in local tap water in coherence with independent chemical analysis using standard methods.

18.4.1.6 Role of Biosensors in Detection of Toxins

Toxins are microbially produced compounds with adverse impacts on human and animal health. These toxins are biomolecules of either small or macroscopic size with the ability to disrupt the plasma membrane of the host by causing leakage in the phospholipid bilayer. Some of the major toxins such as Aflatoxins carry the potential to be a mutagen or carcinogen, adversely affecting the genetic components of the host. Among various toxins, bacterial toxins are major causative agents of severe food-borne illness in humans. The bacterial toxins represent macromolecules having proteinaceous integrity with a high stability and infectious nature. Compared with the conventional methods of toxin detection, biosensors represent a fast, reliable and convenient monitoring tool for the detection of food toxicants. For instance, a biosensor carrying magnetic relaxation along with a fluorescence biosensing ability represented a successful strategy for the detection of ricin B toxin in food-grade oils and potable water samples using fluorescent magnetic nanoparticles (Wang et al. 2022). The presence of ricin B toxin weakened the magnetic and florescent signals. The change in intensity was found to be in correlation with concentrations of toxin showing a linear relationship. Similarly, a detection limit of 0.92 pg mL^{-1} and concentration range of 0.001–80 ng mL^{-1} was provided by a fluorescent biosensor based on the CRISPR/Cas12a complex and MXenes/ssDNA-FAM reporter for the detection of Aflatoxin B1 in peanut samples (Wu et al. 2023). Another study employed the lateral flow nucleic acid biosensor coupled with a multiple loop-mediated isothermal amplification (mLAMP) process for the detection of *Pseudomonas aeruginosa* along with its toxin genes (Chen et al. 2016). The mLAMP process utilized the modified primers for the correct identification of toxin genes *ExoS* and *ExoU* of *Pseudomonas aeruginosa*. Further, the accumulation of gold nanoparticles as a red-coloured band acted as a specific visual signal for the detection of *P. aeruginosa* and its toxin genes. The developed system showed impressive detection capability for a lower concentration value of 20 CFU/mL. In a similar way, saxitoxin was detected at a broad detection range of 10 to 2000 ng/mL and with a LOD of 0.5 ng/mL using an optical biosensor working on biolayer interferometry with aptamers as characteristic receptors (Gao et al. 2017). This biosensor also exhibited a higher selectivity and good stability with reproducible inferences for saxitoxin. Another ultrasensitive colorimetric biosensor was reported to be based on gold nanoparticles and aptamer for specific and quantitative detection of saxitoxin (Qiang et al. 2020). The specificity of aptamers towards the toxin aggregates the nanoparticles resulting in a change of colour. The biosensor represented a lowest saxitoxin detection concentration of 10 fM from sampled seawater with a linear relationship between the absorbance ratio and saxitoxin concentrations.

18.4.2 MOLECULAR BIOLOGY TOOLS AND TECHNIQUES

The ecosystem and biodiversity are continuously interacting entities of the environment that influence the characteristics of each other and, therefore, any natural and biotic cause, including anthropogenic activities, hampering the environment's quality and ecosystem directly affects its biodiversity (Wu et al. 2022). Further, the aspect of biological monitoring refers to assessing changes in the surroundings using the biological responses of ecological biodiversity involving various bioindicators and biomonitors as two key constituents of biomonitoring processes (Sumudumali et al. 2021). However, traditional morphotaxonomy-based methods of biomonitoring along with the present era of biosensors based monitoring mechanisms are falling behind and facing bottlenecks due to their poor sensitivity and selectivity, lack of precise determination, time-consuming nature, and functional and geographical variability in changing environmental conditions (Blattner et al. 2021). This brings up the need for novel molecular approaches that circumvent the drawbacks of conventional methods for monitoring ecosystems. Molecular biology has been a valuable participant in the development of environmental engineering, providing a novel approach for the aspects of limiting pollution, environmental monitoring, and remediation processes for the environment (Sharma et al. 2022). The development of molecular methods such as environmental genomics holds advantages for the non-specific detection of the microscopic world previously inaccessible with conventional methods, easing monitoring processes. Modern genomic technologies remedy the drawbacks of the culture-based approaches previously utilized for the identification of microbial community structure. The next-generation biomonitoring roadmap involves the efficient implementation of molecular techniques for microbial diversity analysis that largely dominates most ecosystems (Cordier et al. 2021). The amplification of genes, such as 16S rRNA gene hypervariable portions, and subsequent sequencing methods are able to provide a comprehensive microbial population assessment. Furthermore, molecular methods are progressively developing as the front-runners for their application for pathogen detection methodologies. In the same direction, environmental DNA (eDNA) analysis is emerging as an important asset in environmental monitoring based on the quantification of time-dependent geological changes for the analysis and forecasting of the variations in the ecosystem (Mathieu et al. 2020). Furthermore, the omics technologies work as an effective apparatus for gene-based analysis of the ecosystem.

18.4.2.1 Molecular Biomarkers

The molecular biomarkers encompasses various biomolecules such as nucleic acids, proteins, fats and metabolic products for the biodiversity analysis. The special physical and chemical characteristics of these biomarkers form the basis of diversity within organisms. The biomarkers such as DNA/RNA are the target molecules for the development of methods utilizing hybridization of target sequences and synthetic oligonucleotides. This leads to the identification of the species-specific gene present in the eDNA. For instance, in the case of water-dwelling pathogens such as bacteria, the DNA or RNA residues of specific species/genus, functional genes or specific antimicrobial genes are potential bioindicator molecules for their detection and quantification, by using specific DNA sequence-targeted polymerase chain reaction (PCR) amplification as most common molecular technique (Zhang et al. 2021). Similarly, proteomic techniques such as mass spectrometry are utilized to identify peptides using specific DNA damage, repair and stress response proteins produced by the organisms under stress conditions (Zhang et al. 2018). A study by Agarwal et al. (2018) reported the amplified expression of different proteins from *Burkholderia* spp. strain SRS-25 and strain SRS-46 with the presence of uranium as the stress compound. Another method of molecular biological monitoring involves the host-specific molecular source tracking (MST) markers representing distinct geographical and temporal distributions of microbial communities. For instance, host-specific intestinal bacterial species were detected by using endpoint and real-time quantitative PCR for the monitoring of water pollution caused by faecal contamination (Ballesté et al. 2020). In a similar way, bacteriophage crAssphage as a highly human-associated bacterial marker was utilized in crAssphage-based qPCR assays, for detection of faecal contamination of water, by Stachler et al. (2018).

18.4.2.2 DNA Metabarcoding

The term DNA metabarcoding refers to the process employing PCR primers on an environmental DNA sample coupled with high-throughput next-generation sequencing (NGS) for identification of the species structure of any environmental sample. The technique's use of metagenomic sequences rather than morphology for assessing overall biodiversity and the presence of an indicator species, by using samples acquired from the sources as water, sediment or air, represents the novel approach of biological monitoring. Therefore, metabarcoding of environmental DNA (eDNA) is an ecosystem scale tool to define community biodiversity along with interactions and ecology of large areas. Furthermore, eDNA metabarcoding has advantages as a fast, low-cost and accurate identification methodology over traditional barcoding process. The ecological biodiversity and communities directly correlates with the dynamic changes of environment and, therefore, acts as a biomonitoring system. Among all living communities of an ecosystem, the microorganisms are preferentially responsive to environmental disturbances due to their higher environmental sensitivity. Further, their significance is supported by their ubiquitous presence, microscopic size, and abundance, providing a potential advantage and ecological relevance for a valid candidate for biomonitoring processes. Environmental genomics is able to provide a cheap and accurate alternative for identifying the microbial habitants using traditional biomonitoring approaches based on morphological indication. Li et al. (2018) utilized eDNA metabarcoding to identify the community structure change of Yangtze River Delta in China identifying a range of diversity in bacteria, metazoan and other smaller eukaryotic populations. It was found that the nutrient levels of the river was influencing the relative abundance of operational taxonomic units (OTUs) acting as indicators. Similarly, more than 500 river sites were studied for the determination and quantification of diatoms based on a short DNA fragment (DNA barcode) present using DNA metabarcoding (Valentin et al. 2019). Along with the microbial world, some microscopic eukaryotes residing in aquatic or terrestrial ecosystems are also major participants of DNA metabarcoding-based community analysis. For instance, Aylagas et al. (2018) used the metabarcoding-based taxonomic characterization of benthic macroinvertebrate populations from 18 different estuarine and shoreline samples. The taxonomic profile obtained with morphology-based analysis accompanied by metabarcoding was able to detect variations in community structure caused by the ecological disturbance. Similarly, Al et al. (2021) utilized high-throughput sequencing of DNA for the determination of the microeukaryotic community structure and their correlation with non-urban and urban polluted waterbodies in Wuhan City of China due to the anthropogenic effects. The study showed that there was noteworthy alteration in the anthropogenically caused environmental condition that led to the change in the microeukaryotic community structure in urban freshwater ecosystems compared to the non-urban reservoirs. Furthermore, the indicator OTU were comparatively more abundant, representing significant and negative correlation with pollution levels. Therefore, several methodologies with 18S rDNA amplicon sequencing, co-existing with interactive and indicator species analysis, provides a throughput direction based on microeukaryotic communities to analyse and predict the quality of urban aqueous ecosystems.

The application of these methods also met some challenges such as these methodologies being entitled to different partialities in terms of richness, abundance and taxonomic profiles responsible for the basic mechanisms of conventional bioindicators and biomonitoring processes. The abundant presence of OTUs lacks a direct link with morphospecies richness, intragenomic differences, or the source of eDNA as from the dead or inactive organisms. Further, the lack of a complete reference database adds to the hurdles faced in implementation of large sequence analysis.

18.4.2.3 Microbial Fingerprinting

The term microbial fingerprinting relates to the identification of microbial diversity based on the abundance, metabolism and functions of microbes of an area by using various molecular techniques. Molecular microbiology has introduced the methods of culture-independent evaluation of microorganisms with numerous techniques such as PCR, denaturing gradient gel electrophoresis (DGGE), fluorescence in situ hybridization (FISH), amplified ribosomal DNA (rDNA) restriction analysis (ARDRA), terminal restriction fragment length polymorphism (TRFLP), rRNA intergenic spacer analysis (RISA), and single-strand conformational polymorphism (SSCP). These techniques selectively utilized for studying microbial ecology offer extraordinary data about microbial community structures and their response in polluted environments. For instance, the gene sequence-dependent three-dimensional structure of rRNA genes utilized by the SSCP technique, coupled with the electrophoretic mobility of single-stranded rRNA genes under nondenaturing conditions, provides the band patterns that are able to differentiate various microbial phylogenic groups. In a study by Kusi et al. (2020), the silver nanoparticles-impacted reduction of microbial activity was detected using community-level physiological profiling (CLPP). The results exhibited 80% reduction of microbial catabolic activity along with a subsequent reduction in substrate richness and diversity under the influence of citrate-AgNP, subsequently reducing the microbial functional diversity. In another study by Fajardo et al. (2020), the viability of *Caenorhabditis elegans* was checked under the influence of heavy metals such as lead, cadmium and zinc using transcriptome analysis and ecotoxicological tests with DNA microarrays. The bacterium was assessed for its potential as a prudent bioindicator representing the nanoremediation of heavy metals.

18.5 CONCLUSION AND FUTURE PROSPECTS

Environmental health and ecological disturbances are to be monitored precisely in order to assess the impacts of hazardous compounds on biological communities and interactions among living beings. Traditional methods offer precise monitoring but are time-consuming and costlier as compared to biological monitoring. Bioindicators are an environmentally friendly, quicker and safer alternative that utilizes visualization or enumeration of affected species or specific metabolites produced by biological agents. A combined approach involving molecular biology tools and techniques along with biological agents and/or their metabolites could improve the detection as well as real-time monitoring of not only pollutants but also the limiting nutrient. A synergistic approach is needed to maintain a healthier indices environment in general and of soil, water and air in particular. Quicker detection of pollutants accelerates environmental cleanup strategies, thereby lowering the chances of potential mixing and incorporation of hazardous contaminants from the environment into the food chain.

REFERENCES

Agarwal, M., Pathak, A., Rathore, R. S., Prakash, O., Singh, R., Jaswal, R. and Chauhan, A. 2018. Proteogenomic analysis of *Burkholderia* species strains 25 and 46 isolated from uraniferous soils reveals multiple mechanisms to cope with uranium stress. *Cells*, 7(12): 269.

Agouridis, C.T., Wesley, E.T.; Sanderson, T.M. and Newton, B. L. 2015. Aquatic macroinvertebrates: Biological indicators of stream health. *Agriculture and Natural Resources Publications*, 175. https://uknowledge.uky.edu/anr_reports/175

Al, M. A., Xue, Y., Xiao, P., Chen, H., Zhang, C., Duan, M. and Yang, J. 2021. DNA metabarcoding reveals the significant influence of anthropogenic effects on microeukaryotic communities in urban waterbodies. *Environmental Pollution*, 285: 117336.

Aslam, M., Verma, D. K., Dhakerya, R., Rais, S., Alam, M. and Ansari, F. 2012. Bioindicator: A comparative study on uptake and accumulation of heavy metals. *Research Journal of Environmental and Earth Sciences*, 4(12): 1060–1070.

Aylagas, E., Borja, Á., Muxika, I. and Rodríguez–Ezpeleta, N. 2018. Adapting metabarcoding–based benthic biomonitoring into routine marine ecological status assessment networks. *Ecological Indicators*, 95: 194–202.

Azzazy, M. F. 2020. Plant bioindicators of pollution in Sadat city, Western Nile Delta, *Egypt. PLoS One*, 15(3): 226–315.

Bacchu, M. S., Ali, M. R., Das, S., Akter, S., Sakamoto, H., Suye, S. I. and Khan, M. Z. H. 2022. A DNA functionalized advanced electrochemical biosensor for identification of the foodborne pathogen *Salmonella enterica* serovar *Typhi* in real samples. *Analytica Chimica Acta*, 1192: 339332.

Baker, A.J.M., Baker, M., Ernst, W.H.O., Ent. A., Malaisse, F. and Ginocchio, R. 2010. Metallophytes: The unique biological resource, its ecology and conservational status in Europe, central Africa and Latin America. In: Batty LC, Hallberg KB (eds) *Ecology of industrial pollution*. Cambridge University Press, Cambridge, pp 7–40.

Ballesté, E., Demeter, K., Masterson, B., Timoneda, N., Sala–Comorera, L. and Meijer, W. G. 2020. Implementation and integration of microbial source tracking in a river watershed monitoring plan. *Science of the Total Environment*, 736: 139573.

Bastida, F., Zsolnay, A., Hernández, T. and García, C. 2008. Past, present and future of soil quality indices: A biological perspective. *Geoderma*, 147(4): 159–171.

Bintsis, T. 2017. Foodborne pathogens. *AIMS Microbiology*, 3(3): 529.

Blattner, L., Ebner, J. N., Zopfi, J. and von Fumetti, S. 2021. Targeted non–invasive bioindicator species detection in eDNA water samples to assess and monitor the integrity of vulnerable alpine freshwater environments. *Ecological Indicators*, 129: 107916.

Burger, J., Gochfeld, M., Niles, L., Powers, C., Brown, K., Clarke, J. and Kosson, D. 2015. Complexity of bioindicator selection for ecological, human, and cultural health: Chinook salmon and red knot as case studies. *Environmental Monitoring and Assessment*, 187(3): 1–18.

Burgos, M. G., Winters, C., Stürzenbaum, S. R., Randerson, P. F., Kille, P. and Morgan, A. J. 2005. Cu and Cd effects on the earthworm *Lumbricus rubellus* in the laboratory: multivariate statistical analysis of relationships between exposure, biomarkers, and ecologically relevant parameters. *Environmental Science & Technology*, 39(6): 1757–1763.

Butterworth, F. M., Gunatilaka, A. and Gonsebatt, M. E. 2012. *Biomonitors and biomarkers as indicators of environmental change: A handbook* (Vol. 56). Springer Science & Business Media.

Cen, S. 2015. Biological monitoring of air pollutants and its influence on human beings. *Open Biomedical Engineering Journal*, 9: 219.

Chen, Y., Cheng, N., Xu, Y., Huang, K., Luo, Y. and Xu, W. 2016. Point-of-care and visual detection of *P. aeruginosa* and its toxin genes by multiple LAMP and lateral flow nucleic acid biosensor. *Biosensors and Bioelectronics*, 81: 317–323.

Cheng J.F. and Sun E.J. 2013. Factors affecting ozone sensitivity of tobacco Bel-W3 seedlings. *Botanical Studies*, 54(1): 21.

Cordier, T., Alonso-Sáez, L., Apothéloz-Perret-Gentil, L., Aylagas, E., Bohan, D. A., Bouchez, A. and Lanzén, A. 2021. Ecosystems monitoring powered by environmental genomics: A review of current strategies with an implementation roadmap. *Molecular Ecology*, 30(13): 2937–2958.

Cotrozzi, L. 2020. Leaf demography and growth analysis to assess the impact of air pollution on plants: A case study on alfalfa exposed to a gradient of sulphur dioxide concentrations. *Atmospheric Pollution Research*, 11(1): 186–192.

Cristani, M., Naccari, C., Nostro, A., Pizzimenti, A., Trombetta, D. and Pizzimenti, F. 2012. Possible use of *Serratia marcescens* in toxic metal biosorption (removal). *Environmental Science and Pollution Research*, 19(1): 161–168.

Das, S. K., & Varma, A. (2010). Role of enzymes in maintaining soil health. In *Soil enzymology* (pp. 25–42). Springer, Berlin, Heidelberg.

De Almeida Rodrigues, P., Ferrari, R. G., Kato, L. S., Hauser–Davis, R. A. and Conte–Junior, C. A. 2021. A systematic review on metal dynamics and marine toxicity risk assessment using crustaceans as bioindicators. *Biological Trace Element Research*, 200(2): 881–903.

Essien, J. P. and Antai, S. P. 2009. *Chromatium* species: An emerging bioindicator of crude oil pollution of tidal mud flats in the Niger Delta mangrove ecosystem, Nigeria. *Environmental Monitoring and Assessment*, 153(1): 95–102.

Fajardo, C., Martín, M., Nande, M., Botías, P., García–Cantalejo, J., Mengs, G. and Costa, G. 2020. Ecotoxicogenomic analysis of stress induced on *Caenorhabditis elegans* in heavy metal contaminated soil after nZVI treatment. *Chemosphere*, 254: 126909.

Follmer, C. 2008. Insights into the role and structure of plant ureases. *Phytochemistry*, 69(1): 18–28.

Forootan, A., Sjöback, R., Björkman, J., Sjögreen, B., Linz, L. and Kubista, M. 2017. Methods to determine limit of detection and limit of quantification in quantitative real-time PCR (qPCR). *Biomolecular Detection and Quantification*, 12: 1–6.

Gao, S., Zheng, X. and Wu, J. 2017. A biolayer interferometry–based competitive biosensor for rapid and sensitive detection of saxitoxin. *Sensors and Actuators B: Chemical*, 246: 169–174.

Geiger, G., Federer, P. and Sticher, H. 1993. Reclamation of heavy metal contaminated soils: Field studies and germination experiments. *Journal of Environmental Quality*. 22: 201–207.

Ghaju Shrestha, R., Tanaka, Y. and Haramoto, E. 2022. A review on the prevalence of *Arcobacter* in aquatic environments. *Water*, 14(8): 1266.

Gil-Sotres, F., Trasar-Cepeda, C., Leirós, M. C. and Seoane, S. 2005. Different approaches to evaluating soil quality using biochemical properties. *Soil Biology and Biochemistry*, 37(5): 877–887.

Grodzińska, K. and Szarek-Łukaszewska, G. 2001. Response of mosses to the heavy metal deposition in Poland—an overview. *Environmental Pollution*, 114(3): 443–451.

Guo, M., Wang, J., Du, R., Liu, Y., Chi, J., He, X. and Xu, W. 2020. A test strip platform based on a whole-cell microbial biosensor for simultaneous on–site detection of total inorganic mercury pollutants in cosmetics without the need for predigestion. *Biosensors and Bioelectronics*, 150: 111899.

Hameed, S., Xie, L. and Ying, Y. 2018. Conventional and emerging detection techniques for pathogenic bacteria in food science: A review. *Trends in Food Science & Technology*, 81: 61–73.

Hanikenne, M. 2003. *Chlamydomonas reinhardtii* as a eukaryotic photosynthetic model for studies of heavy metal homeostasis and tolerance. *New Phytologist*, 159(2): 331–340.

Hassan, S. H. and Oh, S. E. 2010. Improved detection of toxic chemicals by *Photobacterium phosphoreum* using modified Boss medium. *Journal of Photochemistry and Photobiology B: Biology*, 101(1): 16–21.

Hasselbach, L., Ver Hoef, J. M., Ford, J., Neitlich, P., Crecelius, E., Berryman, S. and Bohle, T. 2005. Spatial patterns of cadmium and lead deposition on and adjacent to National Park Service lands in the vicinity of Red Dog Mine, Alaska. *Science of the Total Environment*, 348(3): 211–230.

Hopf, C., Bunting, E., Clark, A. and Childs-Sanford, S. 2022. Survival and release of 5 American Crows (*Corvus brachyrhynchos*) Naturally infected with West Nile Virus. *Journal of Avian Medicine and Surgery*, 36(1): 85–91.

Horvath, H., Kovacs, A.W., Riddick, C. and Presing, M. 2012. Extraction methods for phycocyanin determination in freshwater filamentous cyanobacteria and their application in a shallow lake. *European Journal of Phycology*, 48(3): 278–286.

Hosmani, S. 2014. Freshwater plankton ecology: a review. *Journal of Resource Management and Technology*, 3: 1–10.

Hosmani, S. P. 2013. Fresh Water Algae as Indicators of Water Quality. *Universal Journal of Environmental Research and Technology*, 3(4).

Huang, R., Wen, B., Pei, Z., Shan, X. Q., Zhang, S. and Williams, P. N. 2009. Accumulation, subcellular distribution and toxicity of copper in earthworm (*Eisenia fetida*) in the presence of ciprofloxacin. *Environmental Science & Technology*, 43(10): 3688–3693.

Hui, C. Y., Guo, Y., Liu, L., Zhang, N. X., Gao, C. X., Yang, X. Q. and Yi, J. 2020. Genetic control of violacein biosynthesis to enable a pigment–based whole-cell lead biosensor. *RSC Advances*, 10(47): 28106–28113.

Hui, C. Y., Guo, Y., Li, H., Gao, C. X. and Yi, J. 2022. Detection of environmental pollutant cadmium in water using a visual bacterial biosensor. *Scientific Reports*, 12(1): 1–11.

Jain, A., Singh, B. N., Singh, S. P., Singh, H. B. and Singh, S. 2010. Exploring biodiversity as bioindicators for water pollution. In *National Conference on Biodiversity, Development and Poverty Alleviation*: 50–56.

Jain, P. K., Soloman, P. E. and Gaur, R. K. 2022. Higher plant remediation to control pollutants. *In Biological Approaches to Controlling Pollutants*: 321–363. Woodhead Publishing.

Kalugina, O. V., Mikhailova, T. A. and Shergina, O. V. 2017. Pinus sylvestris as a bio-indicator of territory pollution from aluminum smelter emissions. *Environmental Science and Pollution Research*, 24(11): 10279–10291.

Kargar, M., Jahromi, M. Z., Najafian, M., Khajeaian, P., Nahavandi, R., Jahromi, S. R. and Firoozinia, M. 2012. Identification and molecular analysis of mercury resistant bacteria in Kor River, Iran. *African Journal of Biotechnology*, 11(25): 6710–6717.

Kaur, J. and Gosal, S. K. 2021. Biotransformation of pollutants: A microbiological perspective. In *Rhizobiont in Bioremediation of Hazardous Waste*: 151–*162*.

Kleman–Leyer, K. M., SiiKa–Aho, M., Teeri, T. T. and Kirk, T. K. 1996. The cellulases endoglucanase I and cellobiohydrolase II of *Trichoderma reesei* act synergistically to solubilize native cotton cellulose but not to decrease its molecular size. *Applied and Environmental Microbiology*, 62(8): 2883–2887.

Knoll, A. H and Bauld, J. 1989. The evolution of ecological tolerance in prokaryotes. *Earth and Environmental Science Transactions of The Royal Society of Edinburgh*, 80(3–4):209–223.

Kour, D., Khan, S. S., Kour, H., Kaur, T., Devi, R., Rai, P. K. and Yadav, A. N. 2022. Microbe-mediated bioremediation: Current research and future challenges. *Journal of Applied Biology and Biotechnology*, 10(2): 6–24.

Kumar, S., Verma, N. and Singh, A. K. 2017. Development of cadmium specific recombinant biosensor and its application in milk samples. *Sensors and Actuators B: Chemical*, 240: 248–254.

Kusi, J., Scheuerman, P. R. and Maier, K. J. 2020. Emerging environmental contaminants (silver nanoparticles) altered the catabolic capability and metabolic fingerprinting of microbial communities. *Aquatic Toxicology*, 228: 105633.

Li, F., Peng, Y., Fang, W., Altermatt, F., Xie, Y., Yang, J. and Zhang, X. 2018. Application of environmental DNA metabarcoding for predicting anthropogenic pollution in rivers. *Environmental Science and Technology*, 52(20): 11708–11719.

Li, P., Wang, Y., Yuan, X., Liu, X., Liu, C., Fu, X. and Holmes, D. E. 2021. Development of a whole-cell biosensor based on an ArsR–Pars regulatory circuit from *Geobacter sulfurreducens*. *Environmental Science and Ecotechnology*, 6: 100092.

Li, Y., Yang, F., Yuan, R., Zhong, X. and Zhuo, Y. 2022. Electrochemiluminescence covalent organic framework coupling with CRISPR/Cas12a–mediated biosensor for pesticide residue detection. *Food Chemistry*, 389: 133049.

Liu, X., Cheng, H., Zhao, Y., Wang, Y. and Li, F. 2022. Portable electrochemical biosensor based on laser-induced graphene and MnO2 switch–bridged DNA signal amplification for sensitive detection of pesticide. *Biosensors and Bioelectronics*, 199: 113906

Lovelock, C. E., Wright, S. F. and Nichols, K. A. 2004. Using glomalin as an indicator for arbuscular mycorrhizal hyphal growth: an example from a tropical rain forest soil. *Soil Biology and Biochemistry*, 36(6): 1009–1012.

Ma, Y., Deng, D., Zhan, Y., Cao, L. and Liu, Y. 2022. A systematic study on self–powered microbial fuel cell based BOD biosensors running under different temperatures. *Biochemical Engineering Journal*, 180: 108372.

Mahbub, K. R., Krishnan, K., Naidu, R. and Megharaj, M. 2017. Development of a whole cell biosensor for the detection of inorganic mercury. *Environmental Technology and Innovation*, 8: 64–70.

Makoi, J. H. and Ndakidemi, P. A. 2008. Selected soil enzymes: examples of their potential roles in the ecosystem. *African Journal of Biotechnology*, 7(3).

Makoi, J. H., Chimphango, S. B. and Dakora, F. D. 2010. Elevated levels of acid and alkaline phosphatase activity in roots and rhizosphere of cowpea (*Vigna unguiculata* L. Walp.) genotypes grown in mixed culture and at different densities with sorghum (*Sorghum bicolor* L.). *Crop and Pasture Science*, 61(4): 279–286.

Malik, D. S. and Bharti, U. 2012. Status of plankton diversity and biological productivity of Sahastradhara stream at Uttarakhand, India. *Journal of Applied and Natural Science*, 4(1): 96–103.

Maseko, S. T. and Dakora, F. D. 2013. Rhizosphere acid and alkaline phosphatase activity as a marker of P nutrition in nodulated *Cyclopia* and *Aspalathus* species in the Cape fynbos of South Africa. *South African Journal of Botany*, 89: 289–295.

Mason, L.M., Eagar, A., Patel, P., Blackwood, C.B. and DeForest, J.L. 2021. Potential microbial bioindicators of phosphorus mining in a temperate deciduous forest. *Journal of Applied Microbiology*, 130(1):109–122.

Mateo, P., Leganes, F., Perona, E., Loza, V. and Fernandez-Pinas, F. 2015. Cyanobacteria as bioindicators and bioreporters of environmental analysis in aquatic ecosystems. *Biodiversity and Conservation*, 24:909–948.

Mathieu, C., Hermans, S. M., Lear, G., Buckley, T. R., Lee, K. C. and Buckley, H. L. 2020. A systematic review of sources of variability and uncertainty in eDNA data for environmental monitoring. *Frontiers in Ecology and Evolution*, 8: 135.

McKee, A. M. and Cruz, M. A. 2021. Microbial and viral indicators of pathogens and human health risks from recreational exposure to waters impaired by fecal contamination. *Journal of Sustainable Water in the Built Environment*, 7(2): 03121001.

Mehrotra, P. 2016. Biosensors and their applications–A review. *Journal of Oral Biology and Craniofacial Research*, 6(2): 153–159.

Mei, Y., Li, C., Yang, S., Chen, W., Liu, R. and Xu, K. 2021. A novel biosensor for the rapid and high–sensitive detection of mercury based on cleavable phosphorothioate RNA probe. *Sensors and Actuators B: Chemical*, 337: 129756.

Merino, C., Godoy, R. and Matus, F. 2016. Soil enzymes and biological activity at different levels of organic matter stability. *Journal of Soil Science and Plant Nutrition*, 16(1): 14–30.

Muhsin, S. A., Al–Amidie, M., Shen, Z., Mlaji, Z., Liu, J., Abdullah, A. and Almasri, M. 2022. A microfluidic biosensor for rapid simultaneous detection of waterborne pathogens. *Biosensors and Bioelectronics*, 203: 113993.

Nakatsu, C. H., Carmosini, N., Baldwin, B., Beasley, F., Kourtev, P. and Konopka, A. 2005. Soil microbial community responses to additions of organic carbon substrates and heavy metals (Pb and Cr). *Applied and Environmental Microbiology*, 71(12): 7679–7689.

Nash, I., Thomas, H. and Gries, C. 2002. Lichens as bioindicators of sulfur dioxide. *Symbiosis* 33(1):1–2.

Nguyen, D. K. and Jang, C. H. 2021. An acetylcholinesterase–based biosensor for the detection of pesticides using liquid crystals confined in microcapillaries. *Colloids and Surfaces B: Biointerfaces*, 200: 111587.

Nicholson, P., Osborn, R. W and Howe, C. J. 1987. Induction of protein synthesis in response to ultraviolet light, nalidixic acid and heat shock in the cyanobacterium *Phormidium laminosum*. *FEBS letters*, 221(1):110–114.

Oehl, F., Oberholzer, H. R., van der Heijden, M. G., Laczko, E., Jansa, J. and Egli, S. 2016. Arbuscular mycorrhizal Fungi as bioindicators in agricultural soils. *Agrarforschung Schweiz*, 7(1): 48–55.

Ogunseitan, O. A. 2000. Microbial proteins as biomarkers of ecosystem health. In *Integrated Assessment of Ecosystem Health*, CRC Press, pp. 207–223

Pandey, D., Agrawal, M. and Bohra, J. S. 2014. Effects of conventional tillage and no tillage permutations on extracellular soil enzyme activities and microbial biomass under rice cultivation. *Soil and Tillage Research*, 136: 51–60.

Parmar, T. K., Rawtani, D. and Agrawal, Y. K. 2016. Bioindicators: The natural indicator of environmental pollution. *Frontiers in Life Science*, 9(2): 110–118.

Podgórska, B., Pazdro, K. and Węgrzyn, G. 2007. The use of the *Vibrio harveyi* luminescence mutagenicity assay as a rapid test for preliminary assessment of mutagenic pollution of marine sediments. *Journal of Applied Genetics*, 48(4): 409–412.

Qi, W., Zheng, L., Hou, Y., Duan, H., Wang, L., Wang, S. and Lin, J. 2022. A finger–actuated microfluidic biosensor for colorimetric detection of foodborne pathogens. *Food Chemistry*, 381: 131801.

Qiang, L., Zhang, Y., Guo, X., Gao, Y., Han, Y., Sun, J. and Han, L. 2020. A rapid and ultrasensitive colorimetric biosensor based on aptamer functionalized Au nanoparticles for detection of saxitoxin. *RSC Advances*, 10(26): 15293–15298.

Qureshi, J.A., Collin, H.A., Hardwick, K., and Thurman, D.A. 1981. Metal tolerance in tissue cultures of *Anthoxanthum odoratum*. *Plant Cell Reports*. 1(2):80–82.

Rajapaksha, P., Elbourne, A., Gangadoo, S., Brown, R., Cozzolino, D. and Chapman, J. 2019. A review of methods for the detection of pathogenic microorganisms. *Analyst*, 144(2): 396–411.

Ramić, D., Klančnik, A., Možina, S. S. and Dogsa, I. 2022. Elucidation of the AI–2 communication system in the food-borne pathogen *Campylobacter jejuni* by whole-cell–based biosensor quantification. *Biosensors and Bioelectronics*, 114439.

Ray, M., Ray, A., Dash, S., Mishra, A., Achary, K. G., Nayak, S. and Singh, S. 2017. Fungal disease detection in plants: Traditional assays, novel diagnostic techniques and biosensors. *Biosensors and Bioelectronics*, 87: 708–723.

Ray, S., Senapati, T., Sahu, S., Bandyopadhyaya, R. and Anand, R. 2018. Design of ultrasensitive protein biosensor strips for selective detection of aromatic contaminants in environmental wastewater. *Analytical Chemistry*, 90(15): 8960–8968.

Rietz, D. and Haynes, R. 2003. Effects of irrigation–induced salinity and sodicity on soil microbial activity. *Soil Biology and Biochemistry*. 35: 845–854.

Sadhu, S. and Maiti, T.K. 2013. Cellulase production by bacteria: A review. *British Microbiology Research Journal*, 3(3): 235–258.

Samecka–Cymerman, A., Marczonek, A. and Kempers, A. J. 1997. Bioindication of heavy metals in soil by liverworts. *Archives of environmental Contamination and Toxicology*, 33(2): 162–171.

Sanyal, S. K., Shuster, J and Reith, F. 2019. Cycling of biogenic elements drives biogeochemical gold cycling. *Earth-science Reviews*, 190:131–147.

Sase, H. 2017. Environmental monitoring with indicator plants for air pollutants in Asia. In *Air Pollution Impacts on Plants in East Asia* (pp. 111–121). Springer, Tokyo.

Sharma, P., Pandey, V., Sharma, M. M. M., Patra, A., Singh, B., Mehta, S. and Husen, A. 2021. A Review on Biosensors and Nanosensors Application in Agroecosystems. *Nanoscale Research Letters*, 16(1): 1–24.

Sharma, P., Bano, A., Singh, S. P., Dubey, N. K., Chandra, R. and Iqbal, H. M. 2022. Microbial fingerprinting techniques and their role in the remediation of environmental pollution. *Cleaner Chemical Engineering*, 100026.

Stachler, E., Akyon, B., de Carvalho, N. A., Ference, C. and Bibby, K. 2018. Correlation of crAssphage qPCR markers with culturable and molecular indicators of human fecal pollution in an impacted urban watershed. *Environmental Science and Technology*, 52(13): 7505–7512.

Stürzenbaum, S. R., Georgiev, O., Morgan, A. J. and Kille, P. 2004. Cadmium detoxification in earthworms: from genes to cells. *Environmental Science and Technology*, 38(23): 6283–6289.

Sumampouw, O. J., and Risjani, Y. 2014. Bacteria as indicators of environmental pollution. *Environment: Review. International Journal of Ecosystem* 4(6): 251–258.

Sumudumali, R. G. I. and Jayawardana, J. M. C. K. 2021. A review of biological monitoring of aquatic ecosystems approaches: With special reference to macroinvertebrates and pesticide pollution. *Environmental Management*, 67(2): 263–276.

Sun, D., Fan, T., Liu, F., Wang, F., Gao, D. and Lin, J. M. 2022. A microfluidic chemiluminescence biosensor based on multiple signal amplification for rapid and sensitive detection of *E. Coli* O157: H7. *Biosensors and Bioelectronics*, 212: 114390.

Thakur, R. K., Jindal, R., Singh, U. B. and Ahluwalia, A. S. 2013. Plankton diversity and water quality assessment of three freshwater lakes of Mandi (Himachal Pradesh, India) with special reference to planktonic indicators. *Environmental Monitoring and Assessment*, 185(10): 8355–8373.

Uka, U. N., Hogarh, J. and Belford, E. J. D. 2017. Morpho-anatomical and biochemical responses of plants to air pollution. *International Journal of Modern Botany*, 7(1): 1–11.

Utobo, E. B. and Tewari, L. 2015. Soil enzymes as bioindicators of soil ecosystem status. *Applied Ecology and Environmental Research*, 13(1): 147–169.

Uttah, E.C., Uttah, C., Akpan, P.A., Ikpeme, E.M. and Ogbeche, J. Usip, J.O. 2008. Bio-survey of plankton as indicators of water quality for recreational activities in Calabar River, Nigeria. *Journal of Applied Sciences & Environmental Management,* 12(2): 35–42.

Valentin, V., Frédéric, R., Isabelle, D., Olivier, M., Yorick, R. and Agnès, B. 2019. Assessing pollution of aquatic environments with diatoms' DNA metabarcoding: experience and developments from France Water Framework Directive networks. *Metabarcoding and Metagenomics*, 3: e39646.

Van Aarle, I. M. and Plassard, C. 2010. Spatial distribution of phosphatase activity associated with ectomycorrhizal plants is related to soil type. *Soil Biology and Biochemistry*, 42(2): 324–330.

Walsh, G. E. 1978. *Toxic effects of pollutants on Plankton. Principles of Ecotoxicology.* John Wiley and Sons, Inc., New York, 257–274.

Wang, T., Liu, S., Ren, S., Liu, B. and Gao, Z. 2022. Magnetic relaxation switch and fluorescence dual–mode biosensor for rapid and sensitive detection of ricin B toxin in edible oil and tap water. *Analytica Chimica Acta*, 340471.

White, E. M., Wilson, J. C. and Clarke A. R. 2006. Biotic indirect effects: A neglected concept in invasion biology. *Diversity and Distributions*, 12:443–455.

Wu, D., Xu, C., Wang, S., Zhang, L. and Kortsch, S. 2022. Why are biodiversity-ecosystem functioning relationships so elusive? Trophic interactions may amplify ecosystem function variability. *Journal of Animal Ecology*, 92: 367–376.

Wu, Z., Sun, D. W., Pu, H., and Wei, Q. 2023. A novel fluorescence biosensor based on CRISPR/Cas12a integrated MXenes for detecting Aflatoxin B1. *Talanta*, 252: 123773.

Yang, H., Tong, J., Lee, C. W., Ha, S., Eom, S. H. and Im, Y. J. 2015. Structural mechanism of ergosterol regulation by fungal sterol transcription factor Upc2. *Nature Communications*, 6(1): 1–13.

Yildirim, N., Yildirim, N.C., Tatar, S. and Alp, H. 2019. *Phanerochaete chrysosporium* as a model organism to assess the toxicity of municipal landfill leachate from Elazığ, Turkey. *Environmental Science and Pollution Research*, 26:12807–12812.

Yushin, N., Chaligava, O., Zinicovscaia, I., Vergel, K. and Grozdov, D. 2020. Mosses as bioindicators of heavy metal air pollution in the lockdown period adopted to cope with the COVID–19 pandemic. *Atmosphere*, 11(11): 1194.

Zaghloul, A., Saber, M., Gadow, S. and Awad, F. 2020. Biological indicators for pollution detection in terrestrial and aquatic ecosystems. *Bulletin of the National Research Centre*, 44(1): 1–11.

Zancan, S., Trevisan, R. and Paoletti, M. G. 2006. Soil algae composition under different agro–ecosystems in North–Eastern Italy. *Agriculture, Ecosystems and Environment*, 112(1): 1–12.

Zhang, B., Zhu, Z., Li, F., Xie, X. and Ding A. 2021. Rapid and sensitive detection of hepatitis B virus by lateral flow recombinase polymerase amplification assay. *Journal of Virological Methods*, 291:114094.

Zhang, T., Gaffrey, M. J., Thrall, B. D., & Qian, W. J. 2018. Mass spectrometry-based proteomics for system-level characterization of biological responses to engineered nanomaterials. *Analytical and Bioanalytical Chemistry*, 410(24): 6067–6077.

Zhang, Y., Li, Z., Su, W., Zhong, G., Zhang, X., Wu, Y. and Zheng, L. 2022. A highly sensitive and versatile fluorescent biosensor for pathogen nucleic acid detection based on toehold–mediated strand displacement initiated primer exchange reaction. *Analytica Chimica Acta*, 1221: 340125.

Zhao, F., He, J., Li, X., Bai, Y., Ying, Y. and Ping, J. 2020. Smart plant-wearable biosensor for in–situ pesticide analysis. *Biosensors and Bioelectronics*, 17C: 112636.

Zhelyazkova, I. 2012. Honeybees–bioindicators for environmental quality. *Bulgarian Journal of Agricultural Science*, 18(3): 435–442.

Zhou, M., He G., Liu Y., Yin P., Li Y., Kan H., Fan M., Xue A. and Fan M. 2015. The associations between ambient air pollution and adult respiratory mortality in 32 major Chinese cities, 2006–2010. *Environmental Research*, 137: 278–286.

19 Microbial Community and Management of Soil Health for Environmental Sustainability

Concept, Trends, and Prospects

Shubhra Singh, Kadarkarai Murugan, and Douglas J.H. Shyu

19.1 GENERAL CONCEPTS OF SOIL ORGANISMS: STRUCTURAL AND FUNCTIONAL ROLES AND BEHAVIORAL PATTERN

The core component of the soil ecosystem is the myriad of microorganisms that inhabit naturally, including bacteria, fungus, algae, invertebrates, annelids, plants, and other creatures. The inhabitants are required to play an important role in enzymatic activities that contribute to another complex biochemical phenomenon for maintaining the quality of the soil, considering the health of plants and animals. The assessment of soil quality depends upon its ability to promote the growth and development of plants, minimizing the harmful effects of dangerous chemicals or pollutants produced by industrial, organic waste, and agricultural wastes. Furthermore, soil composition affects the structural and functional organization of microbial and associated communities. The interaction between compositions of the soil matrix with biotic or abiotic factors directly impacts the environment and indirectly on human health. Therefore, the heterogeneity contributes to the high level of biodiversity, referring to the excellent health of the soil. Indeed, the functional role of biological diversity varies widely from one ground to another and from one place to another, which primarily depends upon the organic matter content, soil mineral compositions, pH, and soil-management practices of a particular region. This biodiversity is maintained by the contribution of microorganisms and the circulation of nutrients (micro and macro) to plant life. The fundamental role of soil determines the arrangement and composition of plant communities. Most of the soil communities are responsible for altering the physicochemical characteristics of the environment by transforming nitrogen, phosphorus, and sulfur.

Further, they construct mutualistic relationships with plants that enhance plant growth and development. In this case, it is essential to understand the interaction and dynamicity of above-ground plants with different levels of underground soil flora and fauna. The diversity of land biomass highly affects the distribution and abundance of the soil physicochemical parameters. Earthworms are responsible for maintaining the entire composition and physicochemical properties for proper soil fertility and productivity, although the recent plantation of trees on agricultural lands alters the population of earthworms since all the organic matter is utilized by the trees. The population density and total biomass of earthworms increased during the monsoon and post-monsoon seasons in studies on earthworm communities in rubber plantations of various ages in West Tripura, India. Similarly, seasonal fluctuations have a significant impact on the vertical distribution of earthworms in Pineapple

DOI: 10.1201/9781003408352-22

plantations in Tripura, India. These seasons were distinguished by higher humidity suited for the survival, growth, and reproduction of earthworms. Also, changes in land-use systems had a significant impact on the productivity of earthworms that demonstrated complete synchronization between the above-ground and below-ground environments. Therefore, above-ground habitat maintenance is directly proportional to below-ground productivity (Dey and Chaudhuri, 2016; Gudeta et al., 2022). This indicates the need for amutualistic approach to combine the soil microbiome and the plant agroecosystem. This agroecosystem is responsible for maintaining plant health and textual characteristics from aerial parts to the below-ground root system. All the microorganisms often occupy a niche in the complex root system where they release many organic compounds as exudates and are indirectly associated with the differentiation of minerals resulting in nourishment to the plants. This ecologically balanced system enables microorganisms to interact with the soil, resulting in positive, negative, or neutral systems. The most important parts of this system find the main reason behind this whole. The bacterial microbiome alters the physicochemical characteristics of the plant and environment by forming mutualistic associations that directly affect the transformation of nitrogen, phosphorus, and sulfur (Wolfe and Klironomos, 2005).

19.1.1 MICROORGANISMS: ARCHAEA, BACTERIA, ACTINOMYCETES, FUNGI, AND ALGAE

Microscopic studies of soil microorganisms have recognized the necessary co-existence between microorganisms and the contemporary living world with a deeper understanding of their internal architectures. The main two divisions are eukaryotes and prokaryotes. Eukaryotes are complex structures and this particular type of organisms have true nucleus that includes algae, fungi, and protists. Conversely, prokaryotes constitute fewer complex structures, consisting of two types of microbial communities, eubacteria, and archaea, comprised of a heterogeneous group of microorganisms (Table 19.1). Therefore, the cellular organisms are divided into eukaryotes, eubacteria, and archaea, and further, the eukaryotes can be divided into three categories: plants, animals, and fungi. Additionally, eubacteria can be sectioned based on cell wall formation into purple, green, gram-positive, and gram-negative bacteria. According to nutritional requirements, prokaryotes are divided into photoautotrophs, photoheterotrophs, chemolithoautotrophs, and chemolithoheterotrophs (Table 19.2). Similarly, based on oxygen metabolism, bacteria species have

TABLE 19.1
Comparative Chart of Microorganisms

Microorganisms	Cell wall	Reproduction	Importance
Prokaryotes			
Bacteria	The cell wall is rigid	Reproduce by binary fission or photosynthesis	Component of soil microbial population, regulate atmospheric composition, recycled biomass, a branch of phytoplankton
Archaea	The cell wall is rigid, unusual membrane structure, and photosynthetic membrane	Binary fission and lacks chlorophyll	Some of them are extremophiles, production and consumption of compounds with low molecular weight, dead recycling biomass
Eukaryotes			
Fungi	Rigid cell wall, single cell formation	Budding or multicellular form (hyphae or mycelium)	Recycle biomass, growth, and development of plant and other microorganisms
Algae	Cell wall rigid and photosynthetic	Fragmentation	Member of phytoplankton

TABLE 19.2
Modes of Nutritional Requirements

Microorganisms	Source of energy	Carbon source	Examples
Photoautotrophic	Sunlight	Carbon dioxide	Photosynthetic bacteria, cyanobacteria, extreme halophiles
Photoheterotrophic	Sunlight	Organic compounds	Purple non-sulfur bacteria or green non-sulfur bacteria
Chemolithoautotrophic	Chemical compounds	Carbon dioxide	Nitrosomonas, Nitrobacter
Chemolithoheterotrophic	Chemical compounds	Organic compounds	Most bacteria, fungi, and all animals

been divided into oxybionts and anoxybionts. The diversification in prokaryotes is dependent upon soil pH, temperatures (cold, ambient, or hot), and inorganic salts (Paul, 2006).

Archaea are ancient bacteria, and they were the most primitive microorganisms to have been manifested on the Earth. These bacteria comprise a group of primitive prokaryotes and can survive in highly unpredictable environments like salt marshes and hot sulfur springs. The diverse group of archaea is phylogenetically different from the eubacteria and has several unique traits. They can be identified by the presence of a cell wall, which indicates that they are specifically composed of proteins and non-cellulosic polysaccharides. Additionally, the cell membrane contains lipids with branched-chain structures, which may help them withstand high temperatures and alkaline or basic conditions. The nucleotides containing rRNA also differ from those of the other organisms (Huber et al., 2002). They comprise of two sub-groups as obligate and facultative anoxybiont, and the former can survive only in anaerobic conditions, and aerobic conditions might kill them. They are mostly methanogen and halophile species. On the other hand, facultative anoxybionts are represented by thermoacidophiles and are found in the aerobic conditions as well as anaerobic conditions (Table 19.3) (Kyrpides and Olsen, 1999).

Bacteria are diverse soil microorganisms with numerous shapes and sizes (Table 19.4). Depending on the health of the soil, there are about 2,000 to 10,000 bacterial species found in one gram of soil. Recent advances in the metagenomics approach have identified novel species from different soil types, including agricultural lands, grasslands, arid soils, and forests. It is interesting to note that these microorganisms have such an important role in nutrient recycling and developmental processes, such as maintaining symbiotic relationships with other plant cells and rebuilding the structure and functional properties of soil by forming the rhizosphere. They interact with other organisms and plants and contribute to the containment of diseases. The rhizospheric interaction of plants might be symbiotic or pathogenic, which is strongly associated with the interaction between roots of the plant and soil microorganisms. Bacteria is responsible for the crucial regulation of biogeochemical cycles and mineralization of soils with organic matter, oxidation-reduction of inorganic compounds, and solubilization of minerals. There are different types of bacterial species as follows: nitrogen fixers (diazotrophs or symbiosis with plants), sulfur oxidizers (*Thiobacillus*), sulfates to sulfites and sulfides reducers (sulfate-reducing bacteria), and nitrate producers (nitrifying bacteria), which reduces the contamination of soils with heavy metals such as trace metals and organic compounds such as polycyclic aromatic hydrocarbons (PAHs) by decontamination and mineralization on the partial rate. In addition to these, some of the microorganisms have an innate ability to develop symbiosis with plants that increases the amount of nutrients available for growth and development, for example plant growth-promoters. Most of them consist of extracellular chitinase producers, facultative anaerobes, phosphate solubilizers, antibiotics production, siderophores, or hormone producers (Linderman, 1988).

TABLE 19.3
Types of Archaea

Type	Environment	Habitat	Importance	Examples
Methanogens	Freshwater and marine environments, cold sediments, and hydrothermal vents	Anaerobic	Generate methane as an energy source using carbon dioxide as an electron acceptor	*Methanococcus, Methanosprillum*
Extreme halophiles	Nearby salt or soda lakes and brines	Facultative anaerobic	Pigmented as pink blooms in concentrated saltwater lakes or ponds	*Halobacterium, Halorubrum, Natrinobacterium, Natronococcus*
Methane-generating thermophiles	Near hydrothermal vents	Obligate anaerobic	Grows at 100 degrees	*Methanothermus*
Sulfur and sulfate-reducing hyperthermophiles	Hot springs	Obligate anaerobic	Utilizes sulfur to produce hydrogen sulfide	*Archaeoglobus, Thermoproteus, Thermococcus,, Pyrolobus* and *Pyrodictium*
Sulfur oxidizers	Marine sediments	Anaerobes	Oxidizes sulfur as an energy source using oxygen to produce sulfuric acid	*Sulfolobus*
Thermophilic extreme acidophiles	A hot and acidic environment	Anaerobes	Grow in an extreme hot and acidic environment, mechanism unknown	*Thermophilus, Picrophilus*

TABLE 19.4
Estimation of Microorganisms Found in the Soils

	Counts of microorganisms			Species in culture	
Types	Described species	Estimated species	Total species (%)	Number	Total estimated species (%)
Bacteria	3,000	30,000	10	2,300	7.0
Fungi	69,000	1,500,000	5	11,500	0.8
Algae	40,000	60,000	67	1,600	2.5

Source: Data retrieved from Hawksworth (1992).

Actinomycetes of soil microorganisms can be categorized into specific groups based on distinctive characteristics. They are clustered with other bacterial classes called schizomycetes that are confined to the order actinomycetes. Due to the aerial mycelium, they bear similarities to deuteromycetes, which profusely sporulates and forms a distinct layer in liquid cultures (Benson, 1988). *Streptomyces* is the most frequent actinomycetes accounting for nearly 70% in comparison to less frequent species such as *Nocardia, Micromonospora, Actinoplanes,* and *Streptosporangium* (Rao, 2018; Willey et al., 2020). Decomposed organic matter increases the quantity of actinomycetes, and the number decreases below pH 5.0; therefore, they are intolerant to acidity.

Additionally, waterlogging in soil restricts their growth, whereas arid and semi-arid zones containing sandy soil with low organic matter can promote the growth of actinomycetes populations.

TABLE 19.5
Major Classification of Fungi

Phylum	Examples	Characters	Reproduction	
			Sexual	Asexual
Zymgomycetes	*Rhizopus stolonifera* (black bread mold)	Multicellular, coenocytic mycelia	Developing as zygospores can be dominant in adverse environmental conditions	Spores develop in sporangia on the tips of aerial hyphae
Basidiomycetes	*Agaricus campestris* (meadow mushroom), *Cryptococcus neoformans*	Multicellular, uninucleated mycelia.	Produce basidiospores at the tips of the hyphae	Absent
Ascomycetes	*Neurospora, Saccharomyces cerevisiae*	Unicellular and multicellular with septate hyphae	Ascus formation on specialized hyphae	Common by budding, conidiophores
Deuteromycetes	*Penicillium, Aspergillus*	Mostly human pathogens	Absent/ unknown	Budding

The depth of soil acts as a mediator between multiplication of total microbial populations, and the percentage increases with greater depth. The growth of actinomycetes are promoted at temperatures between 25 and 30 degrees Celsius, while thermophilic cultures, which grow best at 55 and 65 degrees Celsius and primarily come from the genera *Thermoactinomyces* and *Streptomyces*, are frequently found in compost piles.

Fungi represent the most remarkable diversity and dominate among other microorganisms found in the soil. They primarily consist of filamentous mycelium with individual hyphae that might be uninucleate, binucleate, or multinucleate and non-septate or septate. Since the majority of fungi are heterotrophic with nature to reuse the nutrients released by other microorganisms, there are several factors that influence the quality and quantity of fungal flora in the soil, such as the distribution of other microorganisms in the soil with respect to its quantity and quality. In a similar way, organic matter in a variety of soils has a direct impact on the fungal species, and they can dominate the diversity of the soil microbiome since they can survive in acidic soils, unlike other bacteria and actinomycetes. They can also thrive in soils with neutral and alkaline pH, and a few of them can tolerate an exceptional range of pH over 9.0. Fungi are divided into ascomycetes, basidiomycetes, deuteromycetes and phycomycetes (Table 19.5). When it comes to soil moisture levels and depths, some fungal species exhibit selection preferences. For example, species located in the upper layer of soil are rarely found in the lower layer. Similar to this, arable soils have higher moisture contents to support the development and distribution of fungal species. As a result, they are particular, and excess moisture eventually reduces their populations (Bolton et al., 1993).

The distribution of fungi into different phylum are based on their characteristics and ability to reproduce in other environmental conditions. Several fungal species, mostly isolated or obtained from soils, are categorized into fungi imperfecti species due to the production of excess asexual spores, but they lack sexual sequential stages (Lynch, 1987). Later, they were known as deuteromycetes or anamorphic fungi. The key vegetative characteristic of this phyla is septate mycelium, which continuously produces conidiospores. The ability of soil fungus to make a mycelium with polarized growth toward appropriate substrate sources is the key vegetative characteristic. Members of the ascomycetes and basidiomycetes groups are very efficient in degrading intricate organic compounds such as cellulose or lignin present in plant and animal cell membranes along with their ability to associate with plants as root symbionts as mycorrhizas with a mutualistic behavior (Lynch and Hobbie, 1988).

Some examples of root symbionts are *Aspergillus, Botrytis, Penicillium, Spicaria, Trichothecium, Trichoderma, Cladosporium, Fusarium, Alternaria, Mucor, Absidia, Mortierella, Zygorynchus, Cunninghamella, Pythium, Rhizoctonia, Chaetomium* (Newman, 1985; Rao, 2018). Some of the yeasts are grouped as true ascomycetes, such as *Saccharomyces*, that have been isolated from soils, and some belong to deuteromycetes, for example, *Candida albicans*; however, the quantity is comparatively very low as compared to other types of distribution. Some of the fungal species produce humid substances in soil and are responsible for degrading the organic matter, such as *Alternaria, Aspergillus, Humicola, Metarhizium, Cladosporium, Dematium, Gliocladium, Helminthosporium* (Hawksworth, 1992).

Algae are found ubiquitously in surroundings where there is the availability of moisture and sunlight; however, as a member of chlorophyceae, they are dominant in the soil system. Diatoms are photosynthesizing algae found mostly in the aquatic environment, including freshwater and other types of soil with high moisture content. As a result of the presence of chlorophyll in their cells, diatoms are considered photoautotrophic microorganisms, and they can convert carbon dioxide from the atmosphere into oxygen molecules. They are identified as green scum on the surface of water or soil with the naked eye. However, these organisms are microscopic in structure with unicellular (*Chlamydomonas*) or filamentous (*Spirogyra, Ulothrix*) structures. Some of the genera live in the absence of sunlight, below the surface, entering into the different layers of soil. The common examples of green algae belong to the genera *Chlorella, Oedogonium, Chlorococcum, Chlamydomonas, Protosiphon,* and *Chlorochytrium* (Metting B, 1988).

19.1.2 Microfauna: Protozoa, Nematode, Rotifer, and Tardigrade

The microfauna comprises free-living protozoans found in litter or soils belonging to two phyla, *Sarcomastigophora* and *Ciliophora*, mostly characterized as unicellular, colonial, autotrophic, or heterotrophic by nature. There are four divisions based on their ecology:

1. *Flagellates*: Mostly consisting of single or multiple flagella-like structures. The important function of flagellates as an active protozoans are nutrient turnover, with bacteria as their principal prey items (Zwart and Darbyshire, 1991).
2. *Naked amoebae*: Giant soil protozoans are active in different types of soils found in agricultural fields, grasslands, and forests. The common mode of nutrition is phagotrophic with the ingestion of either bacteria or fungi or other suspended, fine particulate organic matter. The plasticity of naked amoeba is one of the distinguishing characteristics that allows them to enter small cavities in soil aggregates, and they survive on undergrounded species that are not accessible by other predators (Clarholm, 1985; Gupta and Germida, 1989).
3. *Testate amoebae*: Testate amoebae are not that widely distributed and are mostly found in a moist, forested areas. They can be easily identified by direct filtration and staining procedures. Researchers favor extracting and quantifying by culture techniques in which serial dilutions are done to count bacterial species.
4. *Ciliates*: They have complex reproductive systems and are confined to moist or occasionally moist habitats. They can also enter either through underground soil cavities caused from cracks etc. and/or pores containing essential organic matter. They utilize bacterial food sources and reproduce at a faster rate. The mode of reproduction is asexually by fission and sexually by conjugation with the help of micronucleus through meiosis in two different cells exchanging haploid gametic nuclei. The resulting individuals are genetically distinct from the pre-conjugant parents.

19.1.3 Mesofauna: Potworm, Collembolan, Mite, Insect Larva, and Symphylan

Soil mesofauna ranges from 0.1 to 2 mm in size and survives on the surface of soil or in the layers below the surface. Mesofauna have important role in the carbon cycle and are highly affected by climate change and adverse environmental conditions. They mostly survive on a variety of organic materials, including tiny soil animals, other microorganisms, live, dead, and decaying matter, and decomposed fungi, algae, lichen, spores, and pollens. Therefore, the common mode of nutrition is decaying plant matter that opens drainage and aeration channels in the soil for the root system. The waste material, such as fecal matter from big invertebrates, is broken down by smaller vertebrates and invertebrates. The re-organization of the soil system is not supported by mesofauna, however; they only prefer to use the existing voids, cavities, and channels to enter and survive in the soil system. In agricultural land, the biological activity at the top layer (about 20 centimeters) is considered to be highly active, and the topmost layer is the organic horizon with the accumulation of animal and plant biomass. Through the process, some forms of nitrogen are fixed with the support of bacterial communities that consume the amino acids and sugars from the residues; consequently major contributors are soil macro and mesofauna where 30% of nitrogen undergoes re-mineralization. Mesofauna mostly contributes to the existing voids filling the free spaces and accounts for a small portion of total porosity in the soil. Therefore, the particle size and sorting of the soil plays an important role as clay has much smaller particles with smaller voids, and organic matter can fill those small voids, which increases nitrogen mineralization since microbes are broken down, and nitrogen is released in the soil system.

19.1.4 Macrofauna: Earthworm, Millipede, Centipede, Woodlice, Snail, Slug, and Insect

Soil macrofauna represents ecosystem engineers that can drastically alter soil microbial communities through the effects of the structure of the soil. Typical members are earthworms, millipedes, centipedes, slugs, snails, and insects. They are one centimeter or more long but smaller than an earthworm and also include spiders, scorpions, and other important predators in soils. Most of them burrow in the soil and help in soil drainage and aeration. Additionally, they are helpful in the movement of organic material through the soil surface. Most of the macrofauna survive on decaying matter and organic debris; however, centipedes, insects, and other predators feed on other soil creatures such as beetles. Some of the species are scavengers and omnivorous. Centipedes mostly survive in moist microhabitats and survive on moist soft-bodied fauna with smaller sizes. On the other hand, millipedes feed on residues of plants that are partially decomposed by fungi and bacteria. These organisms are responsible for altering the physico-chemical and structural properties of their microniches, affecting other biota and ecosystem functions. The most important functions are moving and mixing a large amount of soil and littering from residual plants. Earthworms are hermaphrodite, segmented worms and prefer moist environmental conditions for their movement and activity. They are commonly found in temperate regions such as forests, grasslands, agricultural lands, and tropical forests with the range 50–2000 m^{-2} with 2–10 species with different distributions. Earthworms are differentiated on the basis of three ecological classifications: epigeic, anecic, and endogeic. The first class as epigeic species are found in the layer with dead organic material (litter), and they make small fragments of those debris and promote different stages of decomposition. Anecic species can survive at a depth of 1 meter and feed on the surface of soil litters, and have permanent burrows. They are specialized to form macropores at the depth and pull the litter down their burrows, lining the organic matter evenly at the walls. Endogeic earthworms are found at the upper layer of soil (at least 15 cm deep) with a preference for particulate organic matter. They turn over the burrows and backfill the voids (Coleman et al., 2004).

19.1.5 Megafauna: Large Invertebrate and Vertebrate

In the soil system, megafauna refers to small vertebrates and worms such as moles, mice, hares, rabbits, gophers, snakes, and lizards. Earthworms consume both dirt and organic matter, although the majority of vertebrates eat plant materials, invertebrates, and other tiny vertebrate animals as their primary sources of nutrition. Megafauna is the main force behind soil distribution and turnover, which loosens soil texture, enhances aeration and drainage, and disperses soil microbes. The largest mammals and occasionally the largest birds and reptiles are referred to as megafauna. These animals were frequently the first to go extinct after coming into contact with humans. Larger invertebrates and vertebrates shape the microtopography of the soil landscape. Pocket mounds and thick, rope-like soil fills are created by pocket gophers in the snowbanks' base layers. Skunks, javelinas, coatis, whiptail lizards, roadrunners, and spadefoot toads are examples of vertebrates that mix soils. They burrow during dry spells and emerge when it rains. The horizons and health of the soil are impacted by the excavation of underground passages. For redistribution, pack rats, prairie dogs, skunks, pocket mice, ground squirrels, and ground squirrels dig channels and tunnels, as well as dense underground chambers. In underground chambers, rodents like rats, mice, and gophers store plant materials like seeds and other food-related components. Animal wastes are a source of regional nitrogen, phosphorus, and potassium concentrations. A wide range of animals, such as birds, bats, and rodents, contribute to regional nutrient cycles through their excretions and nesting behaviors. For instance, bats and birds often roost in caves, cliffs, or trees, depositing guano (excrement) that contains valuable nutrients. This guano serves as a natural fertilizer, enriching the soil with organic matter and nutrients. In regions where these animals are abundant, the chemical composition of the soil can be significantly influenced by their presence. Additionally, the nesting habits of certain animals, such as raptors on cliffs, contributes to soil formation and erosion processes. By introducing organic material and disturbing the surface, these megafauna aid in the breakdown of bedrock and the creation of new soil layers that can support vegetation growth.

19.2 CONTRIBUTION OF SOIL AGGREGATION TO ECOSYSTEM FUNCTIONS

19.2.1 Mechanisms of Aggregate Formation and Stabilization

Soil consists of a heterogeneous combination with different types of organisms, including organic as well as inorganic mineral substances present in the form of solids, liquids, and gases. Soil aggregates are formed by the physical forms of forces, such as climatic and topological forces, in addition to the natural grouping of different particles with different arrangements, stabilities, and sizes, which is a basic unit of soil structure. The aggregation of soil particles is affected by several factors, such as soil mineralogy, wetting and drying of soil particles, and the presence of micro and macronutrients, clay, and organic matter, along with the plant root system. The plant root system directly contributes to the stability and management of soil aggregates through the re-organization of structures in organic matter and the production of metabolites that leads to the stimulation of microbial activity and indirectly to the production of monosaccharides and non-carbohydrate substituents such as exopolysaccharides. There is a complex mosaic of gradients that favors the selection of bacterial development based on the geographical variability of the soil. With increasing soil depth, microbial biomass declines, and changes in the microbial community changes the substrate specialization. The distribution of aggregates, such as micro and macro populations, provides different habitats in terms of high or low temperature, compaction, aeration between particles, water retention capacity, and structural movement. Also, soil aggregates have different pore dimensions that affect carbon sequestration and the availability of macro and micronutrients for optimum soil health. The diversification of bacterial communities in the soil is increased by small pore connectivity resulting in low water potential. The expansion of different forms of bacteria's motility, morphologies, and reproductive capacity is directly influenced by three factors: moisture content, pore size, and habitat.

19.2.2 Maintenance of Soil Structure

The stability and structure of soil layers are maintained by the combination of biotic and abiotic components, along with microbial communities, which provides a quantitative assessment of soil health. Biogeochemical processes are the primary factors that influence how bacteria function in their ecosystems. Soil structure and vitality play a crucial role in order for plants and animals to be healthy and productive as well as for water and air quality to be maintained or improved. Soil structure comprises physicochemical and biological characteristics, such as soil taxonomy, climatic conditions, sowing patterns, type of fertilizers, pesticides or insecticides, availability of carbon substrates and micronutrients, toxic elements, and types of microorganism assembly. In order to manage soils sustainably, it is necessary to monitor them, including biological indicators like microbial communities that offer a variety of potentially useful indications for environmental monitoring in response to abiotic or biotic stresses and climatic disturbances. Additionally, modern farm practices have a higher impact on the structure and function of soil aggregation through land-use intensification, disturbances, and biotic and abiotic factors (Souza et al., 2015).

19.2.3 Transformation of Carbon

Macronutrients such as carbon, oxygen, hydrogen, nitrogen, calcium, phosphorus, magnesium, potassium, sulfur, sodium, and chlorine are needed by the plants in greater amounts, and micronutrients such as iron, manganese, boron, copper, molybednum, zinc, iodine, and selenium are required in lower amounts. From the earliest basic stage of solubilization from rock to the fixation of atmospheric nitrogen that is finally recycled by the microorganism, each of these critical substances has a specific role in the growth of plants, their uptake, and the maintenance of soil quality. The influx of carbon in the soil is determined by the relative rate of photosynthesis versus respiration and decomposition. The microorganisms in the soil interact with each other and the plants help to initiate these processes. The mycorrhizal plants grow more through photosynthesis in nutrient- or water-scarce environments. These interactions affect photosynthetic rates by influencing resource competition. Plant respiration rates may be impacted by soil organisms due to root predation or other tissue damages, and it will increase tissue repair requirements and associations with mutualists. Microbial decomposer communities play an important role in the decomposition of plant and animal products, as well as the carbon dioxide emission into atmosphere. Competition among decomposers has an impact on the rate at which material decomposes. The biological reactions to increasing carbon dioxide emissions, and other reactions to an indirect impact on temperature change and climate instability, will decide the responses to environmental changes and human existence. The global carbon cycle has been considerably altered by the huge CO_2 flux into the atmosphere brought on by the burning of fossil fuels and changes in land use. The rate at which carbon enters soils is controlled by soil organisms, and they have made an important contribution to the evolution of carbon in relation to underground carbon storage. The relative contributions of each below-ground microfauna and mesofauna groups to flux carbon effectively and the effects of increasing CO_2 and adverse climate conditions on these fluxes are now the subjects of several studies worldwide. Photosynthetic organisms continuously repair carbon (CO_2) into organic form via the reduction of organic carbon molecules (photosynthesis). Once bonded, the carbon can no longer be used to create new plant life, thus the carbonaceous materials must be broken down and released into the atmosphere; 1.3×10^{14} kg of CO_2 is thought to be fixed each year in the biosphere. With the conversion of so much of the plant's available carbon to organic form each year and the finite amount in the air, it is obvious that the major plant nutrient element would run out in the absence of microbial transformation. To a lesser extent, CO_2 is also fixed through the action of photosynthetic bacteria and other chemolithotrophs. The fixation and regeneration of CO_2 are crucial to the carbon cycle. The only carbon source for green plants is CO_2, and the carbonaceous matter that is created serves to provide carbon to other heterotrophic creatures and animals. Microbes take over as the primary contributors

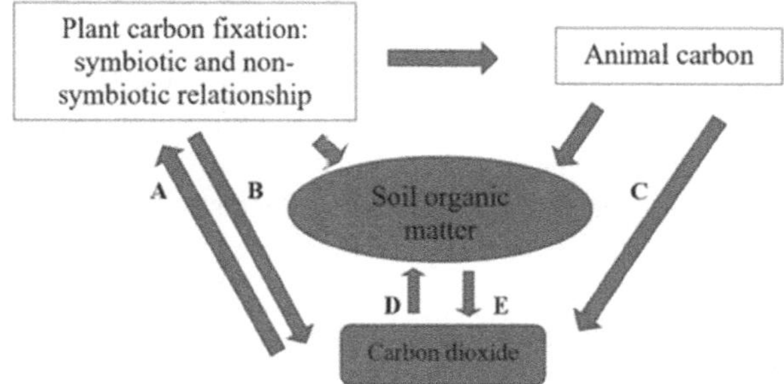

FIGURE 19.1 Carbon utilization: A. Photosynthesis; B. Respiration/plant; C. Respiration/animals; D. Autotrophs; E. Heterotrophs.

Notes: Organic matter derived from photosynthesis provides nutrition for heterotrophs which converts it back to carbon dioxide. Carbon is the backbone of all organic molecules and the most crucial element. Autotrophs, including plants, algae, photosynthetic bacteria, lithotrophs, and methanogens, use carbon dioxide as a carbon source for growth and reduce the molecule to organic cell material. Heterotrophs require organic carbon for growth and convert it back to carbon dioxide.

to the carbon cycle once plants and animals die. The decayed dead tissues are converted into microbial cells, humus, and soil organic fractions. These materials continue to break down, producing CO_2, which is recycled once more (Figure 19.1).

19.2.4 BIOLOGICAL DINITROGEN FIXATION: SYMBIOTIC AND NON-SYMBIOTIC RELATIONSHIP

After carbon fixation through photosynthesis, nitrogen fixation is a fundamental process needed for the production of essential micronutrients from the atmosphere to the Earth's surface. Synthesis of cellular enzymes, chlorophyll, proteins, RNA, and DNA, as well as the production of food and the promotion of plant development, depends on nitrogen. For the nodulating legumes, the essential element nitrogen is supplied and converted into atmospheric nitrogen by nitrogenase with the help of nitrogen-fixing rhizobium. Biological nitrogen fixation accounts for 65% of the nitrogen used in agriculture today, with a considerable impact on developing crop production methods for sustainability and food security. Mutualistic symbionts in plants include nitrogen-fixing bacteria and multifunctional arbuscular mycorrhizal (AM) fungi. Rhizobia, a group of different bacterial taxa, may fix nitrogen in a mutualistic relationship with legume plants. The symbiotic relationship between legumes and nitrogen-fixing organisms drives the majority of the significant biochemical reactions in biological nitrogen fixation. Some of the free-living nitrogen fixers are cyanobacteria, heterotrophic and autotrophic bacteria, with actinomycetes belonging to the genus *Frankia*. The non-symbiotic nitrogen fixers are *Clostridium pasteurianum*, which fix the atmospheric nitrogen by free-living soil bacteria. The major steps in non-symbiotic nitrogen fixation are ammonia production and nitrification.

1. The reduction of nitrogen in the atmosphere results in ammonia and this ammonia is converted into nitrite with the help of organisms such as *Nitrosomonas, Nitrobacter, Nitrosospira* (Equation 1)

$$NH_3^- \text{ (Ammonia)} \rightarrow NO_2^- \text{ (Nitrite)} \tag{1}$$

2. After the conversion of ammonia into nitrite, nitrifying bacteria convert nitrite into nitrate with the help of *Nitrobacter, Nitrospina, Nitrococcus,* and *Nitrospira* (Equation 2)

$$NO_2^- \text{ (Nitrite)} \rightarrow NO_3^- \text{ (Nitrate)} \tag{2}$$

19.2.5 Soil Food Web and Biological Population Regulation

The soil food web comprises a group of organisms that live their entire lives or a portion of them in the soil. It defines a sophisticated living system that coexists with the environment, plants, and animals in the soil. The flow of energy between different level of species in an environment is described by food webs. A food web is more complicated and shows all possible pathways, whereas a food chain just looks at one linear energy pathway through an ecosystem. Sunlight is mostly responsible for this energy transfer. By photosynthesis, plants transform carbon dioxide and minerals into plant material by converting inorganic chemicals into energy-rich organic compounds. Plants release energy-dense nectar from their flowers above ground, and their roots release sugars, acids, and ectoenzymes into the rhizosphere to balance the pH and nourish the food web below ground. Plants generate their own energy and are known as autotrophs. They are also known as producers because they generate energy that may be used by other organisms. Consumers that are heterotrophs are unable to produce their own food. They consume plants or other heterotrophs for energy. Energy goes from producers (plants) to primary consumers (herbivores) and subsequently to secondary consumers in above-ground food webs (predators). The term "trophic level" describes the several tiers or phases in the energy flow. In other words, the primary trophic levels are the producers, consumers, and decomposers. There may be several additional links in this chain of energy transmission from one species to another, but they eventually come to an end. Decomposers, such as bacteria and fungi, convert dead plant and animal matter into straightforward nutrients at the top of the food chain. The characteristics of soil make it challenging to directly observe food webs. There are numerous techniques for extracting soil organisms since their sizes range from smaller than 0.1 mm (nematodes) to more than 2 mm long (earthworms). Metal cores are frequently used to collect soil samples. Smaller nematodes and soil arthropods cannot be manually removed, unlike larger macrofauna like earthworms and insect larvae. There are three different types of food web systems above and below ground level:

1. topological food webs
2. flow webs
3. interaction webs

Early food chains were topological and descriptive and offered a nonquantitative view of consumers, resources, and the connections between them. Pimm, Lawton, and Cohen (1991) described these webs on the basis of eating habits as a map of organisms in a community of different kinds. A topological web is initially built before a flow web is created. Following the selection of the web's members, the biomass of each functional group is determined, often expressed in kilograms of carbon per hectare. Researchers make the assumption that the population of the functional group is in equilibrium in order to compute feeding rates. When the group is at equilibrium, reproduction equals the rate of member loss from natural causes and predation. An interaction web is very similar to a topological web except it shows the movement of energy and materials and the influence of one group onto other. Each link in an interaction food web model has two direct effects: one on the resource and the other on the consumer. The resource has a favorable influence on the consumer (they get to eat), but the consumer has a negative impact on the resource (it is eaten). Some relationships between species in the soil are complex, and some of them are not classified by food webs. Mutualists, ecosystem engineers, and litter transformers all have significant effects on their communities that cannot be classified as top-down or bottom-up. Soil food webs continue to be an effective tool for characterizing ecosystems, despite the fact that they cannot capture all interactions. There is still much to learn about how species interactions in the soil affect decomposition and maintain the stability of soil food webs and the evolution of the species through time. Understanding how food webs impact significant properties like soil fertility depends on this knowledge.

19.3 SOIL MICROBIOME

19.3.1 The Complex Interaction between Biotic and Abiotic Environments

The complex relationship between biotic and abiotic environments above and below the surface plays an important role in the improvement of productivity with less investment. This strategy includes the secluded use of plant microbiome members, probably referred to as a revelation in the "new" green revolution (Lyu et al., 2020). The use of Plant Growth Promoting Bacteria has the ability to improve nutrient absorption for proper growth and development by releasing certain phytohormones and antipeptide-like biotic compounds to combat biotic stress against pathogenic bacteria. The association of these beneficial bacteria is defined by releasing different compounds into the soil, and these unique characteristics reside with other plant-related bacteria maintaining healthy soil and microbiome (Carro and Nouioui, 2017; Lyu et al., 2021). The microbial inoculants can serve as an exogenous source of application to achieve the goal of sustainable and resilient agricultural system without chemicals application (Jiménez-Gómez et al., 2017). Due to frequent climate-changing conditions, we anticipate extreme weather conditions which are difficult for the survival of plants and microbes. The combinatorial role of the soil microbiome has resulted in plant growth under stressful environmental conditions with considerable impact. The integration of plant microbiome members in the form of plant-promoting strains into the agricultural system is a crucial and sustainable way to mitigate climate change. However, there are certain limiting factors in terms of the association between the plant kingdom and the microbiome community. Some of the major factors are soil pH, moisture, temperature, fertility, host specificity, colonization ability, plant-microbe signaling molecules, and microbial competition (Surovy and Islam, 2021). The table shows the plant probiotic role during biotic and abiotic stresses (Table 19.6).

TABLE 19.6
Bacterial and Fungal Species against Biotic and Abiotic Stress Tolerance

Bacterial and fungal species	Biotic or abiotic factors	Crops/plants	References
Stenotrophomonas sp. and *Ochrobacterum* sp.	*Colletotrichum fructicola*	Strawberry vegetative parts and fruits	Mukta et al., 2017; Rahman et al., 2018
Pseudomonas chlororaphis O6 (PcO6) with nanoparticle	Drought tolerance	Wheat roots and shoots	Jacobson et al., 2018
Bacillus amylolequefaciens and *Paraburkholderia fungorum*	Increase antioxidant content	Strawberry fruit	Rahman et al., 2018
Rhizobium laguerreae	Abiotic stress	Spinach roots	Jiménez-Gómez et al., 2018
Pseudomonas strains	*Ralstonia solanacearum* QL-Rs1115 strain	Tomato rhizospheres	Hu et al., 2016; Hu et al., 2021
Pseudomonas plecoglossicida and *Pseudomonas taiwanensis*	Black sigatoka disease	Banana fruits	Marcano et al., 2016
Bacillus subtilis CBR05	Increase antioxidant and carotenoid content	Tomato fruits	Chandrasekaran et al., 2019
Pseudomonas chlororaphis	Microbial pathogens, insects and nematodes	Wheat, tomato, soybean, avocado, green pepper	Anderson and Kim, 2018
Trichoderma virens Gl006 and *Bacillus velezensis* Bs006	Fusarium wilt (*Fusarium oxysporum* f. sp. *physali*)	Cape gooseberry	Izquierdo-García et al., 2020

19.3.2 Conservation of Soil Health Practices

Soil health is essential in providing strength to the plant's physiology in order to produce food, fuel, and other commercial purposes for humans as well as other creatures on Earth. In order to fulfill the needs of the increasing population, the usage of cultivated land has increased by almost 50% in the past five decades, along with a 70% increase in the utilization of fertilizer and pesticides (Banerjee et al., 2019; FAO, 2020). The world's largest agricultural producers and exporters, including the European Union, Brazil, the United States, and China, are major contributors to using high levels of pesticides. According to statistics, these four producers use 827 million, 830 million, 1.2 billion, and 3.9 billion pounds of pesticide consumption, respectively (Donley, 2019). However, this practice has neglected the alarming levels of comprise in soil health leading to it being an ultimate menace for the crop production system. Taking into account the changing climate, the massive use of chemical fertilizers imposes an imbalance in the soil microbiome, which comprises beneficial microorganisms. Some of the other important reasons include improper soil conditions, unsuitable farming practices, and less evolved fertilization methods that are responsible for the continuous increase in intensification of chemical fertilizer application (Guo & Wang, 2021). According to the Food and Agriculture Organization of the United Nations (FAO) *Statistical Yearbook 2020*, global pesticide use went up by one-third between 2000 and 2018, to 4.1 million tonnes, with the major contributor being Asia, followed by the Americas, Europe, Oceania, and Africa. On the other hand, chemical fertilizer use has tremendously increased in almost all regions between 2000 and 2018 (FAO, 2020). The use of fertilizer directly impacts soil health and creates an imbalance in associated microbial communities. Plant microbiomes are always associated with a well-balanced combination of microbes under regulated environmental conditions. Different environmental conditions, such as positive, neutral, or negative environments largely affect the intensity of microbial activities toward host growth and function. The positive and negative interactions play a key role in all physiological processes through microbe-to-plant signaling molecules, either by upregulation or downregulation via the production of phytohormones, signal compounds or peptide formations. The interactions range from natural symbiotic relationships between root system and the inbound rhizosphere microbial population that resides between plant cells to intermittent symbiotic relationships between microorganisms and the plant cell. Therefore, the members belonging to a plant microbiome or other complex microbiome can stimulate growth and development to attain sustainability in an agroecosystem (Lyu et al., 2021).

19.4 TRENDS AND PROSPECTS

19.4.1 Management Goals for Improving Soil Quality

The management of the soil system is dependent upon the support and activities of biomass along with the microorganisms involved. Therefore, the application of growth-promoting microorganisms has a long-term effect on crop productivity in terms of improving quality and quantity based on biological inputs. However, due to its complex nature, a solid application strategy is needed to promote the usage of these micro-consortiums in agriculture and farm systems. Farm practice should be encouraged with more bio-based fertilizers since agrochemicals have a negative impact on the soil microbes and decrease the natural rhythm of microbial populations. Ammonia-rich and sulfur-rich fertilizers can give immediate results but can directly or indirectly hamper the soil pH and increase the risk of pest infestation with a reduction in the quality of the soil. The innate microbial communities have reduced, and the ability of the plant to defend against abiotic and biotic stresses has declined (Gupta et al., 1989; Delcour et al., 2015). Tillage also affects soil porosity and the water-holding capacity of the soil, affecting nutrient exchange capacity. The porosity of soil is determined by the amount of air and water the soil can hold. On the other hand, tillage also affects residual placement, which reduces the temperatures, evaporation rate and nutrient loading, and rate of decay. Soil diversity can be promoted by adding organic matter on a regular basis and diversifying the type of plants in the fields. In other words, farmers should practice crop rotation and reduce the intensive tillage and excessive use of pesticides (Shah et al., 2021). To promote soil rejuvenation, robust screening of novel bacterial and fungal species as probiotic strains with potential ability must be considered for developing a bio-based fertilizer. This screening method may work well with the superior kind of strains along with the overall effect of the pathogen and disease. Precise fertilizer recommendations based on soil mineralogy and dynamics are of utmost importance to improve crop yield and cost-effectiveness. Initially, the in vitro experimental setup should be brought up in the fields by selecting certain districts or counties in a country. A certain number of soil types must then be selected as samples for mineralogy and dynamics and observed for the comparison between conventional types of fertilizers. The quality of soil plays a very important role in the improvement of the quality of most crops. However, for precise and more accurate crop-responsive probiotic application, a suitable formulation might be used to investigate the dynamicity in the system minimizing the use of chemical fertilizers and promoting organic agriculture for a sustainable environment. Moreover, public sector research institutions, fertilizer companies, and agriculture-based industries should work together to develop a research and awareness strategy for the promotion of bio-based fertilizers in agriculture (Menendez and Garcia-Fraile, 2017; Rao, 2018).

19.4.2 TOWARDS SOIL SUSTAINABILITY IN SOIL ECOSYSTEMS

The soil ecosystem is the foundation of multiple functions in the ecosystem and directly or indirectly contributes to the entire livelihood of human health. Soil is the medium to support plant growth, crops, forage, and Sustainable Development Goals (SDGs) 2 and 3, namely "zero hunger" and "good health and well being". The soil ecosystem drives the central processes impacting all the living organisms on Earth. Evidence on the importance of soil biota and its beings is abundant and diverse to support global food production. There are opportunities to save and promote soil biodiversity, which in turn supports the variety of life on Earth, including humans. Numerous activities that promote above-ground biodiversity also promote below-ground biodiversity. Explicit examination of soil biodiversity can offer a holistic strategy to advance many elements of the global sustainability objectives since soil biodiversity is integrated into many features of ecosystems. Utilizing sustainable agriculture methods, preserving already-existing natural areas, rehabilitating damaged habitats, and promoting urban biodiversity are all actions that support and sustain diverse soil communities and the services they offer to all ecosystems. The restoration of soil ecosystems and the use of sustainable agricultural methods, and the adaptation of urban areas for both people and nature, all depend on healthy soil biodiversity.

19.5 CONCLUSIONS

The most important goal of this chapter was to introduce the complexity of soil microorganisms and their influence on the environment along with their structural and functional behavior. The most dominant microbial biomass are fungi and bacteria. The diversity of microorganisms is influenced by climatic conditions and the microscopic world of aggregates in the soil ecosystem. The formation and stabilization of soil aggregates seems to be affected by the soil structure and carbon cycling with respect to the entire food web found in the soil microbiome. The functional and structural interactions of soil microorganisms were targeted to understand the complex relationship between biotic and abiotic factors. Therefore, microorganisms play a crucial role in balancing the natural and artificial environment below and above the soil surface. Many a time, they act as the only tool to achieve agricultural sustainability such as protecting crops from biotic and abiotic stress. Furthermore, soil conservation and preservation must be practiced to achieve ecological balance at ecosystem level.

REFERENCES

Anderson, A. J., and Y. C. Kim. 2018. Biopesticides produced by plant-probiotic *Pseudomonas chlororaphis* isolates. *Crop Protection* 105:62–69. https://doi.org/10.1016/j.cropro.2017.11.009

Banerjee, S., F. Walder, L. Büchi, M. Meyer, A. Y. Held, A. Gattinger, T. Keller, R. Charles, and M. G. A. Van Der Heijden. 2019. Agricultural intensification reduces microbial network complexity and the abundance of keystone taxa in roots. *The ISME Journal* 13:1722–1736. https://doi.org/10.1038/s41396-019-0383-2

Benson D. R. 1988. The genus *Frankia*: Actinimycetes symbionts of plants. *Microbiological Sciences* 5:9–12.

Bolton Jr. H., J. K. Fredrikson, and L. E. Elliot. 1993. Microbiology of the rhizosphere. In: *Soil microbial ecology*, ed. F. B. Metting Jr., 27–63, New York: Marcel Dekker.

Carro, L., and I. Nouioui. 2017. Taxonomy and Systematics of Plant Probiotic Bacteria in the Genomic Era. *AIMS Microbiology* 3: 383–412.

Chandrasekaran, M., S. C. Chun, J. W. Oh, M. Paramasivan, R. K. Saini, and J. J. Sahayarayan. 2019. *Bacillus subtilis* CBR05 for tomato (*Solanum lycopersicum*) fruits in South Korea as a novel plant probiotic bacterium (PPB): implications from total phenolics, flavonoids, and carotenoids content for fruit quality. *Agronomy* 9:838. https://doi.org/10.3390/agronomy9120838

Clarholm, M. 1985. Possible roles for roots, bacteria, protozoa and fungi in supplying nitrogen to plants. In: *Ecological interactions in soil: Plants, microbes and animals (Special publication no. 4 of the British Ecological Society)*, ed. A. H. Fitter, 255–265. Oxford: Blackwell Science Inc.

Coleman, D. C., D. A. Crossley, and P. F. Hendrix. 2004. *Fundamentals of Soil Ecology*. Amsterdam: Elsevier.

de Souza, R., A. Ambrosini, and L. M. P. Passaglia. 2015. Plant growth-promoting bacteria as inoculants in agricultural soils. *Genetics and Molecular Biology* 38:401–419. https://doi.org/10.1590/S1415-4757 38420150053

Delcour, I., P. Spanoghe, and M. Uyttendaele. 2015. Impact of climate change on pesticide use. *Food Research International* 68:7–15. https://doi.org/10.1016/j.foodres.2014.09.030

Dey, A., and P. S. Chaudhuri. 2016. Species richness, community organization, and spatiotemporal distribution of earthworms in the pineapple agroecosystems of Tripura, India. *International Journal of Ecology* 2016:3190182. https://doi.org/10.1155/2016/3190182

Donley, N. 2019. The USA lags behind other agricultural nations in banning harmful pesticides. *Environmental Health* 18:44. https://doi.org/10.1186/s12940-019-0488-0

FAO. 2020. *World Food and Agriculture – Statistical Yearbook 2020*. Rome: FAO.https://doi.org/10.4060/cb1329en

Gudeta, K., A. Bhagat, J. M. Julka, S. A. Bhat, and G. K. Sharma. 2022. Impact of aboveground vegetation on abundance, diversity, and biomass of earthworms in selected land use systems as a model of synchrony between aboveground and belowground habitats in Mid-Himalaya, India. *Soil Systems* 6:76. https://doi.org/10.3390/soilsystems6040076

Guo, Y., and J. Wang. 2021. Spatiotemporal changes of chemical fertilizer application and its environmental risks in China from 2000 to 2019. *International Journal of Environmental Research and Public Health* 18:11911. https://doi.org/10.3390/ijerph182211911

Gupta, V. V. S. R., and J. J. Germida. 1989. Influence of bacterial–amoebal interactions on sulfur transformations in soil. *Soil Biology and Biochemistry* 21:921–930. https://doi.org/10.1016/0038-0717(89)90081-3

Hawksworth, D. L. 1992. Microorganisms. In: *Global Biodiversity,* eds. Groombridge, B, 47–54. Dordrecht: Springer. https://doi.org/10.1007/978-94-011-2282-5_6

Hu, J., Z. Wei, V. P. Friman, S. H. Gu, X. F. Wang, N. Eisenhauer, T. J. Yang, J. Ma, Q. R. Shen, Y. C. Xu, and A. Jousset. 2016. Probiotic diversity enhances rhizosphere microbiome function and plant disease suppression. *MBio* 7:e01790–16. https://doi.org/10.1128/mBio.01790-16

Hu, J., T. Yang, V. P. Friman, G. A. Kowalchuk, Y. Hautier, M. Li, Z. Wei, Y. Xu, Q. Shen, and A. Jousset. 2021. Introduction of probiotic bacterial consortia promotes plant growth via impacts on the resident rhizosphere microbiome. *Proceedings of the Royal Society B: Biological Sciences* 288:20211396. https://doi.org/10.1098/rspb.2021.1396

Huber, H., M. J. Hohn, R. Rachel, T. Fuchs, V. C. Wimmer, and K. O. Stetter. 2002. A new phylum of Archaea represented by a nanosized hyperthermophilic symbiont. *Nature* 417:63–67. https://doi.org/10.1038/417063a

Izquierdo-García, L. F., A. González-Almario, A. M. Cotes, and C. A Moreno-Velandia. 2020. *Trichoderma virens* Gl006 and *Bacillus velezensis* Bs006: A compatible interaction controlling Fusarium wilt of cape gooseberry. *Scientific Reports* 10:6857. https://doi.org/10.1038/s41598-020-63689-y

Jacobson, A., S. Doxey, M. Potter, J. Adams, D. Britt, P. McManus, J. McLean, and A. Anderson. 2018. Interactions between a plant probiotic and nanoparticles on plant responses related to drought tolerance. *Industrial Biotechnology* 14:148–156. https://doi.org/10.1089/ind.2017.0033

Jiménez-Gómez, A., L. Celador-Lera, M. Fradejas-Bayón, and R. Rivas. 2017. Plant probiotic bacteria enhance the quality of fruit and horticultural crops. *AIMS Microbiology* 3:483–501. https://doi.org/10.3934/microbiol.2017.3.483

Jiménez-Gómez, A., J. D. Flores-Félix, P. García-Fraile, P. F. Mateos, E. Menéndez, E. Velázquez, and R. Rivas. 2018. Probiotic activities of *Rhizobium laguerreae* on growth and quality of spinach. *Scientific Reports* 8:295. https://doi.org/10.1038/s41598-017-18632-z

Kyrpides, N. C., and G. J. Olsen. 1999. Archaeal and bacterial hyperthermophiles: Horizontal gene exchange or common ancestry? *Trends in Genetics* 15:298–299. https://doi.org/10.1016/s0168-9525(99)01811-9

Linderman, R. G. 1988. Mycorrhizal interaction with the rhizosphere microflora: The mycorrhizosphere effect. *Phytopathology* 78:366–371. www.apsnet.org/publications/phytopathology/backissues/Documents/1988Articles/Phyto78n03_366.PDF

Lynch, J. M. 1987. Soil biology – accomplishments and potential. *Soil Science Society of America Journal* 51:1409–1412. https://doi.org/10.2136/sssa.1987.03615995005100060004x

Lynch, J. M., and J. H. Hobbie. 1988. *Micro-organisms in action: Concepts and application in microbial ecology.* Oxford: Blackwell Science Inc.

Lyu, D., R. Backer, S. Subramanian, and D. L. Smith. 2020. Phytomicrobiome coordination signals hold potential for climate change-resilient agriculture. *Frontiers in Plant Science* 11:634. https://doi.org/10.3389/fpls.2020.00634

Lyu, D., J. Zajonc, A. Pagé, C. A. S. Tanney, A. Shah, N. Monjezi, L. A. Msimbira, M. Antar, M. Nazari, R. Backer, and D. L. Smith. 2021. Plant holobiont theory: The phytomicrobiome plays a central role in evolution and success. *Microorganisms* 9:675. https://doi.org/10.3390/microorganisms9040675

Marcano, I. E., C. A. Díaz-Alcántara, B. Urbano, and F. González-Andrés. 2016. Assessment of bacterial populations associated with banana tree roots and development of successful plant probiotics for banana crop. *Soil Biology and Biochemistry* 99:1–20. https://doi.org/10.1016/j.soilbio.2016.04.013

Menendez, E., and P. Garcia-Fraile. 2017. Plant probiotic bacteria: Solutions to feed the world. *AIMS Microbiology* 3:502–524. https://doi.org/10.3934/MICROBIOL.2017.3.502

Metting, B. 1988. Micro-algae in agriculture. In: *Micro-algal biotechnology,* eds. Borowitzka M. A. and L. A. Borowitzka, 197–221. Cambridge: Cambridge University Press.

Mukta, J. A., M. Rahman, A. A. Sabir, D. R. Gupta M. Z. Surovy, M. Rahman, and M. T. Islam. 2017. Chitosan and plant probiotics application enhance growth and yield of strawberry. *Biocatalysis and Agricultural Biotechnology* 11, 9–18. https://doi.org/10.1016/j.bcab.2017.05.005

Newman, E. 1985. The Rhizosphere: Carbon Sources and Microbial Populations. *Environmental Science, Biology*, 11. Corpus ID: 82006012.

Paul, E. 2006. *Soil microbiology, ecology, and biochemistry,* 3rd Edition. Amsterdam: Elsevier.

Pimm, Stuart L., John H. Lawton, and Joel E. Cohen. 1991. Food Web Patterns and Their Consequences. *Nature* 350, 669–674.

Rahman, M., A. A. Sabir, J. A. Mukta, M. M. A. Khan, M. Mohi-Ud-Din, M. G. Miah, and M. T. Islam. 2018. Plant probiotic bacteria *Bacillus* and *Paraburkholderia* improve growth, yield and content of antioxidants in strawberry fruit. *Scientific Reports* 8:2504. https://doi.org/10.1038/s41598-018-20235-1

Rao, N. S. S. 2018. *Soil microbiology,* 5th Edition New Delhi: CBS Publishers and Distributors Pvt Limited.

Shah, A., M. Nazari, M. Antar, L. A. Msimbira, J. Naamala, D. Lyu, M. Rabileh, J. Zajonc, and D. L Smith. 2021. PGPR in agriculture: A sustainable approach toincreasing climate change resilience. *Frontiers in Sustainable Food Systems* 5:667546. https://doi.org/10.3389/fsufs.2021.667546

Surovy, M. Z., & Islam, T. (2021). Principle, diversity, mechanism, and potential of practical application of plant probiotic bacteria for the biocontrol of phytopathogens by induced systemic resistance. *Food Security and Plant Disease Management, April*, 75–94. https://doi.org/10.1016/b978-0-12-821843-3.00004-0

Willey, J. M., K. M. Sandman, and D. H. Wood. 2020. *Prescott's microbiology*, 11th Edition. New York: McGraw-Hill Education.

Wolfe, B. E., and J. N. Klironomos. 2005. Breaking new ground: Soil communities and exotic plant invasion. *BioScience* 55:477–487. https://doi.org/10.1641/0006-3568(2005)055[0477:BNGSCA]2.0.CO;2

Zwart, K. B., and J. F. Darbyshire. 1991. Growth and nitrogenous excretion of a common soil flagellate, *Spumella* sp. – a laboratory experiment. *European Journal of Soil Science* 43:145–157. https://doi.org/10.1111/j.1365-2389.1992.tb00126.x

20 Sustainable Approach for Management of Organic Waste Agri-residues in Soils for Food Production and Pollution Mitigation

Suman Chaudhary, Tanvi Bhatia, and Satyavir Singh Sindhu

20.1 INTRODUCTION

The human population is assumed to increase from the current seven billion on the planet to more than 9.5 billion by 2050 (Green et al. 2005). This will greatly influence the food and water demand, which is estimated to be increased by 60% till the year 2050 (IRENA 2015; Ehrlicha and Harteb 2015; Lin et al. 2018). At present, sufficient food is produced to feed the seven billion people by applying intensive agriculture practices. However, the extensive use of agrochemicals (mainly chemical fertilizers and pesticides) in agriculture to increase food production, gradual strengthening of intensive agriculture practices worldwide, and current climatic changes have resulted in poor soil health, reduced microorganisms' biodiversity and an increase in biotic and abiotic stresses along with pollution of the environment causing public health hazards (Tilman 2002; Stamati et al. 2016). In addition, the declining water sources, soil salinization, water pollution and limited availability of cultivated land in the vast majority of countries suggested the need for more sustainable agriculture in the 21st century (Gomlero et al. 2011).

The current population explosion, rapid industrialization, intensification of agriculture and application of high-input agri-technologies has led to a vast growth in the output of agricultural products, but it has also resulted in excess organic waste build-up. The quantities of organic wastes are likely to increase significantly in future due to strengthening of farming systems to increase agricultural production (Obi et al. 2016). Organic wastes are commonly by-products of agricultural, industrial, domestic or municipal actions. In addition, crop-plant residues, waste products from food processing units, animal manure, industrial wastes, municipal wastes and toxic agricultural wastes (i.e., pesticides, insecticides etc.) are also considered as organic wastes. These wastes can be solids, liquids or slurries based on the system and type of agricultural activities. When in soil, the behaviour and fate of the pollutant is controlled by several environmental elements such as the pollutant's physical and chemical characteristics, and the type and extent of its interactions with different constituents of the soil system.

The generation of large quantities of solid organic waste in the agricultural production system has caused the accumulation of around 2.24 billion tonnes of total solid waste globally in 2020. The amount of organic waste produced is further expected to increase significantly by 73% by 2050. According to estimates, the annual production of organic waste from agriculture is 998 million tonnes (MT) (Agamuthu 2009; Obi et al. 2016). In any farm, organic waste can make up to 80% of the total solid wastes produced, with manure production accounting for 5.27 kg/day/1000 kg live weight, on a wet weight basis (Obi et al. 2016). In India, the solid waste generated has increased from 6 MT in 1947 to 48 MT in 1997, and it is further estimated to reach 300 MT by 2047 (Sharholy et al. 2007; Bhat et al. 2018). This continuous increase in organic waste generation is bound to affect water, energy, soil quality and food supplies globally, and has become a major hazard that is causing the environment to deteriorate, resulting in more health-related problems as well as major economic losses. Besides this, agriculture events release greenhouse gases at a rate of 25–33% (Tubiello et al. 2014), and also lead to eutrophication and acidification of natural terrestrial and aquatic ecosystems with agrochemicals (Clark and Tilman 2017). However, organic agriculture is often claimed to have lower environmental consequences than high-input conventional systems as it replaces agrochemicals with organic amendments (Reganold and Wachter 2016).

Traditional techniques for managing solid waste such as thermal processes (incineration, gasification or pyrolysis), chemical degradation and landfilling involve high chemical cost, large energy consumption, and release of greenhouse gases along with toxic chemicals, which have various detrimental effects on the ecosystem. These problems have necessitated the development of sustainable and economic technologies such as composting and anaerobic digestion that are also socially acceptable for the management of solid organic waste, which may help in increasing the economic worth of the rejected solid wastes and may mitigate environmental issues (Goyal and Sindhu 2011; Zhou et al. 2022). Recycling and utilizing organic waste and its by-products is needed to boost crop productivity for fulfilling the high demand for food of the expanding population (Ibrahim et al. 2016). Recently, the biological treatment of organic waste has emerged as a pivotal alternative and these biological technologies utilize diverse soil-inhabiting organisms (such as bacteria, algae, fungi, nematodes, protozoa and arthropods) for organic waste management.

Organic wastes can be potentially used as fertilizers for soil amendment (in the form of compost, biochar or biogas slurry), for energy recovery (such as heat, liquid fuels, biohydrogen or electricity), and for production of certain chemicals such as volatile organic acids and alcohols (Sindhu et al. 2013; Babu et al. 2022; Osman et al. 2022). Animal manures and composts along with municipal biosolids and industrial wastes have been used in agriculture for centuries for improving soil's physical, chemical and biological properties (Westerman and Bicudo 2005; Chojnacka et al. 2022). Composting of agri-wastes (Verma et al. 2014) or converting of agro-residues into vermi-composting (Zhou et al. 2022), or biochar through pyrolysis (Osman et al. 2022) can be a suitable alternative in meeting both environmental and social concerns; and their use in the fields may decrease chemical fertilizers application in organic production systems. Other amendments such as vermi-composting, incorporation of biogas slurry and application of biofertilizers along with organic phyto-stimulants may also contribute towards stimulation of plant growth and increasing crop production. Additionally, such pre-treated organic waste can be used as a potent raw material for different industries as well as for livestock products. These current problems in agriculture practices emphasized the need for eco-friendly organic farming, which combines integrated pest and disease management techniques with biological and natural inputs to make agriculture more sustainable.

20.2 SUSTAINABLE STRATEGIES FOR MANAGEMENT OF ORGANIC WASTE IN SOIL

Intensive mining of mineral nutrients from soil during intensive farming, inappropriate use of agrochemicals and unrestrained burning of crop residues on farms have led to deterioration of agricultural lands as the biological processes that maintained soil health and quality became overtaxed (Ogbonnaya and Semple 2013). However, the soil is capable of restoring its soil fertility and of transforming pollutants to some extent through a combination of physical, chemical and biological processes. Therefore, proper utilization of harvested agricultural residues is needed to use this plant biomass for improving organic matter, nutrient availability and health of the soil (Sindhu et al.

2013). Additionally, it promotes the metabolism and sequestration of organic carbon, and maintains populations of microbes and other beneficial microflora that are important in the processing of essential nutrients (Bargaz et al. 2018). Tejaswini et al. (2022) highlighted the technical improvements in the biodegradation of organic wastes and their sustainable use as secondary resources, and recommended a cutting-edge sustainable approach built on the idea of the circular economy to support the United Nations sustainable development goals (SDGs). Policymakers have recently started to be concerned about agricultural waste management (AWM) for sustainable practices of agriculture (Hai and Tuyet 2010). If handled incorrectly or left untreated, organic wastes, especially animal manure, can significantly degrade the land, water and air quality. One such example is acid rain, which is caused due to the release of odorous gases and ammonia volatilization upon organic waste decomposition (Obi et al. 2016).

The concentration of total solids in organic wastes is the primary factor affecting the handling and treatment of agricultural wastes. It can be increased or decreased according to the treatment. It is increased by adding animal beddings or other on-farm waste like straw or hay; it can be decreased by adding water. Generally, agriculture waste management is a function of six basic steps (Figure 20.1), namely production of organic waste, its collection from various sources, large-scale storage, pre-treatment, transfer and utilization (USDA 2012). Production includes the quantity and quality of waste generated. The act of collecting garbage at its source or place of dumping is referred to as collection. Temporary containment or holding of waste requires storage facilities. Pre-treatment activities like analysing the features of the waste prior to treatment, figuring out the preferred features of the waste after treatment, choosing the type, anticipated dimension, site and installation cost of the treatment setup, and estimating the overall cost of pre-treatment process are all included in treatment. Utilization comprises returning non-reusable waste to the environment and recycling reusable waste (Obi et al. 2016).

Different methods are utilized in the valorization process of turning discarded organic wastes into secondary resources. Composting, vermi-composting, biochar addition etc. are some of the

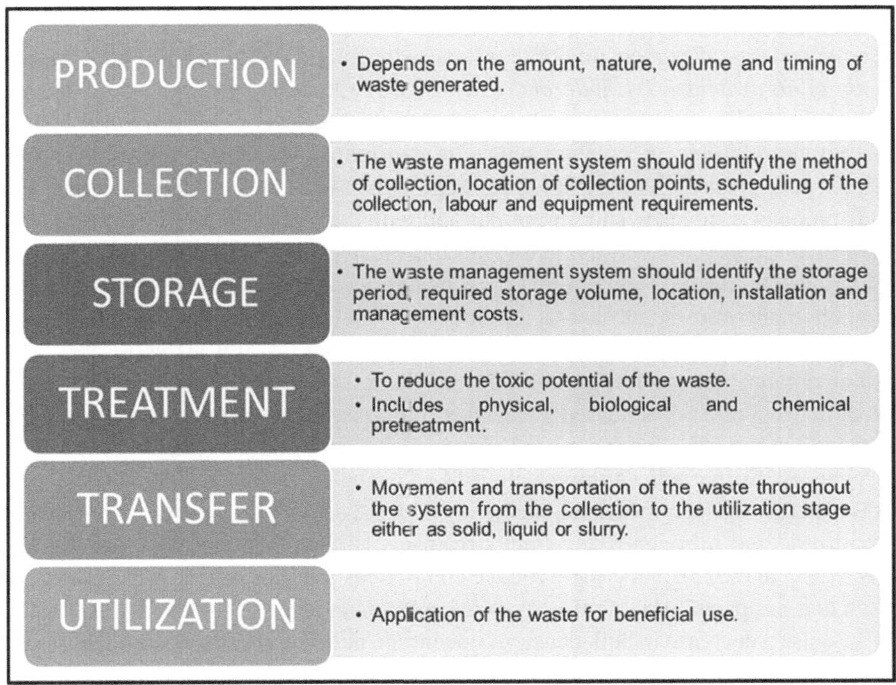

FIGURE 20.1 Different functions of agriculture waste management system.

predominant methods used in recycling of organic wastes, which are subsequently used in sustainable agriculture for improving crop production. These processes are briefly discussed below.

20.2.1 COMPOSTING

Fast decomposition rates and frequent crop cultivation under intensive agricultural practices has resulted in low organic matter content in cultivated soils in tropical climates. Therefore, adding both organic and mineral fertilizers is necessary for sustainable crop production. More than 3000 million tonnes of organic wastes are produced annually in India in the form of crop residues, rural and urban municipal wastes, animal farm wastes, vegetable market wastes, forest wastes and wastes from different industries (Gupta et al. 1998). In addition, municipal wastes are produced at the rate of 125,000 tonnes/day in India, as per the reports of the Central Pollution Control Board (2017) (Avnimelech 1986). The process of composting replenishes required organic matter to the soil and leads to an improvement in the overall soil efficiency and fertility (Tandon 1995). Thus, organic wastes may be effectively recycled in the system itself as organic manure (Verma et al. 2014), and composting the waste organic matter is particularly relevant to agricultural production and management of forestlands. Therefore, the process of composting for disposing of organic waste material is advantageous from an ecological and financial standpoint. Throughout the history of farming, composts have been used to increase agricultural output and soil fertility.

Composting is the process of biodegradation of a combination of organic wastes by a microbial consortium made up of distinct populations of bacteria, fungi and actinomycetes, both under aerobic circumstances and solid state fermentation. Compost is the stabilized and sterilized end result of composting (Insam and Bertoldi 2007; Goyal and Sindhu 2011). Basic nitrogenous and carbonaceous molecules undergo transformation to form comparatively stable complex organic structures that are chemically similar to soil humic substances, during the composting process by the action of microbes (Diacono and Montemurro 2011; Duan et al. 2021; Zhao et al. 2022). The organic components are gradually mineralized by the microbial driven process, releasing nutrients slowly. The decomposition of waste materials often helps in release and subsequent use of minerals and nutrients by the plants and microbiota of the soil. For faster composting systems, microbial inoculants can hasten the process of composting. For instance, inoculation with cellulolytic microorganisms such as *Aspergillus awamori*, *Trichoderma viride*, *Phanerochaete chrysosporium*, *Trichorus spiralis* and *Paecilomyces fusisporus* help to produce composts with a lower C:N ratio from sugarcane trash and rice straw. In addition, inoculation with nitrogen-fixing *Azotobacter*, phosphate-solubilizing fungus *Aspergillus awamori* and cellulolytic fungi along with rock phosphate helped to produce composts from plant residues in shorter periods and with the highest N and P content.

The rate of nutrient release varies based on the organic matter composition, the method and length of composting, the type and timing of raw material, and microbial inoculums application. For instance, amendment of fresh, nutrient-rich residue will result in a quick supply of nutrients, whereas a stabilized residue will sustain a gradual and slow nutrient supply. It is also observed that with the addition of an organic material having low C:N and C:P ratios, its rapid decomposition may result in high rates of nutrient release. Released nutrients will first be immobilized by the growing microbial biomass in similar materials having more C:N and C:P ratios, and only then will they be released from the compost to the soil (Avnimelech 1986). Strong microbial activity during composting improves the quality of the nutrients in the compost by the activity of nitrifiers, ammonia oxidizers, sulphur oxidizers, and nitrogen-fixing microorganisms (Richardson and Simpson 2011; Chander et al. 2018; Xie et al. 2023). They also produce compounds that function as plant growth hormones (Richardson and Simpson 2011). Different composting methods used to decompose organic wastes (Table 20.1) cause a decrease in bulk density, concentrate nutrients, destroy pathogens, reduce odour and have a stabilized product to transport nutrients for application in the farm (Ayilara et al. 2020).

TABLE 20.1
Basic Characteristics of Different Composting Methods

S. No.	Method of composting	Salient Features	Time duration	Advantages	Disadvantages
1.	Indian Bangalore composting	Recommended for the composting of night soil and organic residues	6–8 months	Volume of material is reduced	Laborious and expensive to support
2.	Vessel composting	It is conducted in a closed container or vessel	40 days	Accurate temperature control and monitoring	Forced aeration and mechanical turning is required. Laborious and expensive to support
3.	Windrow composting	It is done in windrows or in long thin piles with timely turning	8–9 months	Rapid and retains heat	Difficult and costly to support
4.	Static composting	A traditional way of composting, in which wastes are composted under aerobic conditions with passive aeration	3–6 months	Simple method with less labour, economy and equipment requirement	Time consuming
5.	Sheet composting	Organic matter is spread directly onto the soil in thin layers as mulch and is allowed to decay	>6 months	The method is simple and economical	Time consuming
6.	Indian Indore composting	Organic wastes are made into a layer of about 15 cm thick until the heap is about one and a half meter high. It is then cut into about 20–25 kg beddings, which are then filled in pits for composting	3–4 months	The process can be controlled	Labor intensive method
7.	Berkley rapid composting	Materials are chopped prior to composting. Harder tissues are needed to be chopped into smaller pieces to enhance decomposition	18 days	Fast method of composting	Some pathogens, weeds and weed seeds are not controlled.

Badillo et al. (2021) characterized leachate obtained from the composting process and tested raw and diluted leachates as culture medium for growth of *Bacillus* species in shake-flask fermentation assays. Leachates were also amended with yeast and whey permeate. They observed that growth of *Bacillus* sp. was better in diluted leachate as compared to that in raw leachate. Also, the growth was enhanced and the rate of sporulation increased when co-substrates were introduced. Co-substrates produced the maximum sporulation rate, causing more than 89% of cells to sporulate. This leachate valorization strategy will result in financial benefits by minimizing the amount of leachate waste to be handled and by assisting in the economical production of biological amendments. In another experiment, Isaac and George (2022) reported that agroforestry systems generate substantial quantities of leaf litters, and these biomass residues may be recycled into composting. Other viable options included biochar production and thermo-chemical digestion. Thus, *in situ* recycling of the leaf litter as a nutrient source may enrich the repository of organic manures in the form of litter compost. The use of these recycled composts will augment organic carbon and other plant nutrients in soil, and will sustain soil health in organic farming. Recently, the production of hydrochar co-compost (HCO) by anaerobic digestion of organic fractions of municipal solid waste (OFMSW) combined with hydrothermal carbonization (HTC) was reported, which may subsequently be used to recover biogas and compost (Al-Naqeb et al. 2022). The cytotoxicity of hydrochar co-compost was comparable to that of regular compost.

20.2.2 Vermi-composting

Traditional composting methods are very tedious and the compost prepared is relatively less effective. In some cases, using semi-decomposed compost may create many nutrient and pest- or weed-related problems. Therefore, vermi-composting is another better alternative to effectively decompose on-farm wastes by combining the action of earthworms with microorganisms to produce quality compost. Not only it is a highly controlled process, it is also less expensive and energy saving; and it also acquires a more efficient biological and recycling process.

Vermi-composting process employs earthworms for decomposing organic wastes. The resulting worm castings are termed as vermi-compost. Compared to traditional composting methods, humus content is higher in vermi-composting with less phytotoxicity and more nitrogen retention (Westerman and Bicudo 2005; Zhou Y et al. 2022). Vermi-composting speeds up the decomposition process by physically breaking down the raw biomass and combining it with a wide variety of bacteria as it passes through the earthworm's gut. The bacteria in the guts of earthworms are extremely capable of breaking down both organic materials and polysaccharides (Chander et al. 2018). However, vermi-composting faces a challenge in consistently sustaining ambient living conditions for earthworms, which is a significant contributor to many on-farm failures. Therefore, designing an automated system utilizing Internet of Things (IoT) technology to regulate temperature and humidity is currently necessary for the efficient production of vermi-compost (Shalini et al. 2022). As a result, the data gathered will significantly aid in preserving favourable pit conditions and enhancing composting. Zhou Y et al. (2022) highlighted recent developments in vermi-composting practices, which include integrated mathematical modelling, novel additives, and intelligence-assisted reactor designs, and have given a preview of future waste management developments using different microbial cultures, microbial consortiums, several additives, bulking agents or a combination of them for process improvement.

20.2.3 Biochar

Biochar is a carbon-rich byproduct obtained from heating organic waste biomass, such as wood, dung, or plant leftovers, in a sealed environment with minimal to no air (Joseph et al. 2009). European explorers in the Amazonian region discovered the "Terra Preta" soils in the 19th century, which contained high quantities of carbon derived from biochar with greater microbial activity, better nutrient accessibility and higher yield in comparison to soils without any biochar addition (Joseph et al. 2009; Ogbonnaya and Semple 2013). Because of its long-lasting stability and recalcitrant behaviour along with nutrient sequestration abilities, biochar is frequently produced under anaerobic conditions, at controlled temperatures. In recent years, this practice has become popular as a soil management practice.

Biochar may be made from various waste biomass sources, such as wood, manure, wheat or paddy straw, sewage sludge and municipal refuse (Sohi et al. 2009). The main components of wood are lignin, cellulose and hemicelluloses. However, the chemical make-up of each species of wood varies, and as a result, the lignin structure in softwood and hardwood differs (Windeisen and Wegener 2008). Subject to the proposed use of end products, organic waste material can be burnt in a reactor through gasification or carbonization at a variety of temperatures and periods. Mass losses, chemical reactions and structural changes occur during heating and are mostly influenced by the heating time, beginning moisture content, kind of wood and treatment temperature (Ogbonnaya and Semple 2013). Three basic products finally produced by the thermal transformation of wood are liquid (liquid hydrocarbon and water), solid (biochar) and gas (Karagöz et al. 2005). In comparison to lower and higher pyrolysis temperatures, wood feedstock with a greater lignin concentration provides the most biochar at 500°C (Fushimi et al. 2003).

Biochar is synthesized by the process of pyrolysis, in which the pyrolysis reactor's temperature typically varies from 300 to 1000°C. Compared to pyrolysis at high temperatures, pyrolysis at low temperatures often results in a higher biochar yield and an improved volatile matter composition (Khan et al. 2020). In addition, the heating temperature has a significant impact on the biochar's physico-chemical characteristics. While having lesser sorptive capabilities than biochar produced at a greater temperature (800°C), biochar produced at an inferior temperature (350°C) may contain significant levels of advantageous nutrients (Ogbonnaya and Semple 2013). This is due to the fact that carbon content and aromaticity rise with temperature, whereas oxygen, hydrogen, and polarity fall as micropore volume eventually increases (Chen et al. 2008). The carbon content of biochar rises as lignin carbonization takes place at temperatures over 250°C, whereas the quantities of oxygen and hydrogen decrease (Ogbonnaya and Semple 2013). This causes structural modifications and condensation processes that result in the formation of lignan (Windeisen and Wegener 2008), as well as the release of CO_2 and VOCs (Ogbonnaya and Semple 2013).

In addition to causing bigger, aromatic and aliphatic molecules to polymerize from changed chemical products at higher temperatures, higher temperatures also result in expansion of surface area and pore volume. It has been found that heating wood at 820°C reduced the amount of micropores and surface area compared to heating it at 700°C. This suggests that each feedstock material may have a peak temperature at which all holes (micro, meso and macro) will open; further increasing the temperature may promote decrease in the distribution of micropores (James et al. 2005). In environmental management, biochar is crucial for reducing greenhouse gas emissions and contaminants (Liu et al. 2019; Khan et al. 2021). Zhu et al. (2022) reported that a common management technique for attaining carbon neutrality in a circular economy is the conversion of agro-residues into biochar by pyrolysis. Agro-residues may be pyrolysed to make biochar, which can then be used as an alternative energy source, a soil conditioner, and a replacement for activated carbon to create biochar that is more environmentally friendly and long-lasting.

Biochar typically contains low-density particles, and applying it to soil can have several beneficial impacts on soil quality and the sorption of inorganic and organic contaminants (Masulili et al. 2010; Khan et al. 2021). A large surface area with many functional groups, a high cation exchange capacity, and great stability are typical characteristics of biochar (Ding et al. 2017). Thus, adding biochar to soil plays a crucial role in promoting pesticide breakdown both chemically and biologically. It also improves the effectiveness of applied fertilizers and increases nutrient availability and nutrient exchange. Foong et al. (2022) presented a thorough examination of the use of biochar made from rice straw (i.e., rice straw-derived biochar; RSB) to treat wastewater and remove various sorts of impurities. The most well-established method for creating biochar to date is pyrolysis. Thus, magnetic biochar and electrochemical deposition have shown promise as methods for biochar modification, whereas acid-modified RSB has the ability to remove metal ions and organic molecules. Biochar application to soil increases soil productivity, stores carbon, mitigates climate change, and filters percolating soil water (Joseph et al. 2009). Soils amended with biochar have better moisture retention, enhanced cation exchange capacity, a more adsorptive nature and higher pH. As a result, biochar may be used more often in soil applications for agriculture, waste management, and restoration of contaminated land (Ding et al. 2017; Osman et al. 2022). However, before it is feasible to precisely evaluate the sustainability of this method to sequester carbon and restore soil function, further research on the long-term effects of adding biochar on soil parameters must be done.

20.2.4 BIOGAS SLURRY

With increases in animal-product demand, a remarkable livestock production growth has resulted worldwide, and with this the question of disposal of animal excreta has emerged (Tang et al. 2020). Anaerobic digestion of this huge amount of excreta in biogas plants is the best approach to dispose of the excreta (Verma et al. 2014). The product of anaerobic digestion is biogas, a renewable energy source and its byproduct is biogas slurry, which is rich in nutrients. The spread of small-scale anaerobic digesters and biogas facilities in several Asian countries has led to a drastic increase in biogas slurry. Approximately 15 tonnes of biogas slurry is discharged daily from an 800 m^3 volume operating biogas plant (Chen et al. 2017).

An efficient approach is the long-term application of biogas slurry to farmland for recycling nutrients in agriculture farms. The biogas slurry is composed of 93% water and 7% dry matter (4.5% organic matter and 2.5% inorganic matter). Considering that it includes significant levels of both macronutrients (N, P, K) and micronutrients (Zn, Mn, B) essential for plant growth, biogas slurry is regarded as a useful source of organic fertilizer. Digested slurry includes readily available plant nutrients, such as organic nitrogen (mostly amino acids), plentiful minerals and bioactive compounds (e.g., vitamins, humic acids and hormones) (Yu et al. 2010).

Seeds soaked in biogas slurry had improved germination rates and greater seedling growth than seeds not given this treatment (Yu et al. 2010; Pathak et al. 1992). Its application in the field increases crop yields and prevents diseases, when sprayed on crop plants (Zhao et al. 2007). As an amendment in soil, it offers a promising soil improvement and it could actually improve crop and fruit qualities (Yu et al. 2010). Biogas slurry can be added directly to the crops or applied along with other organic supplies and synthetic stimulants. Addition of animal urine along with biogas slurry can speed up the composting process because of the increase in nitrogen content, resulting in an increased C:N ratio for easy nutrient accessibility to plants and soil microflora. However, some studies have suggested that long-term application of untreated biogas slurry in croplands results in loss of crop yield and quality (Singla et al. 2014; Yang et al. 2018), and also increases the rate of accumulation of heavy metals or genes for antibiotic resistance in soil (Bian et al. 2014; Pu et al. 2018). It may also cause movement of extra nutrients into groundwater or surface water. Thus, some pre-treatment techniques must be devised before its application in farmlands. Some heavy metal removal techniques from biogas slurry such as integrated membrane technology, including reverse osmosis, microfiltration and ultra-filtration membranes, have been reported (Xu et al. 2019; Zhou et al. 2018; Ruan et al. 2015).

20.2.5 ORGANIC PHYTO-STIMULANTS AND BIOSTIMULANTS

A wide range of substances, which are beneficial to plants and are the most widely utilized stimulants in the agriculture sector, but which do not act as nutrients, insecticides or soil enhancers, are termed as "biostimulant". Due to the biostimulant's unique manner of affecting growth and development of crops and plants, traditional stimulants like fertilizers and pesticides may be readily replaced. The first definition of the term "biostimulants" was given by Zhang and Schmidt in 1997. They described biostimulants as "materials that, in minute concentrations, encourage plant growth and development". The term "minute concentration" distinguish these substances from other soil amendments, such as biofertilizers, that are applied in larger concentrations (Du 2015). The definition was further modified as "biostimulants are compounds, other than fertilizers, that encourage plant development, when administered in low quantities" (Kauffman et al. 2007). Biostimulants come in a wide range of formulations and include a wide range of chemicals; however, their broad classification based on their source and composition is discussed in Table 20.2.

TABLE 20.2
List of Some Major Organic Phyto-stimulants and Their Functions

Biostimulant type	Features	References
Seaweed extract	• Sea weed extract (SWE) may be used directly or in the form of purified products (such as the products of polysaccharides such as laminarin, alginates, and carrageenans). • Micro- and macronutrients, N-containing chemicals (betaines), sterols, and phyto-hormones including auxins, cytokinins, and gibberellins are other components in seaweed extracts that support plant development. • SWE can be applied on soils, as foliar treatment or in hydroponic solutions. • The polysaccharides result in gel formation that helps in water retention. The polyanionic compounds in SWE helps in the fixation and exchange of certain cations and heavy metals for soil remediation. • SWE may also involve some antioxidants or endogenous stress responsive gene regulators.	Arioli et al. 2015; Aremu et al. 2016; Colla et al. 2017a; Chen et al. 2021
Protein hydrolysates	• Protein hydrolysates consist of peptides or amino acids obtained by chemical processes or the enzymatic hydrolysis of agricultural and industrial waste materials (of both plant and animal origin). • They act as biostimulants of plant growth by modulating N uptake and assimilation via enzymes involved and by influencing the signalling system for acquiring nitrogen. • Application of protein hydrolysates also regulates TCA cycle enzymes. • Amino acids such as proline shows chelating effects, hence protecting plants against heavy metals and also helping in micronutrient mobility and acquisition. • Some nitrogenous compounds such as glycine, betaine and proline confer antioxidant activity and contribute to the mitigation of environmental stress. • They also increase soil respiration and soil fertility by increasing microbial biomass activity.	Marfà et al. 2009; Chalamaiah et al. 2012; Colla et al. 2013, 2017b
Fulvic and humic acids	• Humic substances as well as their compounds in soil (humins, humic acid and fulvic acid) are a result of interaction between organic matter, microbes and plant roots. Therefore, the application of humic substances shows inconsistent positive results. • They can be obtained from naturally decomposed organic matter, from composts or vermi-composts or from mineral deposits. • Organic wastes can be subjected to pre-controlled degradation and chemical oxidation, resulting in formation of "humic like substances" as a substitute for natural humic substances. • The effect of humic substances varies with soil environment, the nutrient accepting plant, the dosage and method of application. • They cause increased uptake of nutrients because of improved cation exchange capacity of the soil and also stimulate H^+-ATPases to uptake nitrate and other nutrients. • Humic and fulvic acids also help in stress mitigation by phenylpropanoid metabolism leading to production of phenolic compounds.	Stevenson 1982; Kumar and Chandra 2020; Kulikova and Perminova 2021; Sindhu et al. 2022a

(continued)

TABLE 20.2 (Continued)
List of Some Major Organic Phyto-stimulants and Their Functions

Biostimulant type	Features	References
Chitosan	• The deacetylated form of chitin polymer is known as chitosan. It may be produced naturally or industrially. Its polymers are used in agricultural sectors. • This polycationic compound has the capacity to bind a variety of cellular constituents as well as to specific receptors implicated in the activation of defence genes. For example the purified preparations of laminarin (a form of glucan stored in brown algae) are used in agricultural applications. • As a consequence of chitosan binding to cell receptors with lower specificity, H_2O_2 accumulates and Ca^{2+} leaks into the cell, causing many physiological changes as these are responsible for stress response signalling and growth regulatory mechanisms. • Other applications of chitosan include plant protection against fungal pathogens and abiotic stress (drought, cold and salinity) tolerance, leading to stimulation of plant growth and crop yields.	Ahmed et al. 2020; Bandara et al. 2020; Mujtaba et al. 2020; Rahman and Goswami 2021
Complex biostimulants	• These microbial biostimulants include non-pathogenic bacteria, mycorrhizal fungi and other soil or rhizosphere-inhabiting beneficial microorganisms. • They help the plants by improving their nutrient uptake and abiotic stress tolerance. They also condition the soil and prevent any possible pest invasion. • The symbiotic association of mycorrhiza with plants is widely accepted to promote sustainable agriculture. It helps in absorbing macro- and micronutrients by plants, maintaining water balance and mitigating biotic and abiotic stress. • The tripartite associations between mycorrhizal fungi, plant and rhizobacteria are important for soil and plant health. • Some fungal based biostimulants are used in fields to promote nutrient efficiency, stress tolerance, increased crop yield and quality. These include fungal endophytes such as *Trichoderma* spp., *Phanerocheate* etc. *Trichoderma* spp. has also been used as biocontrol agents and as biopesticides. • As biostimulants, inoculation with mutualistic endosymbionts like *Rhizobium* and plant growth promoting rhizobacteria are mostly favoured.	du Jardin 2015; Backer et al. 2018; Salvador et al. 2022; Xie et al. 2022

20.2.6 BIOFERTILIZERS

To produce adequate food for the ever-increasing human population, farmers are growing high yielding crop varieties and applying modern agricultural biotechnologies to enhance crop yields. But, the growth and yield of crop plants in the soil is limited by the availability of various macro and micronutrients. To supplement the nutrient requirements in soil, there is widespread use of chemical fertilizers and plant growth regulators in securing high crop yields (Tyagi et al. 2022). However, the indiscriminate use of nitrogenous and phosphatic fertilizers along with other agro-chemicals has caused significant soil, air and water contamination. Moreover, these agrochemicals also exert damaging effects on the population of soil microorganisms and influence the soil's fertility, leading to a steady reduction in crop productivity and soil health. In addition, the recent increase in the prices of nitrogen and phosphatic fertilizers has made them unaffordable or unavailable in many countries.

Therefore, adding additional N and P fertilizers to the soil in order to prevent nutritional deficits is exceedingly difficult for farmers and agricultural experts.

Sustainable agriculture is required in order to effectively manage agricultural resources in order to meet changing human requirements while preserving or improving the environmental quality and protecting natural resources. The use of microbial inoculants and their combinations as biofertilizers has proved a sustainable alternative to preserve and improve soil health (Kumar et al. 2022a,b). These beneficial microorganisms are contributing immensely in providing plant nutrients by nitrogen fixation, phosphorous and potassium solubilization or transformation in the soil. Moreover, these beneficial microbes also improve plant growth by production of phytohormones, suppression of plant pathogens and bioremediation of pollutants and heavy metals (Sehrawat et al. 2021; Sindhu et al. 2022b; Chaudhary and Sindhu, 2024). These biofertilizers are affordable to both small and marginal farmers.

These biofertilizers contain specific, competent and living microbial cultures (Dadarwal et al. 1997). Various bacteria, actinobacteria, fungi and cyanobacteria are used in the preparation of biofertilizers (Kumar et al. 2022a,b). These biofertilizers are prepared either in powdered, liquid or granular form, with different sterilized carriers like perlite, charcoal, culture medium, peat and mineral soil. Biofertilizers can also be classified on the basis of nutrients provided by these beneficial microorganisms (Thomas and Singh 2019; Sindhu et al. 2014a,b; Kumar et al. 2022b). On this basis, biofertilizers are categorized as nitrogen-fixing, phosphorous or potassium solubilizing and/or plant growth-promoting microbes (Table 20.3).

TABLE 20.3
Important Groups of Biofertilizers and Their Functions

Types	Functions	Characteristics	References
Nitrogen-fixing microbes	Free-living	• In culture media, *Azotobacter chroococcum* has been found to fix 2–15 mg N/g of carbon source. • Free-living cyanobacteria in rice fields fix approximately 20–30 kg N/ha. • Examples: *Azotobacter, Beijerinkia, Anabena, Nostoc, Chromatium* etc.	Bekele and Yilma 2021; Ladha et al. 2022
	Symbiotic	• Different *Rhizobium* strains possess the capacity to fix nitrogen up to 450 kg N/ha (depending on the bacteria and legume host). • *Frankia* can nodulate the roots of many non-legume actinorhizal plants such as *Alnus, Casuarina* etc. • Some blue green algae like *Anabaena, Nostoc* and *Trichodesmium* contribute approximately 36% of the global nitrogen fixation. • Examples: Rhizobia (*Rhizobium, Allorhizobium, Azorhizobium, Bradyrhizobium, Mesorhizobium and Sinorhizobium*), *Anabaena, Frankia* etc.	Sindhu 2004; Sindhu et al. 2019; Kawaka 2022
	Associative	• These diazotrophic microbes live in intimate association with plant roots. • *Azospirillum* interacts with a variety of annual and perennial plants. Along with nitrogen fixation, it also produces growth promoting substances. • *Acetobacter diazotrophicus* formulation can fix up to 70% sugarcane crop nitrogen requirement. • Examples: *Azospirillum* spp., *Azoarcus* spp., *Acetobacter, Bacillus, Herbaspirilum, Gluconacetobacter* etc.	Sindhu et al. 2021; Sindhu et al. 2022b

(continued)

TABLE 20.3 (Continued)
Important Groups of Biofertilizers and Their Functions

Types	Functions	Characteristics	References
Phosphorous solubilizing microbes	Phosphorous supply	• These microbes secrete organic acids leading to lowering of pH and also produce exopolysaccharides, which help in solubilization of bound phosphates in soil. • Examples: *Bacillus subtilis, Bacillus circulans, Pseudomonas straita, Trichoderma, Rhizobium* etc.	Sindhu et al. 2014b; Dahiya et al. 2021; Rodrigues et al. 2023
Other mineral solubilizing biofertilizers	Potassium, silicate and zinc solubilizing	• Some microbes can hydrolyse bound potassium minerals, silicates and aluminium silicates by producing organic acids and supplying protons that cause hydrolysis. • Some microbes also possess the ability to solubilize insoluble zinc compounds. • Application of potassium or zinc solubilizing bacteria results in higher plant biomass and crop yields. • Examples: *Bacillus subtilis, Pseudomonas, Bacillus edaphicus, Bacillus mucilaginosus, Leclercia adecarboxylata* etc.	Meena et al. 2015; Sindhu et al. 2016; Parmar and Sindhu 2019; Kang et al. 2021; Sharma et al. 2024
Plant growth promoting microbes	Plant growth stimulation	• These microbes synthesize plant growth promoting substances such as phytohormones, as well as siderophores, antibiotics, chitinase, ACC deaminase and other hydrolytic enzymes. • They also provide protection against soilborne diseases and pests infestation in soil. • These microbes also help in mitigation of abiotic stresses. • Examples: *Azotobacter, Bacillus, Pseudomonas, Serratia, Cellulomonas, Enterobacter* etc.	Sharma et al. 2019; Sindhu and Sharma 2020; Vocciante et al. 2022
Mycorrhizal biofertilizers	Plant growth promotion	• Mycorrhizal fungi (MF) form obligate or facultative mutualistic relationship with >80% of land plants. • They help in phosphorous mobilization or phosphate absorption. These fungi also enhance soil quality, water absorption, soil aeration and tolerance to drought and heavy metals. • MF solubilize soil humus nutrients and secretes antimicrobial substances against various plant pathogens. • Examples: Ectomycorrhizal and arbuscular mycorrhizal fungi	Elekhtyar et al. 2022; Kumar et al. 2022a; Salvador et al. 2022; Phour and Sindhu 2024

20.3 EFFECT OF SOIL AMENDMENTS IN IMPROVING PLANT GROWTH AND CROP YIELD

To improve soil fertility and soil health in terms of crop production, soil amendments, including both inorganic and organic materials, are mixed into the field soil. In both conventional and organic farming, biological soil amendments are crucial (Seufert et al. 2012). By adding nutrients to soils, these amendments can boost soil fertility, crop yields, soil structural stability, bulk density, aeration, water-holding capacity, porosity, water movement and nutrient cycling. Repeated applications of soil amendments have also been found to improve soil microbiological diversity and microbial biomass carbon (Ponisio et al. 2015). Soil amendments have been reported to provide a better environment for functioning of soil microflora, development of plants roots and stimulation of plant growth (Shah et al. 2021; Yang et al. 2024).

20.3.1 INCREASING NUTRIENT AVAILABILITY IN SOIL

Organic soil amendments are reported to release nutrients slowly as compared to chemical fertilizers (Reganold and Wachter 2016). However, nutrient losses are faster with the addition of chemical fertilizers, which may leach into the soil with water flow, hence making the nutrients unavailable for uptake by the plant roots. Hence, repeated chemical fertilizer application is required, which in the long run depletes soil fertility, leads to soil desertification and results in pollution of the water bodies. On the other hand, organic amendments such as composts, vermi-composts, biofertilizers and biochar release nutrients slowly and for a longer time. In fact, it is important to coordinate the nutrient demand by crops with the nutrient supply by organic amendments. The increase in nutrient use efficiency by crop plants decreases the loss of nutrients to leaching (Tilman et al. 2002).

Repeated application of organic amendments to farmland resulted in an improved soil nutrient concentration. For example, microbial biomass carbon increased up to 100% by using good quality composts and their enzymatic activity by 30% by sludge addition (Diacono and Montemurro 2011). Similarly, the amount of soil organic-carbon and soil organic-nitrogen contents increased by 90% by storing organic amendments for mineralization in future cropping seasons. It has been observed that a one-time application of compost increased N uptake by tall fescue grass consistently for seven years in a total of 294 to 527 kg/ha. However, nitrogen mobilization and immobilization depends on the degradability and concentration of soil C pools (i.e. C:N ratio) (Sullivan et al. 2003).

Mohamed et al. (2022) suggested that inoculation of kitchen waste with phosphorus-solubilizing bacteria considerably improved enzyme activity and increased the concentration of citric acid phosphorus, Olsen phosphorus, total phosphorus, organic matter decomposition and microbial diversity, thus improving overall biological phosphorus transformation and nutrient content in final composting products. In a similar experiment, Chojnacka et al. (2022) demonstrated that valuable materials, including nitrogen (especially important amino acids), P, K, microelements, and biologically active substances could be recovered with appropriate management of agri-food biological waste. Innovative strategies were outlined for the valorization of different bio-based wastes into biopesticides and biofertilizers. Similarly, lemon plants grown in a pot experiment were treated with chemical fertilizer (CF), and treatments with CFBC (chemical fertilizer combined with biochar) and assessed with no fertilization (CK) treatment (Xie et al. 2023). Compared with CK (control) treatment, potential nitrification rate (PNR) increased by 19.3–34.7% in the rhizosphere; whereas manure (M), MBC (manure combined with biochar), and FMBC treatments decreased PNR about 60.0%. Interestingly, CF and CFBC treatments increased ammonia-oxidizing archaea *amo*A gene copies by 40.2–101.7%, whereas M, MBC and FMBC treatments decreased it by 66.5–81.9%, and five fertilization treatments decreased the ammonia-oxidizing bacteria. Chemical fertilizer and manure combined with biochar affects the population structure of ammonia-oxidizing archaea and bacteria by pH, available phosphorus, carbon to nitrogen ratio, moisture content, and N-content in the rhizosphere.

20.3.2 ENHANCING PLANT GROWTH AND CROP YIELD

Organic amendments in soil have resulted in increased microbial activity, changes in soil microflora community structure and composition, and enrichment of specific beneficial microbes. All these changes have a direct or indirect effect on the growth and yield of plants (Phour and Sindhu 2022, 2023). For example, crop yields increased by 250% by the application of municipal waste compost (Tilman et al. 2002). Amendment of soil with organic matter was found to increase the organic-carbon pool of degraded soils (Lal et al. 2006), which further improved the crop yields. Increases in plant growth and crop yield were correlated with three mechanisms including (i) enhancing the availability of water; (ii) increasing nutrient supply; and (iii) improving the structure and physical properties of soil.

However, the effect of these factors on crop yield depends on several factors such as soil management, previous organic-carbon concentration, frequency and rate of application of chemical fertilizers and other soil amendments. Cherif et al. (2009) showed that wheat yield enhanced by 246% as compared to control, when 80 tonnes/ha of municipal waste compost was applied annually for five years. Similarly, the highest application rate of 23 tonnes/ha/year of biowaste compost increased yields of cereals and potato crop by 10% in comparison to unfertilized control (Erhart et al. 2005). It is also evident that lingering effects of organic amendments on both crop yields and soil characteristics can last for several years. The use of organic soil amendments along with mineral fertilizers also increased crop yield and growth for more than ten years as compared to application of soil amendment alone (Diacono and Montemurro 2011).

Schmidt et al. (2014) reported that organic amendments to a vineyard soil with biochar alone (8t ha^{-1}), aerobic compost (55t ha^{-1}) and biochar-compost (8t ha^{-1} + 55t ha^{-1}) induced negligible effects on plant growth, nutrient uptake, plant health and quality of grapes over the three years. But application of higher amounts of biochar through surface application wasn't found to have any significant economic benefits for vine growing in alkaline, poor fertility and temperate soil. Yu et al. (2019) presented an overview about problem soils (that inhibit or prevent plant growth). Amendment of problem soils with biochars as functional materials, interactions of amended biochars with soil-inhabiting microbes, soil particles, and plant roots were considered in relation to the improvement of problem soil and plant growth. Biochar amendment was recommended as a viable approach for enhancing crop production in problem soils. However, further research is needed on biochar amendment to increase our understanding of the collaborations of biochar with different components of soils, thus further increasing the efficiency of soil remediation and to improve crop production in affected soils.

Ghorbani et al. (2022) compared the nitrogen-use efficiency of wheat in relation to biochar (B) mixed with legume residues, urea and azocompost. Experimental treatments included: control (without biochar and fertilizer), urea @ 46 kg ha^{-1}, azocompost @ 5 tonnes (t) ha^{-1}, legume residue @ 5 t ha^{-1}, biochar @ 10 t ha^{1}, urea–biochar mix, combination of legume residues and biochar, and azocompost–biochar mix. The yield of wheat grain was found to improve by 337% and 312% with urea–biochar and urea–legume residue treatments, respectively. The combined treatments of legume residues and biochar resulted in the highest grain N yield, with an increase of 8.8 times over the control. Legume residues showed the highest N-use efficiency, with a rise of 149% in comparison to the control. The highest N-recovery efficiency and N-harvest index were obtained by using legume residues and biochar treatments with values of 91 and 70%, respectively. In another experiment, Bornø et al. (2022) treated maize plants with straw or wood biochar, which were exposed to different irrigation regimes. Biochar amendments showed positive effects on the leaf water potential and on many photosynthetic factors in the preliminary drought and retrieval phases. In addition, the pH and NH_4^+ content of soil were also affected by biochar and the carbon substrate utilization also improved in the biochar treatments, independently of irrigation status. Biochar in particular was found to increase the organic acids content, and drought had an extra effect on the succinic acid content in the treatment having wood biochar.

In another experiment, four treatments were made in ginger field soil that included: (i) control; no compost or PGPR; (ii) compost; (iii) compost + *Bacillus subtilis*; and (iv) compost + *Bacillus amyloliquefaciens* SQR9 (TC-BA) (Xie et al. 2022). The organic matter of soil was found to increase significantly, with amendment of compost, alkali hydrolysable nitrogen, available phosphorus and available potassium by 17.34, 21.66, 19.56 and 37.20%, respectively in comparison to control CK treatment. Soil enzyme activities viz. urease, sucrase and neutral phosphatase were also increased by 55.89, 57.21 and 35.59%, respectively in treatments having compost. Soil bacterial diversity was significantly improved in TC-BS treatments in comparison to PGPR and compost treatments and the comparative profusion of beneficial bacteria such as *Acidobacteria*, *Proteobacteria*, *Chloroflexi* and *Bacillus* were found to increase. It has been further found that inoculation of *B. subtilis* along with tomato compost in ginger field soil enhanced bacterial diversity, rehabilitated bacterial community structure, supplemented useful microorganisms and encouraged a healthy rhizosphere. Salvador et al. (2022) reported that soybean (*Glycine max*) plants grown in soil containing organic compost and inoculated with *Bacillus subtilis* increased the root dry weight (~50%), while plants cultivated in soil having arbuscular mycorrhizal fungus viz. *Rhizophagus irregularis* showed increased nodule biomass (~30%), chlorophyll content (~20%), symbiotic efficiency (~50%) and nitrogen accumulation (~45%) as compared to unamended soil.

20.3.3 BIOREMEDIATION OF SOIL POLLUTANTS

Bioremediation of soil pollutants is an eco-friendly approach, which may take several years to complete. Various microorganisms possess the capability to degrade many organic contaminants in the presence of aerobic conditions, and on availability of carbon and other nutrients. However, the recalcitrant nature of some pollutants such as PAHs resists microbial degradation. In addition, the microbial capacity of pollutants' degradation is also limited by the soil environment factors such as accessibility of soil organic C, N or other nutrients, aeration conditions, pH, moisture content and temperature (Teng et al. 2010). Additionally, different soil amendments influence the C:N ratio of soil, which in turn affects the bioremediation process. For instance, biodegradation of hydrocarbons requires an optimal C:N:P ratio of 33:5:1 (Zitrides 1983). If one adds an excessive amount of N (i.e., increasing C:N ratio to 1.8:1), it results in increased ammonia toxicity, which inhibits microbial growth resulting in almost no biodegradation. On the other hand, low C:N ratio favours microbial growth and functioning.

Moisture content also influences bioremediation of contaminants. Adding moisture to the soil enhances polyaldehydes' degradation and results in increased mineralization process of phenanthrene, in which maximum mineralization occurred at 20–50% moisture content (Diacono and Montemurro 2011). The percolation of ground water or uncontaminated water saturated with dissolved oxygen and mixed with nutrients resulted in enhanced bioremediation of soil in the presence of acclimated microorganisms, which is termed as bioaugmentation. Low temperature generally slows down the remediation process. In addition, high rates of contaminants' degradation were observed under aerobic conditions, because of the higher rate of bacterial respiration. Under anaerobic conditions, the intermediate end products of bioremediation are formed, which include methane, some amount of CO_2 and trace amounts of hydrogen. Sometimes, intermediate end products are more hazardous than the original compound. For instance, anaerobic degradation of trichloroethylene leads to the production of more hazardous vinyl chloride. Such problems can be avoided by conducting these projects *in situ*. Vinyl chloride can be broken down further under aerobic conditions.

Bioremediation of pesticides using a single or consortium of bacteria or a microbial-algae consortium is used to decontaminate pesticides from the environment and has been found as a promising bioremediation technology for environmental sustainability (Sehrawat et al. 2021; Sheng et al. 2022). Kulikova and Perminova (2021) discussed the interactions between microorganisms and humic substances, elaborating upon enzyme production capable of catalysing the oxidative binding of biological pollutants to humic substances. Thus, humic-reducing microorganisms may be utilized for the biodegradation of organic pollutants and for lowering the metal toxicity (Kulikova and Perminova 2021). Other functional groups of humic acid such as quinones can form covalent bonds with aromatic amines or the same type of organic compounds (Chianese et al. 2020), and humic acids showed a robust capability to sorb cationic and hydrophobic organic contaminants.

20.3.4 BIOLOGICAL CONTROL OF PLANT PATHOGENS BY SOIL ORGANIC AMENDMENTS

Plant pathogens and insects disturb global crop productivity and annually account for 20 to 40% yield reduction in various cereals and legume harvests (Oerke 2006). Pesticides are usually applied to minimize the yield losses, but injudicious use of agrochemicals cause environmental pollution. Some of the soil-inhabiting microorganisms possess the capability to produce antibiotics, siderophores, HCN and hydrolytic enzymes to inhibit the growth of phytopathogens and, thereby, suppress various plant diseases Sindhu et al. 2016; Sehrawat et al. 2022). Researchers are trying their best to find out antagonistic microorganisms effective against plant pathogens to reduce the use of agrochemicals in crop production (Sehrawat et al. 2022). These antagonistic microorganisms along with various organic amendments are currently being exploited for their possible use as bio-control agents in integrated pest management (IPM) programs for improving food safety (Sehrawat and Sindhu 2019; Bellini et al. 2023; Yan et al. 2024).

Ali et al. (2022) evaluated the effects of initial reductive soil disinfestation (RSD) treatment followed by biochar amendment (RSD-BC) along with applications of microbes (i.e., *Trichoderma* and *Bacillus* spp.) and biochar (RSD-SQR-T37-BC) on *Fusarium* wilt suppression in cucumber crop. These treatments were found to decrease nitrate concentration, elevated soil pH and soil organic-carbon as well as ammonium in the treated soils. Under RSD treatment, the applications of *Bacillus* (RSD-SQR), *Trichoderma* (RSDT37) and biochar (RSD-BC) suppressed *Fusarium* wilt incidence by 26.8, 37.5 and 32.5%, respectively, in comparison to non-RSD treatments. *F. oxysporum* populations were decreased greatly in treatments with RSD-SQR-T37-BC and RSD-T37. In another study, the effects of three different organic soil amendments, viz. cassava peel, sawdust and leaves of *Cedrela odorata* L, was assessed on the stalk rot disease of maize (*Zea mays* L.) caused by *Fusarium verticillioides* (Olajumoke et al. 2022). All the soil treatments showed significant growth-promoting effects on plant heights, number of leaves, leaf areas and stem girths as compared to the control plants. Interestingly, maize plants treated with cassava peels combined with *C. odorata* showed significant reduction in disease index and disease severity due to favourable competition with pathogenic fungi *F. verticillioides*. Bellini et al. (2023) reported that IPM strategies statistically reduced *Fusarium* wilt disease severity of lettuce, in both fields and two consecutive years, from 50% to 70% compared to untreated controls. Three treatments were applied, which included: (i) compost enriched with *Trichoderma*; (ii) a combination of *Trichoderma gamsii* + *T. asperellum*, *Bacillus amyloliquefaciens* and potassium phosphite; and (iii) a combination of *T. polysporum* + *T. atroviride*. All these treatments increased microbial diversity and crop yields of lettuce in comparison to untreated controls.

20.4　NEW STRATEGIES FOR SOIL AMENDMENTS: CLOSED-LOOP ORGANIC WASTE MANAGEMENT SYSTEMS IN AGRICULTURE

Traditional agriculture practices (non-judicial exploitation of resources and overuse of chemical fertilizers) have degraded our soil health (Ponisio et al. 2015). Among the various ill-effects caused, a low amount of organic matter is left in most soils. Different conventional sources of organic amendments are either losing their potential or produced in reduced amounts like cattle dung, which has become much less available over time for a variety of reasons (use in biogas production, plastering of the kachha homes and uses as fuel). There are numerous alternatives of soil organic amendments that have a great deal of potential to increase the nutrient level of soil and crop efficiency. There are around 300 million tonnes of alternate sources of soil organic matter available for soil amendments estimated at present. Some of these organic amendments sources are fly ash, tobacco dust, chicken manure, coir pith, crop residue, aquatic and terrestrial weed biomass, green manuring, animal bedding materials, seri-waste, agro industry waste, municipal biosolids, biochar, bamboo sphere and tank silt (Indoria et al. 2018). Use of all these resources and techniques in an interdependent way via closed circular economy techniques may prove promising for the Indian agriculture system. This section discusses some of these organic waste management systems via the circular economy in agriculture (Indoria et al. 2018; Chojnacka et al. 2022).

20.4.1 MANAGEMENT OF ORGANIC WASTE IN A CLOSED-LOOP OR CIRCULAR ECONOMY

Closed-loop technological systems or organic waste management systems may be defined as step by step interdependent stages of managing organic wastes to produce useful products like fuel and fertilizers; thus generating revenue out of organic waste along with mitigation of environmental issues. In today's context, our energy requirements are primarily achieved through fossil fuels, which in turn are getting depleting very quickly. On the other hand, most world markets have been linear and based on the take, make and dispose concept since the start of the industrial age. This results in the continuous extraction of resources for consumption and production as well as a lack of proper planning for reuse and economic regeneration (Kapoor et al. 2020).

By using circular economy ideas on their farms, farmers could increase their income sources and enhance agricultural productivity. However, the mostly rural communities engaged in agriculture are not aware of the advantages of recycling organic waste. Therefore governments and researchers should focus primarily to formulate potential design configurations for closed-loop systems for the management of organic waste and then to make these farmers aware about its ground-level reality (Awasthi et al. 2022). Different techniques used in closed-loop system include anaerobic digestion, pyrolysis, hydroponics, vermifiltration and use of digestate as fertilizer (Figure 20.2).

20.4.1.1 Anaerobic Digestion of Organic Waste

Anaerobic digestion (AD), which supports both circular economy and organic farming, is the method used for valorizing organic waste. In this process, biomass is transformed into digestate with the help of microbes, and this digestate is rich in nutrients and biogas. The application of digestate to the soil may replace chemical fertilizer and small-scale farmers can successfully use AD as a first step in closing the loop (Chojnacka et al. 2022). These techniques may also convert

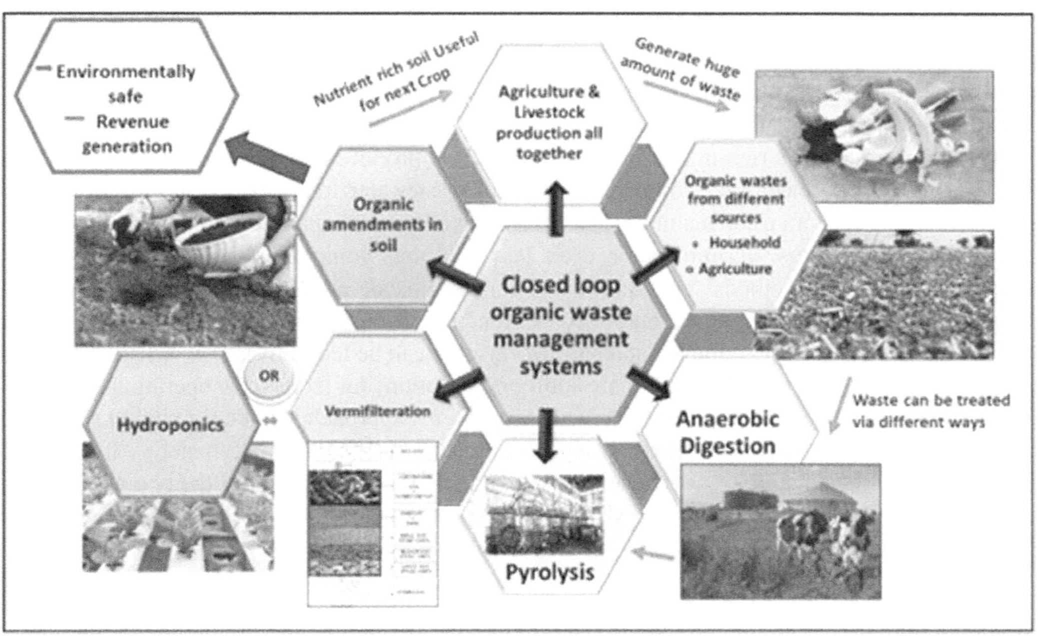

FIGURE 20.2 Various techniques interconnecting in closed-loop organic waste management systems for welfare of both farmers and environment.

Notes: The waste generated from agriculture and livestock production may be converted into simpler inorganic forms which can be further used to improve soil nutrients or direct in crop production (hydroponics) thus reducing dependency on chemical fertilizers and saving our environment.

organic discarded wastes into biogas for food preparation, heating or energy generation as well as organic fertilizer, and they are also reasonably easy to use. Due to its ability to lower nitrous oxide and methane emissions, AD is thought to be more environment-friendly than the direct application of manure to land. Additionally, by offering biogas as a substitute for firewood as a cooking fuel, deforestation can be addressed. Moreover, farmers can save their time consumed in searching for firewood and also less gas will be consumed by this way of cooking. In addition, the population of manure's inhabiting pathogens and odours can be minimized.

For less-suited ligneous wastes like cassava branches in AD, an alternative method is small-scale pyrolysis, which involves heating organic waste under anoxic conditions to create biochar. Infertile soil or areas with limited water supply may get more benefits from the application of biochar causing improvement in soil fertility. It is possible to sequester carbon by adding biochar to the soil, making this method of waste management carbon-negative. Additionally, it reduces fertilizer's nitrous oxide and ammonia emissions. Application of biochar boosts the soil's cation exchange capacity (CEC), which benefits farmers by increasing nutrient retention, preventing soil acidification and enhancing moisture retention (Ding et al. 2017). In addition, biochar increases the soil's microbiological fertility and plants' resistance to disease. These qualities increase the soil's resistance to climatic change and decrease the need for fertilizer, nutrient leaching, seed emergence, crop growth and production (Ghorbani et al. 2022; Osman et al. 2022).

Digestate from AD may act like a good fertilizer, as it contains all of the macro and micronutrients included in the feedstock. Several studies showed that utilization of digestate as fertilizer rather than raw animal manure resulted in comparable or even greater crop growth. Digested animal dung also improves nitrogen availability to plants compared to untreated animal manure, raw or composted crop leftovers. Thus, digestate provides more advantages over chemical fertilizers, including a lower carbon footprint and the ability to withstand drought stress due to its water-retentive properties. Hydroponics is another method for recovering nutrients from left-out digestate, in which plants are grown without soil and nutrients are fed through solution only.

20.4.1.2 Vermifiltration of Organic Waste

An alternative approach involves treatment of digestate by using a vermi-filter, a filtration system in which earthworms assist in the breakdown of nutrients, cleansing the digestate. Vermifiltration also produces compost that is rich in nutrients. Vermifiltration produces little to no sludge, is odourless and uses little to no electricity compared to typical wastewater treatment. Vermifiltration is less expensive to build, operate and maintain, and has similar treatment capabilities to the activated sludge type and trickling filters. Vermifiltration takes less land than competing ecological decentralized wastewater treatment methods. They are useful for small-scale agriculture since they can handle variations in input amount and the output for various irrigation regimes. In addition, vermi-filters provide value by producing compost and earthworms that can be fed to livestock or fish. Utilization of microalgae and struvite precipitation are additional solutions for large-scale operations (Van et al. 2022). In brief, the closed-loop waste management systems approach is not only vital to stop environmental pollution, but it also offers strong economic, social and climate benefits. It positively affects the soil's health and farmers benefit. Thus, suggested approaches have the potential to boost small farmers' income and resilience, while also promoting optimum waste management practices in farms and rural communities.

20.4.2 Organic Amendments with Tobacco Dust

Tobacco is mainly grown in warm climates and cultivated for its use by various cigarette companies. It is first dried and ground into fine powder form, which in turn generates a huge amount of tobacco dust. This agro-industrial waste is usually burnt or incinerated, thus being neglected for years as a potential source of many nutrients. Tobacco dust is rich in many macro and micronutrients including

2.35% N, 1.95% K, 937µg/g P, Fe, Ca, Mg and nicotine. It also has a high amount of organic carbon and, thus, it may be used for soil applications. Tobacco dust amendment is also helpful in increasing the pH of acidic soils, improving soil and iron structure, water intake capacity and maintaining the electrical conductivity and salinity of soils. Its application has been found to improve nutrients and the biomass content of many crops like wheat, paddy, apple and many vegetables including tomatoes. As it contains nicotine, it also has insecticidal properties preventing harmful insect's growth, ants nest in the garden, apple snail and even controlling tobacco mosaic virus infection, which is detrimental for many crops like cucumber, pepper and tobacco itself. So the left-out tobacco dust from industries can be efficiently used for increasing the nutrient value of soil in an eco-friendly manner (Shakeel 2014).

20.4.3 Nanoparticles-enabled Soil Amendment

Nanotechnology may be explored to enhance current or conventional methods of crop feeding and protection, depending on the desired function to increase global food production and sustainability (Dhanker et al. 2022). Nanoparticles are also useful for the improvement of soil, thus further helping the agriculture sector. Due to their high reactivity, ability to decrease and degrade contaminants, and high adsorption capacity, nanomaterials are being recommended as an environmental remediation technology with significant potential. A novel approach of soil restoration has been described that combines the safe cultivation of rice in pentachlorophenol (PCP)-contaminated paddy soil amended with nanoscale zero-valent iron (nZVI). Amendment with 100 mg of nZVI per kg of contaminated soil increased the rice grain output by 47 to 55%, in comparison to rice plants growing in polluted soil with 100 mg PCP per kg of soil lacking nZVI treatment (Liu et al. 2021).

Nanomaterials such as zeolites and nanoclays have been found to control the amount of water retained and released by soil. Additionally, enzymatic and/or microbial activities are also enhanced from changes made by the addition of nanomaterials (Agrawal et al. 2022). Amendment of zinc oxide and copper oxide in the soil may increase overall soil enzymatic activity without affecting the core-inhabiting microbial community. Although phosphorus lessening is a very common problem with iron amendment, the addition of nanoscale iron oxide has been reported to improve the soil's iron content without affecting the plants' ability to bind phosphorus. In the near future, the use of products available in the market will need justifications for the administration of macro and micronutrients as well as nano-enabled delivery methods for active components in existing pesticides and macro/micronutrients. Additionally, nanoscale delivery systems will be needed for the application of novel pesticides, biopesticides or genetic materials that may manage diseases that are not currently under control (Liu et al. 2021). Thus, still more field experiments or treatments are needed for conditioning of the field soil due to the complicated behaviour of the soil system, huge populations of microbial communities and our limited knowledge of the mechanisms governing the persistence, behaviour and functioning of various nanomaterials in soil (Kah et al. 2019).

20.4.4 Fly Ash

Fly ash is a fine-powdered coal-burning spent residue released from thermal power plants all over the world. Solid waste fly ash must be managed properly to protect our environment (Dwivedi and Jain 2014). India is generating around 150 million tonnes of fly ash from different coal-based thermal power plants, which are contributing to approximately 73% of power generation. Fly ash is a combustion residue having a heterogeneous mixture of many elements, such as Ca, Al, Na, Mg, Fe and Si as the principal elements. In addition, it contains a small amount of toxic metals, including Ni, V, As, Hg, Ba, Cr, Zn and Se.

The mixing of fly ash with other organic amendments, such as press mud, paper-factory sludge, cow dung, farmyard manure, organic compost, sewage sludge and crop residues, has been found to improve soil characteristics by reducing heavy metal contents, killing the pathogens of sludge, by increasing porosity, moisture content and nutrient concentrations, giving better texture, lowering the bulk density and enhancing the biological activity of farm soil. Application of fly ash doses in the range 20–100 tonnes/hectare has been found to improve crop yields in agriculture. Also, use of fly ash in ash ponds provide many micronutrients like Zn, Fe, Cu, Mn, Mo and B as well as macronutrients such as P, Ca, K, S and Mg to crops by reducing crust formation. Thus, fly ash can be treated as a valuable resource material to protect our environment (Dwivedi and Jain 2014).

20.4.5 Soil Amendment with Other Potential Organic Wastes

Food waste has become one of the world's major issues in the last decade and it is critical to develop long-term solutions and technologies to minimize its negative effects. Non-judicious dumping of vast quantities of food waste negatively affects human and ecosystem health, and also pollutes water, soil and air. Composting of food waste is confronted with certain technical difficulties, because of the greater moisture content, low C:N ratio, and challenging lignocellulosic contents, which eventually results in lengthy periods of composting and poor product quality. Adjusting the C:N ratio and moisture content with bulking agents like wood shavings, wheat straw and sawdust have been used to enhance the composting of food waste. Use of compost prepared from food waste mixed with left-out waste of bamboo (which is cheap and widely available in South India and China) has been found to increase the seed germination index and increases the soil organic matter. In addition, bamboo waste increased humification of compost by adding more aromatic and humic acid-like compounds (Wu et al. 2021).

Coir industries create huge amounts of coir pith, a spongy lignin containing cementing material made of fibres from coconut husk, which is accounting for 7.5 million tonnes of annual production in India. It is disposed of as unutilized trash in the environment, demanding appropriate and sustainable disposal techniques. The harmful coir pith can be transformed into beneficial vermi-compost and its amendment with green manure improves the vermi-compost fertilization index along with nutrient recovery. The addition of green manure and cow dung also enhances the growth and fertility of earthworms (Karmegam et al. 2021). The development of biorefinery using agricultural waste as feed has been reported recently, seeking towards circular bioeconomy (Awasthi et al. 2022). Therefore, there are still many unexplored organic wastes, which can be used as amendment to improve soil texture and nutrient content, for their subsequent use in improving crop production.

20.5 AGRI-WASTE USE FOR LIVESTOCK AND INDUSTRIAL PRODUCTS THROUGH VALUE ADDITION

Huge amounts of agriculture residue and waste are being generated to produce more food for the ever-increasing human population. An alternative way to manage these agricultural wastes in a sustainable way is their conversion into value-added products. Agricultural wastes can be categorized into four major types of wastes i.e. livestock waste, crop residues, aquaculture and agro-industrial. Cellulose is the main agricultural waste component followed by lignin and hemicellulose. Crop-wastes have traditionally been used for composting, animal fodder, combustion, soil mulching, matchsticks, roof thatching and production of paper (Atinkut et al. 2020). However, lignocellulosic biomass can also be used for the production of biofuel (biogas, bioethanol, biodiesel, biohydrogen) and bioenergy to resolve the issues of fossil fuel scarcity and climate change (Verma et al. 2014; Babu et al. 2022).

Recently, Deivayanai et al. (2022) presented the detailed technology involving pre-treatment strategy, technological advances and perspectives on biological conversion of lignocellulosic biomass into hydrogen, which may be used as biofuel. The increase in production of hydrogen through conversion of lignocellulosic biomass has also been achieved by applying a suitable nano-technological approach. Lignin-containing corn cob were used to convert into cellulose/hemicellulose raw material for a fermentation process using *Clostridium cellulovorans* and peroxidase mimicking $CeFe_3O_4$ nanoparticles at the rate of 4.0 g/L (Singhvi and Kim 2022). Fermentation with *C. cellulovorans* strain yielded maximum 78.45 mL of cumulative hydrogen with hydrogen production rate of 1.55 mL/h using nanoparticles-treated corn cob biomass.

In India, more than 130 million tonnes of rice-husk is produced, half of which is used as animal fodder and the other half of which is discarded in the environment or burnt by the poor farmers. But technology may be developed for the conversion of rice-husk into useful products after value addition including electricity and biofuel. Thus, conversion of lignocellulosic waste has the capacity to impact the economy by generating value-added products such as bio-coal, biofertilizers, bio-bricks, bioplastics, paper, biofuels, bioelectricity, organic acids and industrial enzymes (Babu et al. 2022; Koul et al. 2022).

20.5.1 Agri-waste Use as Livestock Feed

Agri-waste continues to be the primary feed material used to sustain dairy animals in developing countries. Primarily straw and stover retrieved after harvesting the main crop's part of maize, rice, sorghum, wheat and pearl millet can play an important role in meeting the fodder requirement. The leaves of vegetables and other thrown away parts can also be used as animal fodder. Rice straw also serves as an excellent animal feed and is the primary roughage material during the dry season. Rice straw contains 26–34% crude fibre content, 12–16% silica content and 3–5% protein content (Babu et al. 2022). Another major agricultural-based lignocellulosic residues viz. wheat straw possess 30–45% cellulose, 20–25% hemicellulose and 15–20% lignin along with a number of minor organic compounds. On the other hand, legume stover has a higher nutritional value than cereal straw (Wang et al. 2022). Agri-waste obtained from cereals in its crude form is not a high-quality feed because of low digestibility and consumption capability, and therefore, it requires supplementation with other sources as well as pre-treatment to convert it into valuable feed. Furthermore, high lignin contents in wheat straw (amounting to 15%) also prevent its microbial degradation in rumens and reduces its nutrient value resulting in multi-nutritional deficiencies in the animals (Babu et al. 2022).

Thus, pre-treatment of agricultural waste is essential for increasing the nutritional properties. Farm-waste can be pre-treated by one of four ways: chemical, physical, physico-chemical and biological treatment. Physical treatments further include cutting, chopping, soaking, grinding, boiling and irradiation treatment. Feed should be soaked in water after chopping it into smaller parts before feeding, as it increases digestibility and the intake of rice straw. Chemically, waste can be treated with urea, sodium and potassium hydroxide, whereas physico-chemical methods include steam treatment or combination of lime and urea palletization. Indigestible organic residues can also be treated biologically with enzymes or organic acids and also be added with probiotics to enhance nutritional values (Sethupathy 2022). The addition of some other agri-products like sweet potato chips, molasses, fermented edible oil cakes, leaves of *Gliricidia* and *Leucaena* to chopped straw was found to improve its intake and digestibility. Another encouraging recent technique involves the fermentation of agri-residues into useful animal feed using solid-state type fermentation, which results in the manufacture of protein-rich animal feedstock and it also removes anti-nutritional factors from agri-waste (Wang et al. 2022). The feed prepared by the transformation of agri-residue can be kept and used as feed for animals during the lean season. Crop residue can also be used to recover microbial enzymes and proteins for animal feed. Some other products like preparation of silage from cabbage have also been reported as a better alternative approach (Babu et al. 2022).

20.5.2 Industrial Applications of Agri-waste

Cellulose derived from plants is used for the manufacture of electronic parts, flexible displays, drug-transportation and other useful materials. Similarly, rice-husk can be used for the production of different types of enzymes such as xylanase, pectinase, ligninases, amylases and probiotics. Sugarcane bagasse (a byproduct of sugarcane industry), which contains 50% cellulose, 25% hemicellulose and 25% lignin, has the potential for use in the manufacture of hydrophobic or water resistance gel, carbon fibre, bioactive natural products, bioplastics, thermoplastic, foams, coatings and thermosetting materials with broad industrial uses (Li et al. 2021). Agri-waste also has enormous capacity in the paper-pulp and cardboard industries. Currently, pulp of wood is primarily used for production of cardboards, which uses only 4% of the total crop waste. However, the use of wood-pulp is not an environmentally friendly method, because it contributes to deforestation resulting in nature imbalance and climate change. Therefore, scientists should concentrate more on some green initiative options, which have the potential to reduce deforestation and prevent open burning of crop residue in the field.

High N-containing agri-waste chicken manure, a waste generated from poultry farms, can be converted into value-added third generation biofuel product. This chicken manure amended in soil can be used for the production of micro-algal biomass, which is further a potential raw material for the biofuel industry (Rajagopal et al. 2021). Bio-refineries may be established in different random regions that use raw materials, primarily plant biomass, animal and human wastes for further conversion into value-added products, and contributing to the sustainability of the animal, human beings and whole ecosystem (Yaashikaa et al. 2022).

20.6 FUTURE PERSPECTIVE AND CHALLENGES

Organic waste consisting of agri-residues is generated in ample amounts and its proper management has the potential to address many environmental issues. Future research should focus on developing advanced technologies for agri-waste management and the production of other value-added products. Of course, some critical efforts are needed in some neglected areas to make it more practical. Two main goals which need urgent attention include (i) augmented development of sustainable production systems, especially precision farming, and (ii) realization of the financial value of plant biomass, resulting into less accumulation of biowaste. During the planning and assessing stage of a biowaste management strategy, other environmental impacts should also be kept under consideration before execution of specific projects (Jain et al. 2022). Conversion of agri-waste into value-added products is another criterion where scientists, researchers and environmentalists may play a crucial role in the development or formulation of such efficient technologies.

The development of integrated approaches should be given more emphasis to produce manifold products, which will eventually decrease production costs by the efficient utilization of waste as substrates, thus achieving a zero-waste strategy to operate sustainable bioprocess technology (Koul et al. 2022). Currently, many challenges have been posed to scientists and environmentalists regarding the correct management, disposal and use of waste products, which includes: (i) money needed for efficient management of agri-waste, its collection and transportation to sites for treatment of agri-residues; (ii) lack of proper training of farmers regarding advance techniques and consequences of conventional practices; (iii) education awareness gap between scientific community and farmers, and industrialists in relation to environmental protection; (iv) greater scientific input needed from microbiologists regarding the selection of microbial strains, microbial metabolism, characterization of pathways involved and genetic engineering for improving the bioconversion of agri-residues into simpler useful products for use as industrial or commercial level (Yadav et al. 2021; Zhou et al. 2024); (v) low availability of motivational schemes and employment for researchers particularly in developing countries, like almost no patent facility and high registration costs; and (vi) use of wood and agri-residues for production of biodiesel or biofuels. Thus, collaborative efforts are needed from activists, educationalists, environmentalists, governments, the scientific community and farmers for the appropriate disposal of organic waste and its value addition. Currently, the 4R's principle is suggested (i.e. Reduce, Reuse, Recycle, and Recover) to attain sustainable collaborative alternative choices for the appropriate management of organic waste and agri-residues.

20.7 CONCLUSION

Bioeconomic strategies based on agriculture waste management may prevent the under-utilization of livestock manure and the illogical burning of crop residues to safeguard nutrition and health security, and waste transformation to produce value-added products. Organic soil amendments have been found to improve the availability of plant-utilizable nutrients leading to increased plant growth and crop yield. In addition, the organic amendments also contribute towards the bioremediation of pollutants and biocontrol of pathogens. Transformation of agricultural wastes and by-products into useful resources will help in creating employment opportunities and green markets. Furthermore, the appropriate management of agri-waste residues will also reduce emissions of greenhouse gases, minimize the environmental pollution and lessen the use of fossil fuels; thereby contributing to a clean, safe and sustainable environment. Thus, reducing, reusing and recycling organic and agricultural waste is a critical issue to minimize environmental pollution, preventing pressures on microbial biodiversity and global food security. Considering the rapid pace of progress in the field of value addition of organic and agriculture wastes, the recent application of advanced molecular techniques and their impact on plant and human health will help in developing climate smart agricultural technologies for sustainable agriculture. Of course, much more effort and research is needed for the commercialization of value-added products of organic wastes by industries and their acceptance by farmers. These innovative and eco-friendly agri-technologies will deliver long-term solutions for improving crop productivity with available resources and value-added products developed from agricultural and organic wastes.

REFERENCES

Agamuthu, P. 2009. Challenges and opportunities in agro-waste management: An Asian perspective. *Inaugural Meeting of First Regional 3R forum in Asia*, 11–12 November, Tokyo, 1–25.

Agrawal, S., Kumar, V., Kumar, S., and Shahi, S.K. 2022. Plant development and crop protection using phytonanotechnology: A new window for sustainable agriculture. *Chemosphere* 299, 134465. https://doi.org/10.1016/j.chemosphere.2022.134465

Ahmed, K.B.M., Khan, M.M.A., Siddiqui, H., and A. Jahan. 2020. Chitosan and its oligosaccharides, a promising option for sustainable crop production–A review. *Carbohydrate Polymers* 227:115331. doi: 10.1016/j.carbpol.2019.115331

Ali, A., Elrys, A. S., Liu, L., Xia, Q., Wang, B., Li, Y., Dan, X., Iqbal, M., Zhao, J., Huang, X., and Z. Cai. 2022. Deciphering the synergies of reductive soil disinfestation combined with biochar and antagonistic microbial inoculation in cucumber *Fusarium* wilt suppression through rhizosphere microbiota structure. *Microbial Ecology*. https://doi.org/10.1007/s00248-022-02097-3

Al-Naqeb, G., Sidarovich, V., Scrinzi, D., Mazzeo, I., Robbiati, S., Pancher, M., Fiori, L., and V. Adami. 2022. Hydrochar and hydrochar co-compost from OFMSW digestate for soil application: 3. Toxicological evaluation. *Journal of Environmental Management* 320:115910. https://doi.org/10.1016/j.jenvman.2022.115910

Aremu, A.O., Plaèková, L., Gruz, J., Bíba, O., Novák, O., and W.A. Stirk, et al. 2016. Seaweed-derived biostimulant (Kelpak§) influences endogenous cytokinins and bioactive compounds in hydroponically grown *eucomis autumnalis*. *Journal of Plant Growth Regulation* 35:151–162. doi: 10.1007/s00344-015-9515-8

Arioli, T., Mattner, S.W., and P.C. Winberg. 2015. Applications of seaweed extracts in Australian agriculture: past, present and future. *Journal of Applied Phycology* 27:2007–2015. doi: 10.1007/s10811-015-0574-9

Atinkut, H.B., Yan, T., Arega, Y., and M.H. Raza. 2020. Farmers' willingness-to-pay for eco-friendly agricultural waste management in Ethiopia: A contingent valuation. *Journal of Cleaner Production* 261:121211.

Avnimelech, Y. 1986. Organic residues in modern agriculture. In *The role of organic matter in modern agriculture*. Chen, Y., and Y. Avnimelech, eds., Developments in Plant and Soil Sciences, 25, pp. 1–10. Springer, Dordrecht. https://doi.org/10.1007/978-94-009-4426-8_1

Awasthi, M. K., Sindhu, R., Sirohi, R., Kumar, V., Ahluwalia, V., Binod, P., Juneja, A., Kumar, D., Yan, B., Sarsaiya, S. and Z. Zhang. 2022. Agricultural waste biorefinery development towards circular bioeconomy. *Renewable and Sustainable Energy Reviews* 158:112122.

Ayilara, M.S., Olanrewaju, O.S., Babalola, O.O., and O. Odeyemi. 2020. Waste management through composting: Challenges and potentials. *Sustainability* 12(11): 4456.

Babu, S., Rathore, S.S., Singh, R., Kumar, S., Singh, V.K., Yadav, S.K., Yadav, V., Raj, R., Yadav, D., Shekhawat, K., and O.A. Wani. 2022. Exploring agricultural waste biomass for energy, food and feed production and pollution mitigation: A review. *Bioresource Technology* 360:127566. https://doi.org/10.1016/j.biortech.2022.127566

Backer, R., Roken, J.S., Llangumaran, E., Lamont, J., Praslickova, D., Ricci, E., Subramanian, S., and D.L. Smith. 2018. Plant growth-promoting rhizobacteria: Context, mechanism of action and roadmap to commercialization of biostimulants for sustainable agriculture. *Frontiers in Plant Sciences* 9, article 1473. https://doi 10.3389/fpls.2018.01473

Badillo, T. P. S., Pham, T. T. H., Nadeau, M., Allard-Massicotte, R., Jacob-Vaillancourt, C., Heitz, M., and A. A. Ramirez. 2021. Production of plant growth-promoting bacteria inoculants from composting leachate to develop durable agricultural ecosystems. *Environmental Science and Pollution Research* 28:29037–29045. https://doi.org/10.1007/s11356-019-06135-5

Bandara, S., Du, H., Carson, L., Bradford, D., and R. Kommalapati. 2020. Agricultural and biomedical applications of chitosan-based nanomaterials. *Nanomaterials (Basel)* 24;10(10):1903. doi: 10.3390/nano10101903

Bargaz, A., Lyamlouli, K., Chtouki, M., Zeroual, Y., and D. Dhiba. 2018. Soil microbial resources for improving fertilizers efficiency in an integrated plant nutrient management system. *Frontiers in Microbiology* 9:1606.

Bekele, M., and G., Yilma. 2021. Nitrogen fixation using symbiotic and non-symbiotic microbes: A review article. *Biochemistry and Molecular Biology* 6(4):92–98. doi: 10.11648/j.bmb.20210604.12

Bellini, A., Gilardi, G., Idbella, M., Zotti, M., Pugliese, M., Bonanomi, G., and M. L. Gullino. 2023. *Trichoderma* enriched compost, BCAs and potassium phosphite control Fusarium wilt of lettuce without affecting soil microbiome at genus level. *Applied Soil Ecology* 182:104678. https://doi.org/10.1016/j.apsoil.2022.104678

Bhat, R.A., Dar, S.A., Dar, D.A., and G.H. Dar. 2018. Municipal solid waste generation and current scenario of its management in India. *International Journal of Advance Research in Science and Engineering* 7(02):419–431.

Bian, B., Lv, L., Yang, D., and L. Zhou. 2014. Migration of heavy metals in vegetable farmlands amended with biogas slurry in the Taihu Basin, China. *Ecological Engineering* 71:380–383.

Bornø, M. L., Muller-Stover, D. S., and Liu, F. 2022. Biochar modifies the content of primary metabolites in the rhizosphere of well-watered and drought-stressed *Zea mays* L. (maize). *Biology and Fertility of Soils* 58:633–647. https://doi.org/10.1007/s00374-022-01649-6

Chalamaiah, M., Dinesh Kumar, B., Hemalatha, R., and T. Jyothirmayi. 2012. Fish protein hydrolysates: proximate composition, amino acid composition, antioxidant activities and applications: a review. *Food Chemistry* 135:3020–3038. doi: 10.1016/j.foodchem.2012.06.100

Chander, G., Wani, S.P., Gopalakrishnan, S., Mahapatra, A., Chaudhury, S., Pawar, C.S., Kaushal, M., and A.V.R. Rao. 2018. Microbial consortium culture and vermi-composting technologies for recycling on-farm wastes and food production. *International Journal of Recycling of Organic Waste in Agriculture* 7(2):99–108.

Chaudhary, S., and S.S. Sindhu. 2024. Microbiome-mediated remediation of heavy metals: Impact on soil health, crop production, and ecosystem sustainability. In *Microbiome-Assisted Bioremediation: Rehabilitating Agricultural Soils*, J.A. Parray., and W-J. Li, eds., pp. 257–311. Elsevier-Academic Press. https://doi.org/10.1016/B978-0-443-21911-5.00019-2

Chen, B., Zhou, D., Zhu, L., and X. Shen. 2008. Sorption characteristics and mechanisms of organic contaminant to carbonaceous biosorbents in aqueous solution. *Science in China Series B: Chemistry* 51(5):464–472.

Chen, D., Zhou, W., Yang, J., Ao, J., Huang, Y., Shen, D., Jiang, Y., Huang, Z., and H. Shen. 2021. Effects of seaweed extracts on the growth, physiological activity, cane yield and sucrose content of sugarcane in China. *Frontiers in Plant Sciences* 12:659130. doi: 10.3389/fpls.2021.659130

Chen, G., Zhao, G., Zhang, H., Shen, Y., Fei, H., and W. Cheng. 2017. Biogas slurry use as N fertilizer for two-season *Zizania aquatic* Turcz. in China. *Nutrient Cycling in Agroecosystems* 107(3):303–320.

Cherif, H., Ayari, F., Ouzari, H., Marzorati, M., Brusetti, L., Jedidi, N., Hassen, A., and D. Daffonchio. 2009. Effects of municipal solid waste compost, farmyard manure and chemical fertilizers on wheat growth, soil composition and soil bacterial characteristics under Tunisian arid climate. *European Journal of Soil Biology* 45(2):138–145.

Chianese, S., Fenti, A., Iovino, P., Musmarra, D , and S. Salvestrini. 2020. Sorption of organic pollutants by humic acids: A review. *Molecules* 25:918. doi:10.3390/molecules25040918

Chojnacka, K., Moustakas, K., and M. Mikulewicz. 2022. Valorisation of agri-food waste to fertilisers is a challenge in implementing the circular economy concept in practice. *Environment Pollution* 119906. https://doi.org/10.1016/j.envpol.2022.119906

Clark, M., and D. Tilman. 2017. Comparative analysis of environmental impacts of agricultural production systems, agricultural input efficiency, and food choice. *Environmental Research Letters* 12(6):064016.

Colla, G., Svecova, E., Rouphael, Y., Cardarelli, M., Reynaud, H., and R. Canaguier. et al. 2013. Effectiveness of a plant-derived protein hydrolysate to improve crop performances under different growing conditions. *Acta Horticulture* 1009:175–179. doi: 10.17660/ActaHortic.2013.1009.21

Colla, G., Cardarelli, M., Bonini, P., and Y. Rouphael. 2017a. Foliar applications of protein hydrolysate, plant and seaweed extracts increase yield but differentially modulate fruit quality of greenhouse tomato. *HortScience* 52: 1214–1220. doi: 10.21273/hortsci12200-17

Colla, G., Hoagland, L., Ruzzi, M., Cardarelli, M., Bonini, P., Canaguier, R., and Y. Rouphael. 2017b. Biostimulant action of protein hydrolysates: Unraveling their effects on plant physiology and microbiome. *Frontiers in Plant Sciences* 8:2202. doi: 10.3389/fpls.2017.02202

Dadarwal, K.R., Yadav, K.S., and S.S. Sindhu. 1997. Biofertilizer production technology: Prospects. In *Biotechnological Approaches in Soil Microorganisms for Sustainable Crop Production,* K. R. Dadarwal, ed., pp. 323–337. Scientific Publishers, Jodhpur.

Dahiya, A., Kumar, R., and S.S. Sindhu. 2021. Microbial endophytes mediated phosphorus solubilization: Sustainable approach to improve soil fertility and plant growth. In *Endophytes: Mineral nutrients management in the series "Sustainable Development and Biodiversity"*, D.K. Maheshwari, eds., pp. 35–75. Springer Nature, Gewerbestrasse, Switzerland. https://doi.org/10.1007/978-3-030-65447-4_3

Deivayanai, V.C., Yaashikaa, P.R., Senthil Kumar, P., and G. Rangasamy. 2022. A comprehensive review on the biological conversion of lignocellulosic biomass into hydrogen: Pretreatment strategy, technology advances and perspectives. *Bioresource Technology* 365:128166. https://doi.org/10.1016/j.biortech.2022.128166

Dhanker, R., Rawat, S., Chandna. V., Deepa, Das, S., Sharma, A., Kumar, V., and R. Kumar. 2022. Recovery of silver nanoparticles and management of food wastes: obstacles and opportunities. *Environmental Advances* 9, 100303. https://doi.org/10.1016/j.envadv.2022.100303

Diacono, M., and F. Montemurro. 2011. Long-term effects of organic amendments on soil fertility. In *Sustainable agriculture 2.* Lichtfouse, E., Hamelin, M., Navarrete, M., and P. Debaeke, eds., pp. 761–786. Springer, Dordrecht. https://doi.org/10.1007/978-94-007-0394-0_34

Ding, Y., Liu, Y., Liu, S., Huang, X., Li, Z., Tan, X., Zeng, G., and L. Zhou. 2017. Potential benefits of biochar in agricultural soils: A review. *Pedosphere* 27(4):645–661. https://doi.org/10.1016/S1002-0160(17)60375-8

Du Jardin, P. 2015. Plant biostimulants: Definition, concept, main categories and regulation. *Scientia horticulturae* 196:3–14.

Duan, H., Ji, M., Chen, A., Zhang, B., Shi, J., Liu, L., Li, X., and J. Sun. 2021. Evaluating the impact of rice husk on successions of bacterial and fungal communities during cow manure composting. *Environmental Technology Innovation* 24:102084.

Dwivedi, A., and M.K. Jain. 2014. Fly ash-waste management and overview: A Review. *Recent Research in Science and Technology* 6(1):30–35.

Ehrlicha, P.R., and J. Harteb. 2015. To feed the world in 2050 will require a global revolution. *Proceedings of National Academy Sciences, USA* 112(48):14743–14744. www.pnas.org/cgi/doi/10.1073/pnas.1519841112

Elekhtyar, N.M., Awad-Allah, M.M.A., Alshallash, K.S., Alatawi, A., Alshegaihi, R.M., and R.A. Alsalmi. 2022. Impact of arbuscular mycorrhizal fungi, phosphate solubilizing bacteria and selected chemical phosphorus fertilizers on growth and productivity of rice. *Agriculture* 12:1596. https://doi.org/10.3390/agriculture12101596

Erhart, E., Hartl, W., and B. Putz. 2005. Biowaste compost affects yield, nitrogen supply during the vegetation period and crop quality of agricultural crops. *European Journal of Agronomy* 23(3):305–314.

Foong, S. Y., Chan, Y. H., Chin, B. L. F., Lock, S. S. M., Yee, C. Y., Yiin, C. L., Peng, W., and S. S. Lam. 2022. Production of biochar from rice straw and its application for wastewater remediation – An overview. *Bioresource Technology* 360:127588. https://doi.org/10.1016/j.biortech.2022.127588

Fushimi, C., Araki, K., Yamaguchi, Y., and A. Tsutsumi. 2003. Effect of heating rate on steam gasification of biomass. 1. *Reactivity of char. Industrial and Engineering Chemistry Research* 42(17):3922–3928.

Ghorbani, M., Konvalina, P., Neugschwandtner, R.W., Kopecký, M., Amirahmadi, E., Bucur, D., and A. Walkiewicz. 2022. Interaction of biochar with chemical, green and biological nitrogen fertilizers on nitrogen use efficiency indices. *Agronomy* 12:2106. https://doi.org/10.3390/agronomy12092106

Gomlero, T., Pimentel, D., and M.G. Paoletti. 2011. Is there a need for a more sustainable agriculture? *CRC Critical Review in Plant Sciences* 30:6–23.

Goyal, S., and S.S. Sindhu. 2011. Composting of rice straw using different inocula and analysis of compost quality. *Microbiology Journal* 1:126–138.

Green, R.E., Cornell, S.J., Scharlemann, J.P.W., and A. Blamford. 2005. Farming and the fate of wild nature. *Science* 307:550–555.

Gupta, S., Krishna, M., Prasad, R. K., Gupta, S., and A. Kansal. 1998. Solid waste management in India: options and opportunities. *Resource Conservation Recycling* 24:137–154.

Hai, H.T., and N.T.A. Tuyet. 2010. *Benefits of the 3R approach for agricultural waste management (AWM) in Vietnam: Under the framework of joint project on Asia Resource Circulation Research.*

Ibrahim, M.H., Quaik, S. and S.A. Ismail. 2016. An introduction to anaerobic digestion of organic wastes. In *Prospects of organic waste management and the significance of earthworms,* Applied Environmental Science and Engineering for a Sustainable Future, pp. 23–44. Springer, Cham. https://doi.org/10.1007/978-3-319-24708-3_2

Indoria, A.K., Sharma, K.L., Reddy, K.S., Srinivasarao, C., Srinivas, K., Balloli, S.S., Osman, M., Pratibha, G., and N.S. Raju. 2018. Alternative sources of soil organic amendments for sustaining soil health and crop productivity in India-impacts, potential availability, constraints and future strategies. *Current Science* 115(11):2052–2062.

Insam, H. and M. De Bertoldi. 2007. Microbiology of the composting process. In *Waste management series,* Diaz L.F., de Bertoldi M., Bidlingmaier W. and E. Stentiford, eds., 8, pp. 25–48. Elsevier. https://doi.org/10.1016/S1478-7482(07)80006-6

IRENA. 2015. Renewable Energy in the Water, Energy & Food Nexus. International Renewable Energy Agency: 2015. www.irena.org/documentdownloads/publications/irena_water_energy_food_nexus_2015.pdf (accessed Oct, 2022).

Isaac, S.R., and A.M. George. 2022. Recycling of leaf litters: Biowaste management for resource conservation. *Agricultural Reviews.* doi: 10.18805/ag.R-2433

Jain, A., Sarsaiya, S., Awasthi, M.K., Singh, R., Rajput, R., Mishra, U.C., Chen, J., and J. Shi. 2022. Bioenergy and bio-products from bio-waste and its associated modern circular economy: Current research trends, challenges, and future outlooks. *Fuel* 307:121859.

James, G., Sabatini, D.A., Chiou, C.T., Rutherford, D., Scott, A.C., and H.K. Karapanagioti. 2005. Evaluating phenanthrene sorption on various wood chars. *Water Research* 39(4):549–558.

Joseph, S., Peacocke, C., Lehmann, J., and P. Munroe. 2009. Developing a biochar classification and test methods. *Biochar for Environmental Management: Science and Technology* 1:107–126.

Kah, M., Tufenkji, N., and J.C. White. 2019. Nano-enabled strategies to enhance crop nutrition and protection. *Nature Nanotechnology* 14(6):532–540.

Kang, S.-M., Shahzad, R., Khan, M.A., Hasnain, Z., Lee, K.-E., Park, H.-S., Kim, L.-R., and I.-J. Lee. 2021. Ameliorative effect of indole-3-acetic acid- and siderophore-producing *Leclercia adecarboxylata* MO1 on cucumber plants under zinc stress. *Journal of Plant Interactions* 16:30–41, doi: 10.1080/17429145.2020.1864039

Kapoor, R., Ghosh, P., Kumar, M., Sengupta, S., Gupta, A., Kumar, S.S., Vijay, V., Kumar, V., Vijay, V.K., and D. Pant. 2020. Valorization of agricultural waste for biogas based circular economy in India: A research outlook. *Bioresource Technology* 304:123036.

Karagöz, S., Bhaskar, T., Muto, A., Sakata, Y., Oshiki, T., and T. Kishimoto. 2005. Low-temperature catalytic hydrothermal treatment of wood biomass: analysis of liquid products. *Chemical Engineering Journal* 108(1–2):127–137.

Karmegam, N., Jayakumar, M., Govarthanan, M., Kumar, P., Ravindran, B., and M. Biruntha. 2021. Precomposting and green manure amendment for effective vermitransformation of hazardous coir industrial waste into enriched vermicompost. *Bioresource Technology* 319:124136.

Kauffman, G.L., Kneivel, D.P., and T.L.Watschke. 2007. Effects of a biostimulant on the heat tolerance associated with photosynthetic capacity, membrane thermostability, and polyphenol production of perennial ryegrass. *Crop Science* 47(1):261–267.

Kawaka, F. 2022. Characterization of symbiotic and nitrogen fixing bacteria. *AMB Express* 30(12):99. doi: 10.1186/s13568-022-01441-7

Khan, N., Chowdhary, P Ahmad, A., Giri, B.S., and P. Chaturvedi. 2020. Hydrothermal liquefaction of rice husk and cow dung in mixed-bed-rotating pyrolyzer and application of biochar for dye removal. *Bioresource Technology* 309:123294.

Khan, N., Chowdhary, P., Gnansounou, E., and P. Chaturvedi. 2021. Biochar and environmental sustainability: emerging trends and techno-economic perspectives. *Bioresource Technology* 332:125102.

Koul, B., Yakoob, M., and M.P. Shah. 2022. Agricultural waste management strategies for environmental sustainability. *Environmental Research* 206:112285.

Kulikova, N.A., and I.V. Perminova. 2021. Interactions between humic substances and microorganisms and their implications for nature-like bioremediation technologies. *Molecules* 6:2706. https://doi.org/ 10.3390/molecules26092706

Kumar, A., and R. Chandra. 2020. Ligninolytic enzymes and its mechanisms for degradation of lignocellulosic waste in environment. *Heliyon* 6:e03170. https://doi.org/10.1016/j.heliyon.2020.e03170

Kumar, S., Chaudhary, T., Diksha, Sindhu, S.S., Chaudhary, D., and R. Kumar. 2022a. Mycorrhizal fungi: An eco-friendly input for sustenance of soil fertility and plant health. In *Microbes for humanity and its applications*, Malik, D.K., Rathi, M., Kumar, R., and D. Bhatia, eds., pp. 21–74. Daya Publishing House, New Delhi.

Kumar, S., Diksha, Sindhu, S.S., and R. Kumar. 2022b. Biofertilizers: An ecofriendly technology for nutrient recycling and environmental sustainability. *Current Research in Microbial Sciences* 3:100094. https:// doi.org/10.1016/j.crmicr.2021.100094

Ladha, J.K., Peoples, M.B., Reddy, P.M., Biswas, J.C., Bennett, A., Jat, M.L., and T.J. Krupnik. 2022. Biological nitrogen fixation and prospects for ecological intensification in cereal-based cropping systems. *Field Crops Research* 283:108541. https://doi.org/10.1016/j.fcr.2022.108541

Lal, R. 2006. Enhancing crop yields in the developing countries through restoration of the soil organic carbon pool in agricultural lands. *Land Degradation and Development* 17:197–209.

Li, J., Liu, Z., Feng, C., Liu, X., Qin, F., Liang, C., Bian, H., Qin, C., and S. Yao. 2021. Green, efficient extraction of bamboo hemicellulose using freeze-thaw assisted alkali treatment. *Bioresource Technology* 333:125107.

Lin, L., Xu, F., Ge, X., and Y. Li. 2018. Improving the sustainability of organic waste management practices in the food-energy-water nexus: A comparative review of anaerobic digestion and composting. *Renewable and Sustainable Energy Reviews* 89: 151–167.

Liu, X., Ji, R., Shi, Y., Wang, F., and W. Chen. 2019. Release of polycyclic aromatic hydrocarbons from biochar fine particles in simulated lung fluids: implications for bioavailability and risks of airborne aromatics. *Science of the Total Environment* 655:1159–1168.

Liu, Y., Wu, T., White, J.C., and D. Lin. 2021. A new strategy using nanoscale zero-valent iron to simultaneously promote remediation and safe crop production in contaminated soil. *Nature Nanotechnology* 16(2):197–205.

Marfà, O., Cáceres, R., Polo, J., and J. Ródenas. 2009. Animal protein hydrolysate as a biostimulant for transplanted strawberry plants subjected to cold stress. *Acta Horticulture* 842:315–318. doi: 10.17660/ ActaHortic.2009.842.57

Masulili, A., Utomo, W.H., and M.S. Syechfani. 2010. Rice husk biochar for rice based cropping system in acid soil 1. The characteristics of rice husk biochar and its influence on the properties of acid sulfate soils and rice growth in West Kalimantan, Indonesia. *Journal of Agricultural Science* 2(1):39.

Meena, V.S., Maurya, B.R., Verma, J.P., Aeron, A., Kumar, A., Kim, K., and V.K. Bajpai. 2015. Potassium solubilizing rhizobacteria (KSR): isolation, identification, and K-release dynamics from waste mica. *Ecological Engineering* 81:340–347. doi:10.1016/j. ecoleng.2015.04.065

Mohamed, T. A., Wu, J., Zhao, Y., Elgizawy, N., El Kholy, M., Yang, H., Zheng, G., Mu, D., and Z. Wei. 2022. Insights into enzyme activity and phosphorus conversion during kitchen waste composting utilizing phosphorus-solubilizing bacterial inoculation. *Bioresource Technology* 362:127823. https://doi.org/10.1016/j.biortech.2022.127823

Mujtaba, M., Khawar, K.M., Camara, M.C., Carvalho, L.B., Fraceto, L.F., R.E. Morsi, et al. 2020. Chitosan-based delivery systems for plants: A brief overview of recent advances and future directions. *International Journal of Biological Macromolecules* 154:683–697. doi: 10.1016/j.ijbiomac.2020.03.128

Obi, F.O., Ugwuishiwu, B.O., and J.N. Nwakaire. 2016. Agricultural waste concept, generation, utilization and management. *Nigerian Journal of Technology* 35(4):957–964.

Oerke, E.C. 2006. Crop losses to pests. *Journal of Agricultural Sciences* 144:31–43.

Ogbonnaya, U., and K.T. Semple. 2013. Impact of biochar on organic contaminants in soil: a tool for mitigating risk? *Agronomy* 3(2):349–375.

Olajumoke, A., Dorcas, A., Moses, A., Adegboyega, O., and S. Ayodele. 2022. The effect of organic soil amendments on stalk rot of maize caused by *Fusarium verticillioides*. *African Journal of Microbiology Research* 16(10):301–308. https://doi.org/10.5897/AJMR2022.9630

Osman, A. I., Fawzy, S., Farghali, M., El-Azazy, M., Elgarahy, A.M., Fahim, R.A., Abdel Maksoud, M.I.A, Ajlan, A.A., Yousry, M., Saleem, Y., and D.W. Rooney. 2022. Biochar for agronomy, animal farming, anaerobic digestion, composting, water treatment, soil remediation, construction, energy storage, and carbon sequestration: A review. *Environmental Chemistry Letters* 20:2385–2485. https://doi.org/10.1007/s10311-022-01424-x

Parmar, P., and S.S. Sindhu. 2019. The novel and efficient method for isolating potassium solubilizing bacteria from rhizosphere soil. *Geomicrobiology Journal* 36(2):130–136.

Pathak, H., Kushwaha, J.S., and M.C. Jain. 1992. Eyahiation of manurial value of biogas spent slurry composted with dry mango leaves, wheat straw and rock phosphate on wheat crop. *Journal of the Indian Society of Soil Science* 40(4):753–757.

Phour, M., and S.S. Sindhu. 2022. Mitigating abiotic stress: Microbiome engineering for improving agricultural production and environmental sustainability. *Planta* 256:85. https://doi.org/10.1007/s00425-022-03997-x

Phour, M., and S.S. Sindhu. 2023. Soil salinity and climate change: Microbiome-based strategies for mitigation of salt stress to sustainable agriculture. In *Climate Change and Microbiome Dynamics. Climate Change Management,* J.A. Parray, eds., Springer, Cham. https://doi.org/10.1007/978-3-031-21079-2_13

Phour, M., and S.S. Sindhu. 2024. Arbuscular mycorrhizal fungi: An eco-friendly technology for alleviation of salinity stress and nutrient acquisition in sustainable agriculture. In *Advances in Arbuscular Mycorrhizal Fungal Technology for Sustainable Agriculture II: Nutrient and Crop Management,* Parihar, M., Rakshit, A., Adholeya, A., and Y. Chen, eds., Chapter 11. Springer Nature-Singapore Pte Ltd. https://doi.org/10.1007/978-981-97-0300-5_11

Ponisio, L.C., M'Gonigle, L.K., Mace, K.C., Palomino, J., de Valpine, P., and C. Kremen. 2015. Diversification practices reduce organic to conventional yield gap. *Proceedings of Royal Society, B.* 282:20141396.

Pu, C., Liu, H., Ding, G., Sun, Y., Yu, X., Chen, J., Ren, J., and X. Gong. 2018. Impact of direct application of biogas slurry and residue in fields: *in situ* analysis of antibiotic resistance genes from pig manure to fields. *Journal of Hazardous Materials* 344:441–449.

Rahman, L., and J. Goswami. 2021. Recent development on physical and biological properties of chitosan-based composite films with natural extracts: A review. *The Journal of Bioactive and Compatible Polymers* 36(3):225–236. DOI: 10.1177/08839115211014218

Rajagopal, R., Mousavi, S.E., Goyette, B., and S. Adhikary. 2021. Coupling of microalgae cultivation with anaerobic digestion of poultry wastes: Toward sustainable value added bioproducts. *Bioengineering* 8(5):57.

Reganold, J.P., and J.M. Wachter. 2016. Organic agriculture in the twenty-first century. *Nature Plants* 2:15221.

Richardson, A.E., and R.J. Simpson. 2011. Soil microorganisms mediating phosphorus availability: Phosphorus plant physiology. *Plant Physiology (Bethesda)* 156(3):989–996.

Rodrigues, Y.F., Andreote, F.D., Silva, A.M.M., Dias, A.C.F., Taketani, R.G., and S.R. Cotta. 2023. Disentangling the role of soil bacterial diversity in phosphorus transformation in the maize rhizosphere. *Applied Soil Ecology* 182:104739. https://doi.org/10.1016/j.apsoil.2022.104739

Ruan, H., Yang, Z., Lin, J., Shen, J., Ji, J., Gao, C., and B. van der Bruggen. 2015. Biogas slurry concentration hybrid membrane process: Pilot-testing and RO membrane cleaning. *Desalination* 368:171–180.

Salvador, G.L.O., Araujo, F.F., Pereira, A.P. de A., Bonifacio, A., and A.S.F. Araujo. 2022. Rhizobacteria and arbuscular mycorrhizal fungus presented distinct and specific effects on soybean growth when inoculated with organic compost. *Rhizosphere* 22:100513, https://doi.org/10.1016/j.rhisph.2022.100513

Schmidt, H-P., Kammann, C., Niggli, C., Evangelou, M.W.H., Mackie, K.A., and S. Abiven. 2014. Biochar and biochar-compost as soil amendments to a vineyard soil: Influences on plant growth, nutrient uptake, plant health and grape quality. *Agriculture, Ecosystems & Environment* 191:117–123. https://doi.org/10.1016/j.agee.2014.04.001

Sehrawat, A., and S.S. Sindhu. 2019. Potential of biocontrol agents in plant disease control for improving food safety. *Defence Life Science Journal* 4:220–225.

Sehrawat, A., Phour, M., Kumar, R., and S.S. Sindhu. 2021. Bioremediation of pesticides: An eco-friendly approach for environmental sustainability. In *Microbial Rejuvenation of Polluted Environment. Microorganisms for Sustainability,* D.G. Panpatte, Y.K. Jhala, eds. vol. 25, pp. 23–84. Springer Nature, Singapore Pte Ltd. https://doi.org/10.1007/978-981-15-7447-4_2

Sehrawat, A., Sindhu, S.S., and B.R. Glick. 2022. Hydrogen cyanide production by soil bacteria: Biological control of pests and plant growth promotion. *Pedosphere* 32(1):15–38. doi: 10.1016/S1002-0160(21)60058-9

Sethupathy, S., Morales, G.M., Gao, L., Wang, H., Yang, B., Jiang, J., Sun, J., and D. Zhu. 2022. Lignin valorization: status, challenges and opportunities. *Bioresource Technology* 347:126696. doi: 10.1016/j.biortech.2022.126696

Seufert, V., Ramankutty, N., and J.A. Foley. 2012. Comparing the yields of organic and conventional agriculture. *Nature* 485:229–232.

Shah, A., Nazari, M., Antar, M., Msimbira, L. A., Naamala, J., Lyu, D., Rabileh, M., Zajonc, J., and D. L. Smith. 2021. PGPR in agriculture: A sustainable approach to increasing climate change resilience. *Frontiers in Sustainable Food Systems* 5:667546. doi: 10.3389/fsufs.2021.667546

Shakeel, S. 2014. Consideration of tobacco dust as organic amendment for soil: A soil & waste management strategy. *Earth Sciences* 3:117–121.

Shalini, V.B., Maheswari, A.U., Marimuthu, C. and J. Jeshima. 2022. Vermi-Composting using AI in IoT. In *2022 International Conference on Applied Artificial Intelligence and Computing (ICAAIC)* pp. 1489–1493. IEEE.

Sharholy, M., Ahmad, K., Vaishya, R.C. and R.D. Gupta. 2007. Municipal solid waste characteristics and management in Allahabad, India. *Waste management* 27(4):490–496.

Sharma, R., Dahiya, A., and S.S. Sindhu. 2019. Harnessing proficient rhizobacteria to minimize the use of agrochemicals. *International Journal of Current Microbiology and Applied Sciences* 7(10):3186–3197.

Sharma, R., Sindhu, S.S., and B.R. Glick. 2024. Potassium solubilizing microorganisms as potential biofertilizer: A sustainable climate-resilient approach to improve soil fertility and crop production in agriculture. *Journal of Plant Growth Regulation.* https://doi.org/10.1007/s00344-024-11297-9

Sheng, Y., Benmati, M., Guendouzi, S., Benmati, H., Yuan, Y., Song, J., Xia, C., M. Berkani. 2022. Latest eco-friendly approaches for pesticides decontamination using microorganisms and consortia microalgae: A comprehensive insights, challenges, and perspectives. *Chemosphere* 308:136183. https://doi.org/10.1016/j.chemosphere.2022.136183

Sindhu S, Sindhu D., and S.K. Yadav. 2021. Data mining and phylogenetic analysis of NifH protein of *Azospirillum* strain among nitrogen-fixing bacteria using bioinformatics tools. *International Journal of Computer Sciences and Engineering* 9(1):1–10.

Sindhu, S.S. 2004. Biological nitrogen fixation. *In Research methods in plant sciences: Allelopathy, Volume 1, Soil analysis.* Narwal, S.S., Dahiya, S.S., and J.P. Singh., eds., pp. 326–342. Scientific Publishers, Jodhpur.

Sindhu S.S., and R. Sharma. 2020. Plant growth promoting rhizobacteria (PGPR): A sustainable approach for managing soil fertility and crop productivity. In *Microbes for humankind and application*, Malik, D.K., Rathi, M., Kumar, R., and D. Bhatia., eds., pp. 97–130. Daya Publishing House, New Delhi.

Sindhu, S.S., Verma, N., and S. Goyal. 2013. Ecofriendly utilization of crop residues and organic waste material. *In: Human and animal health: environmental perspectives.* Garg, S. R. ed. Satish Serial Publishing House, New Delhi, pp. 385–404.

Sindhu, S.S., Parmar, P., and M. Phour. 2014a. Nutrient cycling: potassium solubilization by microorganisms and improvement of crop growth. In *Geomicrobiology and biogeochemistry,* N. Parmar and A.K. Singh, eds., pp. 175–198. Springer, Berlin, Heidelberg.

Sindhu, S.S., Phour, M., Choudhary, S.R., and D. Chaudhary. 2014b. Phosphorus cycling: prospects of using rhizosphere microorganisms for improving phosphorus nutrition of plants. In *Geomicrobiology and Biogeochemistry.* pp. 199–237. Springer, Berlin, Heidelberg.

Sindhu, S.S., Parmar, P., Phour, M., and A. Sehrawat. 2016. Potassium solubilizing microorganisms (KSMs) and its effect on plant growth improvement. In *Potassium solubilizing microorganisms (KSMs),* Meena, V.S., Verma, J.P., Maurya, B.R., and R.S. Meena, eds., pp. 171–185. Springer- Verlag, Berlin, Heidelberg.

Sindhu, S.S., Sehrawat, A., Sharma, R., and A. Dahiya. 2016. Biopesticides: Use of rhizosphere bacteria for biological control of plant pathogens. *Defence Life Science Journal* 1:135–148.

Sindhu, S.S., Sharma, R., Sindhu, S., and Sehrawat, A. 2019. Soil fertility improvement by symbiotic rhizobia for sustainable agriculture. In *Soil fertility management for sustainable development,* Panpatte, D.G., and Y.K. Jhala, eds., pp. 101–166. Springer Nature, Singapore Pvt Ltd. https://doi.org/10.1007/978-981-13-5904-0_7

Sindhu, S.S., Sehrawat, A., and B.R. Glick. 2022a. The involvement of organic acids in soil fertility, plant health and environment sustainability. *Archives of Microbiology* 204:720.https://doi.org/10.1007/s00 203-022-03321-x

Sindhu, S.S., Sehrawat, A., Phour, M., and R. Kumar. 2022b. Nutrient acquisition and soil fertility: Contribution of rhizosphere microbiomes in sustainable agriculture. In *Microbial biotechnology for sustainable agriculture volume 1. microorganisms for sustainability,* vol 33. N.K. Arora, B. Bouizgarne (eds), Springer, Singapore. pp. 1–41. https://doi.org/10.1007/978-981-16-4843-4_1

Singhvi, M., and B. S. Kim. 2022. Green hydrogen production through consolidated biopro cessing of lignocellulosic biomass using nanobiotechnology approach. *Bioresource Technology* 365:128108. https://doi.org/10.1016/j.biortech.2022.128108

Singla, A., Iwasa, H. and K. Inubushi. 2014. Effect of biogas digested slurry based-biochar and digested liquid on N_2O, CO_2 flux and crop yield for three continuous cropping cycles of komatsuna (*Brassica rapa* var. *perviridis*). *Biology and Fertility of Soils* 50(8):1201–1209.

Sohi, S., Lopez-Capel, E., Krull, E., and R. Bol. 2009. Biochar, climate change and soil: A review to guide future research. *CSIRO Land and Water Science Report* 5(09):17–31.

Stamati, P. N., Maipas, S., Kotampasi, C., Stamatis, P., and L. Hens. 2016. Chemical pesticides and human health: The urgent need for a new concept in agriculture. *Frontiers in Public Health* 4:148.

Stevenson, F.J. 1982. *Humus chemistry.* Wiley, New York

Sullivan, D.M., Bary, A.I., Nartea, T.J., Myrhe, E.A., Cogger, C.G., and S.C. Fransen. 2003. Nitrogen avail- ability seven years after a high-rate food waste compost application. *Compost Science &Utilization* 11(3):265–275.

Tandon, H.L.S. 1995. *Recycling of crop, animal, human and industrial wastes in agriculture.* Fertilizer Development and Consultation Organization, New Delhi, 148 pp.

Tang, Y., Wang, L., Carswell, A., Misselbrook, T., Shen, J., and J. Han. 2020. Fate and transfer of heavy metals following repeated biogas slurry application in a rice-wheat crop rotation. *Journal of Environmental Management* 270:110938.

Tejaswini, M.S.S.R., Pathak. P, and D.K. Gupta. 2022. Sustainable approach for valorization of solid wastes as a secondary resource through urban mining. *Journal of Environment Management* 319:115727. https:// doi.org/10.1016/j.jenvman.2022.115727

Teng, Y., Luo, Y., Ping, L., Zou, D., Li, Z., and P. Christie. 2010. Effects of soil amendment with different carbon sources and other factors on the bioremediation of an aged PAH-contaminated soil. *Biodegradation* 21(2):167–178.

Thomas, L., and I. Singh. 2019. *Microbial biofertilizers: types and applications.* In *Biofertilizers for sustainable agriculture and environment,* pp. 1–19. Springer, Cham.

Tilman, D., Cassman, K.G., Matson, P.A., Naylor, R. and S. Polasky. 2002. Agricultural sustainability and intensive production practices. *Nature* 418(6898):671–677.

Tubiello, F.N., Salvatore, M., CóndorGolec, R.D., Ferrara, A., Rossi, S., Biancalani, R., Federici, S., Jacobs, H., and A. Flammini. 2014. *Agriculture, forestry and other land use emissions by sources and removals by sinks. Rome, Italy.*

Tyagi, J., Ahmad, S., and M. Malik. (2022) Nitrogenous fertilizers: Impact on environment sustainability, mitigation strategies, and challenges. *International Journal of Environmental Science and Technology* 1–24.

USDA. 2012. Agricultural waste management field handbook. United States Department of Agriculture, Soil conservation Service. Accessed from www.info.usda.gov/ viewerFS. aspx?hid=21430> on 10/06/ 2016. 2012.

van der Velden, R., da Fonseca-Zang, W., Zang, J., Clyde-Smith, D., Leandro, W.M., Parikh, P., Borrion, A., and L.C. Campos. 2022. Closed-loop organic waste management systems for family farmers in Brazil. *Environmental Technology* 43(15):2252–2269.

Verma, N., Wati, L., Sindhu, S.S., and S. Goyal. 2014. Organic waste utilization for production of compost and biofuel. In *Microbes in the Service of Mankind: Tiny Bugs with Huge Impact,* R. Nagpal, Ashwani Kumar and R. Singh, eds., pp. 95–124. JBC Press, New Delhi

Vocciante, M., Grifoni, M., Fusini, D., Petruzzelli, G., and E. Franchi. 2022. The role of plant growth-promoting rhizobacteria (PGPR) in mitigating plant's environmental stresses. *Applied Sciences* 12:1231. https://doi.org/10.3390/app12031231

Wang, Z., Xu, Z., Chen, S., Chen, X., Yuan, X., Shen, G., Jiang, X., Liu, S., and M. Jin. 2022. Effects of storage temperature and time on enzymatic digestibility and fermentability of densifying lignocellulosic biomass with chemicals pretreated corn stover. *Bioresource Technology* 347:126359.

Westerman, P.W. and J.R. Bicudo. 2005. Management considerations for organic waste use in agriculture. *BioresourceTechnology* 96(2):215–221.

Windeisen, E., and G. Wegener. 2008. Behaviour of lignin during thermal treatments of wood. *Industrial Crops and Products* 27(2):157–162.

Wu, X., Wang, J., Shen, L., Wu, X., Amanze, C. and W Zeng. 2021. Effect of bamboo sphere amendment on the organic matter decomposition and humification of food waste composting. *Waste Management* 133:19–27.

Xie, J., Wang, Z., Wang, Y., Xiang, S., Xiong, Z., and M. Gao. 2023. Manure combined with biochar reduces rhizosphere nitrification potential and *amoA* gene abundance of ammonia-oxidizing microorganisms in acid purple soil. *Applied Soil Ecology* 181: 104660 https://doi.org/10.1016/j.apsoil.2022.104660

Xie, K., Sun, M., Shi, A., Di, Q., Chen, R., Jin, D., Li, Y., Yu, X., Chen, S., and C. He. 2022. The application of tomato plant residue compost and plant growth-promoting rhizobacteria improves soil quality and enhances the ginger field soil bacterial community. *Agronomy* 12:1741. https://doi.org/10.3390/agronomy12081741

Xu, Z.M., Wang, Z., Gao, Q., Wang, L.L., Chen, L.L., Li, Q.G., Jiang, J.J., Ye, H.J., Wang, D.S., and P. Yang. 2019. Influence of irrigation with microalgae-treated biogas slurry on agronomic trait, nutritional quality, oxidation resistance, and nitrate and heavy metal residues in Chinese cabbage. *Journal of Environmental Management* 244:453–461.

Yaashikaa, P.R., Kumar, P.S., and S. Varjani. 2022. Valorization of agro-industrial wastes for biorefinery process and circular bioeconomy: A critical review. *Bioresource Technology* 343:126126.

Yadav, B., Talan, A., Tyagi, R.D., and P. Drogui. 2021. Concomitant production of value-added products with polyhydroxyalkanoate (PHA) synthesis: A review. *Bioresource Technology* 337:125419.

Yan, W., Liu, Y., Malacrinò, A., Zhang, J., Cheng, X., Rensing, C., Zhang, Z., Lin, W., Zhang, Z., and H. Wu. 2024. Combination of biochar and PGPBs amendment suppresses soil-borne pathogens by modifying plant-associated microbiome. *Applied Soil Ecology* 193:105162. https://doi.org/10.1016/j.apsoil.2023.105162

Yang, H., Yu, D., Zhou, J., Zhai, S., Bian, X. and M. Weih. 2018. Rice-duck co-culture for reducing negative impacts of biogas slurry application in rice production systems. *Journal of Environmental Management* 213:142–150.

Yang, K., Hu, J., Ren, Y., Zhang, Z., Tang, M., Shang, Z., Zhen, Q., and J. Zheng. 2024. Enhancement of soil organic carbon, water use efficiency and maize yield (*Zea mays* L.) in sandy soil through organic amendment (Grass peat) incorporation. *Agronomy* 14:353. https://doi.org/10.3390/agronomy14020353

Yu, F.B., Luo, X.P., Song, C.F., Zhang, M.X., and S.D. Shan. 2010. Concentrated biogas slurry enhanced soil fertility and tomato quality. *Acta Agriculturae Scandinavica: Section B– Soil and Plant Science* 60(3):262–268.

Yu, H., Zou, W., Chen, J., Chen, H., Yu, Z., Huang, J., Tang, H., Wei, X., and B. Gao. 2019. Biochar amendment improves crop production in problem soils: A review. *Journal of Environmental Management* 232:8–21. https://doi.org/10.1016/j.jenvman.2018.10.117

Zhao, S., Guangxin, R., and Y.G. Yang. 2007. Effect of spraying biogas slurry on capsicum. *Acta Agriculturae Borea Liocci Denta Lis Sinice* 16(3):202–203.

Zhao, X., Li, J., Che, Z., and L. Xue. 2022. Succession of the bacterial communities and functional characteristics in sheep manure composting. *Biology* 11:1181. https://doi.org/10.3390/ biology11081181

Zhou, S-P., Ke, X., Jin, L-Q., Xue, Y-P., and Y-G. Zheng. 2024. Sustainable management and valorization of biomass wastes using synthetic microbial consortia. *Bioresource Technology* 395:130391. https://doi.org/10.1016/j.biortech.2024.130391

Zhou, K., Zhang, Y., and X. Jia. 2018. Co-cultivation of fungal-microalgal strains in biogas slurry and biogas purification under different initial CO_2 concentrations. *Scientific Reports* 8(1):1–12.

Zhou, Y., Xiao, R., Klammsteiner, T., Kong, X., Yan, B., Mihai, F-C., Liu, T., Zhang, Z., and M. K. Awasthi. 2022. Recent trends and advances in composting and vermicomposting technologies: A review. *Bioresource Technology* 360:127591. https://doi.org/10.1016/j.biortech.2022.127591.

Zhu, X., Labianca, C., He, M., Luo, Z., Wu, C., You, S., and D.C.W. Tsang. 2022. Life-cycle assessment of pyrolysis processes for sustainable production of biochar from agro-residues. *Bioresource Technology* 360:127601. https://doi.org/10.1016/j.biortech.2022.127601.

Zitrides, T. 1983. Bio-decontamination of spill sites. *Pollution Engineering* 15(11):25–27.